QUANTITATIVE DATA ANALYSIS

QUANTITATIVE DATA ANALYSIS

Doing Social Research to Test Ideas

DONALD J. TREIMAN

JOSSEY-BASS
A Wiley Imprint
www.josseybass.com

Published by Jossey-Bass
A Wiley Imprint
989 Market Street, San Francisco, CA 94103—www.josseybass.com

Jossey-Bass books and products are available through most bookstores. To contact Jossey-Bass directly call our Customer Care Department within the United States at (800) 956-7739, outside the United States at (317) 572-3986, or via fax at (317) 572-4002.

Jossey-Bass also publishes its books in a variety of electronic formats. Some content that appears in print may not be available in electronic books.

Library of Congress Cataloging-in-Publication Data

Treiman, Donald J.
Quantitative data analysis : doing social research to test ideas / Donald J. Treiman.
p. cm.
ISBN 978-0-470-38003-1 (alk. paper)
1. Social sciences—Research—Statistical methods. 2. Sociology—Research—Statistical methods. 3. Sociology—Statistical methods. 4. Social sciences—Statistical methods—Computer programs. 5. Stata. I. Title.
HA29.T675 2008
300.72—dc22

2008038229

FIRST EDITION
PB Printing 10 9 8 7 6 5 4 3 2 1

CONTENTS

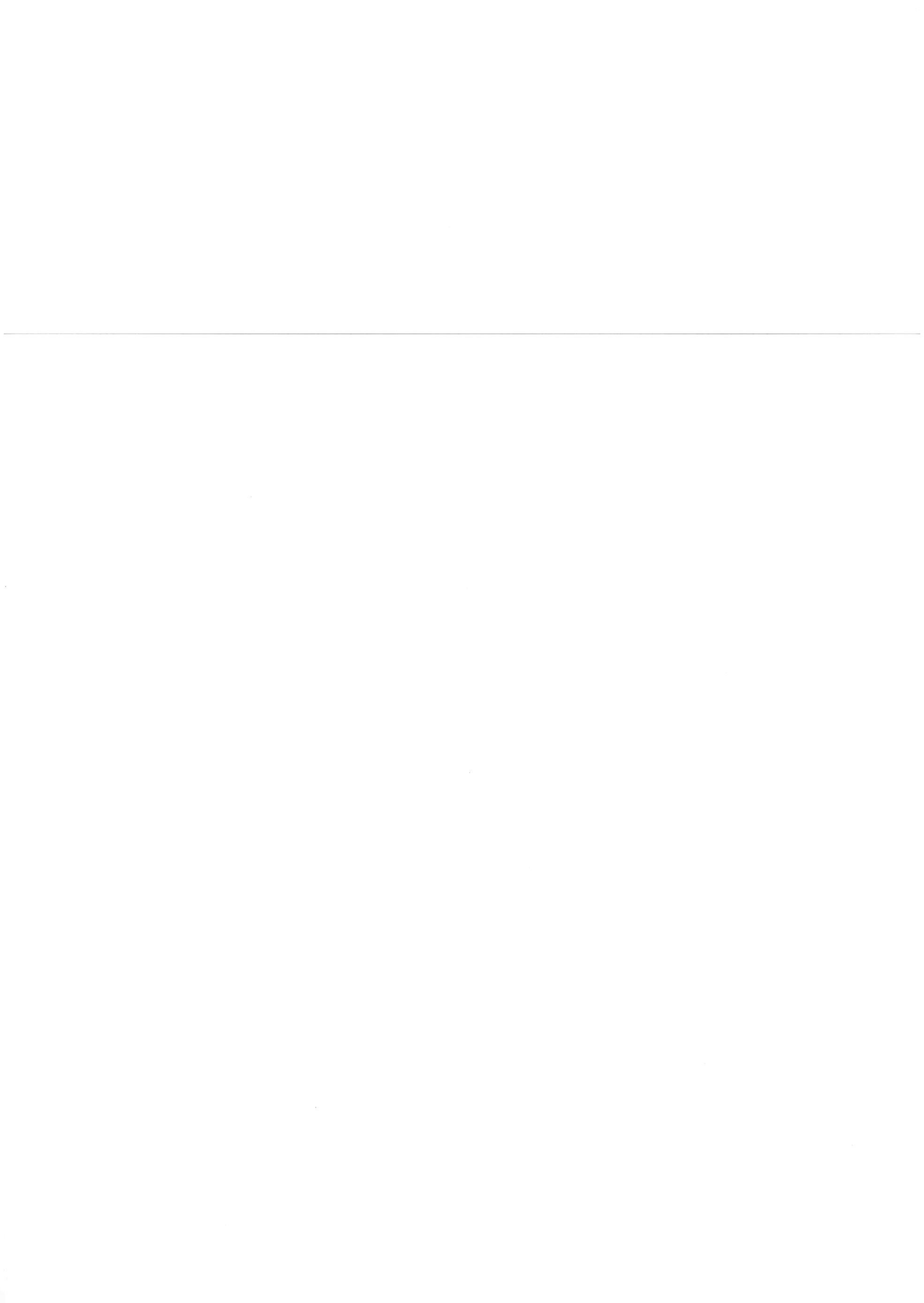

TABLES, FIGURES, EXHIBITS, AND BOXES

TABLES

FIGURES

EXHIBITS

BOXES

To John Pock, who first introduced me to the idea of doing social research to test ideas

PREFACE

This is a book about how to conduct theoretically informed quantitative social research—that is, social research to test ideas. It derives from a course for graduate students in sociology and other social sciences and social science-based professional schools (public health, education, social welfare, urban planning, and so on) that I have been teaching at UCLA for some thirty years. The course has evolved as quantitative methods in the social sciences have advanced; early versions of the course were based on the first half of this book (through Chapter Seven), with additional materials added over the years. Interestingly, I have been able to retain the same format—a twenty-week course with one three-hour lecture per week and a weekly exercise, culminating in a term paper written during the last four weeks of the course—from the outset, which is, I suppose, a tribute to the increasing level of preparation and quantitative competence of graduate students in the social sciences. The book owes much to lively class discussions over the years, of often subtle and complex methodological points.

By the end of the book, you should know how to make substantive sense of a body of quantitative data. That is, you should be well prepared to produce publishable papers in your field, as well as first-rate dissertation chapters. Of course, there is always more to learn. In the final chapter (Chapter Sixteen), I discuss advanced topics that go beyond what can be covered in a first course in data analysis.

The focus is on the analysis of data from representative samples of well-defined populations, although some exceptions are considered. The populations can consist of almost anything—people, formal organizations, societies, occupations, pottery shards, or whatever; the analytic issues are essentially the same. Data collection procedures are mentioned only in passing. There simply is not enough space in an already lengthy book to do justice to both data analysis and data collection. Thus, you will need to look elsewhere for systematic instruction on data-collection procedures. A strong case can be made that you should do this *after* rather than before a course on data analysis because the main problem in designing a data collection effort is deciding what to collect, which means you first need to know how you will conduct your analysis. An alternative method of learning about the practical details of data collection is to become an apprentice (unpaid, if necessary) to someone who is about to conduct a survey and insist that you get to participate in it step-by-step even when your presence is a nuisance.

This book covers a variety of techniques, including tabular analysis, log-linear models for tabular data, regression analysis in its various forms, regression diagnostics and robust regression, ways to cope with missing data, logistic regression, factor-based and other techniques for scale construction, and fixed- and random-effects models as a way to make causal inferences. But this is not a statistics book; the emphasis is on using these procedures to draw substantive conclusions about how the social world works. Accordingly, the book is designed for a course to be taken after a first-year graduate statistics course in the social sciences. Although there are many equations in the book, this is because it is necessary to understand how statistical procedures work to use them intelligently. Because the emphasis is on applications, there are many worked examples, often adapted from my own research. In addition to data from sample surveys I have conducted, I also rely heavily on the *General Social Survey*, an omnibus survey designed for use by the research community and also for teaching. Appendix A describes the main data sets used for the substantive examples and provides information on how to obtain them; they are all available without cost.

The only prerequisites for successful use of this book are a prior graduate-level social science statistics course, a willingness to think carefully and work hard, and the ability to do high school algebra—either remembered or relearned. With only a handful of exceptions (references at one or two points to calculus and to matrix algebra), no mathematics beyond high school algebra is used. If your high school algebra is rusty, you can find good reviews in Helen Walker, *Mathematics Essential for Elementary Statistics*, and W. L. Bashaw, *Mathematics for Statistics*. These books have been around forever. Although more recent equivalents probably exist, school algebra has not changed, so it hardly matters. Copies of these books are readily available at `amazon.com`, and probably many other places as well.

The statistical software package used in this book is *Stata* (release 10). Downloadable command files (`-do-` files in Stata's terminology), files of results (`-log-` files), and ancillary computer files used in the computations are available at www.josseybass.com/go/quantitativedataanalysis. These materials, and updates for Stata 11 and Stata 12, also are available on my personal web page, http://www.ccpr.ucla.edu/dtreiman.

Often the details underlying particular computations are only found in the downloadable `-do-` and `-log-` files, so be sure to download and study them carefully. These files will continue to be updated as new releases of Stata become available.

I use Stata in my teaching and in this book because it has very rapidly become the statistical package of choice in leading sociology and economics departments. This is not accidental. Stata is a fast and efficient package that includes most of the statistical procedures of interest to social scientists, and new commands are being added at a rapid pace. Although many statistical packages are available, the three leading contenders currently are Stata, SPSS, and SAS. As software, Stata is clearly superior to SPSS—it is faster, more accurate, and includes a wider range of applications. SAS, although very powerful, is not nearly as intuitive as Stata and is more difficult to learn (and to teach). Nonetheless, this book can be readily used in conjunction with either SPSS or SAS, simply by translating the syntax of the Stata `-do-` files. (I have done something like this, exploiting Allison's excellent, but SAS-based, exposition of fixed- and random-effects models [Allison 2005] by writing the corresponding Stata code.)

FOR INSTRUCTORS

Some notes on how I have used these materials in teaching may be helpful to you as you design your own course.

As noted previously, the course on which this book is based runs for two quarters (twenty weeks). I have offered one three-hour lecture per week and have assigned an exercise every week. When I first taught the course, I read these exercises myself, but as enrollments have increased, I have enjoyed the services of a T.A. (chosen from among students who had done well in the course in previous years), who assists students with the mysteries of computing and statistics and also reads and comments on the exercises. In recent years, I have offered seventeen lectures and have assigned exercises for all but the last, with the final month of the course devoted to producing two drafts of a term paper—in a marathon session I read the first drafts and write comments, in an attempt to emulate the journal submission process. Thus, in my course, everyone gets a "revise and resubmit" response. I encourage students to develop their term papers in the course of doing the exercises and to complete their drafts in the two weeks after the last exercise is due.

The initial exercises are designed to lead students in a guided way through the mechanics of analysis, and some of the later exercises do this as well. But the exercises increasingly take a free form: "carry out an analysis like that presented in the book." Illustrative answers are provided for those exercises that involve definitive answers—that is, those similar to statistics problem sets.

The course syllabus, weekly exercises, and illustrative answers to those exercises for which I have written illustrative answers are available for downloading from www.josseybass.com/go/quantitativedataanalysis. The course syllabus and exercises, but not the answers, are available on my personal web page, http://www.ccpr.ucla.edu/dtreiman.

ACKNOWLEDGMENTS

As noted earlier, this book has been developed in interaction with many cohorts of graduate students at UCLA who have wrestled with each of the chapters included here and have revealed troubles in the exposition, sometimes by way of explicit comments and sometimes via displays of confusion. The book would not exist without them, as I never imagined myself writing a textbook, and so I owe them great thanks. One in particular, Pamela Stoddard, literally caused the book to be published in its current form by mentioning in the course of a chance airplane conversation with Andrew Pasternack, a Jossey-Bass acquisitions editor, that her professor was thinking of publishing the chapters he used as a course text. Andy contacted me, and the rest is history.

The course on which this book is based first came into being through collaboration with my colleague Jonathan Kelley, when he was a visiting professor at UCLA in the 1970s. The first exercise is borrowed from him, and the general thrust of the course, especially the first half, owes much to him.

My colleague, Bill Mason, recently retired from the UCLA Sociology and Statistics Departments, has been my statistical guru for many years. Often I have turned to him for insights into difficult statistical issues. And much that I have learned about topics that were not part of the curriculum when I was a graduate student has been from sitting in advanced statistics courses offered by Bill. Another colleague, Rob Mare, has been helpful in much the same way. My new colleague, Jennie Brand, who took over my quantitative data analysis course in the fall of 2008, read the entire manuscript and offered many helpful suggestions. Bill Rising, of Stata's Author Support Program, reviewed the downloadable Stata 10 `-do-` files and offered many suggestions for improving and updating them so that they reflect current best practice. The book has benefited greatly from very careful reading by a group of about 100 Chinese students, to whom I gave a special version of the course in an intensive summer session at Beijing University in July 2008. They caught many errors that had gone unnoticed and raised often subtle points that resulted in the reworking of selected portions of the text. A group of graduate students at UCLA (Lan Li and Xi Song) and the Hong Kong University of Science and Technology (Jun Li, Hua Ye, Bingdao Zheng, and Zhuoni Zhang) also carefully read the book in the course of checking a translation into Chinese and found many additional errors.

My understanding of research design and statistical issues, especially concerning causality and threats to causal inference, has benefited greatly from the weekly seminar of the California Center for Population Research, which brings together sociologists, economists, and other social scientists to listen to, and comment on, presentations of work in progress, mainly by visitors from other campuses. The lively and wide-ranging discussion has been something of a floating tutorial, a realization of what I have imagined academic life could and should be like.

Finally, my wife, Judith Herschman, has displayed endless patience, only occasionally asking, "When are you going to finally publish your methods book?"

THE AUTHOR

Donald J. Treiman is distinguished professor of sociology emeritus at the University of California at Los Angeles (UCLA) and was from 2006 through 2008 director of UCLA's California Center for Population Research. He has a BA from Reed College (1962) and an MA and PhD from the University of Chicago (1967). As a graduate student at Chicago, he spent most of his time at the National Opinion Research Center (NORC), where he gained valuable training and experience in survey research. He then taught at the University of Wisconsin, where he decided that he really was a social demographer at heart, and made the Center for Demography and Ecology his intellectual home. From Wisconsin, he moved to Columbia University and then, in 1975, to UCLA, where he has been ever since, albeit with temporary sojourns elsewhere, as staff director of a study committee at the National Academy of Sciences/National Research Council (1978–1981) and fellowship years at the U.S. Bureau of the Census (1987–1988), the Center for Advanced Study in the Behavioral and Social Sciences (1992–1993), and the Netherlands Institute for Advanced Study in the Humanities and Social Sciences (1996–1997).

Professor Treiman started his career as a student of social stratification and status attainment, particularly from a cross-national perspective, and this has remained a continuing interest. He and his Dutch colleague, Harry Ganzeboom, have been engaged in a long-term cross-national project to analyze variations in the status attainment process across nations throughout the world over the course of the twentieth century. To date, they have compiled an archive of more than 300 sample surveys from more than 50 nations, ranging through the last half of the century. In addition to his comparative project, Professor Treiman has conducted large-scale national probability sample surveys in South Africa (1991–1994), Eastern Europe (1993–1994), and China (1996), all concerned with various aspects of social inequality.

His current research has moved in a more demographic direction. In 2008 he conducted a national probability sample survey in China that focused on the determinants, dynamics, and consequences of internal migration. He is now engaged in another survey focused on the consequences of migration for children left behind and children accompanying their migrant parents.

INTRODUCTION

It is not uncommon for statistics courses taken by graduate students in the social sciences to be treated essentially as mathematics courses, with substantial emphasis on derivations and proofs. Even when empirical examples are used—which they frequently are because statistics is, after all, about summarizing data, describing relationships among variables, and making inferences from samples to populations—students often are left with little sense about what to do with what they have learned, that is, how to apply the statistical techniques they have mastered to actual research questions.

Learning to apply statistical techniques is the focus of this book. I assume that you already have had a first year graduate level statistics course, most likely a course taught in a social science department or professional school. In such a course, you have covered basic statistical concepts, at least through ordinary least squares regression and preferably with some exposure to logistic regression. But now you want to become a practicing quantitative researcher, analyzing sample surveys or other quantitative data sets to contribute to the body of substantive knowledge in your field and publishing your results. This book is designed for you. It is based on a two-quarter course I have been teaching at UCLA for the past thirty years, which has been updated year-by-year to keep up with developments in quantitative methods in sociology and other social science fields.

The kinds of quantitative, or statistical, methods used in papers published these days in the leading journals of sociology and other social sciences are vastly different from what they were even a few years ago, and new methods—borrowed largely from econometrics and to some extent from psychometrics and biostatistics, and from developments within mathematical statistics—gain currency every year. Thus, topics that I used to regard as too advanced for a first-year quantitative methods course—for example, fixed- and random-effects models—are now so commonly used in published papers in sociology that I have felt compelled to incorporate them into my course and into this book.

OVERVIEW OF CHAPTERS

Still, this book begins at the beginning, with the most basic approach to analyzing nonexperimental data—percentage tables. **Chapters One** through **Three** describe the logic of cross-tabulations and provide many technical details on how to produce attractive (because they are clear and easy to read) tables. The two central ideas in these chapters are deciding in which direction to percentage tables and understanding statistical controls. It turns out that the first of these is difficult for some students—much more difficult than coping with complex mathematical formulas, which we do later in the book. Thus, even if you think you already know all there is to know about percentage tables, I encourage you to pay careful attention to these chapters. Doing so will pay great dividends.

Chapter Four is an introduction to computing. In this chapter, I show how data are organized for analysis by computer and how analysis is conducted using statistical software. I also provide hints for using Stata, the statistical package used in this book. However, the chapter is written in such a way that it also can serve as an introduction to other statistical packages, such as SPSS and SAS.

Chapters Five through **Seven** consider ordinary least squares correlation and regression, the workhorse of statistical analysis in the social sciences. These procedures provide a way of quantifying the relationship between some quantitative outcome and its determinants—for example, how much of a difference in income should we expect for people who differ by a given number of years in their level of schooling, holding constant other confounding factors? They also provide a way of assessing how good our prediction is—for example, how much of the variability in income can be attributed to differences in education, gender, race, and so on. **Chapter Five** focuses on two-variable correlation and regression to get the logic straight and to consider some common errors in interpretation of correlation and regression statistics. **Chapter Six** considers multiple regression, which is used when there are several predictors of a particular outcome, and introduces the idea of "dummy" or dichotomous variables, which require special treatment. Making use of dummy variables and "interaction terms," I offer a strategy for assessing whether social processes differ across population groups, a frequent question in the social sciences. **Chapter Seven** offers a variety of tricks that permit relatively refined hypotheses to be tested within a regression framework.

Most data sets analyzed by social scientists are plagued by missing data—information on particular variables that is missing for specific individuals. **Chapter Eight** reviews ways to cope with missing data, culminating in a demonstration of how to do multiple imputation of missing data, the current state-of-the-art approach.

Chapter Nine takes up the issue of sampling and its implications for statistical analysis. Whereas the previous chapters assumed simple random sampling, most general population samples are actually complex, multistage samples. Correctly analyzing data from such samples requires that we take account of the "clustering" of observations when we compute standard errors. This chapter introduces *survey estimation* procedures, which do this.

There are many pitfalls to regression that can trap the unwary. As noted, these are first discussed (brie y) in Chapter Five. **Chapter Ten** gives them a fuller treatment, through the introduction of what are known as *regression diagnostics*. These procedures provide protection against the possibility of making false inferences from regression results.

Chapter Eleven shows why and how to construct multiple-item scales, focusing principally on *factor-based* scaling but also introducing *effect-proportional* scaling. Often we want to study concepts for which no one item in a questionnaire provides an adequate measure, for example, "level of living," "liberalism," "type-A personality," and "depression." Summary measures, or *scales*, based on several items usually provide variables that are both more *reliable* and more *valid* than single items. This chapter shows how to create such scales and how to use them.

Chapters Twelve through **Fourteen** provide techniques for considering *limited dependent variables*. Ordinary least squares regression is designed to handle outcome variables that can be treated as continuous, such as income, years of schooling, and so on. But many outcome variables of interest to social scientists are dichotomous (for example, whether people vote, marry, have been victimized by crime, and so on) and others are polytomous (political affiliation in a multiparty system, occupational category, type of university attended, and so on). Log-linear analysis and logistic regression are techniques for dealing with limited dependent variables. **Chapter Twelve** considers log-linear analysis, a technique for making rigorous inferences about the relationships among a set of polytomous variables, that is, inferences about the degree and pattern of associations among cross-tabulated variables. In this sense, log-linear analysis provides a way of doing statistical inference about the kinds of tables we consider in Chapters One through Three. **Chapter Thirteen** introduces binary logistic regression, an appropriate technique for analyzing dichotomous outcomes, and then shows how to use this technique to handle special kinds of cases: progression ratios, where what is being studied are the factors affecting whether people move through a series of steps, say from one level of schooling to the next; discrete-time hazard rate models, where what is being studied is the likelihood that an event (say, first marriage) occurs at a point in time (say, a given age); and case-control models, which provide a way of studying the likelihood of rare events such as contracting diseases, gaining elite occupations, and so on. **Chapter Fourteen** shows how to study still other limited dependent variables: unordered polytomous variables, such as type of place of residence, via *multinomial logistic regression*; ordinal outcomes for which the order of categories is known but not the distance between categories, such as some attitude scales (are you "very happy," "somewhat happy," or "not too happy"), via *ordinal logistic regression*; and "censored" variables, where the range of a scale is truncated, for example, an income variable with the top category "$100,000 per year or more," via *tobit regression*.

When using nonexperimental data, it frequently is difficult to definitively establish that one variable causes another because they both could depend on still a third variable, often unmeasured. **Chapter Fifteen** provides a class of techniques, known as *fixed-effects* and *random-effects* models, for dealing with such problems when one has suitable data—either panel data, in which data are available for the same individuals at more than one point in time, or clustered data, in which observations are available for more than one individual in a family, school, community, or other unit. When appropriate data are available, this is a very powerful approach.

The final chapter (**Chapter Sixteen**) considers techniques that are beyond what I have been able to cover in this book and beyond what usually can be covered in a first-year graduate course in quantitative data analysis. Many of these techniques, now widely used in economics, are ways of coping with various versions of the *endogeneity problem*, the possibility that unmeasured variables affect both predictors and outcomes, resulting in biased estimates. Fixed- and random-effects models provide one way of dealing with such problems, but many other techniques are available, which are reviewed in Chapter Sixteen. I also brie y introduce structural equation modeling, a technique for dealing with complex social processes in which an outcome variable is a predictor variable for another outcome. For example, in status attainment analysis, we would want to study how the social status of parents affects education, how parental status and education affect the status of the first job, and so on. The brief introduction to these advanced techniques is intended to provide guidance in pursuing more advanced training in quantitative analysis. I conclude the chapter with advice on good research practice—hints on how to improve the quality of your work and how to save time and energy in the bargain.

CHAPTER

CROSS-TABULATIONS

WHAT THIS CHAPTER IS ABOUT

In this chapter, we start with an introduction to the elements of quantitative analysis—the material to be covered in this book. Then we deal with the most basic of all quantitative analyst's tools, cross-tabulations or percentage tables. (Strictly speaking, not all percentage tables are cross-tabulations because we can percentage univariate distributions. But the main emphasis of this chapter will be on how to percentage tables involving the simultaneous tabulation of two or more variables.) Although the procedures are basic, they are not trivial. There are clear principles for deciding how to percentage cross-tabulations. We will cover these principles and also their exceptions. In the course of doing this, we will consider the logic of causal argument. Then we will consider other ways, besides percentage tables, of summarizing univariate and multivariate distributions of data, as well as ways of assessing the relative size of associations between pairs of variables *controlling for* or *holding constant* other variables. Take this chapter seriously, even if you have encountered percentage tables before and think you know a lot about them. In my experience, getting right the logic of how to percentage a table proves to be very difficult for many students, much more difficult than seemingly fancier procedures, such as multiple regression.

You will notice that many of the examples in the first three chapters are quite old, drawn from studies conducted as far back as the 1960s. This is because at that time tabular analysis was the "state of the art"—the technique used in most of the articles published in leading journals. Thus, by going back to the older research literature. I have been able to find particularly clear applications of tabular procedures.

INTRODUCTION TO THE BOOK VIA A CONCRETE EXAMPLE

In 1967, Gary Marx published an article in the *American Sociological Review* titled "Religion: opiate or inspiration of civil rights militancy among Negroes?" (Marx 1967a; see also Marx 1967b). The title expressed two competing ideas about how religiosity among Blacks might have affected their militancy regarding civil rights. One possibility was that religious people would be less militant than nonreligious people because religion gave them an other-worldly rather than this-worldly orientation, and established religious institutions have generally had a stake in the status quo and hence a conservative orientation. The other possibility was that they would be more militant because the Black churches were a major locus of civil rights militancy, and religion is an important source of universal humanistic values. Of course, a third possibility was that there would be no connection between religiosity and militancy.

Suppose that we want to decide which of these ideas is correct. How can we do this? One way—which is the focus of our interest here—would be to ask a probability sample of Blacks how religious they are and how militant with respect to civil rights they are, and then to cross-tabulate the answers to determine the relative likelihood, or probability, that religious and nonreligious people say they are militant. If religious people are *less* likely to give militant responses than are nonreligious people, the evidence would support the first possibility; if religious people are *more* likely to give militant responses, the evidence would favor the second possibility; and if there is no difference in the relative likelihoods of religious and nonreligious people giving militant responses, the evidence would favor the third possibility. Of course, evidence favoring an idea does not definitely prove it. I will say more about this later.

This seemingly simple example contains all of the elements that we will be dealing with in this book and that a researcher needs to take account of to arrive at a meaningful and believable answer to any research question. Let us consider the elements one by one.

First, the *idea*: is religion an opiate or inspiration of civil rights militancy? Without an idea, the manipulation of data is pointless. As you will see repeatedly, the nature of the idea a researcher wants to test will dictate the kind of data chosen and the manipulations performed. Without an idea, it is impossible to decide what to do, and the researcher will be tempted to try to do everything and be at a loss to choose from among the various things he or she has done. Ideas to be tested are generally called *hypotheses*; they also will be referred to here and in what follows as *theories*. A theory need not be either grandiose or abstract to be labeled as such. Any idea about what causes what, or why and how two variables are associated, is a theory.

Second is the information, or *data*, needed to test the idea or hypothesis (or theory). In this book, we will be concerned with data drawn from probability samples of populations. A *population* is any definable collection of things. Mostly we will be concerned with populations of people, such as the population of the United States. But social scientists are also interested in populations of organizations, cities, occupations, and so on. A *probability sample* is a subset of the population selected in such a way that the probability that a given individual in the population will be included in the sample is known. Only by using a probability sample is it possible to make inferences from the characteristics of the sample to the characteristics of the population from which the sample is drawn.

That is, if we observe a given result in a probability sample, we can infer within a specified range what the likely result will be in the population.

The sample used by Marx is actually quite complex, consisting of a probability sample of 492 Blacks living in metropolitan areas outside the South, plus four special samples: probability samples of Blacks living in Chicago, New York, Atlanta, and Birmingham. The total number of respondents from the non-Southern urban sample plus the four special samples is 1,119, and Marx treats the combined sample as representative of urban Blacks in the United States. This is not, in fact, entirely legitimate. Later we will explore ways to weight complex samples to make them truly representative of the populations from which they are drawn. Evaluation of the sample used in an analysis is an important part of the data analyst's task. But for now, we will go along with Marx in treating his sample as a probability sample of U.S. urban Blacks.

When our ideas are about the behavior or attitudes of people, a standard way of collecting data is to ask a probability sample chosen from an appropriate population to tell us about their behavior and attitudes by answering a set of specific questions. That is, we *survey* the sample by asking each individual in the sample a set of questions and recording the responses. In most sample surveys, the possible responses are preselected, and the person being surveyed, the *respondent*, is asked to choose the best response from a list (however, see the boxed comment on open-ended questions). For example, one of the questions Marx asked was

What would you say about the civil rights demonstrations over the last few years—that they have helped Negroes a great deal, helped a little, hurt a little, or hurt a great deal?

Helped a great deal	1
Helped a little	2
Hurt a little	3
Hurt a great deal	4
Don't know	5

OPEN-ENDED QUESTIONS Occasionally, questions are worded in a way that requires a narrative response; these are known as *open-ended* questions. Open-ended questions are used when possible responses are too varied or complex to be conveniently listed on a questionnaire or when the researcher doesn't have a very good idea of what the possible responses will be. Open-ended questions must be *coded,* that is, converted into a standard set of response categories, as an editing operation in the course of data preparation. This is very time-consuming and expensive and is avoided whenever possible. Still, some items must be asked in an open-ended format. Both in the decennial census and in many contemporary surveys in the United States, for example, a series of three open-ended questions typically is asked to elicit information necessary to classify respondents according to standard detailed (three-digit) classifications of occupation and industry.

Each response, or *response category*, has a number associated with it, known as a code. The codes are what are actually recorded when the data are prepared for analysis because they are used to manipulate data in a computer. Typically, some respondents will refuse to answer a question or, in a self-administered questionnaire, will choose more than one response. Sometimes, an interviewer will forget to record a response or will record it in an ambiguous way. For these reasons, an extra code is usually designated to indicate nonresponses or uncodable responses. For example, code "9" might be assigned to nonresponses to the preceding question when the data are being prepared for analysis (this topic is discussed further a bit later). How to handle nonresponses, or missing data, is one of the perennial problems of the survey analyst, so we will devote a great deal of attention to this question.

The term *variable* refers to each set of response categories and the associated codes. A *machine-readable data set* (whether stored on computer tape, computer disk, oppy disks, CD-ROMs, thumb drives, or—almost extinct—IBM cards) consists of a set of codes for each individual in the sample corresponding to the response categories for the variables included in the data set. Suppose, for example, that the earlier question on whether civil rights demonstrations have helped Negroes is the tenth variable in a survey. Suppose, also, that the first respondent in the sample had said that demonstrations "helped a little." The data set would then include a "2" in the tenth location for the first individual. To know exactly what is included in a data set and where in the data set it is located, a *codebook* is prepared and used as a map to the data set. In Chapter Four, I will describe how to use a codebook. Here it is sufficient to note that the rudimentary materials necessary to carry out the sort of analysis dealt with in this book are a data set, a codebook for the data set, and documentation that describes the sample. We will not be concerned with problems of data collection or the preparation of a machine-readable data set, except in passing. These topics require full treatment in their own right, and we will not have time for them here.

It is customary to classify variables according to their level of measurement: nominal, ordinal, interval, or ratio. *Nominal variables* consist simply of a set of mutually exclusive and collectively exhaustive categories. Religious affiliation is an example of such a variable. For example, we might have the following response categories and codes:

Protestant	1
Catholic	2
Jewish	3
Other	4
None	5
No answer	9

Note that no order is implied among the responses—no response is "better" or "higher" than any other. The variable simply provides a way of classifying people into religious groups. Note, further, that every individual in the survey has a code, even those who didn't answer the question. This is accomplished by including a residual category, "Other," and a "No answer" category. In properly designed variables, categories are always mutually exclusive and collectively exhaustive—that is, written in such as way that each individual in the sample can be assigned one and only one code. (In Chapter Four, we will discuss various ways of coding missing data.)

Ordinal variables have an additional property—they can be arranged in an order along some dimension: quantity, value, or level. The question on civil rights demonstrations cited previously is an example of an ordinal variable, where the dimension on which the responses are ordered is helpfulness to Negroes. Actually, the variable is a useful example of what we often actually encounter in surveys. Two of the responses, "don't know" and the implicit "no answer" response, are not self-evidently ordered with respect to the other responses. In such situations, the analyst has two choices: either to exclude these responses from the analysis or to assign a position to them by recoding the variable, that is, altering the codes so that they indicate the new order. A plausible argument can be made that a "don't know" response is in between "helped a little" and "hurt a little," essentially a neutral rather than either a positive or a negative response. To treat the question in this way, an analyst would recode the variable by assigning code "3" to "don't know," code "4" to "hurt a little," and code "5" to "hurt a great deal." Whether to do this will depend on the specifics of the research question being investigated; but it is very important to be forthcoming about such manipulations when they are undertaken, reporting them as part of the writeup of the analysis. It would be rather more difficult to make the same sort of plausible case for including "no answer" as a neutral response because the bases for nonresponses are so varied, including simple error, failure to complete the questionnaire, and so on. Hence, there is no way to predict how nonrespondents would have responded had they done so. Therefore, it probably would be wisest to treat "no answer" as missing data.

The important feature of ordinal variables is that they include no information about the distance between categories. For example, we do not know whether the difference between a judgment that civil rights demonstrations "hurt a little" and the judgment that they "helped a little" is greater or smaller than the difference between the judgment that they "helped a little" and that they "helped a great deal." For this reason, some statisticians and social researchers argue that ordinal variables ought to be analyzed using ordinal statistics, which are statistics that make no assumptions about the distance between categories of a variable and use only the order property. This is not the position taken here. In this book, we will mainly consider two kinds of statistics, those appropriate for nominal variables and those appropriate for interval and ratio variables; the latter are known as parametric statistics. There are several reasons for ignoring statistics specifically designed for ordinal variables (with the exception of ordinal logistic regression, which we will consider in Chapter Fourteen). First, parametric statistics are much more powerful and far more mathematically tractable than ordinal statistics and, moreover, tend to be very robust; that is, they are generally quite insensitive to violations of assumptions about the nature of data—for example, that error is normally distributed. Second, ordinal statistics are much less widely used than parametric statistics; moreover, there are many alternatives for accomplishing the same thing and little consensus among researchers about which ordinal statistic to use. Third, many ordinal statistics involve implicit assumptions that are just as restrictive as the assumptions underlying parametric statistics. For example, it can be shown that Spearman's rank order correlation (an ordinal statistic) is identical to the product-moment (Pearson) correlation (the conventional parametric correlation coefficient) when interval or ratio variables are converted to ranks. In effect, then, the Spearman rank order correlation assumes an equal distance between each category rather than making no assumptions about the distance between categories. In sum, we gain little and lose much by using ordinal statistics. (However, if you are interested in such statistics, good discussions can be found in Davis 1971, and Hildebrand and others 1977.)

Interval variables and *ratio variables* are similar in that the distance between categories is meaningful. Not only can we say that one category is higher than another (on some dimension) but also how much higher. Such variables legitimately can be manipulated with standard arithmetic operations: addition, subtraction, multiplication, and division.

SAMUEL A. STOUFFER (1900–1960) was an early leader in the development of survey research. He was born in Sac City, Iowa, and earned a B.A. from Morningside College; earned an M.A. in literature at Harvard; served three years as an editor of the Sac City Sun, a newspaper founded by his father; and then began graduate studies in sociology at the University of Chicago, completing his Ph.D. in 1930. While at Chicago, he came under the tutelage of William F. Ogburn, who introduced him to statistics despite his self-described initial hostility to the subject. He studied statistical methods and mathematics intensively at Chicago and then spent a year as a Social Science Research Council Fellow at the University of London, where he worked with Karl Pearson, among others (see the biographical sketch of Pearson in Chapter Five). Stouffer held academic appointments in statistics and sociology at Wisconsin, Chicago, and Harvard. He was a skilled research administrator, heading a number of large projects designed to provide scientific understanding of major social crises: in the 1930s, a Social Science Research Council project to evaluate the influence of the Depression on social order, which resulted in thirteen monographs; during World War II, a study of soldiers for the Defense Department, which resulted in the classic publication, *The American Soldier* (Stouffer and others 1949); and in the 1950s, a study of the anticommunist hysteria of the McCarthy era, funded by the Ford Foundation's Fund for the Republic, which resulted in *Communism, Conformity, and Civil Liberties* (1955). When he died rather unexpectedly at age sixty after a brief illness, he was in the process of developing for the Population Council a new study on factors affecting fertility in developing nations. He also played an important role in developing the statistical program of the federal government, helping to establish the Division of Statistical Standards in the U.S. Bureau of the Budget. A hallmark of Stouffer's work is that he was strongly committed to using empirical data and quantitative methods to rigorously test ideas about social processes, which makes it fitting that a posthumous collection of his papers is titled *Social Research to Test Ideas* (1962).

Hence, we can compute statistics such as means and standard deviations for them. The difference between the two is that ratio variables have an intrinsic zero point, whereas interval variables do not. We can compare responses to ratio variables by taking the ratio of the value for one respondent (or group of respondents) to the value for another, whereas we can compare responses to interval variables only by taking the difference between them. Examples of interval variables include IQ and occupational prestige. Examples of ratio variables include years of school completed and annual income. It is not meaningful to say that someone's IQ is twice as high as someone else's, but it is meaningful to say that one person's IQ is 10 points higher than another person's IQ or that the within-race variance in IQ is larger than the between-race variance. By contrast, it is meaningful to say both that the incomes of men and women differ by $10,000 per year on the average and that the incomes of men are twice as high on average as those of women.

In this book, we often will treat ordinal variables as if they are interval variables to gain the power of parametric statistics. But we also will deal with procedures for assessing the adequacy of the interval assumption and for treating variables as nominal within the context of a general parametric approach that permits both nominal and interval or ratio variables to be dealt with simultaneously. These procedures involve various forms of regression analysis.

Often concepts of interest cannot be captured fully by single questions. For example, no single item in Marx's questionnaire fully captured what he meant by militancy. Hence, he constructed a multiple-item *scale* to represent this concept. Eight items that were pertinent to the situation in 1964 were used to construct a militancy scale. Individuals were classified as militant if they gave the militant response (shown in parentheses) to at least six of the eight items listed here (Marx 1967b, p. 41):

> *In your opinion, is the government in Washington pushing integration too slow, too fast, or about right?* (Too slow.)
>
> *Negroes who want to work hard can get ahead just as easily as anyone else.* (Disagree.)
>
> *Negroes should spend more time praying and less time demonstrating.* (Disagree.)
>
> *To tell the truth I would be afraid to take part in civil rights demonstrations.* (Disagree.)
>
> *Would you like to see more demonstrations or less demonstrations?* (More.)
>
> *A restaurant owner should not have to serve Negroes if he doesn't want to.* (Disagree.)
>
> *Before Negroes are given equal rights, they have to show that they deserve them.* (Disagree.)
>
> *An owner of property should not have to sell to Negroes if he doesn't want to.* (Disagree.)

There are many advantages to multiple-item scales, including in particular greater *reliability* and *validity* (both defined in Chapter Eleven). There also are many ways to construct multiple-item scales—some clearly superior to others—and some important pitfalls to avoid. Later—in Chapter Eleven—we will devote considerable attention to scale construction and evaluation.

The third element in any quantitative analysis is the *model*, the way we organize and manipulate data to assess our idea or hypothesis. The model has two components: the choice of statistical procedure and the assumptions we make about how the variables in our analysis are related. Given these two components, we can estimate the relative size or strength of the relationships between variables, and thus test our hypotheses (or ideas or theories) by assessing whether our estimates of the size of different effects are consistent with our hypotheses. For the simple example we have been considering, our models are cross-tabulations of militancy by religiosity (with the introduction of successive control variables, which are discussed a bit later), and our expectation (hypothesis) is that a higher percentage of the nonreligious than of the religious will be militant—or, because we have competing hypotheses, that a lower percentage of the nonreligious will be militant. Later in the book we will deal with statistical models that are more sophisticated—mostly variants of the general linear model—but the logic will remain unchanged. How we actually carry out cross-tabulation analysis is the topic of the next section.

CROSS-TABULATIONS

There are several ways to determine whether religious Blacks are more likely (or less likely) to be militant than are nonreligious Blacks. Perhaps the most straightforward approach is to cross-tabulate militancy by religiosity, that is, to count the frequency of persons with each combination of religiosity and militancy. By using four religiosity categories and two militancy categories, there are eight combinations of the two variables. In Marx's sample, the cross-tabulation of militancy by religiosity yields the following joint frequency distribution (Table 1.1).

TABLE 1.1. Joint Frequency Distribution of Militancy by Religiosity Among Urban Negroes in the U.S., 1964.

Religiosity	Militant	Nonmilitant	Total
Very religious	61	169	**230**
Somewhat religious	160	372	**532**
Not very religious	87	108	**195**
Not at all religious	25	11	**36**
Total	**333**	**660**	**993**

Source: Adapted from Marx (1967a, Table 6).

TECHNICAL POINTS ON TABLE 1.1

1) The Total row and Total column are known as marginals. They give the frequency distributions for each variable separately, in other words, the univariate frequency distributions. (Rows are read across and columns are read down.) The total number of cases (or respondents, or individuals) in the table is given in the lower-right cell (or position in the table). Note that this is fewer than the number of cases in the sample (recall that the sample consists of 1,119 cases). The difference is due to missing data; that is, some respondents did not answer all the questions needed to construct the religiosity and militancy scales. Later, we will deal extensively with missing data problems. For the present, however, we ignore the missing data and treat the sample as if it consists of 993 respondents.

2) The eight cells in the interior of the table give the bivariate frequency distribution, that is, the frequency of each combination of religiosity and militancy.

3) The titles of the variables and response categories are given in the table stubs.

4) When constructing a table, it is wise to check the accuracy of your entries by adding up the entries in each row and confirming that they correspond to the column marginal, for example, 61 + 169 = 230, and so on; adding up the entries of each column and confirming that they correspond to the row marginal, for example, 61 + 160 + 87 + 25 = 333, and so on; and adding up the row marginals and the column marginals and confirming that the sum of each corresponds to the table total. It is easy to introduce errors, especially when copying tables, and it is far better to discover them for yourself before committing them to print than for your readers to discover them after you have published. Always double-check your tables.

From this table, can we decide whether religiosity favors or inhibits militancy? Not very well. To do so, we would need to determine the *relative probability* that people of each degree of religiosity are militant. If the probability increases with religiosity, we would conclude that religiosity promotes militancy; if the probability of militancy decreases with religiosity, we would conclude that religion is an opiate. The relative probabilities are to be found by determining the conditional probability of militancy in each religiosity group, that is, the probability of militancy given that one is at a particular religiosity level. These conditional probabilities can be expressed as 61/230, 160/532, 87/195, and 25/36. Although this is a completely correct way of expressing the probabilities, they are more readily interpreted if expressed as percentages: (61/230)*100 = 27, and so on.

In fact, we ordinarily do this initially, by presenting tables of percentages rather than tables of frequencies. This makes direct comparisons of relative probabilities very easy. That is, we ordinarily would never present a table like Table 1.1 but instead would present a table like Table 1.2.

TABLE 1.2. Percent Militant by Religiosity Among Urban Negroes in the U.S., 1964.

Militancy	Very Religious	Somewhat Religious	Not Very Religious	Not at All Religious	Total
Militant	27%	30%	45%	69%	**33%**
Nonmilitant	73	70	55	31	**67**
Total	**100%**	**100%**	**100%**	**100%**	**100%**
N	(230)	(532)	(195)	(36)	(993)

Source: Table 1.1.

TECHNICAL POINTS ON TABLE 1.2

1) Always include the percentage totals (the row of 100%s). Although this may seem redundant and a waste of space, it makes it immediately clear to the reader in which direction you have percentaged the table. When the percentage totals are omitted, the reader may have to add up several rows or columns to figure it out. Using percentage signs on the top row of numbers and again on the Total row also clearly indicates to the reader that this is a percentage table.

2) Whole percentages are precise enough. There is no point in being more precise in the presentation of data than the accuracy of the data warrants. Moreover, fractions of percentages are usually uninteresting. It is hard to imagine anyone wanting to know that 37.44 percent of women and 41.87 percent of men do something; it is sufficient to note that 37 percent of women and 42 percent of men do it. Incidentally, a convenient rounding rule is to round to the even number. Thus, 37.50 becomes 38, but 36.50 becomes 36. Of course, 36.51 becomes 37 and 37.49 also becomes 37. You only want to report more than whole percentages if you have a distribution with many categories and are concerned about rounding error.

3) Always include the number of cases on which the percentages are based (that is, the denominator for the percentages). This enables the reader to reconstruct the entire table of frequencies (within the limits of rounding error) and hence to reorganize the data into a different form. Note that Table 1.2 contains all of the information

that Table 1.1 contains because you can reconstruct Table 1.1 from Table 1.2: 27 percent of 230 is 62.1, which rounds to 62 (within rounding error of 61), and so on. Customarily, percentage bases are placed in parentheses to clearly identify them and to help them stand out from the remainder of the table.

4) Sometimes it is useful to include a Total column, as I have done here, and sometimes not. The choice should be based on substantive considerations. In the present case, about one-third of the total sample is militant (as defined by Marx); hence, the marginal distribution for the dependent variable is reported here. Recall from page 7 that "militants" are those who gave militant responses to at least six of the eight items in the militancy scale. We now see that about one-third of the sample did so. Obviously, if we defined as militant all those who gave at least five militant responses, the percentage militant would be higher.

5) No convention dictates that tables must be arranged so that the percentages run down, that is, so that each column totals to 100 percent. In Table 1.2, the categories of the dependent variable form the rows, and the categories of the independent variable form the columns. If it is more convenient to reverse this, so that the categories of the independent variable form the rows, this is perfectly acceptable. The only caveat is that within each category of the independent variable, the percentage distribution across the categories of the dependent variable must total to 100 percent. Thus, if the categories of the dependent variable form the columns, the table should be percentaged across each row.

The Direction to Percentage the Table

Note that the direction in which this table is percentaged is not at all arbitrary but rather is determined by the nature of the hypothesis being tested. The question being addressed is whether religiosity promotes or hinders militancy. In this formulation, religiosity is presumed to in uence, cause, or determine militancy, not the other way around. (One could imagine a hypothesis that assumed the opposite—we might suppose that militants would tend to lose interest in religion as their civil rights involvement consumed their passions. But that is not the idea being tested here.) The variable being determined, in uenced, or caused is known as the *dependent* variable, and the variables that are doing the causing, determining, or in uencing are known as *independent*, or *predictor*, variables. The choice of causal order is always a matter of theory and cannot be determined from the data.

The choice of causal order then dictates the way the table is constructed. Tables should (almost—an exception will be presented later) always be constructed to express the conditional probability of being in each of the categories of the dependent variable given that an individual is in a particular category of the independent variable(s). (Do not let the fact that the table is expressed in percentages and the rule is expressed in probabilities confuse you. A percentage, which means "per hundred," is just a probability multiplied by 100. Percentages range from 0 to 100; probabilities range from 0 to 1.00.) Thus, in Table 1.2, I show the percentage militant for each religiosity category; that is I show the conditional probability (×100) of being militant, given that an urban Black is, respectively, very religious, somewhat religious, not very religious, or not at all religious. Note that the probability of being militant increases as religiosity decreases. Of the very religious, 27 percent are militant, as are 30 percent of the somewhat religious, 45 percent of the not very religious, and 69 percent of the not at all religious. Thus, given the formulation with which I (and Marx) started, in which religiosity was posited as alternatively an opiate or an inspiration, we are led to conclude that religiosity is an opiate because the more religious people are, the less likely they are to be militant.

It is important to understand this example thoroughly because the logic of which way to compute percentages and which comparisons to make is the same in all cross-tabulation tables.

Control Variables

Thus far, we have determined that the probability of militancy increases as religiosity decreases. Do we want to stop here? To do so would be to accept religiosity as the causative factor, that is, to conclude that religiosity causes people to be less militant. If we had a strong theory that predicted an inverse relationship between religiosity and militancy, regardless of anything else, we might be prepared to accept our two-variable cross-tabulation as an adequate test. Ordinarily, however, we will want to consider whether there are alternative explanations for the relationships we observe. In the present instance, for example, we might suspect that both religiosity and militancy are determined by some third factor. One obvious possibility is education. We might expect well-educated Blacks to be both less religious and more militant than more poorly educated Blacks. If this is so, religiosity and militancy would appear to be inversely related even if there were no causal connection between them. This is known as a *spurious association* or spurious correlation.

How can we test this possibility?

First, we need to determine whether education does in fact reduce religiosity by creating Table 1.3. This table shows that among urban Blacks in 1964, those who are well educated tend to be less religious. Of those with only a grammar school education, 31 percent are very religious, compared to 19 percent of those with a high school or college education. Further, only 1 percent of those with a grammar school education, 4 percent of those with a high school education, and 11 percent of those with a college education are not at all religious. Thus, we can say that education and religiosity are inversely or negatively associated: as education increases religiosity decreases. (Study this table carefully to see why it is percentaged as it is. What would you be asserting if you percentaged the table in the other direction?)

Next we need to determine whether education increases militancy by creating Table 1.4.

From Table 1.4, we see that the higher the level of educational attainment, the greater the percentage militant. Only 22 percent of those with grammar school education, 36 percent of those with high school education, and fully 53 percent of those with college education are militant. Another way of putting this is to say that a positive association exists between education and militancy: as education increases, the probability of militancy increases.

TABLE 1.3. Percentage Distribution of Religiosity by Educational Attainment, Urban Negroes in the U.S., 1964.

	Educational Attainment		
Religiosity	**Grammar School**	**High School**	**College**
Very religious	31%	19%	19%
Somewhat religious	57	54	45
Not very religious	12	24	25
Not at all religious	1	4	11
Total	**101%**	**101%**	**100%**
N	(353)	(504)	(136)

Source: Adapted from Marx (1967a, Table 6).

TABLE 1.4. Percent Militant by Educational Attainment, Urban Negroes in the U.S., 1964.

	Educational Attainment		
Militancy	**Grammar School**	**High School**	**College**
Militant	22%	36%	53%
Nonmilitant	78	64	47
Total	**100%**	**100%**	**100%**
N	(353)	(504)	(136)

Source: Adapted from Marx (1967a, Table 6).

TECHNICAL POINTS ON TABLE 1.3

1) Sometimes your percentages will not total to exactly 100 percent due to rounding error. Deviations of one percentage point (99 to 101) are acceptable. Larger deviations probably indicate computational error and should be carefully checked.

2) Note how the title is constructed. It states what the table is (a percentage distribution), which variables are included (the convention is to list the dependent variable first), what the sample is (urban Negroes in the U.S.), and the date of data collection (1964). The table should always contain sufficient information to enable one to read it without referring to the text. Thus, the title and variable headings should be clear and complete; if there is insufficient space to do this, it should be done in footnotes to the table.

3) In the interpretation of percentage distributions, comparing the extreme categories and ignoring the middle categories is usually sufficient. Thus, we noted that the proportion "very religious" decreases with education, and the proportion "not at all religious" increases with education. Similar assertions about how the middle categories ("somewhat religious" and "not very religious") vary with education are awkward because they may draw from or contribute to categories on either side. For example, the percentage "not very religious" among those with a college education might be larger if either the percentage "somewhat religious" or the percentage "not at all religious" were smaller. But one shift would indicate a more religious college-educated population, and the other shift would indicate a less religious college-educated population. Hence, the "not very religious" row cannot be interpreted alone, and usually little is said about the interior rows of a table. On the other hand, it is important to present the data so that the reader can see that you have not masked important details and to allow the reader to reorganize the table by collapsing categories (discussed later).

4) In dealing with scaled variables, such as religiosity, you should not make much of the relative size of the percentages within each distribution; that is, comparisons should be made across the categories of the independent variable, not across the categories of the dependent variable. In the present case, it is legitimate to note that those with a grammar school education are more likely to be very religious than are those who are better educated, but it is not legitimate to assert that more than half those with a grammar school education are somewhat religious. The reason for this is that the scale is only an ordinal scale; the categories do not carry an absolute value. How religious is "very religious"? All we know is that it is more religious than "somewhat religious." In consequence, it is easy to change the distribution simply by combining categories. Suppose, for example, we summed the top two rows and called the resulting category "religious." In this case, 88 percent of those with grammar school education would be shown as "religious." Consider how this would change the assertions we would make about this sample if we took the category labels seriously.

TECHNICAL POINTS ON TABLE 1.4

1) When you are presenting several tables involving the same data, always check the consistency of your tables by comparing numbers across the tables wherever possible. For example, the number of cases in Table 1.4 should be identical to that in Table 1.3.

Because educated urban Blacks are both less likely to be religious and more likely to be militant than are their less educated counterparts, it is possible that the observed association between religiosity and (non)militancy is determined entirely by their mutual dependence on education and that there is no connection between militancy and religiosity among people who are equally well educated. If this proves true, we would say that education *explains* the association between religiosity and militancy and that the association is spurious because it does not arise from a causal connection between the variables.

To test this possibility, we study the relation between militancy and religiosity within categories of education by creating a three-variable cross-tabulation of militancy by religiosity by education. Such a table can be set up in two different ways. The first is shown in Table 1.5, and the second in Table 1.6.

TABLE 1.5. Percent Militant by Religiosity and Educational Attainment, Urban Negroes in the U.S., 1964.

	Grammar School			High School			College		
Militancy	**V**	**S**	**N**	**V**	**S**	**N**	**V**	**S**	**N**
Militant	17%	22%	32%	34%	32%	47%	38%	48%	68%
Nonmilitant	83	78	68	66	68	53	62	52	32
Total	**100%**	**100%**	**100%**	**100%**	**100%**	**100%**	**100%**	**100%**	**100%**
N	(108)	(201)	(44)	(96)	(270)	(138)	(26)	(61)	(49)

Source: Adapted from Marx (1967a, Table 6).

*V=very religious; S=somewhat religious; N=not very religious or not at all religious.

TECHNICAL POINTS ON TABLE 1.5

1) In this sort of table, education is the control variable. The table is set up to show the relationship between militancy and religiosity within categories of education, that is (synonymously), "controlling for education," "holding education constant," or "net of education." The control variable should always be put on the outside of the tabulation so that it changes most slowly. This format facilitates reading the table because it puts the numbers being compared in adjacent columns. (Sometimes we want to study the relationship of each of two independent variables to a dependent variable, in each case controlling for the other. In such cases, we still make only one table and construct it in whatever way made it easiest to read. If our dependent variable is dichotomous or can be treated as dichotomous, we set up the table in the format of Table 1.6.)

2) Note that the "not very religious" and "not at all religious" categories were combined. This is often referred to as collapsing categories. Collapsing is usually done when there would be too few cases to produce reliable results for some categories. In the present case, as we know from Table 1.1 or 1.2, there are thirty-six people who are not at all religious. Dividing them on the basis of educational attainment would produce too few cases in each group to permit reliable estimates of the percent militant. Hence, they were combined with the adjacent group, "not very religious."

   An additional reason for collapsing categories is to improve clarity. Too much detail makes it difficult for the reader to grasp the main features of the table. Often, it helps to reduce the number of categories presented. On the other hand, if categories of the independent variable differ in terms of their distribution on the dependent variable, combining the categories will mask important distinctions. A fine balance must be struck between clarity and precision, which is why constructing tables is an art.

From Table 1.5, we see that religiosity continues to inhibit militancy even when education is controlled, although the differences in percent militant among religiosity categories tend to be smaller than in Table 1.2 where education is not controlled. (In the next chapter, we will discuss a procedure for calculating the size of the reduction in an association resulting from the introduction of a control variable, the *weighted net percentage difference*.) Among those with grammar school education, 17 percent of the very religious and 32 percent of the not religious are militant; the corresponding percentages for those with high school education are 34 and 47 and for those with college education are 38 and 68. Thus we conclude that education does not completely account for the inverse association between religiosity and militancy.

At this point, we have to decide whether to continue the search for additional explanatory variables. Our decision usually will be based on a combination of substantive and technical considerations. If we have grounds for believing that some other factor might account both for religiosity and militancy, net of education, we probably would want to control for that factor as well. Note, however, that the power of additional factors to explain the association between two original variables (here religiosity and militancy) will depend on their association with previously introduced control variables. To the extent that additional variables are highly correlated with variables already introduced, they will have little impact on the association. This is an extremely important point that will recur in the context of multiple regression analysis. Be sure you understand it thoroughly.

Consider age. What relation would you expect age to have to religiosity and to militancy?

Pause to Think About This

Religiosity is likely positively associated with age—that is, older people tend to be more religious—and militancy is inversely associated with age—younger people tend to be more militant. Hence, we might expect the association between religiosity and militancy to be a spurious function of age. That is, within age categories, there may be no association between religiosity and militancy.

What, however, of the relation between age and education? In fact, from knowledge about the secular trend in education among Blacks, we would expect younger Blacks to be substantially better educated than older Blacks. To the extent this is true, age and education are likely to have similar effects on the association between religiosity and militancy. Hence, introducing age as a control variable in addition to education is not likely to reduce the association between religiosity and militancy by much, relative to the effect of education alone.

Apart from theoretical and logical considerations (is a variable theoretically relevant, and is it going to add anything to the explanation?), there is a straightforward technical reason for limiting the number of variables included in a single cross-tabulation—we quickly run out of cases. Most sample surveys include a few hundred to a few thousand cases. We already have seen that a three-variable cross-tabulation required that we collapse two of the religiosity categories. A four-variable cross-tabulation of the same data is likely to yield so many small percentage bases as to make the results extremely unreliable. The difficulty in studying more than about three variables at a time in a cross-tabulation provides a strong motivation to use some form of regression analysis instead. A substantial fraction of the chapters to follow will be devoted to the elaboration of regression-based procedures.

Table 1.5 also enables us to assess the effect of education on militancy, controlling for religiosity by comparing corresponding columns in each of the three panels. Thus, we note that, among those who are very religious, 17 percent of the grammar school educated are militant compared to 34 percent of the high school educated and 38 percent of the college educated; among those who are somewhat religious, the corresponding percentages are 22, 32, and 48; and among those who are not religious, they are 32, 47, and 68. Hence, we conclude that, at any given level of religiosity, the better educated are more militant.

TABLE 1.6. Percent Militant by Religiosity and Educational Attainment, Urban Negroes in the U.S., 1964 (Three-Dimensional Format).

	Educational Attainment		
Religiosity	**Grammar School**	**High School**	**College**
Very religious	17% (108)	34% (96)	38% (26)
Somewhat religious	22% (201)	32% (270)	48% (61)
Not very or not at all religious	32% (44)	47% (138)	68% (49)

Source: Table 1.5.

TECHNICAL POINTS ON TABLE 1.6

1) Each pair of entries gives the percentage of people who have a trait and the percentage base, or denominator, of the ratio from which the percentage was computed. Thus, the entry in the upper-left corner indicates that 17 percent of the 108 very religious grammar-school-educated people in the sample are militant. From this table, we can reconstruct any of the preceding five tables (but with the two least religious categories collapsed into one), within the limits of rounding error. Try to do this to confirm that you understand the relationships among these tables.

This requires a fairly tedious comparison, however, skipping around the table to locate the appropriate cells. When the dependent variable is dichotomous, that is, has only two response categories, a much more succinct table format is possible and is preferred. Table 1.6 contains exactly the same information as Table 1.5, but the information is arranged in a more succinct way. Tables like Table 1.6 are known as three-dimensional tables.

Compare Tables 1.5 and 1.6. You will see that they contain exactly the same information—all the additional numbers in Table 1.5 are redundant. Moreover, Table 1.6 is much easier to read because we can see the effect of religiosity on militancy, holding constant education, simply by reading down the columns, and can see the effect of education on militancy, holding constant religiosity, simply by reading across the rows.

WHAT THIS CHAPTER HAS SHOWN

In this chapter, we have seen an initial idea formulated into a research problem, an appropriate sample chosen, a survey conducted, and a set of variables created and combined into scales to represent the concepts of interest to the researcher. We then considered how to construct a percentage table that shows the relationship between two variables, with special attention to determining in which direction to percentage tables using the concept of conditional probability distributions—the probability distribution over categories of the dependent variable computed separately for each category of the independent variable(s). This is the most difficult concept in the chapter, and one you should make sure you completely understand.

The other important concept you need to understand fully is the idea of statistical controls, also known as controlling for or holding constant confounding variables, to determine whether relationships hold within categories of the control variable(s). Finally, we considered various technical issues regarding the construction and presentation of tables. The aim of the game is to construct attractive, easy to read tables.

In the next chapter, we continue our discussion of cross-tabulations, considering various ways of analyzing tables with more than two variables and, more generally, the logic of multivariate analysis.

CHAPTER

MORE ON TABLES

WHAT THIS CHAPTER IS ABOUT

In this chapter we expand our understanding of how to deal with cross-tabulations, both substantively and technically. First we continue our consideration of the *logic of elaboration*, that is, the introduction of additional variables to an analysis; second, we consider a special situation known as a *suppressor* effect, when the influences of two independent variables offset each other; third, we consider how variables combine to produce particular effects, drawing a distinction between *additive* and *interaction* effects; fourth, we see how to assess the effect of a single independent variable in a multivariate percentage table while controlling for the effects of the other independent variables via *direct standardization*; and finally we consider the distinction between *experiments* and *statistical controls*.

THE LOGIC OF ELABORATION

In traditional treatments of survey research methods (for example, Lazarsfeld 1955; Zeisel 1985), it was customary to make a distinction between two situations in which a third variable completely or partially accounts for the association between two other variables: *spurious* associations and associations that can be accounted for by an *intervening* variable or variables. The distinction between the two is that when a control variable (Z) is temporally or causally prior to an independent variable (X) and dependent variable (Y), and when the control variable completely or partly explains the association between the independent and dependent variable, we infer that there is no causal connection or only a weak causal connection between the independent and dependent variables. However, when the control variable intervenes temporally or causally between the independent and dependent variables, we would not claim that there is no causal relationship between the independent and dependent variables but rather that the intervening variable explains, or helps explain, how the independent variable exerts its effect on the dependent variable. In the previous chapter we considered spurious associations. In this chapter we revisit spurious associations and also consider the effect of intervening variables.

PAUL LAZARSFELD (1901–1976) was a major force—perhaps the preeminent figure—in the establishment of survey research as the dominant method of data collection in American sociology. Born in Vienna, he earned a doctorate in mathematics (with a dissertation that dealt with mathematical aspects of Einstein's gravitational theory). He came to sociology by way of his 1926 marriage to Marie Jahoda and joined with her (and Hans Zeisel) in the now-classic Marienthal study (see the biosketch of Zeisel later in this chapter). His interest in social research lasted, although his marriage to Jahoda did not. He came to the United States in 1933, with an appointment first at the University of Newark (now part of Rutgers University) and then at Columbia University, where he founded the Bureau for Applied Social Research, the training ground for many prominent social scientists and the organizational base for many innovations in quantitative data collection and analysis. One of the hallmarks of Lazarsfeld's approach was that social research was most effectively accomplished as a collective endeavor, involving a team of specialists. But perhaps most important was his insistence that tackling applied questions was a legitimate endeavor of academic sociology and that answers to such questions could contribute to the theoretical development of the discipline.

Spurious Association

Consider the three variables, X, Y, and Z. Suppose that you had observed an association between X and Y and suspected that it might be completely explained by the dependence of both X and Y on Z. (For a substantive example, recall the hypothesis in the previous chapter that the negative relation between religiosity and militancy was due to the dependence of both on education—Blacks with more education were both less religious and more militant.) Such a hypothesis might be diagrammed as shown in Figure 2.1.

Causal diagrams of this sort are used for purposes of explication throughout the book. They are extensively used in *path analysis*, which is a way of representing and algebraically manipulating structural equation models that was widely used in the 1970s but is less frequently encountered now (see additional discussion of structural equation models and path analysis in Chapter Sixteen). My use of such models is purely heuristic. Nonetheless, I use them in such a way as to be conceptually complete. Hence, the paths from x to X (p_{Xx}) and from y to Y (p_{Yy}) indicate that other factors besides Z influence X and Y.

Now, if the association between X and Y within categories of Z were very small or nonexistent, we would regard the association between X and Y as entirely explained by their mutual dependence on Z. However, this generally does not happen; recall, for example, that the negative association between religiosity and militancy did not disappear when education was held constant. We ordinarily do not restrict ourselves to an all-or-nothing hypothesis of spuriousness—except in the exceptional case where we have a very strong theory requiring that a particular relation be completely spurious; rather, we ask what the association is between X and Y controlling for Z (and what the association is between Z and Y controlling for X). The logic of our analysis can be diagramed as shown in Figure 2.2.

To state the same point differently, rather than assuming that the causal connection between X and Y is zero and determining whether our assumption is correct, we estimate the relation between X and Y holding constant Z and determine its size—which, of course, may be zero, in which case Figure 2.1 and Figure 2.2 are identical.

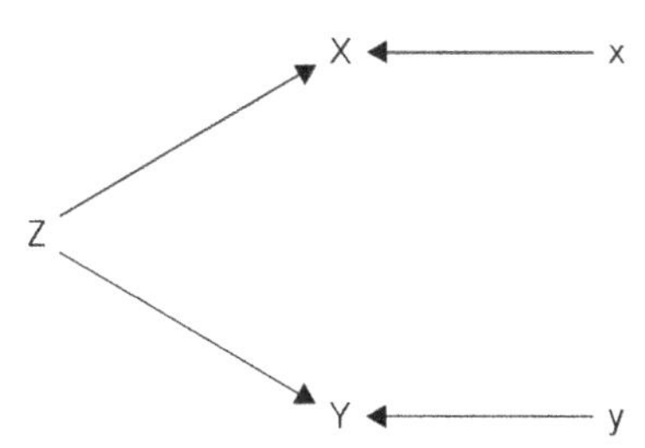

FIGURE 2.1. *The Observed Association Between X and Y Is Entirely Spurious and Goes to Zero When Z Is Controlled.*

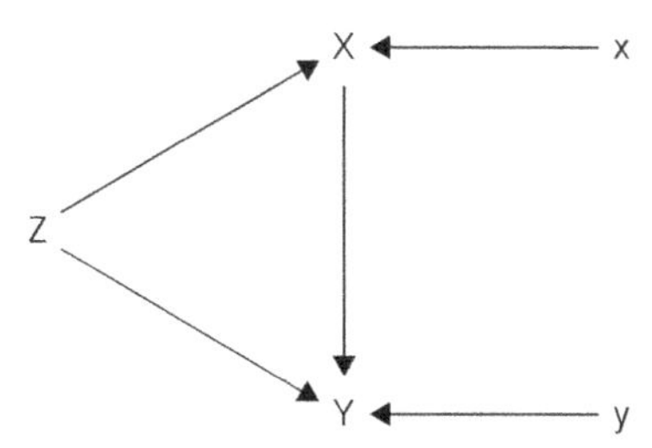

FIGURE 2.2. *The Observed Association Between X and Y Is Partly Spurious: the Effect of X on Y Is Reduced When Z Is Controlled (Z Affects X and Both Z and X Affect Y).*

Intervening Variables

Now let us consider the intervening variable case. Suppose we think two variables, *X* and *Y*, are associated only because *X* causes *Z* and *Z* causes *Y*. An example might be the relation between a father's occupation, son's education, and son's income. Suppose we expect the two-variable association between *X* and *Y*—sometimes called the *zero-order association*, short for zero-order partial association, that is, no partial association—to be positive, but think that this is due entirely to the fact that the father's occupational status influences the son's education and that the son's education influences the son's income; we think there is no direct influence of the father's occupational status on the son's income, only the indirect influence through the son's education. This sort of claim can be diagrammed as shown in Figure 2.3.

But, as before, unless we have a very strong theory that depends on there being no direct connection between *X* and *Y*, we probably would inspect the data to determine the influence of *X* on *Y*, holding constant the intervening variable *Z*, and would also determine the influence of *Z* on *Y*, holding constant the antecedent variable *X*. This can be diagrammed as shown in Figure 2.4

If the net, or *partial*, association between *X* and *Y* proves to be zero, we would conclude that a chain model of the kind described in Figure 2.3 describes the data. Otherwise, we would simply assess the strength and nature of both associations, between *X* and *Y* and between *Z* and *Y* (and, for completeness, the zero-order association between *X* and *Z*).

Notice the similarity between Figure 2.2 and Figure 2.4. With respect to the ultimate dependent variable, *Y*, the two models are identical. The only difference has to do with the specification that *Z* causes *X* or that *X* causes *Z*. There is still another possibility: *X* and *Z* cause *Y*, but no claim is made regarding the causal relation between *X* and *Z*. This can be diagrammed as shown in Figure 2.5.

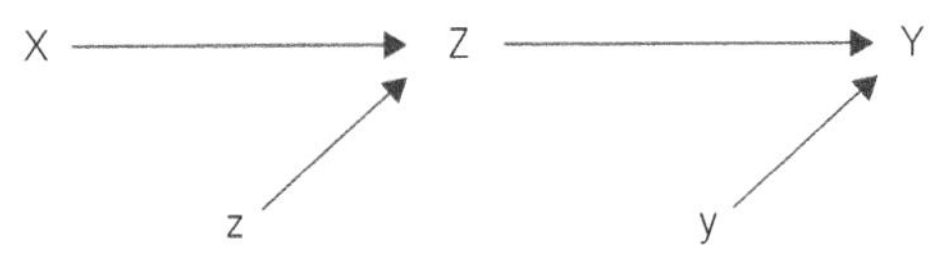

FIGURE 2.3. *The Observed Association Between X and Y Is Entirely Explained by the Intervening Variable Z and Goes to Zero When Z Is Controlled.*

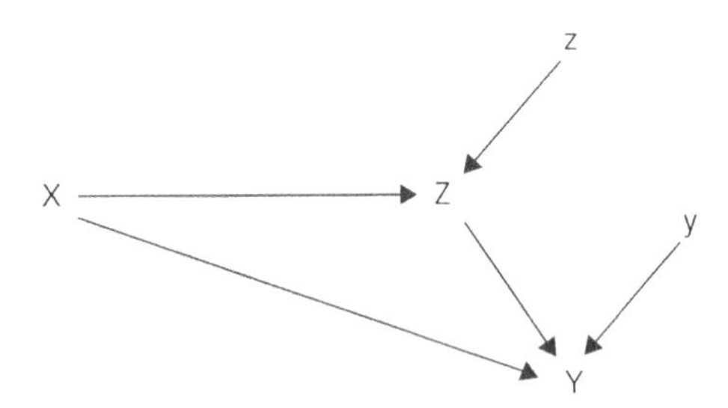

FIGURE 2.4. *The Observed Association Between X and Y Is Partly Explained by the Intervening Variable Z: the Effect of X on Y Is Reduced When Z is Controlled (X Affects Z, and Both X and Z Affect Y).*

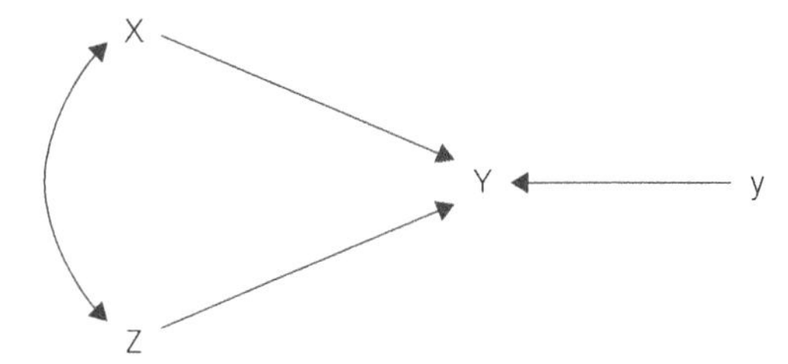

FIGURE 2.5. *Both X and Z Affect Y, but there Is no Assumption Regarding the Causal Ordering of X and Z.*

In almost all of the analyses we undertake—including cross-tabulations of the kind we are concerned with at present, multivariate models in an ordinary least squares regression framework, and log-linear and logistic analogs to regression for categorical dependent variables—the models, or theories, represented by Figures 2.2, 2.4, and 2.5 will be analytically indistinguishable with respect to the dependent variable, Y. The distinction among them thus must rest with the ideas of the researcher, not with the data. From the standpoint of data manipulation, all three models require assessing the net effect of each of two variables on a third variable, that is, the effect of each independent variable holding constant the other independent variable. Obviously, the same ideas can be generalized to situations involving more than three variables.

SUPPRESSOR VARIABLES

One final idea needs to be discussed here, the notion of *suppressor variables*. Thus far, we have dealt with situations in which we suspected that an observed association between two variables was due to the effect of a third, either as an antecedent or an intervening variable. Situations can arise, however, in which there appears to be *no* association between two variables when, in fact, there is a causal connection. This happens when some other variable is related to the two variables in such a way that it *suppresses* the observed zero-order association—specifically, when one independent variable has opposite effects on another independent variable and on the dependent variable, and the two independent variables have opposite effects on the dependent variable. Such situations can be diagrammed as shown in Figure 2.6.

For example, suppose you are interested in the relations among education, income, and fertility. On theoretical grounds, you might expect the following: education will have a positive effect on income; holding constant income, education will have a negative effect on fertility (the idea being that educated people want to do more for their children and regard children as more expensive than do poorly educated people; hence at any given level of income, they have fewer children); holding constant education, the higher the income, the higher the number of children (the idea being that children are generally regarded as desirable so that at any given level of the perceived cost of children, those with more to spend, that is, with higher income, will have more children). These relationship are represented in Figure 2.6, where X = level of education, Z = income, and Y = number of

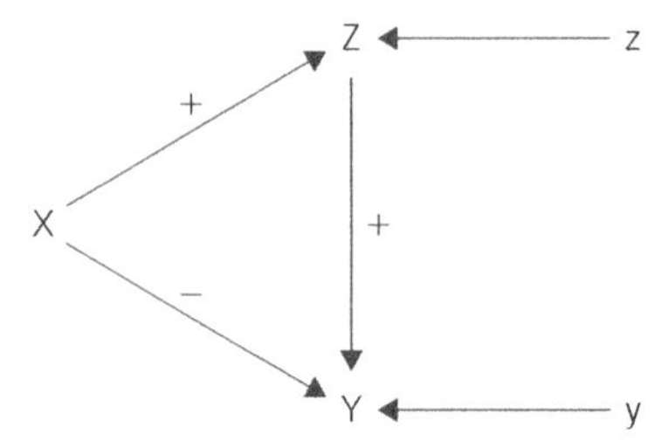

FIGURE 2.6. *The Size of the Zero-Order Association Between X and Y (and Between Z and Y) Is Suppressed When the Effects of X on Z and Y have Opposite Sign, and the Effects of X and Z on Y have Opposite Sign.*

children. The interesting thing about this diagram is that it implies that the gross, or zero-order, relationships between *X* and *Y* and between *Z* and *Y* will be smaller than the net, or first-order partial, relationships and might even be zero, depending on the relative size of the associations among the three variables. To see how this happens, consider the relationship between education and fertility. We have posited that educated people tend to have more income, and at any given level of education, higher-income people tend to have more children. Hence, so far, the relation between education and fertility would be expected to be positive. But we also have posited that at any given income level, better-educated people tend to have fewer children. So we have a positive causal path and a negative causal path at work at once, and the effect of each is to offset or *suppress* the other effect, so that the zero-order relationship between education and fertility is reduced.

ADDITIVE AND INTERACTION EFFECTS

We now consider *interaction effects*, situations in which the effect of one variable on another is contingent on the value of a third variable. To see this clearly, consider Table 2.1.

This table shows that in 1965 educational attainment had no effect on acceptance of abortion among Catholics but that among Protestants, the greater the education, the greater the percentage accepting abortion. Thus, Catholics and Protestants with 8th grade education or less were about equally likely to believe that legal abortion should be permitted under specified circumstances but, among those with more education, Protestants were substantially more likely to accept abortion than were Catholics. Among the college educated, the difference between the religious groups is fully twenty points: about 31 percent of Catholics and 51 percent of Protestants believed that abortion should be permitted.

This kind of result is called an *interaction effect*. Religion and educational attainment *interact* to produce a result different from what each would produce alone. That is, the relationship between education and acceptance of abortion differs for Catholics and Protestants, and the relationship between religion and acceptance of abortion differs by education. Situations in which the relationship between two variables depends on the value of a third, as it does here, are known as interactions. In the older survey analysis literature (for example, Lazarsfeld 1955; Zeisel 1985), interactions are sometimes called *specifications*. Religion *specifies* the relationship between education and beliefs about abortion: acceptance of abortion increases with education among Protestants but not among Catholics.

TABLE 2.1. Percentage Who Believe Legal Abortions Should Be Possible Under Specified Circumstances, by Religion and Education, U.S. 1965 (N = 1,368; Cell Frequencies in Parentheses).

	Educational Attainment			
Religion	**8th Grade or Less**	**Some High School**	**High School Graduate**	**Some College or More**
Catholic	31% (90)	33% (96)	33% (89)	31% (75)
Protestant	29% (287)	36% (250)	43% (256)	51% (225)

Source: Rossi (1966).
Note: Non-Christians omitted.

HANS ZEISEL (1905–1992) is best known for his classic book on how to present statistical tables, *Say It with Figures*, which has appeared in six editions and has been translated into seven languages. Born in what is now the Czech Republic, Zeisel was raised and educated in Vienna, where he and Paul Lazarsfeld were personal and professional collaborators—both leaders of the Young Socialists and, together with Marie Jahoda, authors of the now classic study of the social impact of unemployment on Marienthal, a small Austrian community (1971 [1933]). Zeisel earned a law degree from the University of Vienna in 1927 and practiced law there as well as conducting social research until 1938, when the *Anschluss* forced him to flee to New York. There he carried out market research at McCann-Erickson and then the Tea Council until 1953, when he joined the law faculty at the University of Chicago, where he conducted a number of empirical studies in the area now known as "law and society" research. He was a pioneer in the use of statistical evidence and survey data in legal cases, often serving as an expert witness.

Where we do not have interaction effects we have *additive effects* (or no effects). Suppose, for example, that instead of the numbers in Table 2.1, we had the numbers shown in Table 2.2.

What would this show? We could say two things: (1) The effect of religion on acceptance of abortion is the same at all levels of education. That is, the difference between the percentage of Catholics and Protestants who would permit abortion is 10 percent at each level of education. (2) The effect of education on acceptance of abortion is the same for Catholics and Protestants. For example, the difference between the percentage who would permit abortion among those with some high school education and those who are high school graduates is 10 percent for both Catholics and Protestants. Similarly, the difference between the percentage who would permit abortion among those with a high school education and among those with at least some college is 20 percent for both Catholics and Protestants.

TABLE 2.2. Percentage Accepting Abortion by Religion and Education (Hypothetical Data).

	8th Grade or Less	Some High School	High School Graduate	Some College or More
Catholic	30%	35%	45%	65%
Protestant	40%	45%	55%	75%

The reason this table is *additive* is that the effects of each variable add together to produce the final result. It is as if the probability of any individual in the sample accepting abortion is at least .3 (so we could add .3 to every cell in the table); the probability of a Protestant accepting abortion is .1 greater than the probability of a Catholic accepting abortion (so we could add .1 to all the cells containing Protestants); the probability of someone with some high school accepting abortion is .05 greater than the probability of someone with an 8th grade education doing so (so we would add .05 to the cells for those with some high school); the probability of those who are high school graduates accepting abortion is .15 greater than the probability of those with an 8th grade education doing so (so we would add .15 to the cells for those with high school degrees); and the probability of those with some college accepting abortion is .35 higher than the probability of those with an 8th grade education doing so (so we would add .35 to the cells for those with at least some college). This would produce the results we see in Table 2.2 (after we convert proportions to percentages by multiplying each number by 100).

By contrast, it is not possible to add up the effect of each variable in a table containing interactions because the effect of each variable depends on the value of the other independent variable or variables.

Many relationships of interest to social scientists involve interactions—especially with gender and to some extent with race; but it is also true that many relationships are additive. Adequate theoretical work has not yet been done to allow us to specify very well in advance which relationships we would expect to be additive and which relationships we would expect to involve interactions.

Later you will see more sophisticated ways to distinguish additive effects from interactions and to deal with various kinds of interactions via log-linear analysis and regression analysis.

DIRECT STANDARDIZATION

Often we want to assess the relationship between two variables controlling for additional variables. Although we have seen how to assess partial relationships—that is, relationships between two variables within categories of one or several control variables—it would be helpful to have a way of constructing a single table that shows the average relationship between two variables *net of*, that is, *controlling for*, the effects of other variables. *Direct standardization* provides a way of doing this. Note that this technique has other names, for example, *covariate adjustment*. However, the technique is most widely used in demographic research, so I use the term by which it is known in demography, *direct standardization*. It is important to understand that, even though the same term is used, this procedure has no relationship to standardizing variables to create a common metric. We will consider this subject in Chapter Five.

Example 1: Religiosity by Militancy Among U.S. Urban Blacks

The procedure is most easily explained in the context of a concrete example. Thus we revisit the analysis shown in Tables 1.2 through 1.6 of Chapter One (slightly modified). Recall that we were interested in whether the relationship between militancy and religiosity among Blacks in the United States could be explained by the fact that better-educated Blacks tend to be both less religious and more militant. Because education does not completely explain the association between militancy and religiosity, it would be useful to have a way of showing the association remaining after the effect of education has been removed. We can do this by getting an adjusted percentage militant for each religiosity category, which we do by computing a weighted average of the percent militant across education categories within each religion category but with the weights taken from the overall frequency distribution of education in the sample. (Alternatively, because they are mathematically identical, we can compute the weighted *sum*, using as weights the *proportion* of cases in each category.) By doing this, we construct a hypothetical table showing what the relationship between religiosity and militancy would be if all religiosity groups had the same distribution of education. It is in this precise sense that we can say we are showing the association between religiosity and militancy net of the effect of education. As noted earlier, this procedure is known as direct standardization or covariate adjustment.

Note that the weights need not be constructed from the overall distribution in the table. Any other set of weights could be applied as well. For example, if we wanted to assess the association between religiosity and militancy on the assumption that Blacks had the same distribution of education as Whites, we would treat Whites as the *standard population* and use the White distribution across educational categories (derived from some external source) as the weights. We will see two examples of this strategy a bit later in the chapter.

Now let us construct a militancy-by-religiosity table adjusted, or standardized, for education, to see how the procedure works. We do this from the data in Table 1.6. First, we derive the standard distribution, the overall distribution of education. Because there are 993 cases in the table (= 108 + . . . + 49), and there are 353 (= 108 + 201 + 44) people with a grammar school education, the proportion with a grammar school education is .356 (=353/993). Similarly, the proportion with a high school education is .508, and the proportion with a college education is .137. These are our weights. Then to get the adjusted, or standardized, percent militant among the very religious, we take the weighted sum of the percent militant across the three education groups that subdivide the "very religious" category (that is, the figures in the top row of the table): 17%*.356

TABLE 2.3. Percent Militant by Religiosity, and Percent Militant by Religiosity Adjusting (Standardizing) for Religiosity Differences in Educational Attainment, Urban Negroes in the U.S., 1964 (N = 993).

	Percent Militant	Percent Militant Adjusted for Education	N
Very religious	27	29	(230)
Somewhat religious	30	31	(532)
Not very or not at all religious	48	45	(231)
Percentage spread	21	16	

+ 34%*.508 + 38%*.137 = 29%. To get the adjusted percent militant among the "somewhat religious," we apply the same weights to the percentages in the second row in the table: 22%*.356 + 32%*.508 + 48%*.137 = 31%. Finally, to get the adjusted percent militant among the "not very or not at all religious," we do the same for the third row of the table, which yields 45 percent. We can then compare these percentages to the corresponding percentages for the zero-order relationship between religiosity and militancy (that is, not controlling for education). The comparison is shown in Table 2.3. (The Stata -`do`- file used to carry out the computations, using the command -`dstdize`- and the -`log`- file that shows the results, are available as downloadable files from the publisher, Jossey-Bass/Wiley (www.josseybass.com/go/quantitativedataanalysis) as are similar files for the remaining worked examples in the chapter. Because we have not yet begun computing, it probably is best to note the availability of this material and return to it later unless you are already familiar with Stata.)

STATA -DO- FILES AND -LOG- FILES In Stata, -`do`- files are commands, and -`log`- files record the results of executing -`do`- files. As you will see in Chapter Four, the management of data analysis is complex and is much facilitated by the creation of -`do`- files, which are efficient and also provide a permanent record of what you have done to produce each tabulation or coefficient. Anyone who has tried to replicate an analysis performed several years or even several months earlier will appreciate the value of having an exact record of the computations used to generate each result.

When presenting data of this sort, it is sometimes useful to compare the range in the percentage positive (in this case, the percent militant) across categories of the independent variable, with and without controls. In Table 2.3, we can see that the difference in the percent militant between the least and most religious categories is twenty-one points whereas, when education is controlled, the difference is reduced to sixteen points, a 24 percent reduction (= 1 − 16/21). In some sense, then, we can say that education "explains" about a quarter of the relationship between religiosity and militancy. We need to be cautious about making computations of this sort and only employ them when they are helpful in making the analysis clear. For example, it doesn't make much sense to compute a "spread" or "range" in the percentages if the relationship between religiosity and militancy is not *monotonic* (that is, if the percentage militant does not increase, or at least not decrease, as religiosity declines).

DIRECT STANDARDIZATION IN EARLIER SURVEY RESEARCH Although direct standardization is a conventional technique in demographic analysis, it also appears in the early survey research literature as a vehicle for getting to a "weighted net percentage difference" or "weighted net percentage spread." The really useful part of the procedure is the computation of adjusted, or standardized, rates. The subsequent computation of percentage differences or percentage spreads is only sometimes useful, as a way of summarizing the effect of control variables.

Example 2: Belief That Humans Evolved from Animals (Direct Standardization with Two or More Control Variables)

Sometimes we want to adjust, or standardize, our data by more than one control variable at a time to get a summary of the effect of some variable on another when two or more other variables are held constant. Consider, for example, acceptance of the scientific theory of evolution. In 1993, 1994, and 2000, the NORC *General Social Survey* (GSS) included a question:

> *For each statement below, just check the box that comes closest to your opinion of how true it is. In your opinion, how true is this? . . . Human beings developed from earlier species of animals.*

Table 2.4 shows the distribution of responses for the three years combined. Because there was virtually no difference between years, data from all three years have been combined to increase the sample size. Procedures for assessing variation across years are discussed in Chapter Seven. A description of the GSS and details on how to obtain GSS data sets are given in Appendix A. The Stata -do- file used to create this example and the resulting `-log-` file can be downloaded from the publisher's Web site, noted earlier. In general, the `-do-` and `-log-` files for each worked example are available for downloading.

To me, Table 2.4 is startling. How can it be that the overwhelming majority of American adults fail to accept something that is undisputed in the scientific community?

TABLE 2.4. Percentage Distribution of Beliefs Regarding the Scientific View of Evolution (U.S. Adults, 1993, 1994, and 2000).

Evolutionary explanation is . . .	
Definitely true	15.4%
Probably true	32.3
Probably not true	16.8
Definitely not true	35.5
Total	**100.0%**
N	(3,663)

Perhaps the recent increase in the proportion of the population that adheres to fundamentalist religious beliefs, especially fundamentalist Protestant views in which the Bible is taken as literally true, accounts for this outcome. To see whether this is so, in Table 2.5 I cross-tabulated acceptance of the scientific view of evolution (measured by endorsement of the statement that descent from other animals is "definitely true") by religious denomination, making a distinction between "fundamentalist" and "denominational" Protestants. (For want of better information, I simply dichotomized Protestant denominations on the basis of the proportion of their members in the sample who believe that "The Bible is the actual word of God and is to be taken literally, word for word." Denominations for which at least 50 percent of respondents gave this response—"other" Protestants and all Baptists except members of the "American Baptist Church in the U.S.A." and "Baptists, don't know which"—were coded as fundamentalist; all other Protestant denominations were coded as Denominational Protestants.) Unfortunately, the same distinction, between religious subgroups with and without a literal belief in the holy scriptures of their faith, cannot be made for non-Protestants given the way the data were originally coded in the GSS.

Although there are substantial differences among religious groups in their acceptance of the scientific view of evolution, the fundamentalist-denominational split among Protestants does not seem to be central to the explanation because there is only a 4 percent difference between the two groups. Interestingly, non-Christians appear to be much more willing than Christians to accept an evolutionary perspective, and Catholics appear to be more willing than Protestants to do so.

Given these patterns, it could well be that the observed religious differences are, at least in part, spurious. In particular, educational differences among religious groups—Jews are particularly well educated and fundamentalist Protestants are particularly poorly

TABLE 2.5. Percentage Accepting the Scientific View of Evolution by Religious Denomination (N = 3,663).

	Percentage Accepting the Evolution of Humans from Animals as "Definitely True"	N
Fundamentalist Protestants	8.0	(968)
Denominational Protestants	11.8	(1,222)
Catholics	17.8	(858)
Other Christians	5.6	(18)
Jews	38.6	(83)
Other religion	23.6	(123)
No religion	32.5	(391)

educated—might partly account for religious differences in acceptance of the scientific view. Similarly, age differences among religious groups—the young are particularly likely to reject religion—might provide part of the explanation as well.

To consider these possibilities, we need to determine, first, whether acceptance of the scientific explanation of human evolution varies by age and education and, if so, whether religious groups differ with respect to their age and education. Tables 2.6 and 2.7 provide the necessary information regarding the first question, and Tables 2.8 and 2.9 provide the corresponding information regarding the second question.

Unsurprisingly, endorsement of the statement that humans evolved from other animals as "definitely true" increases sharply with education, as we see in Table 2.6, ranging from 9 percent of those with no more than a high school education to 36 percent of those with post-graduate education. It is also true that younger people are more likely to endorse the scientific explanation of evolution than are older respondents (see Table 2.7): 18 percent of those under age fifty, compared to 7 percent of those seventy and over, say that it is "definitely true" that humans evolved from other animals.

As expected, Table 2.8 shows that Jews are by far the best-educated religious group, followed by other non-Christian groups, and that Fundamentalist Protestants and Other Christians are the least well educated. Also as expected, Table 2.9 shows that those without religion tend to be young. However, members of "other" religious groups also tend—disproportionately—to be young, perhaps because they are mainly immigrants.

TABLE 2.6. Percentage Accepting the Scientific View of Evolution by Level of Education.

	Percentage	N
High school or less	9.2	(1,743)
Some college	11.9	(936)
College graduate	24.8	(561)
Post-graduate education	36.2	(423)

TABLE 2.7. Percentage Accepting the Scientific View of Evolution by Age.

	Percentage	N
18–49	17.5	(2,373)
50–69	13.5	(889)
70+	7.0	(401)

These results suggest that differences among religious groups with respect to age and education might, indeed, explain part of the observed difference in acceptance of the scientific view of evolution.

To see to what extent age and educational differences among religious groups account for religious group differences in acceptance of evolution, we can directly standardize the religion/evolution-beliefs relationship for education and age. We do this by determining the joint distribution of the entire sample with respect to age and education and then use the proportion in each age-by-education category as weights with which to compute, separately for each religious group, the weighted average of the age-by-education-specific percentages accepting a scientific view of evolution. By doing this, we treat each religious group as if it had exactly the same joint distribution with respect to age and education as did the entire sample. This procedure thus adjusts the percentage of each religious group that endorses the scientific perspective on evolution to remove the effect of religious group differences in the joint distribution of age and education.

TABLE 2.8. **Percentage Distribution of Educational Attainment by Religion.**

	High School or Less	Some College	College Graduate	Post-Graduate	Total	N
Fundamentalist Protestants	55.7	26.4	10.7	7.1	99.9	(968)
Denominational Protestants	47.6	26.5	15.2	10.6	99.9	(1,222)
Catholics	45.6	24.1	18.1	12.2	100.0	(858)
Other Christians	61.1	16.7	16.7	5.6	100.0	(18)
Jews	15.7	21.7	31.3	31.3	100.0	(83)
Other religion	36.6	25.2	19.5	18.7	100.0	(123)
No religion	41.4	24.8	16.1	16.6	99.9	(391)
Total	**47.6**	**25.6**	**15.3**	**11.6**	**100.1**	**(3,663)**

TABLE 2.9. **Percentage Distribution of Age by Religion.**

	18–49	50–69	70+	Total	N
Fundamentalist Protestants	59.6	27.9	12.5	100.0	(968)
Denominational Protestants	59.8	25.0	15.1	99.9	(1,222)
Catholics	67.8	24.8	7.3	99.9	(858)
Other Christians	88.9	11.1	0.0	99.9	(18)
Jews	61.4	25.3	13.2	99.9	(83)
Other religion	83.7	13.8	2.4	99.9	(123)
No religion	80.0	15.4	4.6	100.0	(391)
Total	**64.8**	**24.3**	**11.0**	**100.1**	**(3,663)**

To get the necessary weights, we simply cross-tabulate age by education and express the number of people in each cell of the table as a proportion of the total. These proportions are shown in Table 2.10.

We then tabulate the percentage accepting the scientific position on evolution by religion, age, and education. These percentages are shown in Table 2.11. Note that many of these percentages are based on very few cases. This means that they are not very precise, in the sense that they are subject to large sampling variability. We could collapse the education and age categories still further, but that would ignore substantial within-category heterogeneity. As always, there is a balance to be struck between sampling precision and substantive sensibility or—in terminology we will adopt later—between *reliability* and *validity*. In the present case, I might have been better advised to take a more conservative approach, especially because the cell-specific percentages bounce around a lot (exactly as we would expect given the large degree of sampling variability), which makes the differences in the resulting standardized percentages somewhat less clear cut. On the other hand, the weights are very small for the cells based on few cases, which minimizes their contribution to the overall percentages.

Finally, to get the adjusted, or standardized, coefficients, we sum the weighted percentages, where the weights are the proportions in Table 2.10. For example, the adjusted, or directly standardized, percentage of Fundamentalist Protestants who accept the evolutionary viewpoint as "definitively true" is, within rounding error,

$$\begin{aligned} 9.7 = {} & 5.7*.274 + 3.8*.184 + 15.7*.110 + 29.3*.080 \\ & + 4.9*.126 + 3.3*.056 + 25.0*.032 + 40.9*.029 \\ & + 3.3*.076 + 7.7*.015 + 10.0*.012 + 16.7*.007 \end{aligned}$$

The remaining standardized percentages are derived in the same way. They are shown in Table 2.12, with the observed percentages repeated from Table 2.5 to make comparisons easier.

TABLE 2.10. **Joint Probability Distribution of Education and Age.**

	18–49	50–69	70+	Total
High school or less	.274	.126	.076	**.476**
Some college	.184	.056	.015	**.256**
College graduate	.110	.032	.012	**.153**
Post-graduate education	.080	.029	.007	**.116**
Total	**.648**	**.243**	**.110**	**1.001**

TABLE 2.11. Percentage Accepting the Scientific View of Evolution by Religion, Age, and Education (Percentage Bases in Parentheses).

	Fundamentalist Protestants	Denominational Protestants	Catholic	Other Christians	Jewish	Other	None
Age 18–49							
High school or less	5.7 (283)	11.8 (321)	9.1 (220)	[10.0] (10)	[25.0] (4)	21.1 (38)	18.6 (129)
Some college	3.8 (183)	10.1 (208)	20.1 (159)	[0.0] (2)	[28.6] (14)	11.5 (26)	29.3 (82)
College graduate	15.7 (70)	19.3 (119)	30.4 (125)	[0.0] (3)	[41.2] (17)	35.0 (20)	40.4 (47)
Post-graduate education	29.3 (41)	25.3 (83)	37.2 (78)	[0.0] (1)	[62.5] (16)	[36.8] (19)	58.2 (55)
Age 50–69							
High school or less	4.9 (164)	6.2 (146)	11.8 (119)	[0.0] (1)	[75.0] (4)	[0.0] (4)	22.7 (22)
Some college	3.3 (60)	9.0 (78)	6.8 (44)	[0.0] (1)	[25.0] (4)	20.0 (5)	[21.4] (14)
College graduate	25.0 (24)	21.3 (47)	20.0 (25)	– (0)	[20.0] (5)	[25.0] (4)	[41.7] (12)
Post-graduate education	40.9 (22)	20.0 (35)	32.0 (25)	– (0)	[37.5] (8)	[25.0] (4)	[66.7] (12)

(Continued)

TABLE 2.11. Percentage Accepting the Scientific View of Evolution by Religion, Age, and Education (Percentage Bases in Parentheses). *(Continued)*

	Fundamentalist Protestants	Denominational Protestants	Catholic	Other Christians	Jewish	Other	None
Age 70 or more							
High school or less	3.3	1.7	5.8	–	[20.0]	[33.3]	[27.3]
	(92)	(115)	(52)	(0)	(5)	(3)	(11)
Some college	[7.7]	5.3	[0.0]	–	–	–	[0.0]
	(13)	(38)	(4)	(0)	(0)	(0)	(1)
College graduate	[10.0]	10.0	[20.0]	–	[0.0]	–	[50.0]
	(10)	(20)	(5)	(0)	(4)	(0)	(4)
Post-graduate education	[16.7]	[16.7]	[0.0]	–	[50.0]	–	[100.0]
	(6)	(12)	(2)	(0)	(2)	(0)	(2)

Note: Percentages based on fewer than twenty cases (shown in square brackets) should be interpreted with caution.

TABLE 2.12. Observed Proportion Accepting the Scientific View of Evolution, and Proportion Standardized for Education and Age.

	Percentage Accepting Scientific View of Evolution as "Definitely True"	Percentage Accepting Scientific View, Standardized by Age and Education	N
Fundamentalist Protestants	8.0	9.7	(968)
Denominational Protestants	11.8	12.2	(1,222)
Catholics	17.8	16.6	(858)
Other Christians	5.6	2.7	(18)
Jews	38.6	36.0	(83)
Other religion	23.6	19.9	(123)
No religion	32.5	30.2	(391)

As you can see, despite the association between religious group affiliation and, respectively, age and education, and the association of age and education with acceptance of an evolutionary account of human origins, standardizing for these variables has relatively little impact on religious group differences in acceptance of the claim that humans evolved from other animals. The one exception is the non-religious, whose support for a scientific view of evolution appears to be due, in part, to their relatively young age. Despite minor shifts in the expected direction for Fundamentalist Protestants and for Jews and those with other religions, the dominant pattern is one of religious group differences in acceptance of an evolutionary view of the origins of mankind that are *not* a simple reflection of religious differences in age and education but presumably reflect, instead, the theological differences that distinguish religious categories.

Example 3: Occupational Status by Race in South Africa

Now let us consider another example: the extent to which racial differences in occupational attainment in South Africa can be explained by racial differences in education (the data are from the *Survey of Economic Opportunity and Achievement in South Africa*, conducted in the early 1990s [Treiman, Lewin, and Lu 2006]; the Stata `-do-` and `-log-` files for the worked example are available as downloadable files; for information on the data set and how to obtain it, see Appendix A). From the left-hand panel of Table 2.13, it is evident that there are strong differences in occupational attainment by race. Non-Whites, especially Blacks, are substantially less likely to be managerial, professional, or technical workers than are Whites and are substantially more likely to be semiskilled or unskilled manual workers. Moreover, Blacks are far more likely to be unemployed than are members of any other group. It is also well known that substantial racial differences exist in educational attainment in South Africa, with Whites by far the best educated, followed, in order, by Asians (who in South Africa are mainly descendants of people brought as indentured workers from the Indian subcontinent), Coloureds (mixed-race persons), and Blacks (these are the racial categories conventionally used in South Africa); and also that in South Africa, as elsewhere, occupational attainment depends to a considerable degree on educational attainment (Treiman, McKeever, and Fodor 1996). Under these circumstances, we might suspect that racial differences in occupational attainment can be largely explained by racial differences in educational attainment. Indeed, this is what Treiman, McKeever, and Fodor (1996) found using the International Socioeconomic Status Index (ISEI) (Ganzeboom, de Graaf, and Treiman 1992; Ganzeboom and Treiman 1996) as an index of occupational attainment. However, it also is possible that access to certain types of occupations, such as professional and technical positions, depends heavily upon education, whereas access to others, such as managerial positions, may be denied on the basis of race to those who are educationally qualified.

To determine to what extent, and for which occupation categories, racial differences in access can be explained by racial differences in education, I adjusted (directly standardized) the relationship between race and occupational status by education. Here I used the White distribution of education, computed from the weighted data, as the standard distribution to determine what occupational distributions for each of the non-White groups might be expected were they able to upgrade their levels of educational attainment so that they had the same distributions across schooling levels as did Whites.

The results are shown in panel B of Table 2.13. They are quite instructive. Bringing the other racial groups to the White distribution of education (and assuming that doing so would not affect the relationship between education and occupational attainment within each group), racial differences in the likelihood of being a professional would entirely disappear. Indeed, Blacks would be slightly more likely than members of the other groups to become professionals. By contrast, the percentage of each race group in the managerial category would remain essentially unchanged, suggesting that it is not education but rather norms about who is permitted to supervise whom that account for the racial disparity in this category. The remaining large changes apply to only one or two of the three non-White groups: Asians would not be very substantially affected except for a reduction in the proportion semiskilled; Coloureds would increase the proportion in technical jobs and reduce the proportion in semiskilled and unskilled manual jobs and farm labor; and Blacks would increase the proportion in clerical jobs and reduce the proportion in all manual categories. If all four racial groups had the same educational distribution as Whites, the dissimilarity (measured by Δ; see Chapter Three) between the occupational distributions of Whites and Asians would be reduced by about 30 percent (from 29.2 to 20.5) as would the dissimilarity in the occupational distributions of Whites and Coloureds (from 37.9 to 26.5), whereas the dissimilarity in the occupational distributions of Whites

TABLE 2.13. Percentage Distribution of Occupational Groups by Race, South African Males Age 20–69, Early 1990s (Percentages Shown Without Controls and also Directly Standardized for Racial Differences in Educational Attainment;[a] N = 4,004).

	Without Controls				Adjusted for Education			
	White	Asian	Coloured	Black	White	Asian	Coloured	Black
Managers	18.4	9.7	3.3	0.7	18.1	10.5	5.2	1.5
Professionals	13.7	7.0	5.3	3.3	13.2	13.6	11.9	16.5
Technical	17.1	11.6	5.0	2.0	16.9	13.3	8.5	2.4
Clerical	7.2	13.4	7.1	5.8	7.2	13.0	9.1	8.5
Service	4.0	3.9	4.0	6.8	4.0	2.8	4.3	6.7
Sales	2.2	8.2	2.8	3.5	2.2	9.4	2.5	2.6
Farm	4.6	0.5	2.3	0.8	4.5	0.3	1.1	0.3
Skilled manual	18.5	21.1	22.2	14.4	18.6	18.8	19.3	8.2
Semiskilled	3.2	12.9	10.4	16.7	3.2	7.6	6.6	9.9
Unskilled manual	0.6	3.6	14.4	13.5	0.6	2.6	7.6	6.2
Farm laborers	0.0	0.2	7.8	2.3	0.0	0.0	3.1	0.4
Occupation unknown[b]	9.8	5.6	11.4	13.0	9.7	6.2	16.6	20.1
Unemployed	0.9	2.2	4.0	17.2	0.9	1.8	4.4	16.6
Total	**100.2**	**99.9**	**100.0**	**100.0**	**99.1**	**99.9**	**100.0**	**100.0**
N	(1236)	(412)	(396)	(1960)	(1236)	(412)	(396)	(1960)
Dissimilarity from Whites (Δ)[c]		29.2	37.9	52.4		20.5	26.5	46.1

[a]The standard population is the educational distribution of the White male population—computed from the survey data weighted to conform to the census distributions of region by urban versus rural residence.

[b]Because coding of occupation data to detailed occupations had not been completed when this table was prepared, I have included a separate category "occupation unknown."

[c]Index of Dissimilarity = 1/2 the sum of the absolute values of the differences between the percentage of Whites and the percentage of the other racial group in each occupation category. See Chapter Three for further exposition of this index.

and Blacks would be reduced only about 12 percent (from 52.4 to 46.1). The substantial remaining dissimilarity in the occupational distributions of the four race groups net of education suggests that Treiman, McKeever, and Fodor's (1996) conclusion that education largely explains occupational *status* differences between race groups in South Africa does not tell the whole story.

Example 4: Level of Literacy by Urban Versus Rural Residence in China

Now consider a final example—the relationship among education, urban residence, and degree of literacy in the People's Republic of China. In a 1996 national sample of the adult population (Treiman, Walder, and Li 1996), respondents were asked to identify ten Chinese characters (see Appendix A regarding the properties of this data set and how to obtain access to it). The number of correct identifications is interpreted as indicating the degree of literacy (Treiman 2007a). Obviously, literacy would be expected to increase with education. Moreover, I would expect the urban population to score better on the character recognition task just because urban respondents tend to get more schooling than do rural respondents. The question of interest here is whether educational differences between the rural and urban population entirely explain the observed mean difference in the number of characters correctly identified, which is 1.8 (as shown in Table 2.14). To determine this, I adjusted (directly standardized) the urban and rural means by assuming that both populations have the same distribution of education—the distribution for the entire adult population of China, computed from the weighted data. Note that in this example it is not percentages that are standardized but rather means. The procedure is identical in both cases, although if this is done by computer (using Stata), a special adjustment needs to be made to the data to overcome a limitation in the Stata command—the requirement that the numerators of the "rates" to be standardized (what Stata calls `-charvar-`) be integers. To see how to do this, consult the Stata `-do-` and `-log-` files for this example, which are included in the set of downloadable files for this chapter.

TABLE 2.14. Mean Number of Chinese Characters Known (Out of 10), for Urban and Rural Residents Age 20–69, China 1996 (Means Shown Without Controls and Also Directly Standardized for Urban-Rural Differences in the Distribution of Education;[a] N = 6,081).

	Without Controls	Adjusted for Education	N
Urban residents	4.8	4.0	(3,079)
Rural residents	3.0	3.4	(3,002)
Difference	1.8	0.6	

[a]The standard population is the entire population of China age 20–69, computed from the survey data weighted to reflect differential sampling rates for the rural and urban populations and to correct for variations in household size. Nine cases for which data on education were missing were omitted.

The results are quite straightforward and require little comment. When education is standardized, the urban-rural gap in the mean number of characters correctly identified is reduced from 2.8 to 1.6. Thus, about 43 percent (= 1 − 1.6/2.8) of the urban-rural difference in vocabulary knowledge is explained by rural-urban differences in the level of educational attainment.

Although the four examples presented here all standardize for education, this is purely coincidental. Many other uses of direct standardization are imaginable. For example, it probably would be possible to explain higher crime rates among early twentieth-century immigrants to the United States than among natives simply by standardizing for age and sex. Immigrants were disproportionately young males, and young males are known to have higher crime rates than any other age-sex combination.

A FINAL NOTE ON STATISTICAL CONTROLS VERSUS EXPERIMENTS

In describing the logic of cross-tabulations, I have been describing the logic of nonexperimental data analysis in general. True experiments are relatively uncommon in social research, although they are widely used in psychological research and increasingly in microeconomics (for a very nice example of the latter, see Thomas and others [2004]). A true experiment is a situation in which the objects of the experiment are randomly divided into two or more groups, and one group is exposed to some treatment while the other group is not, or several groups are exposed to different treatments. If the groups then differ in some outcome variable, the differences can be attributed to the differences in treatments. In such cases we can unambiguously establish that the treatment caused the difference in outcomes (although we may not know the exact mechanism involved). (Of course, this claim holds only when differences between the experimental and control groups are not inadvertently introduced by the investigators as a consequence of design flaws or of failure to rigorously adhere to the randomized trial design. For a classic discussion of such problems, see Campbell and Stanley [1966] or a shorter version by Campbell [1957] that contains the core of the Campbell and Stanley material.)

When experiments are undertaken in fields such as chemistry, sampling is not ordinarily a consideration because it can be confidently assumed that any batch of a chemical will behave like any other batch of the same chemical; only when things go wrong do chemists tend to question that assumption. In the social and behavioral and many of the biological sciences, by contrast, it cannot be assumed that one subject is just like another subject. Hence, in experiments in these fields, subjects are randomly assigned to treatment groups. In this way, it becomes possible to assess whether group differences in outcomes are larger than would be likely to occur by chance because of sampling variability. If so, we can say, subject only to the uncertainty of statistical inference, that the difference in treatments caused the difference in outcomes.

In the social sciences, random assignment of subjects to treatment groups is often—in fact usually—impossible for several reasons. First, both ethical and practical considerations limit the kind of experimentation that can be done on human subjects. For example, it would be neither ethical nor practically possible to determine whether one sort of schooling was pedagogically superior to another by randomly assigning children to different schools and several years later determining their level of educational achievement. In addition, many phenomena of interest to social scientists are simply not experimentally manipulable, even in principle. The propensity for in-group solidarity to increase in wartime, for example, is not something that can be experimentally confirmed, nor can the proposition that social stratification is more pronounced in sedentary agricultural societies than in hunter-gatherer societies.

Occasionally, "natural experiments" can be analyzed. Natural experiments are situations in which different individuals are exposed to different circumstances, and it can be reasonably assumed that the circumstance to which individuals are exposed is essentially random. A very nice example of such an analysis is the test by Almond (2006) of the "fetal origins hypothesis." He showed convincingly that individuals in utero during the few months in which the 1918 flu pandemic was raging suffered reduced educational attainment, increased rates of physical disability, and lower income in midlife relative to those in utero in the few months preceding and following the epidemic. Because there is no basis for expecting the exact month of conception to be correlated with vulnerability to the flu virus, the conditions of a natural experiment were well satisfied in this elegant analysis. Natural experiments have become increasingly popular in economics as the limitations of various statistical fixes to correct for "sample selection bias" have become more evident. We will return to this issue in the final chapter. (For additional examples of natural experiments that are well worth reading, see Campbell and Ross [1968], Berelson [1979], Sloan and others [1988], and the papers cited in Chapter Sixteen.)

Given the limited possibilities for experimentation in the social sciences, we resort to a variety of statistical controls of the sort discussed here and later. These procedures share a common logic: they are all designed to hold constant some variable or variables so that the net effect of a given variable on a given outcome can be assessed.

THE WEAKNESS OF MATCHING AND A USEFUL FIX

Sometimes survey analysts attempt to simulate random assignment by matching comparison groups on some set of variables. In its original form, this practice was inherently unsatisfactory. When attempting to match on all potentially relevant factors, it is difficult to avoid running out of cases. Moreover, no matter how many variables are used in the match, it is always possible that the experimental and control groups differ on some nonmatched factor that is correlated with the experimental outcome. However, combining matching with statistical controls can be a useful strategy, especially when the adequacy of the match is summarized via a "propensity score" (Rosenbaum and Rubin 1983). For recent treatments of propensity score matching, see Smith (1997), Becker and Ichino (2002), Abadie and others (2004), Brand (2006), Brand and Halaby (2006), and Becker and Caliendo (2007). Harding (2002) is an instructive application. Propensity score matching is also discussed in Chapter Sixteen.

Compared to experiments, statistical controls have two fundamental limitations, which make it impossible to definitively prove any causal theory (although definitive *disproof* is possible). First, no matter how many control variables we introduce, we can never be sure that the remaining net relationship is a true causal relationship and not the spurious result of some yet-to-be-introduced variable.

Second, although we speak of *holding constant* some variable, or set of variables, what we usually do in practice is simply reduce the within-group variability for these variables. This is particularly obvious when we are dealing with cross-tabulations because we generally divide the sample into a small set of categories. In what sense, for example, can we be said to "hold education constant" when our categories consist of those with less than a high school education, those with some sort of high school experience, and those with some sort of college experience? Although the within-category variability in educational attainment obviously is smaller than the total variability in the sample as a whole, it is still substantial. Hence, if two other variables both depend on educational attainment, they are likely to be correlated within educational categories as gross as these, as well as across educational categories. As you will see in more detail later, using interval or ratio variables in a regression framework will not solve the problem but merely transform it. Although the within-category variability generally will be reduced, the very parsimony in the expression of relationships between variables that regression procedures permit will generally result in some distortion of the true complexities of such relationships—discontinuities, nonlinearities, and so on, only some of which can be represented succinctly.

Our only salvation is adequate theory. Because we can seldom definitively establish causal relationships by reference to data, we need to build up a body of theory that consists of a set of plausible, mutually consistent, empirically verified propositions. Although we cannot definitively prove causal relations, we can determine whether our data are *consistent with* our theories; if so, we can say that the proposition is tentatively empirically verified. We are in a stronger position when it comes to *disproof*. If our data are *inconsistent* with our theory, that usually is sufficient grounds for rejecting the theory, although we need to be sensitive to the possibility that there are omitted variables that would change our conclusions if they were included in our cross-tabulation or model. In short, to maintain a theory, it is *necessary* but not *sufficient* that the data be as predicted by the theory. Because consistency is necessary to maintain the theory, inconsistency is sufficient to require us to reject it—provided we can be confident that we have not omitted important variables. (On the other hand, as Alfred North Whitehead is supposed to have said, never let data stand in the way of a good theory. If the theory is sufficiently strong, you might want to question the data. I will have more to say about this later, in a discussion of concepts and indicators.)

WHAT THIS CHAPTER HAS SHOWN

In this chapter we have considered the logic of multivariate statistical analysis and its application to cross-tabulations involving three or more variables. The notion of an interaction effect—a situation in which the effect of one independent variable depends on the value or level of one or more other independent variables—was introduced. This is a very important idea in statistical analysis, and so you should be sure that you understand it thoroughly. We also considered suppressor effects, situations in which the effect of one independent variable offsets the effect of another independent variable because the two effects have opposite signs. In such situations, the failure to include both variables in the model can lead to an understatement of the true relationships between the included variable and the dependent variable. We then turned to direct standardization (sometimes called covariate adjustment), a procedure for purging a relationship of the effect of a particular variable or variables. Direct standardization can be thought of as a procedure for creating "counterfactual" or "what if" relationships—for example, what would be the relationship between religiosity and militancy if we adjusted for the fact that well educated Blacks in the 1960s tended to be both less religious and more militant than less well educated Blacks. Having discussed the logic of direct standardization, we considered several technical aspects of the procedure to see how to standardize data starting not only from tables but also from individual records; to standardize percentage distributions; and to standardize means. We concluded by considering the limitations of statistical controls, in contrast to randomized experiments.

In the following chapter, we complete our initial discussion of cross-tabulation tables by considering how to extract new information from published tables; then note the one circumstance in which it makes sense to percentage a table "backwards"; touch for the first but not for the last time on how to handle missing data; consider cross-tabulation tables in which the cell entries are means; present a measure of the similarity of percentage distributions, the Index of Dissimilarity (Δ); and end with some comments about how to write about cross-tabulations.

CHAPTER

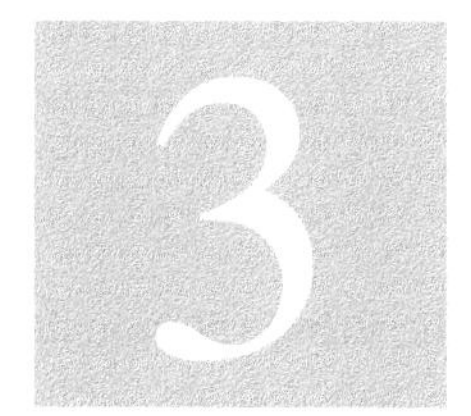

STILL MORE ON TABLES

WHAT THIS CHAPTER IS ABOUT

In this chapter we wrap up our discussion of cross-tabulations for now. After spending some time learning to love the computer—a very brief time, actually—and then delving into the mysteries of regression equations, we will return to cross-tabulations and discuss procedures for making inferences about relations embodied in them via log-linear analysis.

We begin this chapter with a discussion of how to extract new information from published tables; then note the one circumstance in which it makes sense to percentage a table "backwards"; touch for the first but not for the last time on how to handle missing data; consider cross-tabulation tables in which the cell entries are means; present a measure of the similarity of percentage distributions, the Index of Dissimilarity (Δ); and end with some comments about how to write about cross-tabulations.

REORGANIZING TABLES TO EXTRACT NEW INFORMATION

Often when working with published data, reading research papers, and so on, we wish that data had been presented differently. Sometimes sufficient information is presented to enable us to reorganize or recompute the table to draw the inferences that we, as opposed to the original author, are interested in. Two such possibilities are reviewed here.

Collapsing Dimensions

Suppose that you are interested in the gross relationship between acceptance of abortion and religion, and you have available only a table such as Table 2.1 in Chapter Two. How could you extract the two-variable table, "Percent accepting abortion (that is, thinking legal abortions should be possible under specified circumstances) by religion"? The procedure is quite straightforward. First, you convert the percentage table into a table of frequencies: 31 percent of 90 = 27.9, which rounds to 28; 33 percent of 96 = 31.68, which rounds to 32; and so on. This yields Table 3.1. (The total row is, of course, formed simply by adding up each column. A check on your arithmetic can be conducted by adding across the total row, adding the frequencies in Table 2.1, and determining whether the sums agree. They do, both yielding a sample size of 1,368.) The total row can then be percentaged in the usual way, to yield 32 percent (= 100*[112/(112 + 238)]) of Catholics and 39 percent (= 100*[398/(398 + 620)]) of Protestants accepting abortion.

The same result can be obtained more efficiently by simply computing the weighted average of the percentage accepting abortion across the four education groups, separately for Catholics and Protestants, in just the same way as we compute weighted averages to obtain directly standardized rates: for example, for Catholics [(31%)(90) + (33%)(96) + (33%)(89) + (31%)(75)]/(90 + 96 + 89 + 75) = 32%. The advantage of computing the full table of frequencies is, first, that it provides a better check on the accuracy of your com-

TABLE 3.1. Frequency Distribution of Acceptance of Abortion by Religion and Education, U.S. Adults, 1965 (N = 1,368).

	Catholic		Protestant	
Education	Accept	Don't	Accept	Don't
8th grade or less	28	62	83	204
Some high school	32	64	90	160
High school grad	29	60	110	146
Some college or more	23	52	115	110
Total	**112**	**238**	**398**	**620**

Source: Table 2.1.

of your computations and, second, that it permits other tabulations to be constructed, for example, the zero-order relation between education and acceptance of abortion.

Although many other examples could be given, they all follow the same logic. You should get in the habit of manipulating tables to extract information from them. Not only is it a useful skill but it also gives you a better understanding of how tables are constructed.

Collapsing Categories to Represent New Concepts

Sometimes we want to view a variable in a manner entirely different from that envisioned by the original investigator, in which case we may want to reorder the categories. We already have seen one example of this, in our discussion of how to treat "no answer" in our consideration of nominal variables in Chapter One. "No answer" may be thought of as a neutral response and hence as lying between the least positive and the least negative response; or "no answer" might be thought of as not on the continuum at all, and hence best treated as missing data.

Another example can be drawn from the U.S. Congress. In the late 1970s, the *New York Times*, the *Washington Post*, and similar rags took to calling conservative Democrats "boll weevils" and liberal Republicans "gypsy moths" (fads come and go; you never hear these terms any more). Suppose we were conducting a study of members of the U.S. House of Representatives and initially classified every member into one of the following four categories:

1. Standard Republicans
2. Gypsy moths
3. Boll weevils
4. Standard Democrats

This four-category classification can be collapsed into three distinct two-category classifications, each of which represents a different theoretical construct. If we were interested in studying party politics and wanted to know which party controlled the House, we would combine category 1 with category 2, and combine category 3 with category 4:

Standard Republicans
Gypsy moths } Republicans

Boll weevils
Standard Democrats } Democrats

If we were interested in distinguishing between liberals and conservatives, we would combine category 1 with category 3 and combine category 2 with category 4:

Standard Republicans
Boll Weevils } Conservatives

Gypsy moths
Standard Democrats } Liberals

If we were interested in studying party loyalty and wanted to know what proportion of congressmen are party loyalists, we would combine category 1 with category 4 and combine category 2 with category 3:

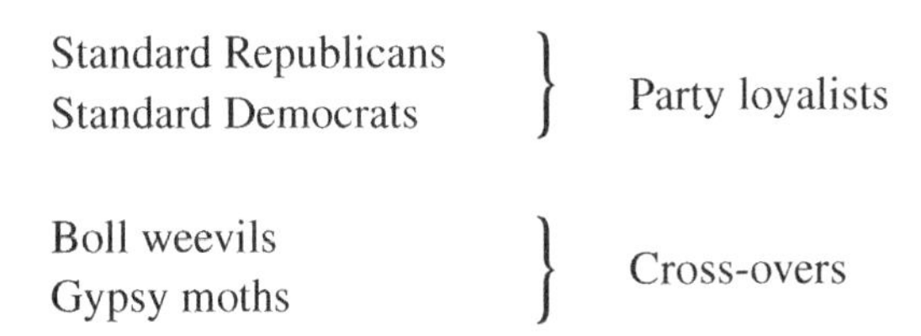

The point of all this is that nothing is sacrosanct about the way a variable is originally constructed. You can and should recode variables freely to get the best representation of the concept you are interested in studying.

A very important corollary of this point is that when you are designing or executing a data collection effort, you should always conserve as much detail as possible. In the early days of survey research, the technology of data manipulation encouraged researchers to pack as many variables as possible onto one IBM card; hence highly aggregated classifications were adopted to save space (and the tedium of manipulation). The technology has changed. Today there is—with one exception—no reason not to preserve as much detail as possible in the initial coding of your variables. (The exception is that you need to design your data-collection instrument in a way that minimizes respondent, interviewer, and coder error. For example, in a survey with data collection done by face-to-face interviews, a lengthy and complicated coding scheme for a variable is likely to increase interviewer error.) You never know when you will get a new idea that will require recoding one or several variables; and if you lack imagination, the next user of the same data set may not. Every experienced survey analyst has faced great frustration on countless occasions because detail that could have been preserved by those who initially collected the data was not. Collapsing, or aggregating, variables into a smaller number of categories is an easy computer operation; disaggregating variables is impossible, at least without going back to the original questionnaire and usually not then either.

WHEN TO PERCENTAGE A TABLE "BACKWARDS"

There is one exception to the rule that tables should be percentaged so that the categories of the dependent variable add to 100 percent. This is in cases where the sample is not representative of the population "at risk" of falling into the various categories of the dependent variable. Sometimes samples are stratified on the dependent variable rather than the independent variables or variables; that is, sometimes they are chosen on the basis of their value on the dependent variable. Various hard-to-find populations are typically sampled in this way: convicted criminals, university students, political activists, cancer patients, and so on.

TABLE 3.2. Social Origins of Nobel Prize Winners (1901–1972) and Other U.S. Elites (and, for Comparison, the Occupations of Employed Males 1900–1920).

	Father's Occupation			
Elite	**Professional**	**Manager or Proprietor**	**Other**	**Total%**
Supreme Court justices	56	34	10	**100%**
Nobel laureates	54	28	18	**100%**
Admirals and generals	45	29	26	**100%**
Business leaders	15	57	28	**100%**
Diplomats	32	36	32	**100%**
Senators	24	35	41	**100%**
Employed males				
1900	3	7	90	**100%**
1910	4	8	88	**100%**
1920	4	8	88	**100%**

Source: Adapted from Zuckerman (1977, 64).

For example, Table 3.2 shows the social origins of various American elites. In this case the table is percentaged to show the distribution of fathers' occupations for each of a number of elite groups, and also for the U.S. labor force as a whole for selected years roughly corresponding to when the fathers were in the labor force. The point of the table is, of course, to show that elites come from elite origins: much higher percentages of the members of these elites are from professional or managerial origins than would be expected if their fathers' occupations corresponded to the distribution of professionals and managers in the labor force. The table is percentaged in this direction, contrary to the usual rule that percentages express the conditional probability of some outcome given some causal or antecedent condition, because it is constructed from information obtained from samples of elites (plus some general labor force data), and therefore is not representative of the social origins of the population. It would not be sensible to use data from a representative sample of the population to study the likelihood that the children of professionals become Supreme Court justices, Nobel laureates, and so on, because we would virtually never find any cases with these outcomes unless we obtained data from the entire population—the outcomes are simply too rare. Thus, in such cases, we rely on *response-based samples*, and percentage the table to show the distribution of the independent variable for each response category—in the present case, the social origins of various elites compared to the general population.

CROSS-TABULATIONS IN WHICH THE DEPENDENT VARIABLE IS REPRESENTED BY A MEAN

When the dependent variable is an interval or ratio variable, it often is useful to display the means of the dependent variable within categories formed by a cross-classification of independent variables. For example, suppose you are interested in the relationship among gender, education, and earnings, perhaps because you suspect that women get smaller returns on their education than do men. Table 3.3 shows the mean annual income in 1979 for full-time workers by level of educational attainment and gender, computed from the 1980 NORC *General Social Survey*.

TABLE 3.3. **Mean Annual Income in 1979 Among Those Working Full Time in 1980, by Education and Gender, U.S. Adults (Category Frequencies Shown in Parentheses).**

	Male	Female	Total
Post-graduate studies	31,864 (57)	14,113 (21)	**26,890** (78)
College graduate	27,227 (46)	11,789 (35)	**20,512** (81)
Some college	19,222 (68)	13,003 (52)	**16,540** (120)
High school graduate	16,288 (131)	10,324 (105)	**13,536** (236)
Less than twelve years	15,855 (78)	8,399 (33)	**13,489** (111)
Total	**20,415** (380)	**11,135** (246)	**16,654** (626)

TECHNICAL POINTS ON TABLE 3.3

1. Note that the format of this table is identical to that of Table 1.6 from Chapter One, except that percentages are presented in Table 1.6, and means are presented here. The tables are read in the same way.

2. In this table levels of educational attainment are presented in descending order. Either descending or ascending order is appropriate; the choice should depend on which makes the discussion easier.

3. Note that this table includes only 626 cases, out of a total sample of 1,468. This reflects the fact that many individuals do not work full time, particularly women, and also that information on education and income is missing for some individuals. Sometimes it is useful to catalog the missing cases, especially when there are many missing cases or when their distribution has substantive importance. In such cases, a footnote can be added to the table or an addition made at the bottom of the table, for example,

Number of cases in table	626
No information on income	57
No information on education	1
No information on education and income	1
Total working full time	685
Men not working full time	235
Women not working full time	549
Total in full sample	1,469

   The reason this tabulation shows 1,469 when there are 1,468 cases in the sample is due to rounding error. Because of errors in the execution of a "split ballot" procedure in the 1980 GSS, the data have to be weighted to be representative of the population (Davis, Smith, and Marsden 2007). We will consider weighting issues in Chapter Nine.

   Even when you do not present the information shown in the tabulation, it is wise to compile it for yourself, as a check on your computations. In fact, in the course of creating the preceding accounting of missing cases, I discovered a computing error I had made that resulted in incorrect numbers in Table 3.3 (since corrected).

4. An alternative way to display these data, which would make the point of the table more immediately evident to the reader, is to show, in the rightmost column, female means as a percentage of male means, rather than the total means. Table-making is an art, and the aim of the game is to make the message as clear and easy to understand as possible.

Inspecting Table 3.3, you see that in 1980, women earned much less than equally well-educated men, although for both men and women income tended to increase as the level of education increased. The gender difference in incomes is striking: on average, women earned just over half of what men did, and the best educated women (those with post-graduate training) earned less on average than did the least educated men (those who did not complete high school).

To provide an easily grasped comparison of male and female average incomes for each level of education, we can compute the ratio of female to male means. Ordinarily, these would simply be included in an additional column in the table or as a substitute for the total column.

Educational Attainment	*Mean Female Income Expressed as a Percentage of Male Mean Income*
Post-graduate training	44
College graduate	43
Some college	68
High school graduate	63
Less than 12 years	53
Total	55

The computations here are just the ratios multiplied by 100, which yield the female means expressed as percentages of the male means. They show that within education categories, women on average earn between two-fifths and two-thirds what men do. You might be curious whether things have changed since 1980. To find out, you can construct the same table from a more recent GSS.

SUBSTANTIVE POINTS ON TABLE 3.3

The ratio of female to male incomes shown here (55 percent) is somewhat lower than the ratio typically estimated from census data (for example, Treiman and Hartmann 1981, 16), which is about 60 percent. The discrepancy may reflect differences in the definition of full-time workers. Most of the computations based on census (or Current Population Survey [CPS]) data define "full-time year-round" workers as those employed at least thirty-five hours in the week preceding the survey and employed at least fifty weeks in the previous year. The GSS question, by contrast, asks whether people were working in the previous week and if so, how many hours, or, if they had a job but were not working in the previous week, how many hours they usually worked. It may be that the GSS table includes a substantial number of people who did not work full time in the previous year and therefore had lower incomes than those employed full time, whereas these people would be excluded from computations based on census or CPS data. Because women tend to have more unstable employment histories than men, it is probable that those included in the GSS but not the census definition of "full time" would be mainly women, which would lower the GSS ratio relative to ratios computed from census or CPS data. Note that there is a certain amount of slipperiness to the analysis using either the GSS or the census definitions of full-time workers: information on hours worked per week at the time of the survey is related to income computed for the previous calendar year. There is no help for this because the alternative is to ask about hours per week typically worked last year, which is bound to be highly error prone, or to ask about current salary or wage—which is also highly error prone because income is highly variable over the course of the year. The convention, which is the convention because it is thought to yield the best data, is to ask about hours worked in the past week but to ask the weeks worked and income questions with respect to the past calendar year.

Another possible reason for the discrepancy between the GSS and census estimates of the ratio of female to male incomes is that the GSS figures are subject to substantial sampling error. We will take up statistical inference in survey analysis in Chapter Nine.

The point of this note is to emphasize that whenever your results differ from those reported by others, especially those that are widely cited, it is important to attempt to account for the differences as best you can, and to eliminate candidate explanations that prove to be incorrect. Your papers should be filled with comments of this sort; they give the reader confidence that you have thought through the issues and are aware of what is going on—in your data and in the literature.

Inferences from Information on Missing Data

Note that the catalog of sources of missing data presented in the technical note on Table 3.3 can be combined with information in the table to get an approximate estimate of sex differences in labor force participation rates. The row marginal of the table tells us that there are 380 males and 246 females employed full time for whom complete information is available. From the information in the technical note, we see that there are 235 males and 549 females who are not employed full time. If we are willing to ignore the 59 people who are employed full time but for whom information is missing on education or income, we can estimate that 62 (= [380/(380 + 235)]*100) percent of the males in the sample and 31 (= [246/(246 + 549)]*100) percent of the females in the sample were employed full time during the week of the survey. Of course, because we have the data, we could get these estimates directly and would not have to ignore the 59 missing cases. But if we had only the published table and the accounting of sources of missing data, we could use them to estimate labor force participation rates, even though the table was not presented with this in mind.

Still Another Way of Presenting the Same Data

Sometimes it is useful to present standard deviations as well as means in tables such as Table 3.3. When you need to present standard deviations as well as means, a useful way to avoid overcrowding your tables is to present several panels, as in Table 3.4. The point of presenting the standard deviations is both to enable the reader to do statistical inference computations from the data in the table (the standard deviations are needed to compute confidence intervals for tests of the significance of the difference between means) and to provide substantive information. For example, it is informative to note—from the rightmost column—that the heterogeneity in income is more than three times as great for men with post-graduate training as for women with post-graduate training—a ratio that is much larger than for any of the other levels of education. This gives us a hint as to why the average income of women with post-graduate training is so low—unlike their male counterparts, some of whom get extremely high-paying jobs, these women appear to be locked into a set of jobs with a very narrow range of incomes. We could take this further by investigating the properties of such jobs—but we will not do so here.

A serious shortcoming in the comparison of means across groups is that means, unlike medians, are sensitive to *outliers*—extreme observations. Thus, for example, the inclusion of a few very high-income people in a sample can substantially affect the computed means. This is equally a problem when the data are coded into a set of categories with a *top code* for incomes higher than some value, as is the case for the income measures used in the GSS. In 1980 the top code for income was $50,000. To compute a mean, a value has to be assigned to each category. This is not much of a problem for most categories; it is conventional, and reasonably accurate, to simply assign the midpoint of the range included. For example, the bottom category, "under $1,000," would be assigned $500, and so on. But for the top code, any decision is likely to be arbitrary. One possibility is to use a Pareto transformation to estimate the mean value of the top code (Miller 1966, 215–220), but this depends on rather strong assumptions regarding the shape of the distribution. In the analysis shown here I thus, rather arbitrarily, assigned $62,500 to the top code. Had I assigned, say, $75,000, the male-female income differences for well-educated people would have been larger, and the male standard deviations would have been larger as well. In the case of *skewed* (asymmetrical) distributions where one tail is longer than the other, of which income is perhaps the most common example, it makes more sense to compute *medians* for descriptive purposes, although for analytic purposes most analysts resort to a transformation of income, usually by taking the natural log of income because medians are very algebraically intractable. Table 3.5 is the equivalent of Table 3.3 except that medians are substituted for means. (If an analyst wants an analog to a standard deviation, the interquartile range is commonly used.) In this case the means and the medians yield similar interpretations, but often this is not the case.

TABLE 3.4. Means and Standard Deviations of Income in 1979 by Education and Gender, U.S. Adults, 1980.

Education	Males	Females	Female as Percent of Male
	Means		
Post-graduate training	31,864	14,113	44
College graduate	27,227	11,789	43
Some college	19,222	13,003	68
High school graduate	16,288	10,324	63
Less than 12 years	15,855	8,399	53
Total	**20,415**	**11,135**	**55**
	Standard Deviations		
Post-graduate training	17,541	5,019	29
College graduate	14,618	6,794	46
Some college	12,912	9,704	75
High school graduate	8,935	7,573	85
Less than 12 years	11,488	6,280	55
Total	**13,790**	**7,750**	**56**
	Frequencies		
Post-graduate training	57	21	
College graduate	46	35	
Some college	68	52	
High school graduate	131	105	
Less than 12 years	78	33	
Total	**380**	**246**	

TABLE 3.5. Median Annual Income in 1979 Among Those Working Full Time in 1980, by Education and Gender, U.S. Adults (Category Frequencies Shown in Parentheses).

	Male	Female	Total
Post-graduate studies	37,500 (57)	13,750 (21)	**21,250** (78)
College graduate	23,750 (46)	11,250 (35)	**18,750** (81)
Some college	16,250 (68)	11,250 (52)	**13,750** (120)
High school graduate	16,250 (131)	9,000 (105)	**11,250** (236)
Less than 12 years	13,750 (78)	6,500 (33)	**11,250** (111)
Total	**16,250** (380)	**11,250** (246)	13,750 (626)

INDEX OF DISSIMILARITY

Thus far we have studied the association between two or more variables by comparing percentages, means, or medians across categories of the independent variable or variables. As we have noted already, there are situations in which this strategy does not yield particularly informative results. In particular, when there are large numbers of categories in a distribution, comparing the conditional percentages in any one category ignores most of the information in the table.

Suppose you are interested in knowing whether the labor force is more segregated by sex or by race. You might investigate this by cross-tabulating occupation by sex and race, as in Table 3.6. Visually, the table is of little help—it is not obvious whether the distributions of the two racial groups or the two gender groups are more similar. To decide this, you can compute the *Index of Dissimilarity* (Δ), given by

$$\Delta = \frac{\sum | P_i - Q_i |}{2} \qquad (3.1)$$

where P_i equals the percentage of cases in the ith category of the first distribution and Q_i equals the percentage of cases in the ith category of the second distribution. This index can be interpreted as the percentage of cases in one distribution that would have to be shifted among categories to make the two distributions identical. If the two distributions are identical, Δ will of course be 0. If they are completely dissimilar, as would be, for example, the distribution of students by gender in an all-girls school and an all-boys school, Δ will be 100.

From Table 3.6 we can compute Δ for each pair of columns. For example, the Δ for White males and White females (which gives us the extent of occupational segregation by sex among Whites) is computed as $42.1 = (|15.6 - 16.4| + |14.9 - 6.8| + \cdots + |1.5 - 0.9|)/2$. In the present case, four of the six comparisons are of interest:

Occupational segregation by sex among	
Whites	42.1
Blacks and others	41.3
Occupational segregation by race among	
Men	24.3
Women	18.2

From these computations, we see that more than 40 percent of White women would have to change their major occupation group to make the occupational distribution of White females identical to that of White males, and similarly for Black and other women relative to Black men (note that the coefficient is symmetrical, so we could as easily discuss the extent of change required of males to make their distributions similar to those of females). By contrast, less than one-quarter of Black males would have to change major occupation groups to make the Black male distribution identical to the White male distribution, and among women, the corresponding proportion is less than one-fifth. Thus, we conclude that occupational segregation by sex is much greater than occupational segregation by race. Although it is not common to report tests of significance for Δ, it is possible to do so. (See Johnson and Farley [1985] and Ransom [2000] for discussions of the sampling distribution of Δ.)

One important limitation of the Index of Dissimilarity is that it tends to increase as the number of categories increases (Δ cannot get smaller if the categories of a distribution are disaggregated into a larger number of categories; it can only get larger or remain unchanged). Hence, comparisons of Δs are legitimate only when they are computed from distributions based on identical categories. For example, it would not be legitimate to use Δ as a measure of the degree of occupational sex segregation in different countries because occupational classifications tend to differ from country to country (unless, of course, the distributions were recoded to a standard classification, for example, the *International Standard Classification of Occupations* [International Labour Office 1969, 1990]) or some aggregation of this classification.

TABLE 3.6. Percentage Distribution Over Major Occupation Groups by Race and Sex, U.S. Labor Force, 1979 (N = 96,945).

Major Occupation Group	White Men	Black and Other Men	White Women	Black and Other Women
Professional and technical workers	15.6%	10.5%	16.4%	14.2%
Managers and administrators	14.9	6.9	6.8	3.4
Sales workers	6.4	2.5	7.4	3.1
Clerical workers	6.0	7.6	35.9	29.0
Craft and kindred workers	22.0	16.6	1.9	1.2
Operatives, except transport	11.2	15.4	10.2	14.7
Transport equipment operatives	5.6	8.5	0.8	0.6
Non-farm laborers	6.7	12.7	1.3	1.6
Private household workers	0.0	0.1	2.0	6.8
Other service workers	7.7	15.7	16.1	24.6
Farmers and farm managers	2.5	0.6	0.4	0.1
Farm laborers and supervisors	1.5	2.9	0.9	0.7
Total	**100.1%**	**100.0%**	**100.1%**	**100.0%**
N (thousands)	(50,721)	(5,779)	(35,304)	(5,141)

Source: Adapted from Treiman and Hartmann (1981, 26).

WRITING ABOUT CROSS-TABULATIONS

In writing about cross-tabulations, or for that matter, quantitative relationships of any kind, the aim of the game is *clarity,* not *elegance*. You should try to say enough about what the table shows to guide the reader through it but not so much as to confuse or bore. Strive for economy of prose. Hemingway is a good model. Among quantitative social scientists, Nathan Keyfitz (who adds charm to simplicity) and Paul Lazarsfeld (a good example because his native language was German, not English, and he was said to write his drafts a half dozen times or more before being satisfied with the prose) are worth emulating. Robert Merton is not a quantitative sociologist but is a good negative role model anyway. He is excessively ornate and uses erudition to finesse sticky points. Too many social scientists are simply turgid. Howard Becker's book, *Writing for Social Scientists* (1986), is a wonderful primer on writing good social science, but he does not pay much attention to writing about quantitative data. However, two recent books by Jane Miller (2004, 2005) do this very well, providing much useful advice. It would be well worth your time to consult both of these texts, the first of which focuses on cross-tabulations and the second on multivariate models. The following are some specific pointers for writing about the sort of data we are concerned with here:

- **Describe tables mainly in terms of their substantive implications.** Cite numbers only as much as is necessary to make clear what the table shows, and then state the conclusions the numbers lead you to. The point of presenting data is to test ideas, so the data should be discussed in terms of their implications for the ideas (hypotheses) being tested. Simply citing the numbers is not sufficient. On the other hand, you need to cite enough numbers to guide the reader through the table because most readers—including most professional social scientists—are more or less illiterate when it comes to reading tables.

- **Strive for simplicity.** Try to state your argument and describe your conclusions in terms your ancient grandmother or your cousin the appliance salesman would understand. There is no virtue in obscurity. Obscurity and profundity are not synonyms; obscurity and confusion are, at least in this context. As our brethren in the physical sciences know, truly elegant explanations are almost always simple.

- **Avoid phrases that add no meaning.** For example, instead of "We now investigate what inference we can make as to whether A might be said to have an effect on B," write "Does A affect B?"

- **Avoid passive constructions.** "It is found that X is related to Y" tells us no more than "X is related to Y." Avoid "A scale of support for U.S. foreign policy was constructed." Who constructed it, God? Write "I constructed a scale of support for U.S. foreign policy" or "I used the University of Michigan Internationalism Scale to measure support for U.S. foreign policy."

- **Avoid jargon when it does not help.** Note that I did not suggest avoiding jargon altogether. Jargon, the technical terms of a particular discipline or craft, has a clear function—economy. Use jargon terms when they enable you to convey a point in a sentence that otherwise would require a paragraph. But if ordinary

English will work just as well, use it. Unnecessary jargon does not make your writing more professional or more scientific, only more ponderous.

- **Avoid abbreviations.** Abbreviations are unavoidable in tables because often there actually is not much room. But the space you save in the text is not worth the cost of annoying or confusing your readers. Psychologists are particularly guilty on this score, and their papers provide good negative role models ("negative role model" is an example of a jargon term that is efficient). Why say "twenty-seven Rs were male" when you can say "twenty-seven respondents were male" or, better yet, "we have data for twenty-seven men," and so on.

- **Do not say "we" when you mean "I."** It is pretentious in a solo-authored paper to say "We constructed a scale." It invites the retort, "What do you mean 'we'? Got a mouse in your pocket?" However, it is acceptable to use "we" when you mean yourself and the reader, for example, "As we see in Table 3. . . ." As you will notice, I do this a lot.

- **"Data" is a plural word.** Never say "data is" but always "data are." The fact that semiliterate professors (and writers of computer manuals, for example, Stata) frequently violate this rule does not make the violations any more acceptable.

- **The terms "association" and "correlation" describe relationships between variables, not particular categories of variables.** These terms are never appropriately applied to particular cells or particular rows or columns of a table. Useful phrases for describing association include "there is a positive correlation between A and B," "when A increases, B tends to increase," and "when A is high, B tends to be high." Do not say, "there is a positive correlation of A with high levels of B." Also do not say "A correlates 82 percent with B"—this is incorrect because it is not what the coefficient means.

To break through the writing barrier, which we all experience to one degree or another, find a friend who is at about the same career stage as you are and edit each other's drafts. The advantage of the symmetry is that it creates an equal exchange of what is a major investment of time and energy and minimizes the defensiveness that all writers feel, because you can give as good as you get. As a writer you will benefit from having someone you trust look at your prose with a detached eye, and as an editor you will discover all sorts of interesting things about what makes prose work and not work—things you can put to work in your own writing (it is the same principle as "the best way to learn something is to teach it to others").

Here are some aphorisms to pin on the wall next to your word processor, for comfort when the writing is getting you down:

- Don't get it right, get it written.

- Write, don't read.

- Don't let the perfect become the enemy of the possible.

- Have the courage to be simple minded.
- Anything worth doing is worth doing superficially (with thanks to John Tukey).
- The last 10 percent of the work takes half the time.
- The first 10 percent of the work also takes half the time.
- You can't write the second draft until you have written the first draft.
- Write honest first drafts. (Show your friends your first drafts, not your fifth drafts passed off as first drafts. It is much more efficient to get others to tell you what is wrong—and what is right—with your prose than to try to figure it out for yourself.)
- There is no such thing as good writing, only good rewriting.
- Accept criticism gracefully, even though it feels like rape, castration, or some similar violation of your person; it happens to everybody, and everybody feels the same way.

WHAT THIS CHAPTER HAS SHOWN

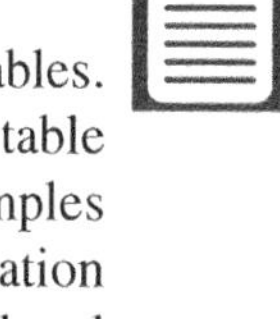

In this chapter we have seen how to extract new information from published tables. Then we noted the one circumstance in which it makes sense to percentage a table "backwards"—when we analyze data derived from "response-based" samples, samples stratified on the dependent variable. We saw why it is necessary to provide information on cases in the sample but excluded from a table, and how to do this. We considered how to construct and interpret cross-tabulations in which the cell entries are means (and standard deviations). We learned how to compute the Index of Dissimilarity (Δ), a measure of the similarity of percentage distributions. And we considered how to write about cross-tabulations.

All of our work so far has been based on paper-and-pencil operations, involving at most a hand calculator. In the next chapter we enter the world of modern social research by learning how to construct cross-tabulations from data on individuals via computer software designed for statistical analysis, focusing on the statistical package Stata, which we will use in the remainder of this book.

CHAPTER

4

ON THE MANIPULATION OF DATA BY COMPUTER

WHAT THIS CHAPTER IS ABOUT

In this chapter we consider how to manipulate data by computer to produce cross-tabulations. The same logic of data manipulation will apply when we get to regression analysis, so this chapter serves also as an introduction to statistical analysis by computer. We consider how data files (of the kind that are of interest in this book) are organized and how to extract data from them; we consider ways to transform variables to make them represent the concepts of interest to us; and we again address the nagging problem of how to handle missing data.

INTRODUCTION

Most statistical analysis by social scientists is done using what are known as "package programs." These are large general-purpose computer programs that enable the researcher to do a variety of data manipulations and statistical analyses on any data set. Although they are not very efficient from a computing point of view (that is, they take longer to perform their tasks than would a program custom-designed to accomplish a particular task), this is no longer important as computers have become more powerful. More to the point, they are very efficient for the researcher. Researchers are usually in the position of writing new commands for each new analysis and hence need an easy way to instruct the computer to perform any particular operation. Statistical package programs meet this need very well. Among the most widely used programs are SPSS and SAS. Among academic users, Stata is rapidly becoming the statistical package of choice. These three

A HISTORICAL NOTE ON SOCIAL SCIENCE COMPUTER PACKAGES

For many years, SPSS was the package of choice among sociologists, probably because it was written by and for social scientists. The manual accompanying the early versions (until SPSS-X appeared on the scene) was a model technical reference and also doubled as an excellent introduction to the statistical techniques available in SPSS. The fact is that it was the manual that sold SPSS because as a computer program per se, it was inferior to a number of others. The SPSS-X manual, and subsequent versions of SPSS, were written by computer programmers and are unfortunately both much less clear as computing manuals and less adequate as introductions to statistics. But the original manual is no longer generally available. Moreover, SPSS has become more and more oriented to business users. Although still widely used by social scientists in Europe and Asia, it has largely lost its market in leading U.S. research universities. As social scientists became more statistically adept and computationally literate, many researchers switched to SAS, a much more powerful package. The main difficulty with SAS is that its manuals are very difficult to read and its commands are often counter-intuitive. It is not an easy language to teach or to learn. Fortunately, Stata, which began life as an econometrics package with very little data management capability, has gotten better and better over time, so that by now it can happily be used as a general-purpose package. Stata is powerful and fast, which makes it viable to carry out analysis on a PC using Stata, even with very large data sets (for example, a 1 percent sample of the Chinese census). It is capable of doing most of the things required in modern data analysis. The commands are generally simple and straightforward. And Stata data sets can be used on any platform. Overall, Stata is a very good choice for our kind of work. Appendix 4.A, at the end of this chapter, provides tips for carrying out data analysis using Stata.

programs are available both for Unix platforms and for PCs. Of the three, Stata is the most nearly identical across platforms. There are, of course, many other statistical packages as well. Many of these packages are worth exploring but only after you have mastered the material in this book. As you will discover, the *logic* of data analysis by computer is fairly standard, although the *command syntax* differs somewhat from program to program. Having once mastered the basic logic, it is easy to apply it to other data sets and other statistical package computer programs.

HOW DATA FILES ARE ORGANIZED

The easiest way to think of the organization of data in a computer is to imagine a matrix in which the rows are cases and the columns (or sets of columns) are variables. Specifically, consider a data set that contains 257 variables (but 422 columns of data because some variables require more than one column; for example, suppose the respondent identification number requires 4 columns and age requires 2 columns) and 1,609 cases (and hence 1,609 rows of data). For the sake of concreteness, regard this data set as containing information from a representative sample of the U.S. population. In such a data set, the information might be organized as in Exhibit 4.1.

To manipulate these data, we need a map to the data set, which tells us where in the matrix particular information is located and what the information means. Such a map is known as a *codebook*. In the present example, we might have a codebook something like what is shown in Exhibit 4.2.

Armed with the information in the codebook, we now know exactly what the data set contains. It consists of one record per respondent for each of 1,609 respondents. High-quality codebooks also generally provide information about the characteristics of the sample on which the data set is based and other documentation necessary to process the data. Data sets that are readable by a computer are said to be *machine readable* or

EXHIBIT 4.1 Illustration of How Data Files Are Organized.

Variable Name	*IDNO*				*SEX*	*AGE*			*POLICY*
Variable number	1				2	3		...	257
Column number	1	2	3	4	5	6	7	...	422
1	0	0	0	1	1	2	7	...	4
2	0	0	0	2	2	4	1	...	1
3	0	0	0	3	2	-	1	...	9
.	.	.	.	.	.	.	.	...	.
1,609	1	6	0	9	1	5	3	...	5

EXHIBIT 4.2 A Codebook Corresponding to Exhibit 4.1.

Variable Number	*Column*	*Variable Name*	*Variable Label and Code*
1	1–4	idno	Respondent identification number
2	5	sex	Sex of respondent
			1 Male
			2 Female
3	6–7	age	Age (exact year)
			99 99 or older
			−1 No answer or uncodable
.	.	.	. .
257	422	policy	The policies of the president are
			1 Wonderful
			2 OK
			3 Not so hot
			4 God awful
			8 Who knows and who cares
			9 No answer or uncodable

computer readable and are known as *files*. In the present example, the first four columns give the identification number for each respondent. Usually this is of little interest at the data analysis stage, but it is vitally necessary to keep track of the data and is crucial if ever we want to add additional data to the file—for example, if we have conducted another survey of the same respondents and want to merge the data from the two surveys, or if we want to supplement interview responses with information from organizational records, and so on. Column 5 gives the sex of the respondent, columns 6 and 7 give the age, and column 422 gives the response to a question about the policies of the president.

Using the response categories indicated in the codebook, we see that the first respondent is a twenty-seven-year-old male who thinks the president's policies are god awful and that the second respondent is a forty-one-year-old female who thinks the president's policies are wonderful. The third respondent is a woman for whom no information is available regarding either her age or her judgment of presidential policies. Perhaps she refused to answer these questions in the interview or gave nonsensical responses, or perhaps there was some sort of editing error that destroyed the information; in any event, it is unavailable to the data analyst. (Note that there is no "n/a" code for sex. It is rare to find a "no answer code" for sex, at least in interview surveys, because the interviewer usually records this information.) Some codebooks give the frequency distribution (the *marginals*) for each variable. This is a very useful practice, and if you construct a codebook, you should include the marginals (the -codebook- command in Stata accomplishes this). Their inclusion permits better initial judgments as to suitable cutting-points for variables as well as a standard against which to check your computer output for accuracy. It is very easy to make mistakes when specifying computer runs, so you should check each run for consistency with previous runs and the marginals.

Suppose we wanted to ascertain whether men and women differ in their support for presidential policies. To do this, we might cross-tabulate the presidential policy question by sex, percentaging the table so that the judgment of presidential policies is the dependent variable. Thus we have to instruct the computer where to find each variable, to do the cross-tabulation, and to percentage the table in the appropriate direction. We also have to instruct the computer what to do about the "no answer" category in the presidential policy variable.

There are two ways to specify how to locate data in a file, and computer programs differ as to whether either or both is permissible. Some programs use instructions that point to particular columns in the file—for example, "cross-tabulate column 422 by column 5." More commonly, programs require that the analyst first specify where in the data set each variable is located and then use variable names to command particular manipulations—for example, "The variable *SEX* is in column 5 and the variable *POLICY* is in column 422. Cross-tabulate *POLICY* by *SEX*." A variant of this approach is to require a map numbering the variables sequentially and specifying their location, for example:

Variable	*Columns*
001	1–4
002	5
003	6–7
.	.
257	422

"Cross-tabulate *VAR257* by *VAR003*." In most current programs, including Stata, SAS, and SPSS, such maps are created in the course of creating *system files*; as part of the preparation of the file, *variable names* (usually restricted to eight characters, although no longer so in Stata beginning with Version 6.0), *variable labels*, and *value labels* (indicating the meaning of each response category) are attached to the file, and variables are then identified by name. In instructions to the program, which are known as *commands*, the analyst uses the names of the variables and need not be concerned about their location in the file. For example, the Stata command

```
tab policy sex,col
```

tells the computer to cross-tabulate *POLICY* (the row variable) by *SEX* (the column variable) and compute column percentages. Note that in Stata, variable names are case sensitive. Thus Stata regards *sex*, *SEX*, and *Sex* as three different variables. (Although in this book, variable names appear in *ALLCAPS*, to make it easier to distinguish variable names from other words in a sentence, in my Stata command files [-do- files—see the following discussion], I always name files with lowercase names to avoid extra typing and the error that accompanies it.)

A Digression on Card Decks and Card-Image Computer Files

Computers began to be used extensively in the social sciences in the mid-1960s but did not become ubiquitous until the 1970s. As a consequence, many data sets still of interest were created for use with pre-computer analytic technology, specifically with machinery that reads IBM punch cards (see Figure 4.1). Although the logic of data organization is similar to that used for analysis by computer, the technology dictated several important differences. Whereas there is in principle no limit to the number of variables that can be included in a single computer record (although there are limitations as to how many variables a program can handle), an IBM card contains eighty columns. Because the machinery for manipulating IBM cards could handle only one card at a time (such machines were known as *unit record* equipment, where the record was one card length), there was a premium on packing as many variables as possible onto a single card.

HERMAN HOLLERITH (1860–1929) was the inventor of the punch card, which revolutionized data processing, radically reducing the processing time for the 1890 census (where it was first used). Born in Buffalo, New York, he entered the City University of New York in 1875; at the age of fifteen but transferred to the Columbia University School of Mines, where he earned an EM (Engineer of Mines) degree in 1879. After completing his education, he worked on the 1880 census, where he was exposed to the laborious and error-prone hand-counting methods in use at the time. His design for a mechanical data processing system (cards and card reader) won a competition sponsored by the Census Bureau. For his invention, he was granted a Ph.D. in 1890 by the Columbia School of Mines, without having matriculated in the doctoral program. (Those were the days!) In 1896 he established his own company, which, after merging with other companies, became IBM (the International Business Machine Company). Hollerith made many subsequent inventions that improved data processing, as well as inventions in other fields.

A card data set consists of one or more cards per respondent. For example, to represent all of the data contained in our illustrative 257-variable, 422-column data set would require 6 cards per respondent (= 422/80, rounded up) times 1,609 respondents, or 9,654 cards. The information shown for the first respondent might be represented on an IBM card as in Figure 4.1, where the response to presidential policies is contained in column 64, but otherwise the columns correspond to those in Exhibit 4.1.

An analyst wanting to cross-tabulate responses to presidential policies by sex would pass the deck through a counter-sorter, which would physically divide the deck into two subdecks by reading the holes punched in a designated column, column 5 in this case. Cards with a "1" punch would fall into the 1 pocket in the machine and cards with a "2" punch would fall into the 2 pocket. Each of these subdecks would then be passed through

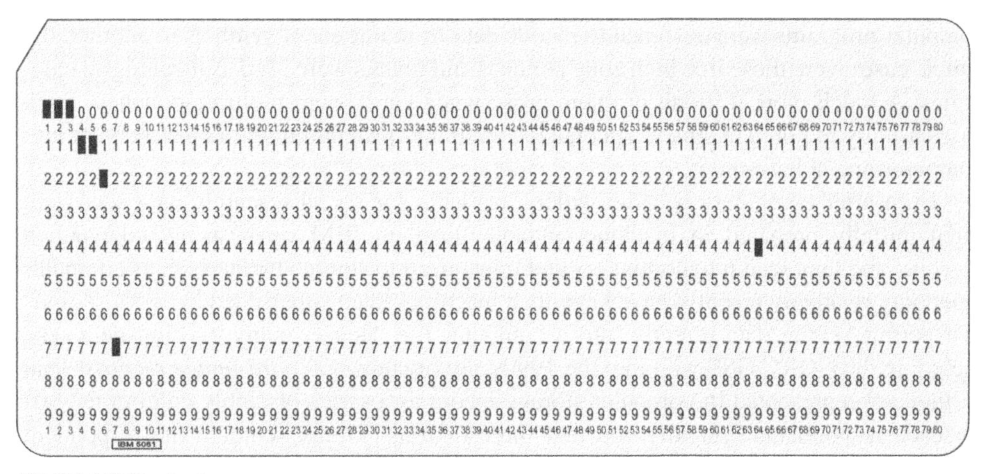

FIGURE 4.1. *An IBM punch card.*

the machine a second time, and the distribution of punches in column 64 would be counted and displayed for the analyst to copy by hand onto paper. These counts would generate the bivariate frequency distribution of judgments regarding presidential policy by sex, which would then be percentaged in the usual way (using a desk calculator).

This technology had several important consequences for data organization and data analysis. First, it discouraged the use of statistical methods other than cross-tabulations because all it could do was generate the counts needed as input to statistical procedures—the algebraic manipulation still had to be carried out by hand. Second, it discouraged the retention of detailed information; there was, indeed, a great premium on squeezing the response categories into a single column if at all possible because a two-column variable was tedious to manipulate (it required much more card handling because the variable had to be sorted on the first digit, and then each of the resulting categories had to be sorted on the second digit) and produced more detail than could be used effectively in a cross-tabulation. This resulted in the use of what are known as zone punches, the locations on an IBM card above the numerical columns, which also were used for "+" and "−" (sometimes called "x" and "y" punches), and also the use of blanks (no punch) as a meaningful category. Thus, for example, it would be unlikely for age to be represented by single years in data sets designed for use with unit record technology; rather, a set of age categories would be predesignated. Third, in the interest of getting as many variables as possible on a single card—because it was impossible to include in a single tabulation variables located on different cards—some analysts resorted to putting more than one variable into a single column. Consider variables 2 and 257 in the preceding example (Exhibit 4.1). Because there are only two possible responses to the sex item and six to the presidential policy item, they could be included in a single column simply by using punches 4–9 for the presidential policy response categories. A device on the counter-sorter machine made it possible to suppress some punches and sort on others. Columns of this kind were known as *multiple-punched* columns.

All of these devices for packing as much data as possible into a single IBM card played havoc when the shift to data analysis by computer occurred. Because most computer programs were designed to recode data from one set of symbols to another, the simple cases were those in which zone punches and blanks were used as meaningful categories. A much more difficult problem arose when cards were multiple-punched. Such cases usually required extensive specialized computer programming to convert them into computer-readable form.

Even after computers became widely available for social research, data sets were often initially prepared in machine-readable form on IBM cards using a keypunch machine and then read into computers and transferred to storage media such as computer tape; only relatively recently have keypunch machines been superseded by work stations that permit keying data directly into a computer file. Hence, many existing data sets, including the NORC GSS well into the 1990s, are organized as *card-image records*. That is, they are represented in computer storage media as a series of eighty column records for each respondent. Typically, the first three or four columns contain the respondent identification number and column 80 contains the record number, or *deck ID*. This organization of data has no consequences for analysis, but it does affect the way the computer is instructed to read the data. The specific details vary depending on the program you use, but you should be aware of this alternative mode of data organization, in addition to the specification of one long record per respondent with which we began our discussion.

THE WAY THINGS WERE Before electronic data entry terminals became available, command files also were input to computers via IBM punch cards. The analyst wrote out his command file, and then each line of the command file was keypunched into a separate IBM card (by a keypunch operator or, in the case of underfunded graduate students, by the analyst). The resulting "deck" of IBM cards was then transported to the university central computing center and either submitted to a clerk or fed directly into a card reader. Eventually the command file was executed ("the job ran"), often after a delay of several hours, and the printed output plus the box of cards were returned to the analyst. If there were errors, the entire process was repeated. This technology limited the number of computer runs to two or three per day, which made the completion of any particular analysis a very time-consuming proposition by current standards—but did at least have the salutary feature of allowing more time to think while waiting for the job to run.

TRANSFORMING DATA

As noted several times in previous chapters, data are not always initially represented in a form that is suitable to our research needs. For many purposes, we may want to alter the codes of a single variable or to combine several variables. These operations are known as *data transformations*, and each of the major statistical package programs has a set of procedures for accomplishing them. Here are some examples, to convey the range of possibilities. Facility at transforming variables to a form that most closely expresses theoretical concepts is an important skill of the quantitative data analyst.

Recoding

Recoding is the term used for changing the values of a variable to a different set of values. Recoding has many uses, some of which we have already seen.

One is to collapse categories of a variable into a smaller number of categories, for example, when I created the leftmost column of Table 2.3 from Table 1.1. To see how this procedure works, let us consider this example in detail. I started with a religiosity scale composed of the following categories:

1. Very religious
2. Somewhat religious
3. Not very religious
4. Not at all religious

(For the moment, ignore the possibility of missing data.) To combine the last two categories, I simply changed, or *recoded*, category 4 to category 3, which yields a new variable:

1. Very religious
2. Somewhat religious
3. Not religious

Although some computer programs permit a variable to be "written over"—that is, to be replaced with a new variable—this is very poor practice. Rather, you should create a new variable containing the transformed values. The reason for this should be obvious: both to protect against error and to permit you to transform a variable more than once in the same computer run, you should preserve the original coding of a variable as well as any recoded or otherwise transformed versions of the variable. Typically, statistical package computer programs operate line-by-line; each line of code operates on the data in whatever form they appear after the previous operation. Hence, it is all too easy to transform a variable and then inadvertently transform it again, unless a new variable is created in the course of the transformation.

A second use of recoding is to redefine a variable by creating a new set of categories representing a new dimension. You have seen an example of this also, in our discussion of property space in Chapter Three. Recall our classification of U.S. congressmen into

1. Standard Republicans
2. Gypsy moths
3. Boll weevils
4. Standard Democrats

To create a classification according to party membership, we can recode 2 to 1 and 3 to 4, yielding a new variable with values of 1 (= Republican) and 4 (= Democrat). To create a classification of congressmen as liberal or conservative, we can recode 2 to 4 and 3 to 1, again yielding a new variable with values of 1 (= conservative) and 4 (= liberal).

Note, however, that when variables are recoded to dichotomies, the convention is to code one category 1 and the other category 0 and to name the variable by the category coded 1. Thus, in the first example, the convention would be to create a new variable, named "Republican," in which 1 and 2 in the original variable are coded 1, and 3 and 4 in the original variable are coded 0. As we will see in later chapters, the 0–1 recoding convention facilitates the use of dichotomous variables in both OLS and logistic regression.

A third use of the recode operation is to assign scale scores to the categories of a variable. For example, we might have a variable measuring educational attainment, which is initially coded as follows:

1. No schooling
2. 1–4 years of elementary school
3. 5–7 years of elementary school
4. 8 years of elementary school
5. 1–3 years of secondary school
6. 4 years of secondary school
7. 1–3 years of college
8. 4 years of college
9. 5 or more years of college
10. No information

For many purposes, it is useful to treat years of school completed as a ratio variable. By doing so, it is possible to compute the mean number of years of school completed by different subgroups of the population, to use years of school completed in regression equations, and so on. To do this, we might recode the original variable by assigning the midpoint or another estimate of the years of school completed by individuals in each category:

Original Code	*Recode*
1	0
2	2.5
3	6
4	8
5	10
6	12
7	14
8	16
9	18
10	−1

In making recodes of this sort, it is important to justify your choice of values rather than assigning them arbitrarily. For example, the decision to assign "18 years" to the category "5 or more years of college" rather than 17 years or 19 years must be justified, not simply asserted.

Note the special treatment of category 10, "no information." In carrying out the analysis, we want either to exclude this category or otherwise give it special treatment. We thus give it a special code, which we can either define as *missing data* (see the discussion later in the chapter) or otherwise modify. It is convenient to use negative numbers to flag categories that we are going to treat as missing data because doing so minimizes the likelihood of inadvertently treating them as substantively meaningful. (A useful alternative, now available in Stata, is to use the code "." to specify missing values when we have no need to distinguish between different types of missing value, and to use the codes ".a," ".b," . . . ".z" when we want to distinguish different types of nonresponse—again, see the discussion of missing data later in the chapter). For example, suppose we recoded the "no information" category as 99. If we subsequently decided to analyze those with at least some college education, we might instruct the computer to select all cases with years of school completed greater than or equal to 14, forgetting that category 99 means no information. This, of course, would result in the inclusion in the highest education category of those for whom education is not known along with the college educated.

TREATING MISSING VALUES AS IF THEY WERE NOT

There is a famous example in the literature of the treatment of missing data codes as substantive quantities. Guillermina Jasso (1985, 237) concluded that the frequency of sexual intercourse per month increases with a wife's age—contrary to all expectations! Alas, as discovered by Kahn and Udry (1986, 736), she failed to notice four outliers, cases erroneously coded 88 rather than 99, the specified missing data code. When these four cases are omitted, the positive effect of wife's age disappears. Kahn and Udry also omitted four other outliers, prompting a lively response from Jasso (1986) about what should be regarded as an outlier. We will return to a discussion of outliers when we consider regression diagnostics in Chapter Ten.

A final use of the recode operation is to convert data from old surveys that use zone punches and blanks into a form that permits numerical manipulation. This typically can be done by reading the data in an alphanumeric format and converting them to a floating-point decimal format.

Arithmetic Transformations

Sometimes we want to transform variables by performing arithmetic operations on them. Such transformations will be particularly important when we get to regression analysis because it is sometimes possible to represent nonlinear relationships by linear equations involving nonlinear variables. For example, it is well known that the relationship between income and age is curvilinear—income increases up to a certain age and then declines.

This relationship can be represented by constructing a regression equation of the following form:

$$\hat{Y} = a + b(A) + c\,(A^2) \tag{4.1}$$

that is, income (= Y) is taken to be a linear function of age and the square of age. To estimate this equation, we need to create a new variable, the square of age. So we simply compute

$$AGESQ = AGE*AGE \tag{4.2}$$

and then regress Y on *AGE* and *AGESQ*. Most of the statistical package programs provide extensive transformation capabilities. It generally is possible to transform a variable using any arithmetic operator or any of a number of specialized functions, such as the square root function.

Contingency Transformations

A final way to transform variables is to use "if" specifications in your commands. "If" specifications are an alternative to recode commands and are much more flexible in some ways because they make it possible to specify complex contingency relationships involving several variables. For example, if we wanted to distinguish those who were upwardly mobile from those who were not, we might identify as upwardly mobile those who had jobs that were of higher status than those of their fathers. This can be accomplished by specifying the following: if *PRESTIGE* is greater than *PRESTIGE-OF-FATHER*, construct a new variable, *MOBILITY*, and give it the value 1; otherwise, give it the value 0. Although the syntax of the computer command required to do this will vary depending on the program used, the logic is, as usual, straightforward: a *dichotomous variable* is created, scored 1 for those individuals who are upwardly mobile, and scored 0 otherwise (where "upward mobility" is defined as having an occupation of higher prestige than the occupation of one's father).

Another kind of contingency transformation is to create a variable consisting of a count of the number of responses to a specified set of other variables that meet specific criteria. For example, we might create a scale of acceptance of abortion by counting the number of "pro-choice" ("accepting") responses to a set of questions about the circumstances under which abortion should be permitted.

Contingency statements are used not only to transform variables but also to select subsamples for analysis. For example, if we were interested in analyzing completed fertility, we might want to restrict our sample to women age forty-five or older. This is accomplished in some computer packages, such as SPSS, by selecting a subsample and doing all the subsequent operations only on the subsample. Other packages, such as Stata, select cases as part of each command, although selection of a subsample is possible in Stata as well.

Missing Data

Often, substantive information on certain variables is missing from a data set. The sources of missing data are nearly endless. In data sets derived from interviews with a sample of people, the information may never have been elicited from the respondent, either in error or as a matter of design (some questions are "not applicable," for example, spouse's education for the never-married; and sometimes questions are asked of random subsets of respondents to increase the length of the questionnaire without increasing the respondent burden—the GSS often does this). The respondent may have refused to answer certain questions, may have responded to some questions by claiming not to know the answer or not to have an opinion, or may have given logically inconsistent answers (for example, responding "never married" to a question on marital status but providing an answer to a question on "age at first marriage"). Interviewers may have failed to record responses or may have recorded them incorrectly. Errors may have been introduced in the process of preparing data for analysis—as when narrative responses are incorrectly assigned to code categories by coding clerks or when correctly assigned codes are incorrectly keyed in the course of data entry. Similar problems plague other sorts of data sets. Bureaucratic records are often incomplete and frequently contain inconsistent information.

PEOPLE GENERALLY LIKE TO RESPOND TO (WELL-DESIGNED AND WELL-ADMINISTERED) SURVEYS

Interestingly, refusals to answer specific questions are rare when questionnaires are well written. By and large, people are flattered that they are asked their opinions and asked to talk about themselves. There is a famous story from the lore of survey analysis about the Indianapolis Fertility Survey, one of the earliest surveys that asked explicitly about sexual behavior. One of the analysts went out with considerable trepidation to conduct a pretest of the questionnaire, not knowing how women would respond to "intimate" questions. As it happened, the interview went off without a hitch—until the very end, when the interviewer got to the routine demographic questions and asked the respondent her age, at which the lady drew herself up indignantly and said, "Now you're getting personal!"

The exception to the general willingness to respond is with respect to information that people fear might put them in jeopardy, such as income, which they suspect might find its way to the tax authorities.

In high-quality surveys, great pains are taken to minimize the extent of error. In the course of readying data sets for analysis, they are *cleaned*, that is, edited to identify and if possible correct *illegal* codes (codes not corresponding to valid response categories) and logically impossible combinations of codes. For example, when a respondent who claims never to have been married gives his age at first marriage, sometimes it is possible to decide which is the correct and which the incorrect response by inspecting other responses given by the same individual. When this is not possible, the respondent might be contacted and asked to resolve the inconsistency.

Of course, error may be *introduced* in the editing process as well as corrected. For instance, I observed the field editing operation undertaken in conjunction with the Madison, Wisconsin, 1968 pretest for the 1970 census. In this operation, middle-class housewives, taken on as temporary employees by the Census Bureau, "corrected" census returns in which a woman's marital status was left unanswered but children were reported by changing the marital status response to "married." The likelihood, of course, is that in 1968, a mother failing to report her marital status was, in fact, an unmarried mother who finessed the question by ignoring it. This sort of editing to make the world more orderly—as perceived by middle-class Wisconsin ladies—is not supposed to occur, but it does.

In the course of the editing process, explicit codes are assigned to the various categories of nonsubstantive responses, in accordance with the principle that the set of categories for each variable be mutually exclusive and collectively exhaustive—that is, designed in such a way that each case is assigned one and only one code on each variable. For example, a code may be assigned to "don't know," another to "refused," another to "not applicable," and still another to responses left blank when a code was expected. In general, it is desirable to distinguish between different forms of nonsubstantive answers to allow flexibility in their treatment in the course of data analysis. For example, as noted earlier, a "don't know" response to an attitude question might alternatively be regarded as a nonresponse or as a neutral response, standing in between positive and negative responses. If "don't know" is combined with "no answer" in the coding process, it becomes impossible to make such a distinction. In short, for the same reason that it is important to preserve detail in the coding of substantive responses (discussed in the previous chapter in the section "Collapsing Categories to Represent New Concepts"), it is important to preserve detail in the coding of nonsubstantive responses.

Analyzing Surveys with Missing Data

Presuming that the data are coded in such a way as to preserve all relevant distinctions, the analyst is confronted with a set of decisions about how to define and how to treat "missing data." The first issue is a substantive one: which responses should be regarded as substantively interesting and which as representing noninformation? What to do about "don't know" responses to attitude questions, mentioned already, is the archetypical example here. Another case that arises frequently in tabular analysis is what to do about very small categories, for example, "other" in a question on race for a U.S. sample (meaning neither White nor Black). The temptation is to leave such categories out of the analysis. This is not a good idea—data should be presented for the entire population of interest. Thus if you are studying the adult population of the United States, your tables should refer to the entire adult population, not just to Whites and Blacks. The solution in this case is simply to create a residual category, "other," and to include it in the table but not bother to discuss it. It is included for the sake of completeness—which, among other things, enables the reader to reorganize the table for other purposes—but is not discussed because the small number of cases and the heterogeneity of residual categories generally make the specific numbers uninteresting.

A more difficult problem arises when there is genuinely no information for one or more variables in the table, for example, when respondents fail to report their education or their income. Again, one alternative is to include a "no answer" category in each row and column of the table. If there are many missing cases, this is wise. If there are only a few missing cases, the increased size of the table probably is not warranted. In this case it is sufficient simply to report how many cases are missing, in a footnote to the table.

When our variables are continuous, we must either exclude missing values from the analysis or in some way impute the values. Chapter Eight is entirely devoted to the treatment of missing data.

Most statistical package programs allow the analyst to specify which codes are to be treated as *missing values* (and indeed *require* it in the sense that any codes not specified as missing values are included in the computation whether you intend it or not). Typically, statistical package programs are not completely consistent across procedures (commands) in the way they handle missing data, so it is very important to understand exactly what each procedure does and to design your analysis accordingly. In designing any analysis, you must know how the procedure will treat each logically possible code in your data, including in particular those codes you designate as missing values; otherwise you inevitably will get into trouble.

In the example on education discussed earlier, "no information" was assigned a code of −1. When computing a mean, we ordinarily would declare −1 to be a missing value for education. In SPSS syntax, missing values are explicitly declared: "missing values educ (−1)"; in Stata, as noted previously, missing values may be excluded automatically by assigning one of several "missing value" codes, or may be explicitly excluded from a procedure by limiting the sample with an *if* qualifier: . . . if educ~=−1 (that is, if *EDUC* is not equal to −1). These statements tell the computer to omit all individuals for whom education is coded −1 (or assigned the missing value code) from the computation of the mean. Neglecting to so inform the computer results in an incorrect mean because any individuals who are coded in the data as having −1 years of schooling are included in the computation. Errors of this sort are very common, which is why it is imperative to check and recheck the logic of your commands. A useful check is to work through the logic of your computer commands *line by line* for specified values of your original variables to see how the computer transforms them at each step in the process. You will make some surprising discoveries.

One of the things that typically happens to novice data analysts is that they do some computation and discover that their computer printout shows no cases or a very small number of cases. Usually this turns out to be the result of a logical error in the specification of data transformations. For example, consider an income variable originally coded in a set of categories representing ranges of income, for example, 1 = under $3,000 per year, 2 = $3,000 to $4,999, and so on, but where 97, 98, and 99 are used to specify various kinds of nonresponses. If the analyst recodes the income categories to the midpoints of their ranges, for example, recodes 1 to 1,500, 2 to 4,000, and so on, but then forgets this and specifies as missing values all codes greater than or equal to 97, all the cases will be excluded because all cases for which income was reported have been recoded to values in the thousands of dollars, that is, greater than 97. If you do not think this will happen to you, wait until you try it! It happens to all of us. The trick is to catch logically similar, but more subtle, errors before you construct entire theoretical edifices upon them.

WHAT THIS CHAPTER HAS SHOWN

This chapter has been an introduction to statistical computing by computer, with some attention to the history of social science computing, and more attention to the logic of data manipulation and the treatment of missing data. The chapter thus serves as a foundation that should make it easy to learn any statistical package program—Stata, which we will use for the remainder of the book, or any other package such as SPSS or SAS.

In the next chapter we turn to the general linear model, with a gentle introduction via a discussion of bivariate correlation and regression.

APPENDIX 4.A DOING ANALYSIS USING STATA

TIPS ON DOING ANALYSIS USING STATA

This appendix offers some simple tips that will greatly enhance the ease and efficiency with which you use Stata for analysis. In addition, the appendix lists some particularly useful commands that are easily overlooked.

Do Everything with `-do-` *Files*

You should from the outset develop the habit of carrying out *all* your analysis by creating command files, known in Stata parlance as "`-do-` files." Doing so has two major advantages: it makes it easy to repeat your analysis until you get it right, and it makes it easy to document your work. Keeping a log of your analysis is not an adequate substitute (although, of course, you must create a `-log-` file to save your output) because a log faithfully records all of your errors and false steps, making it difficult to follow the direct path to successful execution and tedious to repeat your analysis. Here is an example of part of one of my `-do-` files, shown to suggest a standard format you might want to adopt. I use this set of commands at the beginning of each `-do-` file I create. The commands in the file are shown in `Courier New` type, and my comments in square brackets are shown in Times New Roman type (the standard font for the text).

```
capture log close
```

[This command closes any `-log-` file (see the next command) it finds open. The `-capture-` prefix to a command is very useful because it instructs Stata not to stop if an "error" is encountered—which would be the case if it could not find a `-log-` file to close.]

```
log using class.log,replace
```

[This command tells Stata to keep a file of commands and the results of the commands, called a "`-log-` file," and to replace any previous versions of the `-log-` file. The `-replace-` part of the command is crucial because otherwise when you execute the -do- file, fix an error, and try to execute it again, Stata will complain that a previous version of the -log- file exists.]

```
#delimit;
```

[This command tells Stata to end all subsequent commands whenever a ";" is encountered. I find this the most convenient way to handle long lines. The default in Stata is to regard a carriage return (the computer command that ends a line) as the end of the command, which means that, unless the carriage return is "commented out" (see below), commands are restricted to one line. Of course, the line may be very long, extending well beyond the width of a page, but this makes your file difficult to read.]

```
version 10.0;
```

[This command tells Stata for which version of Stata the file was created. Stata always permits old -do- files to run on more recent versions of the software, if the version is specified.]

```
set more 1;
```

[This command tells Stata not to stop at the end of every page of the output. When executing a -do- file, you want the program to run completely without stopping. The way to inspect the output is to read the log file.]

```
clear;
```

[This command clears any data left over from a previous attempt to execute the program or any other Stata command. Stata is good about warning you against inadvertently destroying data you have created. But the fact that Stata warns you means that you need a way to override the warning, which is what this command does.]

```
program drop _all;
```

[This command drops any existing programs that you might have created in a previous execution of the -do- file. Failing to do this causes Stata to stop if you have included any programs in your -do- file.]

```
set mem 100m;
```

[This command tells Stata to reserve 100 MB of memory. Space permitting, Stata reads all data into memory and does its analysis on these data, which is why it is so fast. If you specify too little memory, Stata will complain that it has no room to add variables or cases.]

```
*CLASS.DO (DJT initiated 5/19/99,
last revised 2/4/08);
```

[I always name my -do- file, and because I work often with others, indicate the author, the initial date of creation, and the last date of revision. This is very useful in identifying different versions of the same -do- file, which might exist because my coauthors and I have both revised the same file, or because I have made a revision on my office computer and have forgotten to update the version on my home computer, and so on. Note that *comments* are distinguished from *commands* by an asterisk in the first column.]

```
*This -do- file creates computations for
a paper on literacy in China.;
```

[I always include a description of the analysis the -do- file is carrying out. Because I often work on a paper over an extended period, the description is extremely helpful in jogging my memory and helping me to locate the correct file.]

```
use d:\china\survey\data\china07.dta;
```

[This command loads the data into memory. The remainder of the -do- file then consists of commands that perform various operations on the data and produce various computations.]

```
...

log close;
```

[This command closes the -log- file, so that it can be opened by my editor or word processor.]

The basic procedure for creating and successfully executing a -do- file is to (1) open a new file (either in the Stata editor or in your favorite word processor or ASCII editor), remembering that -do- files must include the extension "do"; (2) insert a front end, of the sort just outlined (which I always copy from my previous file to my current file to minimize typing); (3) create a set of commands to carry out the first task; (4) save the file (as an ASCII file); (5) toggle to Stata and execute the -do- file by typing `do <file name>`; note the error that stops the file from reaching the end, if any (and there surely will be an error most of the time); (5) toggle back to the editor; (6) correct the error or add to the analysis; and (7) repeat the process until your -do- file includes all the necessary steps in the analysis and runs to the end.

This process is known as *debugging* your program. When this happens, you will have a -log- file with a set of results and a -do- file that (1) provides a clean and complete record of how you got the results shown in the -log- file and (2) can be rerun at any time—which you may want to do if, as often happens, you discover an error in the logic of your analysis or update your data set. Also, despite our best intentions, it is very difficult to fully describe our computing operations in a research paper. This means that if you submit a paper for publication and get an invitation to "revise and resubmit" the paper, the best way to recover exactly what computations you did and what assumptions you made several months, or even years, ago is to consult your -do- file. The availability of a -do- file will also greatly speed your revision work.

Build In Extensive Checks of Your Work

It is *extremely* easy to make errors—of both a logical and a clerical kind—when doing computer-based data analysis. The only way to protect yourself from happily making up stories about results produced in error is to *compulsively* check your work. You can do this in two ways. First, check the logic of each set of data transformation commands by working through—as a pencil and paper operation—how each value of a variable being transformed is affected by each command. Second, tabulate or summarize each new variable and *actually look at the output*. You will be surprised how many errors you discover by taking these two simple steps!

Document Your `-do-` File Exhaustively

You should make extensive notes in your `-do-` file about the purpose of each set of commands and the underlying logic—especially in the case of data transformations. Including comments summarizing the outcome of each set of commands makes it clear why I carry out the next step of the analysis. The `-do-` file then becomes a document summarizing my entire analysis. *I cannot emphasize strongly enough the importance of adequate documentation.* It is typical in our field to work on several problems at once and to return to a problem after months or even years. In addition, the editorial review process often takes a very long time. If you have not documented your work, you may have a great deal of trouble remembering why you have done what you have done. This is inefficient and can be highly embarrassing—as when a journal editor asks you to do some additional analysis, and you have no idea why you made particular computations, much less what the chain of reasoning was, and cannot reproduce the previous results. This happens much more often than most of us want to admit.

Include "Side" Computations in Your `-do-` File

This is a corollary to the point about exhaustive documentation made in the previous section. Often we do "side" computations in the course of writing papers to make points or add illustrations to the text—for example, computing the ratio of two coefficients in a table we have made or computing a correlation coefficient between two variables listed in some other publication. The way to make your `-do-` file a comprehensive document of all your computations is to use Stata, rather than a hand calculator or spreadsheet, to do the work; or, at minimum, to include both the data and the results as comments in your `-do-` file. More than once, I have produced a paper with a well-documented `-do-` file but have failed to include side computations in the `-do-` file, and then have discovered months later that I had no idea how the side coefficients reported in the paper were derived.

Rerun Your `-do-` File as a Final Check

At the point you have completed a paper and are about to submit it for a course, for posting in an online paper series, or for publication, you should make a point of executing your `-do-` file in a single step and then checking *every* coefficient in the paper against the corresponding coefficients in your resulting log file. You likely will be startled to discover how many discrepancies there are. Because `-do-` files often are developed over an extended period of time and often are executed in pieces, it is extremely easy for inconsistencies to creep in. If you have a `-do-` file that will run from beginning to end, without interruption, and will produce *every* result you report in the paper, you will have met the gold standard for documentation. You also will be a happy camper months or years later when you need to make a minor change that affects many results. You will discover that the change usually can be made in a matter of minutes—although updating your tables by hand is usually a much more tedious business.

Make Active Use of the Stata Manual

The only way to become facile at any statistical program, including Stata, is to make a point of continuously improving your skills. Each time you are unsure how to carry out a task, look for a solution in the manual. You will find the improvement in facility very rewarding. After you become reasonably facile at Stata, you should then take advantage of Stata's `-net-` commands, which link you to the Stata user community and the most up-to-date applications. Of course, to use the `-net-` commands you must be connected to the Internet.

SOME PARTICULARLY USEFUL STATA 10.0 COMMANDS

Here is a list of key data manipulation and utility commands. It is to your advantage to study the descriptions of these commands in the Stata manual—in addition to reading through the *User's Guide*. The time you spend gaining familiarity with these commands, and with the logic of Stata procedures, will be more than repaid by improvements in the efficiency of your work. I have included few of the commands for carrying out estimation procedures because they will be introduced in later chapters.

`adjust`	Gets adjusted values for means and proportions.
`append`	Combines two data sets with identical variables but different observations. (See also `-merge-`.)
`by`	Repeats a Stata command on subsets of data.
`capture`	Captures return code (that is, allows Stata to continue whether the condition is true or not).
`cd`	Changes directory.
`codebook`	Produces a codebook describing the data.
`collapse`	Produces aggregate statistics, such as means, for subsets of data. Particularly useful for making graphs. Similar to the "aggregate" command in SPSS.
`compress`	Compresses variables to make a data set smaller but without altering the logical character of any variable. Useful when your data will not fit into memory.
`count`	Gives the number of observations satisfying specified conditions. `-count-` without a suffix gives the number of observations in the data set.
`#delimit`	Changes the delimit character.

`describe`	Describes the contents of a data set.
`dir`	Displays file names in the current directory.
`display`	Substitutes for a hand calculator.
`do`	Executes commands from a `-do-` file.
`drop`	Drops variables or observations from the file.
`edit`	Allows you to edit your file cell by cell. Useful in inspecting the content of your file or correcting errors in the file.
`egen`	Extensions to the `-generate-` command.
`encode`	Permits recoding of string variables to numeric variables.
`foreach`	Repeats Stata command for a list of items (variables, values, or other entities). Similar to "do repeat" in SPSS but more powerful.
`forvalues`	Repeats Stata command for a set of consecutive values.
`generate`	Creates or changes the contents of a variable.
`help`	Obtains online help. (See also `-search-` and `-net search-`.)
`infile`	Reads data into Stata. (See also `-infix-` and `-insheet-`.)
`input`	Inputs data from the keyboard.
`inspect`	Useful summary of numerical variables, particularly when you are not familiar with a data set. Reports the number of negative, zero, and positive values; the number of integers and nonintegers; the number of unique values; and the number of missing values; and produces a small histogram.
`keep`	Keeps variables or observations in the file—that is, drops everything not specified. Useful when you want to create a new file containing a small subset of variables.
`label`	Creates or modifies value and variable labels.
`list`	Lists values of variables.
`log`	Creates a log of your session.
`mark`	Marks observations for inclusion (a way of maintaining consistency regarding missing values throughout an analysis).
`merge`	Merges two data sets with corresponding observations but different variables. (See also `-append-`.)
`net`	Installs and manages user-written additions from the net.
`netsearch`	Searches the Internet for installable commands.
`numlabel`	Combines numerical values with labels so that both are displayed. This often is very convenient.
`notes`	Puts notes into data set.
`order`	Reorders variables in a data set.
`predict`	Obtains predictions after any estimation command.
`preserve`	Preserves data. Use this before a command that will alter the data set, such as `-collapse-`. Then use `-restore-` to restore the preserved data set.
`quietly`	Performs Stata command without showing intermediate steps.
`recode`	Recodes variables.
`rename`	Renames variables.

`replace`	Replaces values of a variable with new variables if specified conditions are satisfied.
`reshape`	Converts data set from wide to long format and vice versa.
`restore`	Restores a previously preserved file.
`#review`	Reviews previous commands. Helpful when you start playing around rather than making a `-do-` file, and then want to capture your commands to create a `-do-` file.
`save`	Saves a data set.
`search`	Searches Stata documentation for a particular term or phrase. Particularly useful when `-help-` fails to produce what you want.
`sort`	Sorts the data set on one or more variables.
`summarize`	Produces summary statistics for continuous variables.
`table`	Produces tables of summary statistics.
`tabulate`	Produces one- and two-way tables of frequencies and percentages.
`update`	Updates Stata from the net. (See also `-ado-` and `-net-`.)
`use`	Opens a data set for analysis. A particularly useful feature of the `-use-` command is to specify the `-using-` option to limit the number of variables brought into memory. This dramatically speeds up the work for computers that have relatively small amounts of memory. The syntax is `use <variable list> using <file name>`.
`version`	Specifies which version of Stata applies (for a given `-do-` file or command).
`xi`	Interaction expansion. Forms new variables that are products of specified variables.

CHAPTER

INTRODUCTION TO CORRELATION AND REGRESSION (ORDINARY LEAST SQUARES)

WHAT THIS CHAPTER IS ABOUT

So far we have been dealing with procedures for analyzing categorical data. We now turn to a powerful body of techniques that can be applied when the dependent variable is an interval or ratio variable: ordinary least-squares regression and correlation analysis. In this chapter we deal with the two-variable case, where we have a dependent variable and a single independent variable, to illustrate the logic. In the following two chapters we deal with multiple regression, which is used when we want to explore the effects of several independent variables on a dependent variable, the typical case in social science research.

INTRODUCTION

Suppose we have a set of data arrayed like this:

Father's Years of Schooling	*Respondent's Years of Schooling*
2	4
12	10
4	8
13	13
6	9
6	4
8	13
4	6
8	6
10	11

What can we say about the relationship between father's education and respondent's education? Not much. Visual inspection of the two arrays is quite uninformative. However, if we *plot* the two variables in two-dimensional space, the nature of the relationship is revealed. When you inspect the plot (Figure 5.1), it is immediately evident that the children of highly educated fathers tend to be highly educated themselves. In this situation, we say that the father's and the respondent's education are *positively correlated.*

Although we can see that the father's and respondent's education are positively correlated, we want to quantify the relationship in two respects. First, we want a way to

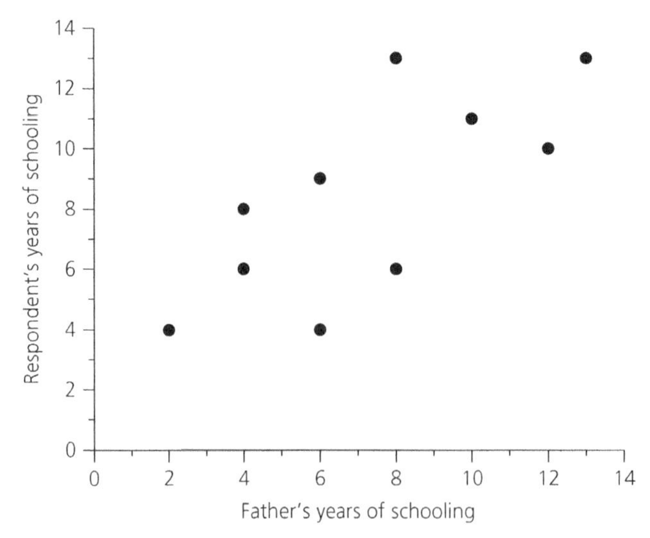

FIGURE 5.1. *Scatter Plot of Respondent's Years of Schooling by Father's Years of Schooling (Hypothetical Data, N = 10).*

describe the *character* of the relationship between the father's and respondent's years of schooling. How large a difference in the dependent variable, years of schooling, would we expect on average for a person whose father's schooling (the independent variable) differs by one unit (one year)? What level of schooling would we expect, or predict, on average for each person, given that we know how much schooling his or her father has? Second, we want a way to characterize the *strength* of the co-relation, or *correlation*, between the respondent's and father's years of schooling. Can we get a precise prediction of the respondent's level of education from the father's level of education or only an approximate one?

QUANTIFYING THE SIZE OF A RELATIONSHIP: REGRESSION ANALYSIS

The conventional and simplest way to describe the character of the relationship between two variables is to put a straight line through the points that "best" summarizes the average relationship between the two variables. Recall from school algebra that straight lines are represented by an equation of the form

$$Y = a + b(X) \tag{5.1}$$

where a is the *intercept* (the value of Y when the value of X is zero) and b is the *slope* (the change in Y for each unit change in X). Figure 5.2 shows the coefficients a and b for our example involving respondent's years of education (Y) and father's years of education (X). The figure is a graphic representation of the equation:

$$\hat{E} = 3.38 + .687(E_F) \tag{5.2}$$

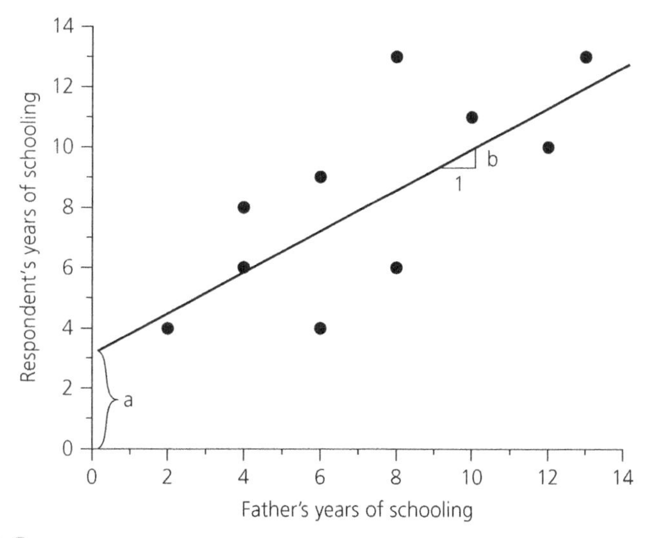

FIGURE 5.2. *Least-Squares Regression Line of the Relation Between Respondent's Years of Schooling and Father's Years of Schooling.*

Here $\hat{E}$ indicates the *expected* number of years of school completed by people with each level of father's years of schooling (E_F) on the assumption that the relationship is *linear*, that is, that each increase in the father's education produces a given increase in the respondent's education regardless of the initial level; 3.38 is the *intercept*, that is, the expected years of schooling for people whose fathers had no schooling at all; and .687 is the *slope*, that is, the expected increase in years of schooling for each one-year increase in the father's schooling. From this equation, we would predict that those whose fathers have 10 years of schooling would have 10.25 years of schooling because 3.38 + 10*.687 = 10.25. Similarly, we would predict that the children of university graduates would have 2.75 more years of schooling, on average, than the children of high school graduates because .687*(16 – 12) = 2.75. Estimating the value of the dependent variable in a regression equation for given values of the independent variable is known as *evaluating* the equation.

So far we have said nothing about how we derive the values for the coefficients shown in Equation 5.2. The criterion for putting a line through a set of points is that we minimize the sum of the squared errors of prediction—that is, we minimize the sum of the squared differences between the observed and predicted values. Lines derived in this way are known as *ordinary least-squares regression* lines. Figure 5.3 illustrates this criterion. The term e_i ($= E_i - \hat{E}_i$, or the actual number of years of schooling for the *i*th person minus the expected number of years of schooling for that person given his or her father's years of schooling) shown in the figure is the error of prediction between the specified point and the regression line. If we square each of these errors of prediction (which are also called *residuals*) and sum them, there is one and only one line for which this sum of squares is smallest. This is the ordinary least-squares (OLS) regression line.

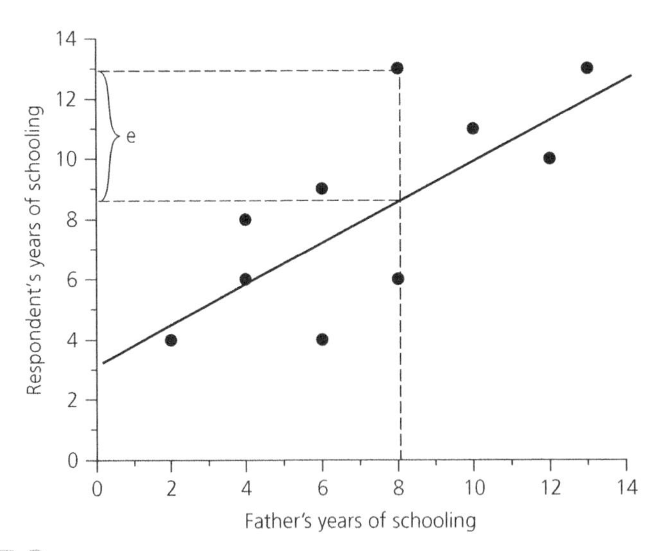

FIGURE 5.3. *Least-Squares Regression Line of the Relation Between Respondent's Years of Schooling and Father's Years of Schooling, Showing How the "Error of Prediction" or "Residual" Is Defined.*

WHY USE THE "LEAST SQUARES" CRITERION TO DETERMINE THE BEST-FITTING LINE? Note that "least squares" is not the only plausible criterion of "best fit." An intuitively more appealing criterion is to minimize the sum of the absolute deviations of observed values from expected values. Absolute values are mathematically intractable, however, whereas sums of squares have convenient algebraic properties, which is probably why the inventors of regression analysis hit upon the criterion of minimizing the sum of squared errors. The consequence is that observations with unusually large deviations from the typical pattern of association can strongly affect regression estimates; because the deviations are squared, such observations have the greatest weight. The presence of atypical observations, known in this context as high leverage points, can therefore produce quite misleading results. We will discuss this point further in the upcoming paragraphs and in Chapter Ten.

It can be shown, via algebra or calculus, that the following formulas for the slope and intercept satisfy the least squares criterion:

$$b = \frac{\text{cov}(X,Y)}{\text{var}(X)} = \frac{\sum(X-\bar{X})(Y-\bar{Y})}{\sum(X-\bar{X})^2} = \frac{N\sum XY - (\sum X)(\sum Y)}{N\sum X^2 - (\sum X^2)} \qquad (5.3)$$

and

$$a = \bar{Y} - b(\bar{X}) = \frac{\sum Y}{N} - b\frac{\sum X}{N} \qquad (5.4)$$

ASSESSING THE STRENGTH OF A RELATIONSHIP: CORRELATION ANALYSIS

Now that we have seen how regression lines are derived and how they are interpreted, we need to assess how good the prediction is. Our criterion for goodness of prediction or *goodness of fit* is the fraction or proportion of the variance in the dependent variable that can be attributed to variance in the independent variable. We define

$$r^2 = 1 - \frac{\sum(Y-\hat{Y})^2/N}{\sum(Y-\bar{Y})^2/N} \qquad (5.5)$$

That is, r^2, which is just the square of the Pearson correlation coefficient, is equal to 1 minus the ratio of the variance around the regression line to the variance around the mean of the dependent variable. (The Pearson correlation coefficient is, of course, the correlation coefficient you have encountered in introductory statistics courses. It has the advantage of ranging from −1 to +1 depending on whether two variables move together or in opposite directions. But it is not as readily interpretable as its square.) When the variance around the regression line is just as large as the variance around the mean of the dependent variable—that is, when knowing the value of the independent variable does not help us predict the value of the dependent variable (in which case, the mean of the dependent variable is the least squares prediction of each value)—the ratio is 1 and $r^2 = 0$; this case is illustrated in (a) of Figure 5.4. When knowledge of the value of the independent variable allows perfect prediction of the value of the dependent variable, the ratio is 0, and hence $r^2 = 1$; this case is illustrated in (b) of Figure 5.4.

Note that OLS regression finds the best *linear* relationship between two variables, even when the actual functional form of the relationship is nonlinear. For example, the correlation between X and Y in (c) of Figure 5.4 is zero, even though it is obvious that the two variables are perfectly (curvilinearly) related. See also Figure 10.1, which reproduces a set of graphs constructed by Anscombe (1973) to show that a given correlation may be associated with very different relationships between two variables. Linear regression provides an adequate summary of a relationship only when it correctly represents the

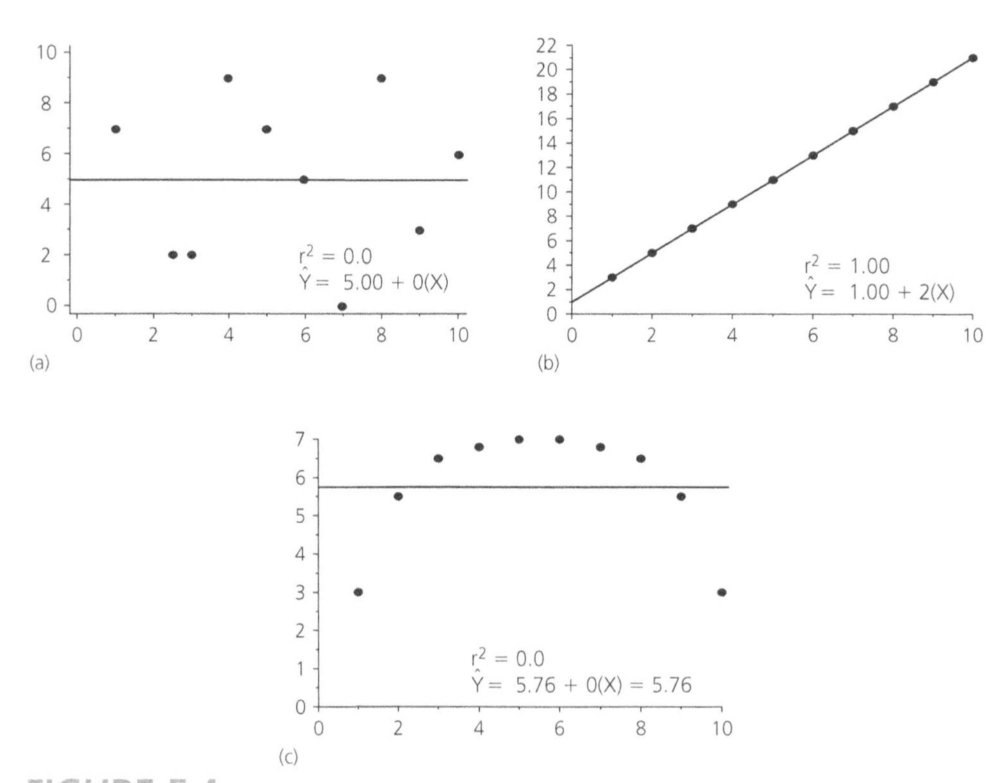

FIGURE 5.4. *Least-Squares Regression Lines for Three Configurations of Data: (a) Perfect Independence, (b) Perfect Correlation, and (c) Perfect Curvilinear Correlation—a Parabola Symmetrical to the x-Axis.*

KARL PEARSON (1857–1936) established the discipline of mathematical statistics and was the principal developer of linear regression and correlation; in recognition of this, the product moment, or ordinary least squares, correlation coefficient, r, is also known as Pearson's r. Pearson's work on classifying probability distributions forms the basis for classical (frequentist) statistical theory and underlies the general linear model. But his contributions are very extensive—for example, he invented the standard deviation and the χ^2 test. He founded the journal Biometrika in 1901 and edited it until his death; he also founded the journal Annals of Eugenics (now Annals of Human Genetics) in 1925. Pearson was born in London to a family of religious dissenters. He studied mathematics at Cambridge but then studied medieval and sixteenth-century German literature at the Universities of Berlin and Heidelberg and became enough of an expert to be offered a Germanics post at Kings College, Cambridge, which he declined. Instead, he read law (his father was a barrister) but never practiced, returning to mathematics. In his youth he also became a feminist and a socialist (the transformation of his birth name, Carl, to Karl, was said to have resulted initially from the way his name was spelled by a clerk when he enrolled at Heidelberg but supposedly was adopted by him in tribute to Karl Marx, whom he apparently had met). He eventually became universally known as KP. In 1884 he was appointed to the Goldsmid Chair of Applied Mathematics and Mechanics at University College, London, and, in 1891, to the chair in Geometry at Gresham College, London. There he met W.F.R. Weldon, a zoologist interested in evolutionary theory who posed a number of research problems that stimulated Pearson to think about statistical distributions; their collaboration lasted until Weldon's untimely death in 1906.

character of the relationship. When it fails to do so, additional variables need to be included in the model. You will see how to do this in the next chapter.

Returning to our example about intergenerational continuity in educational attainment, we note that $r^2 = .536$, which tells us that the variance around the regression line is about half the size of variance around the mean of the dependent variable, and therefore that about half of the variance in educational attainment is explained by the corresponding variability in father's education. As social science results go, this is a very high correlation.

A USEFUL COMPUTATIONAL FORMULA FOR *r* The following is a useful computational formula for the correlation coefficient, r, which comes in handy when you have to do hand calculations:

$$r = \frac{\text{cov}(X,Y)}{\sqrt{\text{var}(X)}\sqrt{\text{var}(Y)}} = \frac{N\Sigma XY - (\Sigma X)(\Sigma Y)}{\sqrt{N\Sigma X^2 - (\Sigma X)^2}\sqrt{N\Sigma Y^2 - (\Sigma Y)^2}}$$

THE RELATIONSHIP BETWEEN CORRELATION AND REGRESSION COEFFICIENTS

Suppose we were to *standardize* our variables before computing the regression of Y on X, by, for each variable, subtracting the mean from the value of each observation and dividing by the standard deviation. Doing this produces new variables with mean = 0 and standard deviation = 1. Then we would have a regression equation of the form

$$\hat{y} = \beta(x) \tag{5.6}$$

(The convention adopted here, which is widely but not universally used, is to represent standardized variables by lowercase Latin symbols and the coefficients of standardized variables by Greek rather than Latin symbols.) There is no intercept because the regression line must necessarily pass through the mean of each variable, which for standardized variables is the (0,0) point. We interpret β as indicating the number of *standard deviations* by which we would expect two observations to differ on Y that differ by one standard deviation on X. (This follows directly from the fact that for standardized variables, the standard deviation is one. Thus, one standard deviation on X is one unit on x; and the same for Y and y.) It can be shown, through a simple manipulation of the algebraic computational formulas for the coefficients, that in the two-variable case, $r = \beta$. It is also true that r is invariant under linear transformations. (A linear transformation is one in which a variable is multiplied [or divided] by a constant and/or a constant is added [or subtracted] . Consider two variables, Y and Y', with $Y' = a + b(Y)$. In this case, $r_{xy} = r_{xy}'$.) So the correlation between standardized variables and unstandardized variables is necessarily perfect.

A convenient pair of formulas for moving between b and β (which also holds for *multiple regression* coefficients) is

$$\beta = b\left(\frac{s_X}{s_Y}\right) \Rightarrow \quad b = \beta\left(\frac{s_Y}{s_X}\right) \tag{5.7}$$

$$a = \bar{Y} - b(\bar{X}) \tag{5.8}$$

where s_X and s_Y are the the standard deviations of X and Y, respectively.

FACTORS AFFECTING THE SIZE OF CORRELATION (AND REGRESSION) COEFFICIENTS

Now that we see how to interpret correlation and regression coefficients, we need to consider potential troubles—factors that affect the size of coefficients in ways that may lead to incorrect interpretation and false inferences by the unwary.

Outliers and Leverage Points

As noted, correlation and regression statistics are very sensitive to observations that deviate substantially from the typical pattern. This is a consequence of the least squares criterion—because "errors" (differences between observed and predicted values on the

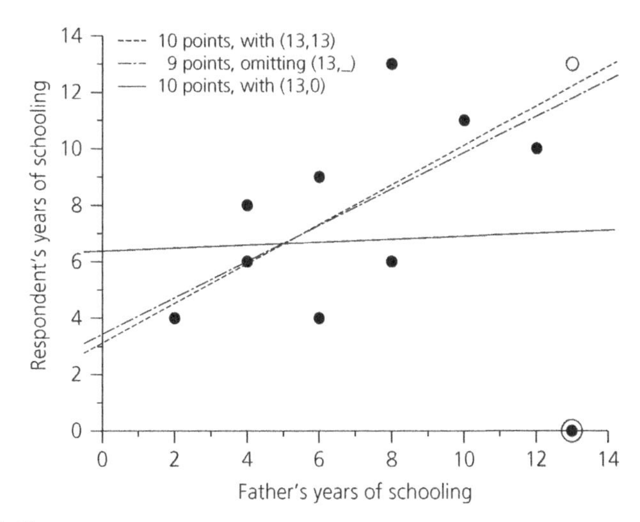

FIGURE 5.5. *The Effect of a Single-Deviant Case (High Leverage Point).*

dependent variable) are squared, the larger the error, the more it will contribute to the sum of squared errors relative to its absolute size. Thus, correlation coefficients can be substantially affected by a few deviant observations, with regression slopes pulled strongly toward them, producing misleading results. To see this, consider the following example, illustrated in Figure 5.5. Suppose that in our example about intergenerational educational transmission, the fourth case had values (13,0) (shown as a solid circle surrounded by an open circle) instead of (13,13) (shown as an open circle). That is, suppose that in the fourth case the child of a man with thirteen years of schooling had no education instead of thirteen years of schooling—perhaps because the child was mentally impaired. The alteration of just one point, from (13,13) to (13,0), dramatically changes the regression line and misrepresents the typical relationship between the father's and respondent's education, making it appear that there is no relationship at all (the regression equation for the ten points with (13,0) as the fourth value is $\hat{E} = 6.74 + .0491(E_F)$; $r^2 = .002$).

This example illustrates the condition under which deviant cases are influential—that is, have high "leverage." This is when points are far away from the center of the multivariate distribution. Outliers close to the center of the distribution, for example, the (8,13) point in Figure 5.5, have less influence because, although they can pull the regression line up or down, they have relatively little effect on the slope. We will consider this distinction further in Chapter Ten.

The most straightforward solution is to omit the offending case. When this is done, the regression line through the remaining nine points is very close to the regression line through ten points with (13,13). However, this generally is an undesirable practice because it creates the temptation to start "cleaning up" the data by omitting whatever cases tend to fall far from the regression surface. Two better strategies, which will be elaborated in Chapters Seven and Ten, are (1) to think carefully about whether the outliers might have been generated by a different process from the remainder of the data and, when you suspect that possibility, to explicitly model the process; or (2) to use a robust regression procedure that downweights large outliers. Fortunately, the damage done by outliers diminishes as sample sizes increase. However, even with large samples extreme outliers can be distorting—for example, incomes in the millions of dollars. One simple way to deal with extreme values on univariate distributions is to truncate the distribution, for example, in the United States in 2006 by specifying $150,000 for incomes of $150,000 or above (this is what the GSS does; in 2006, just over 2 percent of the GSS sample had incomes this high); but this creates its own problems, as we will see next. A better way, which you will see in Chapter Fourteen, is to use interval regression (an elaboration of tobit regression) to correctly specify the category values.

Truncation

Analysts are sometimes tempted to divide their study population into subgroups on the basis of values on the independent or dependent variable or on variables substantially correlated with the independent or dependent variable. For example, an analyst who suspects that income depends more heavily on education among those with nonmanual occupations than among those with manual occupations might attempt to test this hypothesis by correlating income with education separately for nonmanual and manual workers. This is a bad idea because income is correlated with occupational status; thus, dividing the population on the basis of occupational status will truncate the distribution of the dependent variable, which, all else equal, will reduce the size of the correlation. Moreover, if one subgroup, say manual workers, has a smaller variance with respect to income than does the other subgroup, say nonmanual workers (and this is likely to be true in most societies), the size of the correlation will be more substantially reduced for manual than for nonmanual workers, thus leading the analyst to—mistakenly—believe that the hypothesis is confirmed.

To see this, consider a highly stylized example, shown as Figure 5.6. To keep the example simple, imagine that all manual workers in the sample have less than seven years of schooling and that all nonmanual workers have more than seven years of schooling. Note that in the example, there is exactly the same income return to an additional year of education for nonmanual and manual workers. Note further that each point is an equal distance from the regression line. Now, suppose the correlation between income and education were computed separately for manual and nonmanual workers. The correlation for both groups would be smaller than the correlation computed over the total sample, and the correlation would be smaller for manual than for nonmanual workers. This follows directly from Equation 5.5 because, from the way the example was constructed, the variance around the regression line is identical in all three cases, but the variance around the mean of the dependent variable is smaller for nonmanual workers than for the total sample and smaller for manual workers than for nonmanual workers. Although, for the sake of clarity, the example is highly stylized, the principle holds generally: when distributions are truncated the correlation tends to be reduced. This, by the way, is the main reason GRE scores are weak predictors of grades in graduate school courses: graduate

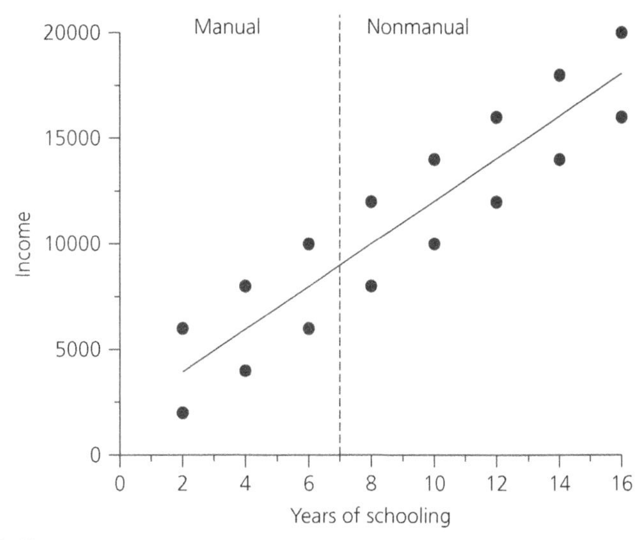

FIGURE 5.6. *Truncating Distributions Reduces Correlations.*

departments do not admit people with low GREs, thereby truncating the distribution of GRE scores. But this does not imply that GRE scores should be ignored in the admissions process, as statistically illiterate professors argue from time to time.

A "REAL DATA" EXAMPLE OF THE EFFECT OF TRUNCATING THE DISTRIBUTION Analyzing the U.S. sample for the *Political Action: An Eight Nation Study, 1973–1976* (Barnes and Kaase 1979) some years ago, I was puzzled to discover an extremely low correlation between education and income (less than .1, whereas in U.S. surveys the typical correlation between these two variables is on the order of .3). Further investigation revealed that the low end of both the education and income distributions were severely truncated, presumably as a result of inadequacies in either the sampling or the field work procedures. When the data were weighted to reproduce the bivariate distribution of education and income observed in the U.S. census for 1980 (the year closest to the survey), the estimated correlation approximated that typically found in U.S. surveys.

Regression Toward the Mean

The consequences of truncation actually are worse than just suggested, because of a phenomenon known as "regression toward the mean." When two measurements are made at different points in time, for example, pre-test and post-test measurements in a randomized experiment or scores on the GRE, it is typical to observe that those cases with high values on the first observation tend, on average, to have lower values on the second observation, and that those cases with low values on the first observation tend to have higher values on the second observation. That is, both the high and the low values move toward (or "regress toward") the mean. This is true even when there is no change in the *true* value between the two measurements.

The reason for this is that observed measurements consist of two components: a true score and a component representing error in measurement of the underlying true score. For example, consider the GRE. The observed score for each individual can be thought of as consisting of a component measuring the candidate's "true" (or underlying or constant) ability to do the kind of work measured by the test and a random component comprised of variations in the exact questions asked in that administration of the test, the candidate's level of energy and mental acuity, level of confidence (Steele 1997), and so on. It then follows that those who have high scores in any given administration of the test will disproportionately include those who have high positive random components, and those who have low scores will disproportionately include those who have low random components. But because the second component *is* random, those who have high random components on the first test will tend, on average, to have lower random components on the second test and those who have low random components on the first test, will tend, on average, to have higher random components on the second test. The result is that the correlation between the two tests will be less than perfect and also that the regression coefficient relating the second to the first test will be less than 1.0. This is true even if the means and standard deviations of the two tests are identical.

An important implication of this result is that a researcher who targets for special intervention a low-scoring group (those who did poorly on a practice GRE, those with low grade point averages, and so on) will be bound to conclude, incorrectly, that the intervention was successful. Of course, if that same researcher chose the high-scoring group for the same intervention, he or she would be forced to conclude that the intervention was completely unsuccessful—indeed, that it was counterproductive. All of this is a simple consequence of analyzing a nonrandom subset of the original sample.

Exactly the same phenomenon—measurement error—has the effect of lowering the correlation between separate phenomena, for example, education and income, the heights of fathers and sons, and so on. This kind of observation is what led Francis Galton, one of the founders of correlation and regression analysis, to conclude in the late nineteenth century that a natural phenomenon of intergenerational transmission was a "reversion" (or "regression") toward "mediocrity"—hence the term "regression analysis" to describe the linear prediction procedure discussed here. But what Galton failed to notice is that there is also, and for exactly the same reason, a tendency for values near the mean to move *away* from the mean. The result is that the variance of the *predicted* (but not the *observed*) values and the slope of the regression line are reduced in proportion to the complement of the correlation between the variables. (For a book-length treatment of this topic, see Campbell and Kenney [1999].)

Aggregation

Students who have spent some time studying the behavior of populations of individuals usually conclude that we live in a stochastic world in which nothing is very strongly related to anything else. For example, in the United States, typically about 10 percent of

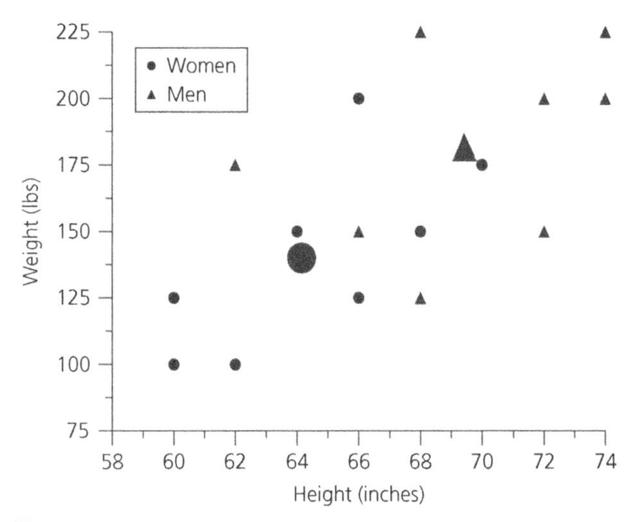

FIGURE 5.7. *The Effect of Aggregation on Correlations.*

the variance in income can be attributed to variance in education ($r \approx .3 \Rightarrow r^2 \approx .09$). Students are then puzzled when they discover that seemingly comparable correlations computed over aggregates, for example, the correlation between mean education and mean income for the detailed occupational categories used by the U.S. Bureau of the Census, tend to be far larger (in the present example, $r \approx .7 \Rightarrow r^2 \approx .49$). Why is this so? The explanation is simple. When correlations are computed over averages or other summary measures, a great deal of individual variability tends to "average out." In the extreme case, where there are only two aggregate categories, the correlation between the means for the two categories will necessarily be 1.0, as you can see in Figure 5.7 (where the large circle represents the mean height and weight for women, and the large triangle represents the mean height and weight for men); but the principle holds for more than two categories as well.

CORRELATION RATIOS

So far we have been discussing cases where we have two interval or ratio variables. Sometimes, however, we want to assess the strength of the association between a categorical variable and an interval or ratio variable. For example, we might be interested in whether religious groups differ in their acceptance of abortion. Or we might be interested in whether ethnic groups differ in their average income. The obvious way to answer these questions is to compute the mean score on an abortion attitudes index for each religious group or the mean income for each ethnic group. But if we discover that the means differ substantially enough to be of interest, we still are left with the question of how strong the relationship is. To determine this we can compute an analog to the (squared) correlation coefficient, known as the (squared) correlation ratio, η^2 (eta squared). η^2 is defined as

$$\eta^2 = 1 - \frac{\textit{Variance around the subgroup means}}{\textit{Variance around the grand mean}}$$
$$= 1 - \frac{\textit{Within group sum of squares}}{\textit{Total sum of squares}} \qquad (5.9)$$
$$= 1 - \frac{\sum_j \sum_i (Y_{ij} - \bar{Y}_{.j})^2}{\sum_j \sum_i (Y_{ij} - \bar{Y}_{..})^2}$$

where Y is the dependent variable, there are j groups, and i cases within each group. Thus, $\bar{Y}_{.j}$ is the mean of Y for group j, and $\bar{Y}_{..}$ is the grand mean of Y. From Equation 5.9, it is evident that if all the groups have the same mean on the dependent variable, knowing which group a case falls into explains nothing; the variance around the subgroup means equals the variance around the grand mean, and $\eta^2 = 0$. At the other extreme, if the groups differ in their means, and if all cases within each group have the same value on the dependent variable—that is, there is no within-group variance—then the ratio of the within-group sum of squares to the total sum of squares is 0, and $\eta^2 = 1$. From this we see that η^2, like r^2, is a *proportional reduction in variance* measure.

Let us explore the religion and abortion acceptance example with some actual data. In 2006 (and for most years since 1972) the GSS asked seven questions about the acceptability of abortion under various circumstances:

. . . should [it] be possible for a woman to obtain a legal abortion . . .

- if there is a strong chance of serious defect in the baby?
- if she is married and does not want any more children?
- if the woman's own health is seriously endangered by the pregnancy?
- if the family has a very low income and cannot afford any more children?
- if she became pregnant as a result of rape?
- if she is not married and does not want to marry the man?
- if the woman wants it for any reason?

From these items I constructed a scale by counting the positive responses, excluding all cases with any missing data. The scale thus ranges from 0 to 7. Table 5.1 shows the mean number of positive responses by religion. All those who specified religions other than Protestant, Catholic, or Jewish or said they had no religion were included in the "Other and None" category. From the table, it is evident that Jews and other non-Christians are much more accepting of abortion than are Christians (Protestants and Catholics). But how important is religion in accounting for acceptance of abortion? To see this, we compute $\eta^2 = .070$. (The Stata computations to create Table 5.1 and to obtain η^2 are shown in the downloadable `-do-` and `-log-` files for the chapter.)

TABLE 5.1. Mean Number of Positive Responses to an Acceptance of Abortion Scale (Range: 0–7), by Religion, U.S. Adults, 2006.

Religion	Mean Number of Positive Responses	Standard Deviation	N
Protestants	3.7	2.5	(923)
Catholics	3.8	2.5	(420)
Jews	5.6	2.5	(26)
Other or none	5.3	2.2	(395)
Total	**4.1**	**2.5**	**(1,764)**

Clearly, religious affiliation does not explain much of the variance in abortion attitudes. How can this be, given the substantial size of the mean differences? The answer is simple. Jews and "Others" differ substantially from Protestants and, especially, Catholics in their acceptance of abortion. But these groups are quite small, especially Jews. Hence, no matter how deviant they are from the overall average, they are unlikely to have much impact; when more than half of the population is included in one group, as is the case here with Protestants, a large fraction of the variance in abortion acceptance is bound to be within-group variance rather than between-group variance.

A second use of the correlation ratio is to test assumptions of linearity. We will take this up in Chapter Seven.

WHAT THIS CHAPTER HAS SHOWN

In this chapter we have considered simple (two-variable) ordinary least-squares (OLS) correlation and regression, as a way of seeing the conceptual basis of OLS regression, the workhorse of modern statistical analysis. We also considered how the size of correlation and regression coefficients is affected by the bivariate distribution of cases—specifically, how results are affected by high-leverage outliers, by truncation, by regression to the mean, and by aggregation. It is important that you understand these effects thoroughly because many confused claims are made by those who fail to understand them. We then considered a variant on correlation coefficients, the squared correlation ratio, which is an analog to correlation when we have an interval or continuous dependent variable but a categorical independent variable. In the next chapter we extend our discussion to multiple correlation and regression, the analogous OLS technique when we have two or more independent variables.

CHAPTER

INTRODUCTION TO MULTIPLE CORRELATION AND REGRESSION (ORDINARY LEAST SQUARES)

WHAT THIS CHAPTER IS ABOUT

In this chapter we consider the central technique for dealing with the most typical social science problem—understanding how some outcome is affected by several determining variables that are correlated with each other. We begin with a conceptual overview of multiple correlation and regression, and then continue with a worked example to illustrate how to interpret regression coefficients. We then turn to consideration of the special properties of categorical independent variables, which can be included in multiple regression equations as a set of dichotomous ("dummy") variables, one for each category of the original variable (except that to enable estimation of the equation, one category must be represented only implicitly). In the course of our discussion of dummy variables, we develop a strategy for comparing groups that enables us to determine whether whatever social process we are investigating operates in the same way for two or more subsegments of the population—males and females, ethnic categories, and so on. We conclude with an alternative way of choosing a preferred model, the Bayesian Information Coefficient (*BIC*).

INTRODUCTION

For most social science purposes, the two variable regressions we encountered in the previous chapter are not very interesting, except as a baseline against which to compare models involving several independent variables. Such models are the focus of this chapter. Here we generalize the two-variable procedure to many variables. That is, we predict some (interval or ratio) dependent variable from a *set* of independent variables. The logic is exactly the same as in the case of two-variable regression, except that we are estimating an equation in many dimensions.

Let us first consider the case where we have two independent variables. Extending the ten-observation example from the previous chapter, suppose we think that education depends not only on the father's education but also on the number of siblings. The argument is that the more siblings one has, the less attention one receives from one's parents (all else equal), and hence, in consequence, the less well one does in school and, therefore, the less education one obtains, on average (for examples of studies of sibship-size effects in the research literature, see Downey [1995], Maralani [2004], Lu [2005], and Lu and Treiman [2008]). Suppose, further, that we have information on all three variables for our sample of ten cases:

Father's Years of Schooling	*Respondent's Years of Schooling*	*Number of Siblings*
2	4	3
12	10	3
4	8	4
13	13	0
6	9	2
6	4	5
8	13	3
4	6	4
8	6	3
10	11	4

Note that the first two columns are simply repeated from the example in the previous chapter (see page 88).

To test our hypothesis that the number of siblings negatively affects educational attainment, we would estimate an equation of the form:

$$\hat{E} = a + b(E_F) + c(S) \quad (6.1)$$

(Note that I use generic symbols, for example, X and Y, to indicate variables in equations of a general form, but mnemonic symbols, for example, E, E_F, and S, to indicate variables in equations that refer to specific concrete examples. I find it much easier to keep track of what is in my equation when I use mnemonic symbols for variables.)

Equations such as Equation 6.1 are known as *multiple regression* equations. In multiple regression equations the coefficients associated with each variable measure the expected difference in the dependent variable associated with a one-unit difference in the given independent variable, *holding constant each of the other independent variables*. So in the present case, the coefficient associated with the number of siblings tells us the expected difference in educational attainment for each additional sibling among those whose fathers have exactly the same years of education. Correspondingly, the coefficient associated with the father's education tells us the expected difference in years of education for those whose fathers differ by one year in their education but who have exactly the same number of siblings. In the three-variable case (that is, when we have only two independent variables), but not when we have more variables, we can construct a geometric representation that illustrates the sense in which we are *holding constant* one variable and estimating the *net* effect of the other.

In multiple regression, as in two-variable regression, we use the least-squares criterion to find the "best" equation—that is, we find the equation that minimizes the sum of squared errors of prediction. However, whereas in bivariate regression we think in terms of the deviation between each observed point and a line, in multiple regression the analog is the deviation between each observed point and a k-dimensional geometric surface where $k = 1 +$ the number of independent variables. Thus, where there are two independent variables, the least-squares criterion minimizes the sum of squared deviations of each observation from a *plane*, as shown in Figure 6.1.

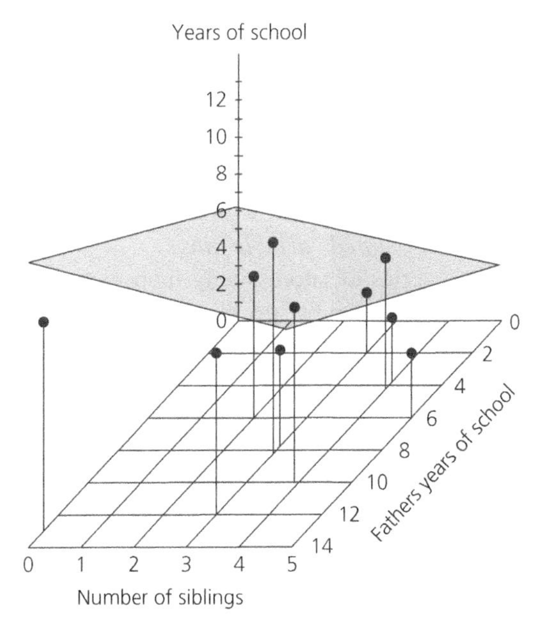

FIGURE 6.1. *Three-Dimensional Representation of the Relationship Between Number of Siblings, Father's Years of Schooling, and Respondent's Years of Schooling (Hypothetical Data; N = 10).*

Metric Regression Coefficients

The coefficients associated with each independent variable are known as *regression coefficients*, or *net regression coefficients* (or sometimes *raw* or *metric regression coefficients*, to distinguish them from *standardized coefficients*, about which you will learn more later). In the present case, the estimated regression equation is

$$\hat{E} = 6.26 + .564(E_F) - .640(S) \tag{6.2}$$

This equation tells us that a person who had no siblings and whose father had no education would be expected to have 6.26 years of schooling; that among those whose fathers had the same amount of education, the expected cost of each additional sibling would be nearly two-thirds of a year of schooling (precisely, 0.640 years); and that among those with the same number of siblings, those whose fathers' schooling differed by one year would be expected to differ in their own schooling by just over one half year (precisely, 0.564 years).

Note that the coefficient associated with father's education in Equation 6.2 is smaller than the corresponding coefficient in Equation 5.2 (.564 and .687, respectively). This is because the father's education and the number of siblings are correlated; r is, in fact, $-.503$ in this example). Thus, in Equation 5.2, part of the observed effect of the father's education on the respondent's education is due to the fact that poorly educated fathers tend to have more children, and those from large families tend to go less far in school. Equation 6.2 takes account of this association and gives the effect of the father's education *net of* (or holding constant or controlling for) the number of siblings. The implication of this result is very important: if a variable that is a true cause of the dependent variable is left out of the equation, the coefficients of the variables included in the equation will be biased—that is, will overstate or understate the true causal relation between the given independent variable and the dependent variable (except in the limiting case that the left-out variable is uncorrelated with the variables in the equation). This is known as *specification error* or *omitted variable bias*.

Some analysts present a series of successively more complete multiple regression models and discuss changes in the size of specific coefficients resulting from the inclusion of additional variables. This is a sensible strategy under one specific condition: when the analyst wants to consider how the effect of one or more variables is modified by the inclusion of another variable (or variables). That is, in a manner closely analogous to the search for spurious or intervening relationships in tabular analysis (see Chapters Two and Three), the analyst might want to investigate whether a particular relationship is explained, or partly explained, by another factor. For example, it may be observed that Southerners are less tolerant of social deviants than are persons living outside the South. However, the analyst may want to assess the possibility that this relationship is entirely (or largely) spurious, arising from the fact that Southerners tend to be less well educated and less urban than others, and that education and urban residence increase tolerance. In such a case it would be appropriate to present two models—one regressing tolerance on Southern residence and a second regressing tolerance on Southern residence, education, and size-of-place—and then to discuss the reduction in the size of the coefficient associated with Southern residence that occurs when education and size-of-place are added to the equation. However, absent specific hypotheses regarding spurious or mediating effects, there is no point in estimating successive equations (except for models involving sets of dummy variables, discussed in the next section, or variables that alter functional forms, discussed in the next chapter); rather, all relevant variables should be included in a single regression equation. However, even in this case the analyst should present a table of zero-order (two-variable) correlation coefficients between pairs of variables, plus means and standard deviations for all interval and continuous variables and percentage distributions for all categorical variables. These descriptive statistics help the reader to understand the properties of the variables being analyzed. In addition, as noted earlier, the zero-order correlations provide a baseline for assessing the size of net effects when other variables are controlled.

Testing the Significance of Individual Coefficients

It is conventional to compute and report the standard error of the coefficient of each independent variable—although, as you will soon see, standard errors have limited utility in the case of dummy variables or interaction terms. The convention is to interpret coefficients at least twice the size of their standard error as statistically significant. This convention arises from the fact that the sampling distribution of regression coefficients follows a t-distribution and that, with 60 *d.f.* (where the degrees of freedom is computed as $N - k - 1$, with k the number of independent variables), $t = 2.00$ defines the 95 percent confidence interval around the value $b = 0$. It is important to understand that the t-statistics indicate the significance of each coefficient net of the effect of all other coefficients in the model. Thus, when several highly correlated variables are included in the model, it is possible that no one of them is significantly different from zero, although as a group they are significant (see also the following boxed comment on multicollinearity).

Some analysts estimate regression models involving several independent variables, drop the variables with nonsignificant coefficients (this is known as *trimming the regression equation*), and reestimate the model, on the ground that to leave coefficients in the model that have nonsignificant effects biases the estimates of the other variables. However, other analysts argue that the best estimate of the dependent variable is obtained by including all possible predictors, even those for which the difference from zero cannot be established with high confidence. The latter strategy is preferable because it provides the best point estimate based on a set of variables that the analyst has an *a priori* basis for suspecting affect the outcome.

Standardized Coefficients

A question that naturally arises when there are multiple determinants of some dependent variable is which determinant has the greatest impact. We cannot directly compare the coefficients associated with each independent variable because they typically are expressed in different metrics. Is the consequence of a difference of one year of schooling completed by the father greater or smaller than the consequence of a difference of one sibling? Although the question can, of course, be answered—as we saw earlier, the cost of each additional

MULTICOLLINEARITY When independent variables are highly correlated, a condition known as *multicollinearity*, regression coefficients tend to have large standard errors and to be rather unstable, in the sense that quite small changes in the distribution of the data can produce large changes in the size of the coefficients. As Fox notes (1991, 11; see also Fox 1997, 337–366), the inflation in the sampling variance of an independent variable, *j*, due to multicollinearity, is given by $1/(1 - R_j^2)$, where R_j^2 is the coefficient of determination (discussed later in this chapter) associated with the regression of variable *j* on the remaining independent variables; this is known as the *variance inflation factor* and can be computed in Stata by using the `-estat vif-` command after the `-regress-` command. (See Fox and Monette [1992] for a generalization to sets of independent variables, such as a set of dummy variables or a variable and its square; see also the discussion in Chapter Seven of this book, in the section on "Nonlinear Transformations.")

Clearly, the independent variables must be quite highly correlated for multicollinearity to be an important problem. For example, if $R_j^2 = .75$, the error variance will be quadrupled, and the standard error will thus be doubled. Because R_j^2's as large as .75 are quite uncommon, multicollinearity is not often a problem in the social sciences, mainly arising in situations in which alternative measures of the same underlying concept are included in a single model and most commonly when aggregated data, such as properties of occupations, cities, or nations, are analyzed. In such situations, a reasonable solution is to combine the measures into a multiple-item scale (see Chapter Eleven).

Some analysts attempt to minimize multicollinearity by employing what is known as *stepwise regression*, in which variables are selected into (or out of) a model one at a time, in the order that produces the greatest increment (or the smallest decrement) in the size of the R^2. Such methods are generally misguided, both because they are completely atheoretical and because the order in which variables are selected can be quite arbitrary, given the previously noted instability in regression coefficients when variables are highly correlated.

sibling is somewhat greater than the gain from each year of the father's education—the answer does not tell us which variable has the stronger effect on the dependent variable because the variance in the number of siblings is much smaller than the variance in the father's years of schooling. If it is not obvious why the size of the variance matters, consider the effect of education and income on the value of the car a person drives. Suppose that for a sample of U.S. adults, we estimate such an equation and obtain the following:

$$\hat{V} = 15{,}000 + .5(I) - 500(E) \qquad (6.3)$$

We would hardly want to conclude from this that the effect of education is 1,000 times as large as the effect of income, or to measure income in units of $100 and then to conclude that the effect of education is 10 times that of income. Actually, the equation indicates that a year of education reduces the (expected) value of a person's car by \$500, net of income, whereas a \$1,000 increment in income increases the (expected) value of a person's car by \$500, net of education. In this precise sense, a year of education exactly offsets \$1,000 in income. However, a more general way to compare regression coefficients is to transform them into a common metric.

The conventional way this is done is to express the relationship between the dependent and independent variables in terms of standardized variables—that is, variables transformed by subtracting the mean and dividing by the standard deviation. Because such variables all have standard deviation = 1, the regression coefficients associated with standardized variables indicate the number of standard deviations of difference on the dependent variable expected for a one standard deviation difference on the independent variable, net of the effects of all other independent variables. In the present example, the equation relating the standardized coefficients—that is, the standardized counterpart to Equation 6.2—is

$$\hat{y} = .601(e_f) - .260(s) \tag{6.4}$$

(Reminder: As noted in the previous chapter, there is no intercept because standardized variables all have mean = 0 and a regression surface must pass through the mean of each variable.) From inspection of the coefficients in Equation 6.4, we conclude that the father's education has a greater effect on educational attainment than does the number of siblings—a greater effect in the precise sense that a one standard deviation difference in the father's years of schooling implies an expected difference of .60 of a standard deviation in the respondent's years of schooling, whereas a one standard deviation difference in the number of siblings implies only a –.26 standard deviation expected difference in the respondent's years of schooling.

Note that in practice we do not ordinarily standardize the variables and recompute the regression equation but rather instruct the software to report standardized coefficients (usually in addition to metric coefficients). Because standardized coefficients often are not reported, particularly in the economics literature, we also can make use of the relation $\beta_{YX} = b_{YX}(s_X/s_Y)$—that is, the fact that the standardized coefficient relating independent variable X to dependent variable Y is equal to the metric coefficient multiplied by the ratio of the standard deviations of the independent and dependent variables—to convert metric to standardized coefficients (or vice-versa). (Recall Equation 5.7 and 5.8.)

There is some controversy regarding standardized regression coefficients. The conventional wisdom in sociology and other social sciences is that they are useful for the purpose just described—to assess the relative effect size of each of a set of independent variables in determining some outcome—but that they are inappropriate for assessing the relative effect size of a given variable in different populations, precisely because the standardized coefficients will differ if the relative standard deviations differ in the populations even if the metric regression coefficients are identical (Kim and Mueller 1976). (Note the analogy to the truncation of correlation coefficients discussed in the previous chapter. In fact, as noted in the previous chapter, the correlation coefficient is identical to the standardized regression coefficient in the two-variable case.) For example, suppose we wanted to compare the effect of the number of siblings on education among Blacks and Whites in the United States. Suppose, further, that the metric regression coefficient relating years of schooling to the number of siblings is identical for Blacks and Whites, that the standard deviation of years of schooling for Blacks and Whites is identical, but that the standard deviation of number of siblings is larger for Blacks than for Whites. Under these circumstances the standardized coefficient relating the number of siblings to years of schooling would also be larger for Blacks than for Whites (this follows directly from the mathematical relation between the standardized and metric regression coefficients shown in Equations 5.7 and 5.8). Would we really want to conclude that the number of siblings has a stronger effect for Blacks than for Whites in determining how much schooling they get if the "cost" (in terms of years of schooling) of each additional sibling is identical for Blacks and Whites? Probably not. However, there are those (for example, Hargins 1976) who argue that it would be meaningful to say that sibship size matters more for Blacks precisely because there is more variability in the sibship sizes of Black families.

Additional light may be shed on this point by comparing the interpretation of standardized and unstandardized coefficients *within* a single sample. In an analysis of educational attainment in a 1962 U.S. national representative sample, Beverly Duncan (1965, 60, 65) showed that the cost of coming from a nonintact family was very high—about a year of schooling, net of a large number of other factors. However, the standardized coefficient relating education to family intactness was relatively weak, about .09, far from the largest standardized coefficient. How can these two results be reconciled? The fact is that there is nothing inconsistent about them. The metric coefficient indicates that for the relatively few persons from non-intact families (these are data from 1962), the cost was very substantial. But the standardized coefficient indicates that family intactness was not a very important determinant of variance in educational attainment, precisely because only a relatively small fraction of the sample was from non-intact families. Given the near constancy of the family intactness variable, it could hardly explain much of the variability in educational attainment.

REMINDER REGARDING THE VARIANCE OF DICHOTOMOUS VARIABLES "Family intactness" is a dichotomous variable. As you will recall from elementary statistics, the standard deviation of such variables is given by the proportion "positive" (whichever category is so defined). That is, $s_p = \sqrt{p(1-p)}$. Thus, the more skewed the distribution, that is, the further it departs from .5 positive, .5 negative, the smaller the standard deviation and hence the smaller the standardized coefficient. Because for dichotomous variables the size of standardized coefficients depends not only on the size of the metric coefficient but also on the proportion of the sample with the "positive" attribute, it generally is unwise to make much of standardized coefficients for such variables.

Coefficient of Determination (R^2)

How well does Equation 6.2 explain the variance in educational attainment? We determine this via an exact analogy to r^2, known as R^2, or the *coefficient of determination*, which gives us the proportion of variance in the dependent variable explained by the entire set of independent variables. Just as for r^2, $R^2 = 1$ – the ratio of the error variance (the variance

A FORMULA FOR COMPUTING R^2 FROM CORRELATIONS A convenient formula for computing R^2 from a matrix of correlations and standardized regression coefficients is

$$R^2_{Y.X_1 \ldots X_k} = \sum_{i=1}^{k} r_{YX_i} \beta_{YX_i}$$

That is, R^2 can be computed as the sum of the products of the correlations between each of the independent variables and the dependent variable and the corresponding standardized regression coefficients.

ADJUSTED R^2 When the number of variables included in a model is large relative to the number of cases in the sample, the explained variance is necessarily large because the amount of information used in the explanation approaches the amount of information to be explained. To correct for this, most computer programs report the "Adjusted R^2" as well as the ordinary R^2. The formula for Adjusted R^2 is

$$R^2_{adj} = 1 - (1 - R^2)\left(\frac{N - 1}{N - (k + 1)}\right)$$

where N is the number of cases and k is the number of independent variables. It is clear that as k approaches N, R^2_{adj}. gets small; indeed, it can become negative. The problem of overfitting the data only arises when samples are quite small; but in such cases the Adjusted R^2 should be taken seriously. However, the ordinary R^2 should be used in tests of the significance of the increment in R^2 (discussed later in the chapter, in the section "A Strategy for Comparisons Across Groups").

around the regression surface) to the total variance in the dependent variable. In the present case, $R^2 = .586$. Note that the R^2 for Equation 6.2 is not much larger than the r^2 for Equation 5.2 (= .536). This is another consequence of the fact that the two independent variables are correlated: either variable taken alone includes its unique effect *plus* its joint effect with the other independent variable. The difference between the R^2 for Equation 6.2 and the r^2 for Equation 5.2 thus gives the unique effect of the number of siblings on educational attainment. Later in the chapter, in the section "A Strategy for Comparisons Across Groups," we will see how to use this fact to compare different regression models. For now let us note the substantive implication: variables added to regression models increase the explained variance to the extent that they are uncorrelated with variables already included in the model. This simple fact places strong constraints on theory building.

Here is a case in point. When graduate students first encounter the literature on status attainment, they sometimes are disappointed by what they perceive to be "low R-squares" (the most competent studies of the U.S. population—for example, Featherman and Hauser [1978, 235]—explain only about 40 percent of the variance in occupational attainment, about 30 percent of the variance in educational achievement, and about 20 percent of the variance in annual earnings), and immediately start proposing other variables that were left out of the analysis and ought to be included. However, the burden of the preceding point is that for the new variables to have much effect on the explained variance, they must be relatively uncorrelated with variables already in the model. It is not easy to think of variables that influence educational attainment that are uncorrelated with parental education, father's occupation, number of siblings, and so on. I stand with Dudley Duncan in celebrating rather than deploring low R^2s. In the context of the analysis of status attainment, at least, and—I would venture to say—many other phenomena as well, they are not so much an indictment of our science as a tribute to our society. As Duncan notes (Blau and Duncan 1967, 174),

> *Sociologists are often disappointed in the size of the residual, assuming that this is a measure of their success in "explaining" the phenomenon under study. They seldom reflect on what it would mean to live in a society where nearly perfect explanation of the dependent variable could be secured by studying causal variables like father's occupation or respondent's education. In such a society it would indeed be true that some are "destined to poverty almost from birth . . . by the economic status or occupation of their parents." Others, of course, would be "destined" to affluence or to modest circumstances. By no effort of their own could they materially alter the course of destiny, nor could any stroke of fortune, good or ill, lead to an outcome not already in the cards.*

Indeed, it is striking that the one society with substantial R^2s for models predicting status attributes is *apartheid*-era South Africa, a society with an extremely rigid racially based stratification system (see, for example, Treiman, McKeever, and Fodor 1996, and Treiman 2007b).

Standard Error of Estimate (Root MSE)

Another useful measure of the adequacy of a regression model is the standard error of estimate (which Stata calls the "root mean square error"—root MSE). This is given by

$$s.e.e. = \sqrt{\frac{1}{N-k}\sum_{i=1}^{N} e_i^2} \tag{6.5}$$

where N is the number of cases, k is the number of independent variables, and e_i is the error, the discrepancy between the actual and predicted value on the dependent variable for the ith individual. The standard error of estimate is thus the root of the sum of squared errors divided by the degrees of freedom, which is $N - k$. It is interpretable as the average deviation of observations from the regression surface, or average size of the residuals (Fox 1997, 104). On the assumption that errors are normally distributed (the conventional assumption of OLS regression), we can construct a confidence interval for the standard error of estimate. For example, a 95 percent confidence interval is given by $\hat{Y} \pm 1.96(s.e.e.)$. This tells us that we can expect 95 percent of the observations to lie within 1.96 *s.e.e.* from the regression surface (Hanushek and Jackson 1977, 57).

An advantage of the *s.e.e.* over R^2 is that the *s.e.e.* is not sensitive to the relative size of the variances of the independent or dependent variables—that is, it is not affected by truncation in the way both r^2 and R^2 are. Usually it is desirable to report both the R^2 and the *s.e.e.*

A WORKED EXAMPLE: THE DETERMINANTS OF LITERACY IN CHINA

Let us work through the presentation and interpretation of a multiple regression equation of the kind that might appear in a published article (the Stata `-do-` and `-log-` files that created the coefficients discussed can be downloaded from the course Web site). A 1996 survey of the population of the People's Republic of China (for details, see Appendix A and also Treiman 1998) included a short-form vocabulary test, in which respondents were asked to identify ten Chinese characters of varying degrees of difficulty. Here we study the factors that account for variation in the number of characters correctly identified.

Obviously, the most important determinant of the ability to recognize characters is years of schooling. Because written Chinese is an ideographic language, one of the prime goals of schooling is to increase knowledge of Chinese characters, and literacy in China is, in fact, measured in terms of the number of characters recognized. This is quite analogous to the expectation that vocabulary will increase with additional schooling in Western nations.

Apart from education, a number of other factors might be expected to contribute to literacy. First, coming from a literate family may increase literacy even when account is taken of the respondent's own level of schooling. Knowledge acquired at home may enhance or substitute for knowledge learned at school. The home environment is measured by two variables: the years of schooling completed by the father and a measure of family "cultural capital"—a scale of the extent to which reading was important in the family when the respondent was age fourteen.

Urban origins (residing in a city or town rather than in a rural village at age fourteen) may enhance literacy both because the quality of schools tends to be better in urban areas in China and because in China, as in the remainder of the world, urban environments tend to offer more opportunities for exposure to written materials. Newspapers are more readily available and more sophisticated, libraries are more likely to exist and to contain wider selections of books, and so on.

Employment in a nonmanual as against a manual occupation should enhance literacy net of years of schooling because words are the tools of the trade in nonmanual occupations. To a much greater extent than manual occupations, nonmanual occupations require reading documents and preparing memoranda. Thus, knowledge of characters learned in school is likely to be more strongly reinforced in nonmanual jobs than in manual jobs.

Finally, China, in common with many East Asian societies, is strongly male-dominant—manifest in a strong preference for sons, higher educational attainment of males than of females, and patrilocal residence. It is quite plausible that the advantages of males over females extend to their participation in literate discourse, both at work and outside of work, thus reinforcing more for males than for females knowledge of characters learned in school; for example, males are more likely to hold supervisory posts and, within the broad nonmanual and manual sectors, are more likely to have high-status jobs requiring more sophisticated use of written language. To test this hypothesis, a variable distinguishing males and females is included.

Table 6.1 shows the means, standard deviations, and correlations among all the variables included in the analysis. Chinese adults were able to correctly identify an average of 3.6 of the 10 characters asked in the survey. The average adult had about 6.5 years of schooling (compared to an average of about 3 years of schooling for their parents); 18 percent were employed in nonmanual jobs; 18 percent grew up in urban areas; 56 percent of the sample is male (note that this is a labor force sample), and the average respondent had "cultural capital" about a quarter of that of the most richly endowed respondents. (It is a labor force sample because I excluded those missing data on occupation. I also excluded those missing data on any other variable, but the bulk of the exclusions were of those lacking occupations. In Chapter Eight we will see how to conserve cases when some variables have substantial amounts of missing data.) Unsurprisingly, the number of characters correctly identified is far more highly correlated with the number of years of schooling than with any other variable.

ALWAYS PRESENT DESCRIPTIVE STATISTICS A table showing descriptive statistics for the variables included in the analysis should always be presented. Such statistics are very informative to the reader and also permit the estimation of alternative models by the reader. The current tendency to start with the presentation of complex models is unfortunate because it often leaves the reader in the dark about the most basic aspects of the data being analyzed. For example, you might gain a rather different impression of the determinants of literacy in a society, such as China, where the average person has less than seven years of schooling, where more than 80 percent work at manual jobs, and more than 80 percent grew up in rural villages, than in a society such as the United States where the average person has about thirteen years of schooling, most people grew up in urban areas, and a majority of the population works at nonmanual jobs.

TABLE 6.1. Means, Standard Deviations, and Correlations Among Variables Affecting Knowledge of Chinese Characters, Employed Chinese Adults Age 20–69, 1996 (N = 4,802).

	V	E	E_F	N	U	M	C
V: Number of characters correctly identified (of 10)[a]		.819	.372	.372	.331	.247	.495
E: Years of schooling			.397	.400	.341	.216	.514
E_F: Father's years of schooling				.226	.307	.013	.574
N: Nonmanual occupation[b]					.368	.030	.327
U: Urban origin						.003	.399
M: Male							.030
C: Level of cultural capital (range 0–1)[c]							
Mean	3.60	6.47	3.07	.177	.180	.558	.227
Standard deviation	2.22	4.10	3.70	.381	.384	.497	.224

[a]The items, in increasing order of difficulty, are *yiwan* (ten thousand), *xingming* (full name), *liangshi* (grain), *hanshu* (function), *diaozhuo* (carve), *sinue* (wreak havoc or wanton massacre), *chuanmiu* (erroneous), *qimao* (octogenarian), *chichu* (walk slowly), and *taotie* (glutton).

[b]Variables *N, U,* and *M* are dichotomies, scored 1 for those in the category and scored 0 for those not in the category.

[c]This scale is the mean of standardized scores for five variables measuring the behavior of parents when the respondents were age fourteen: the number of books in the home, the presence of children's magazines in the home, the frequency with which parents read a newspaper, the frequency with which parents read serious nonfiction, and the presence of an atlas in the home. If information was missing for an item, that item was excluded from the average. The resulting scale was transformed to a 0–1 metric—that is, the lowest score is 0 and the highest score is 1.

Table 6.2 confirms the importance of years of schooling because the standardized coefficients for years of schooling in both models are far larger than the standardized coefficients for any other variable. Each additional year of schooling produces an expected increase of about .4 in the number of characters correctly identified, net of all other factors. This means, for example, that a university graduate (sixteen years of schooling) would be expected to identify about two more characters than would an otherwise similar vocational or technical school graduate (eleven years of schooling).

TABLE 6.2. Determinants of the Number of Chinese Characters Correctly Identified on a Ten-Item Test, Employed Chinese Adults Age 20–69, 1996 (Standard Errors in Parentheses).[a]

Variable	Model 1	Model 2
Metric regression coefficients		
E: Years of schooling	.407 (.006)	.393 (.006)
E_F: Father's years of schooling	.030 (.006)	.009 (.007)
N: Nonmanual occupation	.246 (.057)	.211 (.057)
U: Urban origin	.255 (.053)	.177 (.054)
M: Male	.370 (.045)	.385 (.044)
C: Level of cultural capital (range 0–1)		.866 (.118)
Intercept	.579 (.039)	.546 (.039)
R^2	.683	.687
s.e.e.	1.25	1.24
Standardized regression coefficients		
E: Years of schooling	.750	.725
E_F: Father's years of schooling	.049	.015
N: Nonmanual occupation	.042	.036
U: Urban origin	.044	.030
M: Male	.082	.086
C: Level of cultural capital (range 0–1)		.087

[a]All variables are significant at or beyond the .001 level except for father's education in Model 2 ($p = .195$).

TECHNICAL POINT ON TABLE 6.2

Note that both models in Table 6.2 are based on exactly the same cases, the number of cases shown in Table 6.1. A common error analysts make is to present successive models based on different cases—all the cases for which complete information is available for the variables included in that model. This is ill advised because it makes it impossible to determine whether differences in the coefficients for successive models are due to the inclusion of additional variables or are due to variation in the samples. Moreover, formal comparisons of the increment in explained variance resulting from the inclusion of additional variables (presented in the next section) are not correct unless the models are based on the same cases. Stata has a command, `-mark-`, which makes it easy to ensure that all models being compared are based on the same cases.

Model 1 predicts the number of characters identified from all variables except "cultural capital." In this model all coefficients are significant at the .001 level. Net of other factors, those with nonmanual occupations score about a quarter of a point higher than manual workers, those from urban origins score about a quarter of a point higher than those from rural origins, and males score more than a third of a point higher than females. Clearly, all these effects are real but, with the exception of education, are modest in size. Interestingly, the father's years of schooling significantly increases knowledge of characters, net of all other factors, although the effect is very small (the expected difference between those with the most-educated and least-educated fathers is only about half a character—precisely .54 = .030*18). Together, the factors in Model 1 explain more than two-thirds of the variance in vocabulary knowledge, which is a very strong relationship. Also, the standard error of estimate for Model 1, 1.25, tells us that 95 percent of the actual vocabulary scores lie within 2.45 points (±1.96*1.25) of the regression surface. It is instructive to note how large the error is. Even with a very high R^2 by social science standards, the cases are distributed over nearly half the range of the dependent variable. This suggests the need to exercise considerable caution in interpreting regression estimates.

The intercept, .579, is interpreted as the expected vocabulary score for those with a score of zero on each of the independent variables—that is, for rural origin females working at manual jobs without any schooling whose fathers had no schooling. This is not a very meaningful value. Although in China there are people who fit this description, in many nations a person with 0 scores on all variables would be beyond the range of the observed data. To achieve a meaningful intercept, it often is useful to reexpress the continuous independent variables as deviations from their mean. If this is done, the intercept is then interpretable as the expected value on the dependent variable for people who are at the mean with respect to each of the continuous variables (and, of course, have scores of 0 with respect to each dichotomous variable). In the present case such a reexpression would give us the expected vocabulary score—in this case 3.30—for rural females working at manual jobs (the 0 values on each of the dichotomous variables) who have average education and whose fathers have average education. Note that such a reexpression of the independent variables has no effect on the regression coefficients, the standard errors, the R^2, or the standard error of estimate. Only the intercept is affected.

Model 2 includes "cultural capital" as an additional factor. The associated coefficient indicates that, net of all other factors, people raised in households with the highest cultural capital, that is, maximally involved with reading, score almost a full point higher in their knowledge of vocabulary than do people raised in households minimally involved with reading. Although the explained variance is significantly increased (in the next section we consider how to assess the significance of the increment in R^2), the increase is hardly important from a substantive point of view. What is important is that the introduction of "cultural capital" reduces the effect of father's education to nonsignificance. This makes clear the reason why, net of the respondent's own education, knowledge of vocabulary is enhanced by the father's education: households with educated fathers tend to be more involved in reading than other households. After the "cultural capital" of the household is taken into account, the father's education has no additional effect on vocabulary skill. The "cultural capital" variable also reduces the size of the "urban origin" effect, which indicates that part of the advantage of urban origins is the tendency of urban households to be more involved with reading than are otherwise similar rural households. None of the other coefficients is much affected by the introduction of cultural capital.

Graphic Representation of Results

Sometimes, for ease of exposition, it is useful to graph the net relationship implied by the model between a given independent variable and the dependent variable. This is easy to do. The trick here is to simplify the estimation equation by substituting the means, or other appropriate values, for the remaining independent variables except the one of interest and collecting them into the constant. This yields the expected value on the dependent variable at each level of the independent variable, holding constant all other independent variables at the specified values. The same procedure can be extended to show separate graphs for each category of a categorical variable—for example, if we were interested in how the relationship between literacy scores and years of schooling implied by Model 2 differed for males and females. For continuous variables the mean is a good choice of the value to substitute into the equation. For dichotomous variables we could substitute either the mean or some suitable value—for example, nonmanual workers from urban origins. Of course, for dichotomous variables, the mean is just the proportion that is "positive" with respect to the variable. Thus, if we substitute means for dichotomous variables, we are not evaluating the equation for any actual person—after all, one cannot be 18 percent urban or 56 percent male; rather, we are evaluating what are, in some sense, the typical circumstances of the population.

To see how the procedure works, let us evaluate the equation in two ways: for nonmanual workers from urban origins, and for the mean values of these variables. In each case, we evaluate the equation separately for males and females to create graphs that show separate lines for males and females. Considering first an equation evaluated for nonmanual workers from urban origins, we have, for females

$$\hat{V} = a + b(E) + c(\bar{E}_F) + d(N) + e(U) + f(M) + g(\bar{C})$$
$$= .546 + .393(E) + .009(3.07) + .211(1) + .177(1) + .385(0) + .866(.227) \quad (6.6)$$
$$= 1.158 + .393(E)$$

and for males

$$\hat{V} = a + b(E) + c(\bar{E}_F) + d(N) + e(U) + f(M) + g(\bar{C})$$
$$= .546 + .393(E) + .009(3.07) + .211(1) + .177(1) + .385(1) + .866(.227) \quad (6.7)$$
$$= 1.543 + .393(E)$$

Having arrived at a pair of bivariate equations, differing only by a constant (= .385, the coefficient associated with "male"), we can simply graph the equations. Figure 6.2 shows the graph, which makes clear the relative magnitude of the education and gender effects net of all other determinants of vocabulary knowledge in China. Clearly, education is far more important than gender, although within levels of education there is a small difference favoring males.

Now, suppose that instead of evaluating the equation for nonmanual workers from urban origins, we evaluated the equation at the means of each of the independent variables (except, of course, education and gender, because we want to display the effects of these two variables). Our equation for females is then

$$\hat{V} = a + b(E) + c(\bar{E}_F) + d(\bar{N}) + e(\bar{U}) + f(M) + g(\bar{C})$$
$$= .546 + .393(E) + .009(3.07) + .211(.177) + .177(.180) + .385(0) + .866(.227)$$
$$= .839 + .393(E) \quad (6.8)$$

and for males is

$$\hat{V} = a + b(E) + c(\bar{E}_F) + d(\bar{N}) + e(\bar{U}) + f(M) + g(\bar{C})$$
$$= .546 + .393(E) + .009(3.07) + .211(.177) + .177(.180) + .385(1) + .866(.227)$$
$$= 1.224 + .393(E) \quad (6.9)$$

Note that the only differences between Equations 6.6 through 6.7 and Equations 6.8 through 6.9 are in the intercepts and also that the difference in the intercepts between each

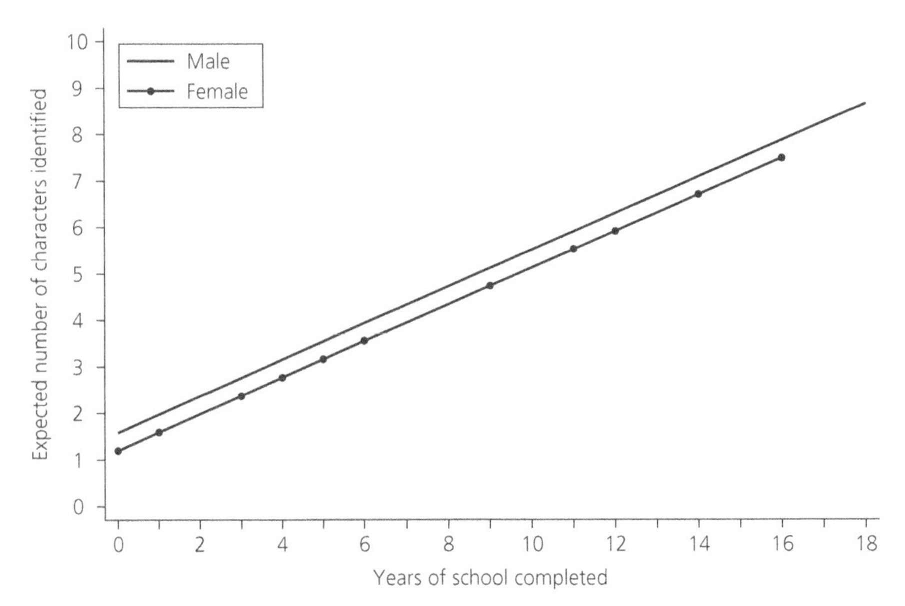

FIGURE 6.2: *Expected Number of Chinese Characters Identified (Out of Ten) by Years of Schooling and Gender, Urban Origin Chinese Adults Age 20 to 69 in 1996 with Nonmanual Occupations and with Years of Father's Schooling and Level of Cultural Capital Set at Their Means (N = 4,802).*

Note: The female line does not extend beyond 16 because there are no females in the sample with post-graduate education.

pair of equations is identical. Thus, a graph of Equations 6.8 through 6.9 would be almost identical to Figure 6.2 except that both lines would be shifted down. For this reason I do not bother to show a graph of Equations 6.8 through 6.9. Whether to substitute means or other specific values when evaluating an equation is a matter of judgment that should be decided by the analyst's substantive concerns.

DUMMY VARIABLES

Often situations arise in which we want to analyze the role of categorical variables, such as religious affiliation, marital status, or political party membership, in determining some outcome. Moreover, typically we want to combine categorical variables with interval variables, to study the effect of each controlling for the other. Thus, we need a way to include categorical variables within a regression framework.

To see how this is done, let us revisit the problem we considered in the final section of Chapter Five, on correlation ratios. Recall that we were interested in the relation between religious affiliation and acceptance of abortion, and we analyzed this by estimating the mean number of positive (accepting) responses to a seven-item scale for each of four religious groups (Protestants, Catholics, Jews, and those with other or no religion) from the 2006 *General Social Survey* (GSS) data. Here we explore a similar substantive problem, but this time using data from the 1974 GSS, because the results for that year are particularly clear-cut and hence more suitable for exposition of the method. (As an exercise, you might want to carry out a similar analysis using the 2006 data). We start by converting the religious denomination variable into a set of four dichotomous variables, one for each religious group, with each variable scored 1 for persons with that religion and scored 0 otherwise. That is, we define a set of new variables (see the downloadable `-do-` or `-log-` file):

R_1 = 1 if the respondent is Protestant, and = 0 otherwise

R_2 = 1 if the respondent is Catholic, and = 0 otherwise

R_3 = 1 if the respondent is Jewish, and = 0 otherwise

R_4 = 1 if the respondent has another religion, no religion, or failed to respond, and = 0 otherwise

Variables of this kind are known as *dichotomous* or *dummy variables*. Using these variables, we can estimate a multiple regression equation of the form:

$$\hat{A} = a + \sum_{i=2}^{4} b_i R = a + b_2 R_2 + b_3 R_3 + b_4 R_4 \tag{6.10}$$

where A is the number of "pro-choice" responses—that is, positive responses to questions about the circumstances under which legal abortions should be permitted (in 1974 six such questions were asked, so the scale ranges from 0 to 6), and the R_i are as specified in the preceding paragraph.

Note that it is necessary to omit one category from the regression equation to avoid a linear dependency (the situation in which any one independent variable is an exact function of other independent variables); because of the way dummy variables are constructed, with each individual scored 1 on one dummy variable and 0 on all the remaining variables in the set, knowing the value on all but one of the dummy variables allows perfect prediction of the value of the remaining dummy variable. In such situations OLS equations cannot be estimated. Any category may be omitted, but because, as we will see, the coefficients of the dummy variables included in the equation are interpreted as deviations from the value for the *omitted*, or *reference*, category, it is best to choose the reference category on substantive grounds—the category against which the analyst wants to contrast other categories. The only exception to the substantive criterion is that very small categories should not be chosen as the omitted category because doing so may create a near-linear dependency among the remaining categories, which could result in unstable numerical estimates of the coefficients.

Estimating Equation 6.10, we have

$$\hat{A} = 3.98 - .33(R_2) + 1.61(R_3) + .88(R_4); \quad R^2 = .045 \tag{6.11}$$

Now, let us derive the predicted values for each category:

For Protestants:

$$\hat{A} = a + b_2(0) + b_3(0) + b_4(0) = a = 3.98 \tag{6.12}$$

For Catholics:

$$\hat{A} = a + b_2(1) + b_3(0) + b_4(0) = a + b_2 = 3.98 - .33 = 3.65 \tag{6.13}$$

For Jews:

$$\hat{A} = a + b_2(0) + b_3(1) + b_4(0) = a + b_3 = 3.98 + 1.61 = 5.59 \tag{6.14}$$

For Other and None:

$$\hat{A} = a + b_2(0) + b_3(0) + b_4(1) = a + b_4 = 3.98 + .88 = 4.86 \tag{6.15}$$

From Equations 6.12 through 6.15, it is evident that the intercept, a, gives the expected value for the omitted, or reference, category and that the coefficients associated with each of the dummy variables, the b_i, give the *difference* in the expected value between that category and the omitted, or reference, category. In the present example, Protestants have an expected score of 3.98 on the dependent variable—that is, on average they find abortion acceptable for about four of the six conditions. Catholics, by contrast, have an expected score of 3.65, .33 less than Protestants. Jews on average endorsed nearly all six items (precisely, 5.59), 1.61 more than Protestants. Finally, the remainder of the sample, the residual category, "Other and None," falls midway between Protestants and Jews in their average level of acceptance of abortion.

WHY YOU SHOULD INCLUDE THE ENTIRE SAMPLE IN YOUR ANALYSIS Although the "Other and None" category is a catch-all residual category and is thus rather uninteresting, it is desirable to include such categories in the analysis rather than restricting the analysis to those with "interpretable" religions. The reason for this is that we ordinarily want to generalize to an entire population, not a subset of the population with definable characteristics. If we omit the residual category from the analysis, our estimates of the average population characteristics are likely to be biased, and, what is worse, biased in unknown ways. Moreover, the regression coefficients may also be biased. See the discussion of "sample selection bias" in Chapter Sixteen.

Note that the R^2 is identical to the correlation ratio, η^2, that we encountered in the previous chapter. Moreover, the predicted values on the abortion attitudes scale are just the means for each religious group shown in Table 5.1. This follows from the fact that, absent any additional information, the mean is the least-squares prediction of the value of an observation. Thus, the "least-squares" estimates for each religious category are just the subgroup means. So far we seem to have no more than a complicated approach to estimating subgroup means and correlation ratios.

The real value of dummy variables is when they are used in combination with other variables to test the effect of group membership on the dependent variable net of the effects of other variables and also to assess the effect of group membership on the relationship between other variables and the dependent variable (and the effect of other variables on the relationship between group membership and the dependent variable)—that is, to assess interactions between the group categories and other variables. To see this, let us continue with our example. Suppose we are interested in assessing the effect of education on acceptance of abortion. Suppose that in addition we want to assess the possibility that religious groups differ in their acceptance of abortion and, moreover, in the relation between education and abortion acceptance—with Catholics tending to oppose abortion regardless of their education, Jews tending to accept abortion regardless of their education, and the remaining two groups becoming more accepting as their education increases. To test these claims, we estimate three successively more complicated regression equations:

$$\hat{A} = a + bE \tag{6.16}$$

$$\hat{A} = a' + b'E + \sum_{i=2}^{4} c_i R_i \tag{6.17}$$

$$\hat{A} = a'' + b''E + \sum_{i=2}^{4} c_i' R_i + \sum_{i=2}^{4} c_i R_i E \tag{6.18}$$

The first model (Equation 6.16) posits an effect of education but no effect of religion. This model assumes that all religious groups are alike in their acceptance of abortion. The second model (Equation 6.17) posits an across-the-board, or constant, difference between religious groups in their acceptance of abortion but assumes that the relation between education and acceptance of abortion is the same for all religious groups. The third model (Equation 6.18) posits an *interaction* between education and religion in the acceptance of abortion, or, to put it differently, assumes that the religious groups differ in the way education affects abortion acceptance. (The conventional way to represent an interaction in a regression framework is to construct a variable that is the product of the two [or more] variables among which an interaction is posited—although other nonlinear functional forms are sometimes posited.)

A STRATEGY FOR COMPARISONS ACROSS GROUPS

Our first task is to decide among the models represented by Equations 6.16 through 6.18. In this situation, where we are assessing whether groups differ with respect to some social process, we would generally prefer the most parsimonious model (except when we have a strong theoretical reason for positing differences between groups or, as noted in the section on Metric Regression Coefficients earlier in the chapter, when we suspect the possibility of omitted variable bias). That is, in general we should prefer a more complicated model only if it does a significantly better job of explaining variability in our dependent variable, acceptance of abortion. We decide on the preferred model by comparing the variance explained by each model. If a more complicated model explains a significantly larger amount of variance in the dependent variable, we accept that model; if it does not, we accept the simpler model. (This is the classical, or frequentist, approach. The next section provides an alternative approach to model assessment based on Bayesian notions, a comparison of *BIC*s.)

We begin by comparing the first and third models. That is, we contrast a model that assumes that there are no religious differences in acceptance of abortion but only educational differences with a model that assumes that the relation between education and acceptance of abortion differs across religious groups. To assess the significance of the difference in R^2, we compute an F-ratio:

$$F = \frac{(R_B^2 - R_A^2)/m}{(1 - R_B^2)/(N - k - 1)} \tag{6.19}$$

where R_B^2 is the variance explained by the larger model (here, Equation 6.18); R_A^2 is the variance explained by the smaller model (here, Equation 6.16); N is the number of cases; k is the total number of independent variables in the larger model; m is the difference in the number of independent variables between the larger and smaller model; the numerator degrees of freedom $= m$; and the denominator degrees of freedom $= N - k - 1$. In our numerical example we have

$$F = \frac{(.097 - .053)/6}{(1 - .097)/(1481 - 7 - 1)} = 11.96 \tag{6.20}$$

with 6 and 1,473 degrees of freedom. To determine whether this F-ratio is significant, we find the p-value corresponding to the numerical value of the F-ratio with the specified numerator and denominator degrees of freedom. If our p-value is smaller than some critical value (.05 is conventional), we reject the null hypothesis (Model 1) in favor of the alternative hypothesis (Model 3). In the present case that is what we are led to do because $F_{(6;1,473)} = 11.96$, which implies $p < .0000$).

GETTING *p*-VALUES VIA STATA Until fairly recently, the way *p*-values were obtained for hand computations such as Equation 6.20 was to look them up in statistical tables of the kind found at the back of many statistics texts. This is no longer necessary and is now something of an old-fashioned approach. Stata provides a set of built-in statistical tables, including a table of probabilities associated with specific *F*-ratios. The probability for a given *F*-ratio can be computed by executing the command, `-display fprob(df_1, df_2,F)-`, where *df_1* is the numerator degrees of freedom, *df_2* is the denominator degrees of freedom, and *F* is the calculated *F*-ratio).

USING STATA TO COMPARE THE GOODNESS-OF-FIT OF REGRESSION MODELS The F-test for the increment in R^2 is equivalent to the Wald test that the coefficients for the set of variables that are included the larger model but not in the smaller model are not significantly different from zero. Thus software that implements the Wald test as a post-estimation command (for example, `-test-` and `-testparm-` in Stata) can be used to carry out the *F*-test. It also can be shown that when a single variable is added to a regression equation, the t-ratio for the additional variable equals the square root of the *F*-ratio for the increment in R^2, and the *t*- and *F*-ratios have identical probability distributions. Thus, when two equations differ by a single variable, they can be contrasted simply by inspecting the significance of the *t*-ratio, which is routinely provided as part of the regression output.

Having determined that we cannot posit a single model of the relation between education and abortion acceptance for all religious groups, we next investigate whether it is necessary to posit religious group differences in the relation between education and abortion acceptance or whether there are simply across-the-board differences between the religious groups in their acceptance of abortion but a similar relation between education and abortion acceptance for all groups; that is, we ask whether both the slopes and intercepts differ across religious groups or whether only the intercepts differ? To answer this question, we contrast the R^2 for Model 3 (Equation 6.18) and Model 2 (Equation 6.17), estimating an *F*-ratio using Equation 6.19. For our current numerical example, we get

$$F = \frac{(.097 - .089)/3}{(1 - .097)/(1{,}481 - 7 - 1)} = 4.35 \qquad (6.21)$$

with 3 and 1,473 degrees of freedom. Because $F_{(3;1473)} = 4.35$, which implies $p = .0046$, we reject the null hypothesis that the relationship between education and abortion acceptance

R. A. (RONALD AYLMER) FISHER (1890–1962) was a British statistician with a strong interest in biology (he was a founder, with Sewall Wright—see the biosketch of Wright in Chapter Sixteen—and J.B.S. Haldane, of theoretical population genetics). He was responsible for major advances in experimental design, introducing the notion of random assignment of cases to different treatments and showing how to use analysis of variance, which he invented—the *F*-distribution is named after Fisher—to assess the contribution of each of several factors in determining an outcome, a procedure that greatly enhanced the power of experimental designs. He also invented the concept of the maximum likelihood and made major contributions to statistical procedures for assessing small samples. His text *Statistical Methods for Research Workers*, first published in 1925, was very widely used, especially as a handbook for the design and analysis of experiments, and ran through fourteen editions, the latest published in 1970.

Fisher was born in London, the son of an art dealer and auctioneer. He was a precocious student, winning the Neeld Medal (a competitive essay in mathematics) at Harrow School at the age of sixteen. (Because of his poor eyesight, he was tutored in mathematics without the aid of paper and pen, which developed his ability to visualize problems in geometrical terms, as opposed to using algebraic manipulations. He became legendary in being able to produce mathematical results without setting down the intermediate steps.) Fisher studied mathematics at Cambridge, spent some time after graduation working on a farm in Canada, tried to enlist in the army during World War I but was rejected because of his poor eyesight, and then spent several years teaching mathematics in secondary schools. At the end of the war, he was offered a position at the Galton Laboratory by Karl Pearson (see the biosketch in Chapter Five) but because of his growing rivalry with Pearson rejected it in favor of a position at a small agricultural experimental station (Rothamsted), where he remained until being appointed Professor of Eugenics at University College London in 1933 and then to the Balfour Chair of Genetics at Cambridge in 1943. After retiring from Cambridge in 1957, he spent the last three years of his life as a senior research fellow at the Commonwealth Scientific and Industrial Research Organization in Adelaide, Australia. Fisher's important contributions to both genetics and statistics are emphasized by the remark of the well-known statistician, Leonard J. Savage (1976): "I occasionally meet geneticists who ask me whether it is true that the great geneticist R. A. Fisher was also an important statistician."

is the same for all religious groups but that the groups differ in their across-the-board acceptance of abortion in favor of the alternative hypothesis, that the relationship between education and abortion acceptance differs across religious groups. Thus, in sum, our preferred model is one that posits that both education and religion affect abortion attitudes and that the effect of education varies by religion (and, necessarily as well, that the effect of religion varies by education).

TABLE 6.3. Coefficients of Models of Acceptance of Abortion, U.S. Adults, 1974 (Standard Errors Shown in Parentheses); N = 1,481.

	Model 1	Model 2	Model 3	Model 3′
E: Education	.136 (.015)	.125 (.015)	.155 (.018)	.155 (.018)
R_2: Catholic		−.373 (.111)	1.059 (.455)	−.371 (.111)
R_3: Jewish		1.341 (.282)	3.202 (1.015)	1.547 (.310)
R_4: Other or None		.747 (.184)	.535 (.687)	.702 (.187)
R_2*E			−.121 (.037)	−.121 (.037)
R_3*E			−.140 (.072)	−.140 (.072)
R_4*E			.014 (.053)	.014 (.053)
Intercept	2.402 (.183)	2.533 (.182)	2.185 (.219)	4.015 (.059)
R^2	.053	.089	.097	.097

Note: Model 3′ is identical to Model 3 except that in Model 3′ years of schooling (education) is expressed as a deviation from the mean years of schooling.

The conventional practice is to report the estimated coefficients for each model, not merely the preferred model. These are shown in Table 6.3.

Let us see how to interpret each of these models. Model 1 is just a two-variable regression equation of the sort we encountered in the previous chapter; nothing further needs to be said here. As we have noted, Model 2 posits the same relationship between education and abortion acceptance for all religious groups but across-the-board differences between religious groups in the level of acceptance of abortion among those who have a given level of education. What is meant by an "across-the-board" difference is clarified by writing out Equation 6.17 separately for each religious group. For Protestants, we have

$$\hat{A} = a + b(E) \tag{6.22}$$

For Catholics, we have

$$\hat{A} = a + b(E) + c_2 = (a + c_2) + b(E) \tag{6.23}$$

For Jews, we have

$$\hat{A} = a + b(E) + c_3 = (a + c_3) + b(E) \tag{6.24}$$

For Others, we have

$$\hat{A} = a + b(E) + c_4 = (a + c_4) + b(E) \tag{6.25}$$

From Equations 6.22 through 6.25 it is evident that Model 2 (Equation 6.17) implies that the religious groups differ in their intercepts but not in the slopes relating education to abortion acceptance. If Model 2 were our preferred model, we could conclude that each year of education resulted, on average, in a .125 increase in abortion acceptance for people of all religions, so that, for example, college graduates (sixteen years of schooling) would be expected to accept abortion for one reason more than would those of the same religion with only eight years of schooling. And we would expect Jews to agree that abortion ought to be permitted for 1.3 reasons more than Protestants, on average, and Catholics to agree .4 times less on average than Protestants. In short, interpretation of the coefficients for Model 2 is straightforward, and the net effects of education and of religious group membership can be assessed separately. However, although the *size* of the coefficient for each religious group can be interpreted individually, it generally is not meaningful to assess the significance of individual coefficients because each coefficient indicates the difference between the expected value for the given category and the expected value for the omitted category, net of all other factors. Therefore, a significant t-ratio merely indicates that a coefficient is significantly different from the implied coefficient of zero for the omitted category, and which coefficients are shown as significant in one's computer output is entirely dependent upon the choice of omitted, or reference, category. Thus, the appropriate procedure is to assess the significance of the entire set of dummy variables representing a given categorical variable by computing an F-test of the increment in R^2 for models including and excluding the *set* of dummy variables corresponding to a single classification (or equivalently, the Wald test that the set of coefficients is jointly = 0).

HOW TO TEST THE SIGNIFICANCE OF THE DIFFERENCE BETWEEN TWO COEFFICIENTS There may be occasions in which an analyst wants to assess the significance of the difference between two specific categories of a dummy variable classification. In this case, it is possible to make use of the formula:

$$t = (b_i - b_j) / (\text{var}(b_i) + \text{var}(b_j) - 2\text{cov}(b_i b_j))^{1/2}$$

where b_i and b_j are the two coefficients being compared. Most statistical packages permit the estimation of the variance-covariance matrix of coefficients. Of course, in these days of high-speed computing, it probably is easier to simply reestimate the model, redefining the reference category. Stata provides an even easier way to compare coefficients, by computing a Wald test that $b_i = b_j$.

When interaction terms are involved, the requirements are even more stringent. Not only must the significance of all of the associated coefficients be assessed simultaneously, but the coefficients themselves must be interpreted together rather than individually. Consider Model 3, which includes interaction terms between education and religious group membership. It helps to write out Equation 6.18 separately for each religious group.

For Protestants, we have

$$\begin{aligned} \hat{A} &= a + b(E) \\ &= 2.18 + .155(E) \end{aligned} \tag{6.26}$$

For Catholics, we have

$$\begin{aligned} \hat{A} &= a + b(E) + c_2 + d_2(E) \\ &= (a + c_2) + (b + d_2)E \\ &= (2.18 + 1.06) + (.155 - .121)E \\ &= 3.24 + .034(E) \end{aligned} \tag{6.27}$$

For Jews, we have

$$\begin{aligned} \hat{A} &= a + b(E) + c_3 + d_3(E) \\ &= (a + c_3) + (b + d_3)E \\ &= (2.18 + 3.20) + (.155 - .140)E \\ &= 5.38 + .015(E) \end{aligned} \tag{6.28}$$

For Others, we have

$$\begin{aligned}\hat{A} &= a + b(E) + c_4 + d_4(E) \\ &= (a + c_4) + (b + d_4)E \\ &= (2.18 + .53) + (.155 + .014)E \\ &= 2.71 + .141(E)\end{aligned} \tag{6.29}$$

Again, it is evident from Equations 6.26 through 6.29 that Equation 6.18 allows both the slopes and the intercepts to vary between groups. The coefficients associated with the dummy variables, the c_is, indicate the differences in the intercept between the reference category and each of the explicitly included categories whereas the coefficients associated with the interaction terms, the d_is, indicate the differences in the effect of education (the slope) between the reference category and each of the explicitly included categories.

From Equations 6.26 through 6.29 it should be clear that for equations of the form of Equation 6.18, no overall summary of the effect of a variable (in this case education or religious group membership) is possible but only the effect of each combination of education and religious group membership. Specifically, the coefficient .155 in Model 3 of Table 6.3 (and in Equation 6.23) does not refer to the overall effect of education but rather to the effect of education among Protestants; and so on for the other coefficients.

Because Equation 6.18 is a *saturated model*—it includes all possible interactions among the independent variables (in the present case, all possible interactions between education and the religious group categories)—it is mathematically equivalent to estimating separate equations for each religious group. Equations 6.26 through 6.29 show this equivalence: the coefficients resulting from rewriting Equation 6.18 as Equations 6.26 through 6.29 and rearranging terms are identical to the coefficients obtained by estimating the equation separately for each religious group (you might want to persuade yourself of this by carrying out the computation both ways). The advantage of estimating Equation 6.18 is that it permits an explicit test of the hypothesis that the groups differ, through the F-test shown as Equation 6.19. Because the conversion of Equation 6.18 to Equations 6.26 through 6.29 is a fairly tedious hand operation, especially when the number of variables is large, we often estimate equations of the form of Equation 6.18 to obtain the R^2 required for the F-test (or the coefficients required for the Wald test) but then estimate separate equations for each group and report them in a separate table.

From Table 6.3, and even from Equations 6.26 through 6.29, it is difficult to interpret the relationships among religion, education, and abortion attitudes. With equations of this sort, graphing the relationships is often useful. Figure 6.3, which can be constructed in the same way as Figure 6.2 (see the downloadable `-do-` or `-log-` file for details), shows the level of abortion acceptance expected from education and religious group membership. Inspecting the graph, it is evident that Jews are highly accepting of abortion regardless of their level of education, that Catholics are relatively unaccepting of abortion regardless of their level of education, and that abortion acceptance varies strongly by education for Protestants and "Others," with the poorly educated similar to Catholics and the well educated similar to Jews.

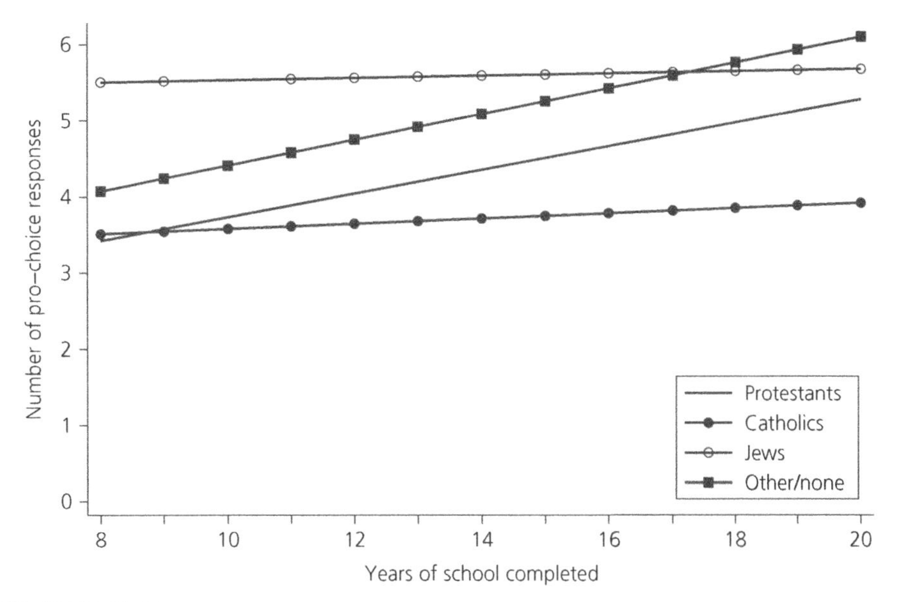

FIGURE 6.3. *Acceptance of Abortion by Education and Religious Denomination, U.S. Adults, 1974 (N = 1,481).*

Reexpressing Variables as Deviations from Their Means

Apart from graphing the relations among variables, as in Figure 6.3, we can use one other device to render the coefficients in models containing interaction terms more readily interpretable—we can reexpress the continuous variables as deviations from their means—just as we saw in the earlier example predicting knowledge of Chinese characters. The advantage of doing this here is that the group effects (the main effects of the dummy variables for groups) can then be interpreted as indicating the expected differences among groups with respect to the dependent variable for persons who are *at the average* with respect to the interval-level independent variables or variables.

In the present context, this means reexpressing years of school completed by subtracting the sample mean from each observation. (The reexpressed coefficients are shown in Table 6.3 as Model 3′.) The intercept then gives the expected value on the abortion scale among Protestants with average education (where the mean is computed over the entire sample, not just Protestants). The coefficients associated with each of the dummy variables then give the difference in the expected level of pro-choice sentiment between Protestants and the specified category among persons with average education. Note that the slope associated with years of schooling (which, as noted, gives the effect of schooling for Protestants) is unchanged, as are the coefficients associated with the interaction terms; only the group intercepts change. However, the interpretation of the coefficients is greatly facilitated: we see that among those with average education, Protestants endorse about 4 of the 6 abortion items on average, Jews an additional 1.5 items, "Others" an additional .7 items, and Catholics about .4 items fewer. We also see that each additional year of schooling increases the expected endorsements of Protestants by .155, and of those without religion by about the same amount because the difference in the slopes is only .014; by contrast, education matters little for Jews and Catholics because the deviations from the Protestant slope are negative and almost as large as the Protestant slope.

Testing Additional Hypotheses: Constraining Coefficients to Zero or to Equality

Inspecting Figure 6.3, we might be led to infer that education has no effect on abortion attitudes for Catholics and Jews and the same effect for Protestants and "Others." How can we formally test the correctness of our inference? We can do this by estimating an equation of the form:

$$\hat{A} = a + \sum_{i=1,3,4} b_i R_i + c(ER_1 + ER_4) \quad (6.30)$$

where, in this case, Catholics are the omitted category. To see how this equation represents the particular hypothesis of interest, we can again write out Equation 6.30 separately for each religious group.

For Protestants:

$$\hat{A} = (a + b_1) + c(E) \quad (6.31)$$

For Catholics:

$$\hat{A} = a \quad (6.32)$$

For Jews:

$$\hat{A} = a + b_3 \quad (6.33)$$

For Others:

$$\hat{A} = (a + b_4) + c(E) \quad (6.34)$$

As is evident from inspection of Equations 6.31 through 6.34, under the specification of Equation 6.30 each religious group differs in the intercept; the slope relating education to abortion acceptance is zero for Catholics and Jews; and the slope is identical for Protestants and "Others." To test whether this constrained specification is an adequate representation of the data, we cannot compute the increment in R^2 for Model 3 relative to Model 3′ because the two models do not stand in a hierarchical relationship to each other: there is no main effect for education in the constrained model. So, what to do? Fortunately, a solution is available.

A BAYESIAN ALTERNATIVE FOR COMPARING MODELS

We can exploit an alternative way of contrasting models, the Bayesian Information Criterion (*BIC*), introduced into the sociological literature by the statistician Adrian Raftery for log-linear analysis (1986) and generalized to a variety of applications in an important article in *Sociological Methodology* (1995a; see also the critical comment by Gelman and Rubin [1995], the appreciative comment by Hauser [1995], and Raftery's reply to both [1995b], and also the February 1999 issue of *Sociological Methods and Research*, which is devoted entirely to an assessment of *BIC*). In a sense *BIC* operates on the opposite principle from classical tests of significance. It is a likelihood ratio measure that tells us which model is most likely to be true given the data (for a brief introduction to maximum likelihood estimation, see Appendix 12.B); classical inference, by contrast, tells us how likely it is that the observed data could have been generated by sampling error given that some theoretical model (the null hypothesis) is true.

BIC has three important advantages over the *F*-test introduced previously. First, unlike the *F*-ratio, *BIC* can be used to compare nonhierarchical models. Any two models purporting to describe the same phenomenon can be contrasted. Second, *BIC* builds in a correction for large samples whereas if the sample is large enough virtually any increment in R^2 will be significant, no matter how small and substantively unimportant. A larger increment in R^2 is required to generate a particular *BIC* value for large samples than would be required for small samples. Thus, *BIC* re ects the conventional advice to choose a smaller probability value when the sample is large. Third, *BIC* penalizes large models. That is, if it takes the introduction of many additional variables to generate much of an increase in R^2, *BIC* is more likely than the *F*-test to lead us to prefer the simpler model. There are several specific ways to calculate *BIC*, depending on the particular statistic being analyzed. To compare regression models, we can use Raftery's Equation 26:

$$BIC_k = N[\ln(1 - R_k^2)] + p_k[\ln(N)] \tag{6.35}$$

where R_k^2 is the value of R^2 for Model *k*, p_k is the number of independent variables for Model *k*, and N = the number of cases being analyzed. A negative value of *BIC* indicates that a specified model is more likely to be true than the baseline model of no association between the independent variables and the dependent variable. To compare two models, we estimate *BIC* for each of them and choose the model with the more negative *BIC*. Raftery (1995a, Table 6) gives a rule of thumb for comparing *BIC*s: a *BIC* difference of 0 to 2 constitutes "weak" evidence for the superiority of one model over another; a difference of 2 to 6 constitutes "positive" evidence; a difference of 6 to 10 constitutes "strong" evidence; and a difference of >10 constitutes "very strong" evidence. However, because *BIC* tends to increase as the sample size increases, Raftery's rule of thumb is most appropriate for relatively small samples.

To see how *BIC* is used, let us compute *BIC* values for the three models shown in Table 6.3. For Model 1, we have

ALTERNATIVE WAYS TO ESTIMATE *BIC* Even for a given statistic, there are alternative versions of *BIC*. I prefer Raftery's formulas because they build in a comparison to a baseline model. Thus, I have written a small `-do-` file, `-bicreg.do-`, to calculate *BIC* following Raftery:

```
*BICREG.DO (Updated for Stata 7.0 11/11/01.)
version 7.0
*Compute BIC from saved results from regression.
gen bic=e(N)*ln(1-e(r2))+e(df_m)*ln(e(N))
*Note: BIC is the same for all observations. Thus, I can
*list BIC for any observation.
list bic in 1
drop bic
```

Invoke `-bicreg-` immediately following your regression command. However, Stata 10.0 now offers *BIC* as a post-estimation statistic. To have Stata calculate *BIC* without using my `-do-` file, invoke the command `-estat ic-` immediately following your regression command. The numerical value of each *BIC* will differ from mine, but the difference between the *BIC*s for alternative models will be identical regardless of which version of *BIC* is calculated.

$$BIC_1 = 1{,}481{*}\ln(1 - .053) + 1{*}\ln(1{,}481) = -73.4 \qquad (6.36)$$

For Model 2, we have

$$BIC_2 = 1{,}481{*}\ln(1 - .089) + 4{*}\ln(1{,}481) = -108.4 \qquad (6.37)$$

For Model 3, we have

$$BIC_3 = 1{,}481{*}\ln(1 - .097) + 7{*}\ln(1{,}481) = -100.3 \qquad (6.38)$$

From a comparison of the *BIC*s for the three models, we are led to conclude that the data are most consistent with Model 2, which posits the same effect of education on abortion attitudes for all religious groups and an across-the-board difference (that is, a difference that holds at each level of education) in abortion acceptance for the various religious groups. From the size of the *BIC* differences, we conclude that the data "very strongly" favor Model 2 over Model 1 and "strongly" favor Model 2 over Model 3.

Note that these results are inconsistent with the results we obtained previously through a comparison of R^2s via an F-test. What are we to make of this? There is no definitive answer. My advice is, first, go with theory. If you have a theoretical reason to prefer one model over the other, choose that one. This advice is consistent with one of Weakliem's [1999] criticisms of *BIC*—that *BIC* assumes a "unit prior." *BIC* is an approximation of the Bayes factor, which involves a comparison of the posterior likelihood of models, where "the posterior likelihood is simply the product of the data likelihood and the researcher's prior. The researcher then chooses that model with the greatest likelihood; that is, the model that has the highest probability of being the true model given the researcher's priors and the data" (Winship 1999a, 356). If there is no clear reason to expect a departure from the null hypothesis, a "unit prior"—which amounts to saying we have little information about the likely outcome—is appropriate. But if we have strong theoretical reasons to expect a relationship, *BIC* can be too conservative. In this case, classical inference would seem to be the preferred tool unless we were to modify *BIC* in ways that go beyond this course. We will discuss likelihoods in Chapters Twelve and Thirteen.

Absent a strong theory, go for parsimony, which is what *BIC* generally does. In the present case, I would be inclined to prefer Model 3 because I think there are good reasons to expect Catholics and Jews to have consistent reactions to abortion regardless of their level of education (Catholics because abortion is prohibited by the Church and Jews because—still in 1974 even if less so today—the Jewish community was socially liberal, and Jews lacking education tended to be immigrants who had the values of educated persons) and to expect Protestants and Others to be more accepting if they are better educated (because of the increasing sophistication that education brings). But if I did not have a strong, coherent, explanation for the religious difference, I would then prefer Model 2.

We can, of course, also compute *BIC* for the constrained model derived from the data:

$$BIC_{3'} = 1{,}481*\ln(1 - .096) + 4*\ln(1{,}481) = -121.0$$

which is more negative than the *BIC* for any of Models 1 through 3 and thus "very strongly" suggests that, for these data, the constrained model is to be preferred.

INDEPENDENT VALIDATION

Note that I said that *for these data* the constrained model is to be preferred. This is because we arrived at a new preferred model based on our inspection of the data rather than from a priori theory. Thus, we are vulnerable to the possibility that we are simply capitalizing on sampling error. To arrive at a definitive preference for the constrained model, we need to show that it is the preferred model in an independent data set. If our sample size permitted, we would want to carry out all of our exploratory analysis using half of the data and then to reestimate our final model (and its competitors) using the other half of the data. The GSS provides a close approximation to this ideal because it repeats identical questions in successive surveys conducted using the same sampling procedures. Thus, it is reasonable to treat adjacent surveys as independent samples drawn from the same population, at least for phenomena not subject to short-term uctuation. The implication of this is that we can carry out all of our exploratory analysis for one year and then use the data from the previous or subsequent year to validate our conclusions.

TABLE 6.4. Goodness-of-Fit Statistics for Alternative Models of the Relationship Among Religion, Education, and Acceptance of Abortion, U.S. Adults, 1973 (N = 1,499).

Model	*BIC*	R^2	F	d.f.	p
Model 1	−149.9	.0995			
Model 2	−197.7	.1405			
Constrained model	−205.1	.1448			
Model 3	−191.1	.1493			
Contrasts					
Model 3 vs. Model 1	−41.2	.0498	14.52	6; 1491	.0000
Model 3 vs. Model 2	6.6	.0088	5.11	3; 1491	.0016
Constrained vs. Model 2	−7.4				
Model 3 vs. Constrained	14.0				

Here we can exploit the GSS in just this way, reestimating the four models of pro-choice attitudes using data from the 1973 GSS. Insofar as we can assume that abortion attitudes did not change in the population between 1973 and 1974, reestimating the models using the data from 1973 constitutes an independent test of the claim that the "constrained" model is the preferred model. Table 6.4 shows *BIC* and R^2 values based on the 1973 data for all four models and contrasts between models wherever meaningful and appropriate. The outcomes are, in fact, just the same as for 1974: Model 3 is preferred to Model 1 and Model 2 by the criteria of classical statistical inference, whereas by the *BIC* criterion Model 2 is preferred to Model 3; and by the *BIC* criterion the constrained model is the most preferred. Thus, we can conclude that our preference for the constrained model, derived from inspection of the data, is valid.

WHAT THIS CHAPTER HAS SHOWN

In this chapter you have learned how to carry out multiple regression and correlation analysis and how to interpret the resulting coefficients, considering a worked example on the determinants of literacy in China. We then focused on the manipulation of dummy variables (sets of dichotomous variables that represent categorical variables), including especially interactions between dummy variables and other variables, using as a worked example an analysis of attitudes regarding the acceptability of abortion. We also considered how to use a goodness-of-fit measure, *BIC*, as an alternative to the conventional goodness-of-fit measures used in classical statistical inference. Using these tools, we explored a strategy for comparisons of subgroups, which enables us to decide whether a particular social process is similar for different groups (for example, those based on race, religion, gender, and so on). Finally, we touched on the desirability of validating results using an independent data set, and saw how to do this when using the GSS, which is a set of replicated cross-sectional surveys.

In the next chapter we consider various ways of specifying multiple regression equations to increase their exibility and hence the precision with which they can be used to explore substantive ideas in the social sciences.

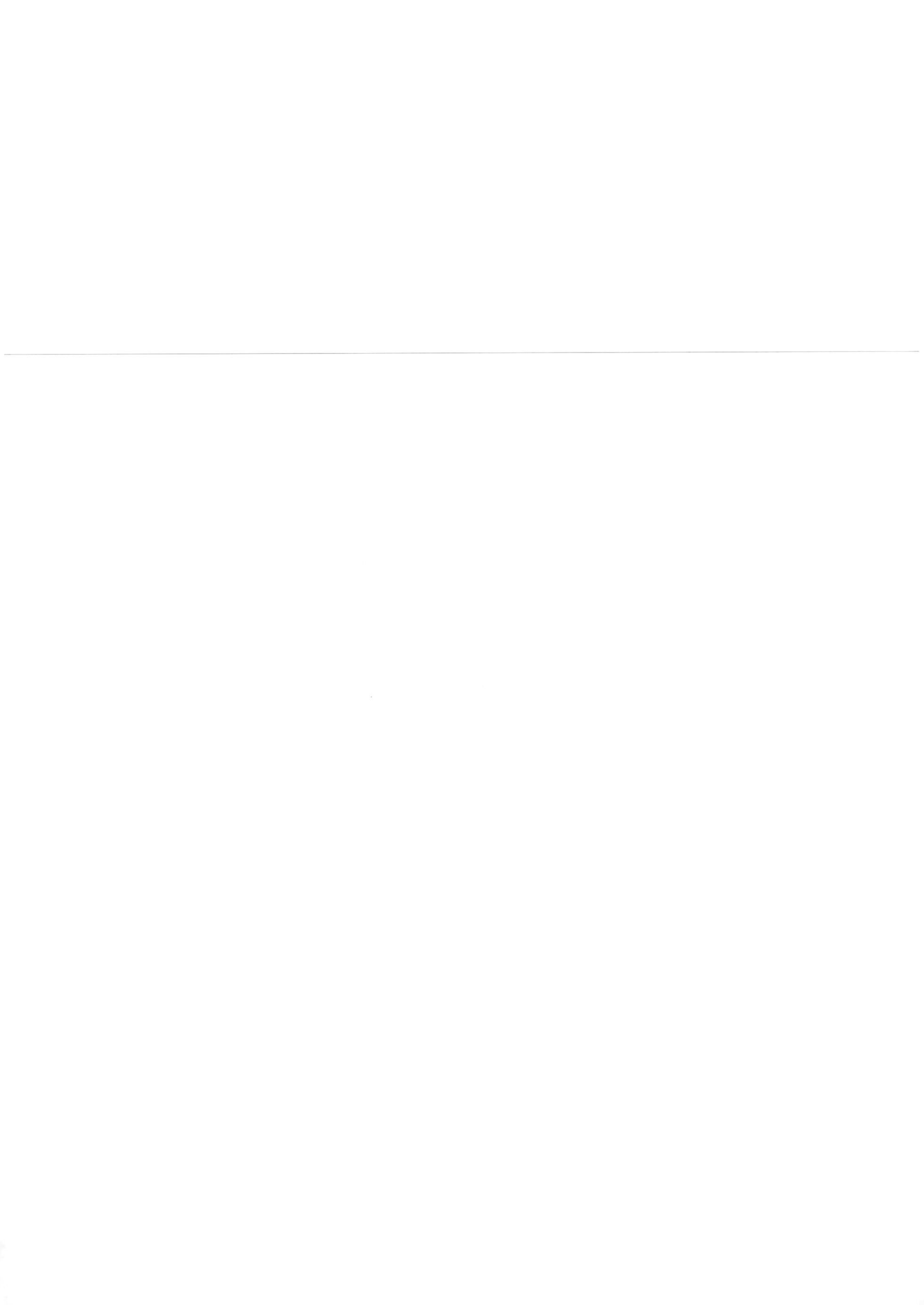

CHAPTER

7

MULTIPLE REGRESSION TRICKS: TECHNIQUES FOR HANDLING SPECIAL ANALYTIC PROBLEMS

WHAT THIS CHAPTER IS ABOUT

This chapter presents various "tricks" for dealing in a multiple regression framework with specific analytic problems faced by social researchers. The Stata -do- and -log- files for all the worked examples in this chapter are available as downloadable files. Specifically, we consider nonlinear transformations of both dependent and independent variables; ways to test the equality of coefficients within an equation; how to assess the assumption of linearity in a relationship, with a trend analysis as a worked example; how to construct and interpret linear splines as a way of representing abrupt changes in slopes; alternative ways of expressing dummy variable coefficients; and a procedure for decomposing the difference between two means.

NONLINEAR TRANSFORMATIONS

Often when doing regression analysis, we have reason to suspect that the relationship between particular independent variables and the dependent variable is nonlinear. Hence, an estimate of the linear relationship between the independent and dependent variables would not properly represent the relationship in the sample under study. You have seen an example of this kind in (c) of Figure 5.4 in Chapter Five, which shows a perfect parabolic relationship between two variables but which produces a slope and correlation of zero when estimated by a linear regression equation. Fortunately, there is a simple solution to problems of this kind—you can transform one or more variables so that the dependent variable is a linear function of the independent variables. Here are several examples, together with some interpretive tricks.

Curvilinear Relationships: Age and Income

In cross-sectional data income commonly increases with age up to a point in the middle of the career and then begins to fall. A reasonable way to represent this is to estimate an equation of the form

$$\hat{Y} = a + b(A) + c(A^2) \tag{7.1}$$

where Y = annual income, A = age, and $A^2 = A*A$.

In the 2004 *General Social Survey* (GSS), the estimated values for this equation are (for people age 20 to 64 with information on personal income; N = 1,573; the open-ended upper interval—$110,000 per year or more—was recoded to $150,000; the remaining income intervals were recoded to their midpoints):

$$\hat{Y} = -49{,}139 + 3{,}777(A) - 35.95(A^2); \quad R^2 = .084 \tag{7.2}$$

which can be represented graphically, as shown in Figure 7.1.

WHY THE RELATIONSHIP BETWEEN INCOME AND AGE IS CURVILINEAR

There are several possible explanations for the curvilinearity of the relationship between income and age, of which the two major ones are the following:

- Economists argue that productivity increases with age up to a point and then falls; sociologists sometimes make similar arguments but also point out that various institutional factors, such as the greater difficulty older workers have in returning to work after layoffs, result in the same observed pattern.
- The cross-sectional observation may simply be an artifact of a cohort progression of earnings, with successive cohorts earning more at any given age than their seniors, and the earnings of all workers continuing to rise throughout the career.

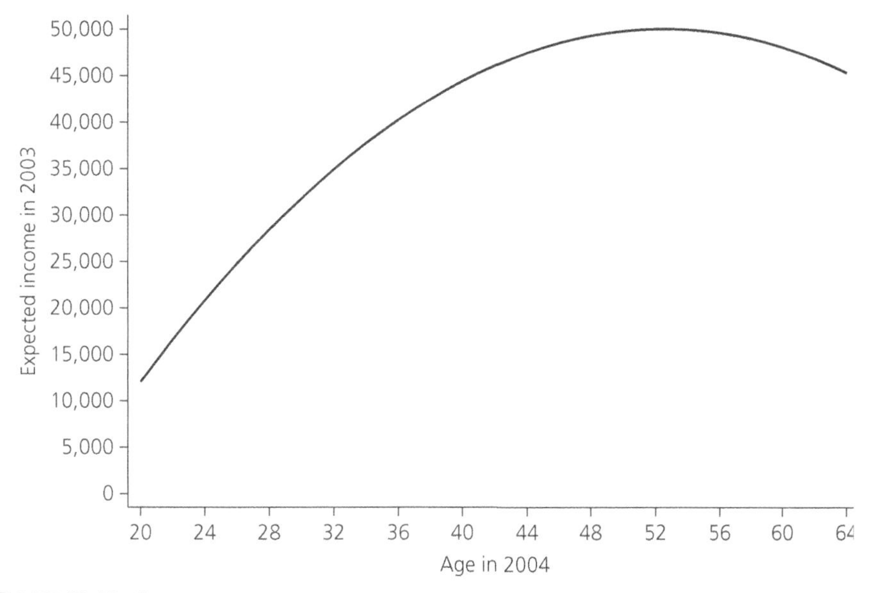

FIGURE 7.1. *The Relationship Between 2003 Income and Age, U.S. Adults Age Twenty to Sixty-Four in 2004 (N = 1,573).*

From the graph we see that in 2003, people age twenty tended to earn about \$12,000 per year, and that earnings were highest among those around age fifty-two, peaking at about \$50,000 per year. Without the graph, however, interpretation of Equation 7.2 is difficult because the coefficients themselves do not have a straightforward substantive interpretation. It is possible, however, to rewrite Equation 7.1 in such a way as to yield a direct interpretation. It can be shown that in the equation

$$\hat{Y} = m + c(F - A)^2 \tag{7.3}$$

where $m = a - b^2/4c$ and $F = -b/2c$ (with the coefficients on the right side taken from Equation 7.1), m is the maximum income, and F is the age at which the maximum income is attained. In the present case, the numerical estimate of Equation 7.3 is

$$\hat{Y} = 50{,}066 - 35.95(52.53 - A)^2; \quad R^2 = .084 \tag{7.4}$$

Equations 7.2 and 7.4, of course, yield the same graph because they are equivalent expressions. But Equation 7.4 also tells us precisely that the peak income is \$50,066 and that this peak is attained between fifty-two and fifty-three years of age (precisely, 52.53).

An equivalent transformation is possible for an equation containing additional independent variables. Consider an equation of the form

$$\hat{Y} = a + b(A) + c(A^2) + d(Z) \tag{7.5}$$

where Z is some other independent variable, and the remaining variables are as before. We could then represent the relation between A and Y, net of Z, by substituting the mean of Z, $\bar{Z}$, so that

$$\hat{Y} = (a + d(\bar{Z})) + b(A) + c(A^2) \tag{7.6}$$

or, equivalently,

$$\hat{Y} = m + c(F - A)^2 \tag{7.7}$$

where, in this case,

$$m = (a + d(\bar{Z})) - b^2/4c \tag{7.8}$$

and F is as before.

Semilog Transformations: Income

A useful transformation when predicting income is the semilog transformation; that is, instead of predicting income, we predict the natural log of income. This has three advantages.

First, economic theories about what generates income tend to make predictions in terms of log income. Specifically, human capital theory takes income as determined by an investment process (Mincer and Polachek 1974). Hence, insofar as we take such theories seriously or are interested in testing them seriously, we probably should predict income in its log form.

Second, income tends to be distributed lognormally in the United States and other advanced industrial societies, so the log of income is distributed normally, a convenient property.

Third, and most important, when the dependent variable is in (natural) log form, the metric regression coefficients can be interpreted as indicating approximately the proportional increase in the dependent variable associated with a one-unit increase in the independent variable, for b less than about 0.2. To see this, consider the equation

$$\widehat{\ln(Y)} = a + b(X) \tag{7.9}$$

Now consider two individuals who differ by one unit with respect to X; that is, $X_1 = X_2 + 1$. Then

$$\widehat{\ln(Y_1)} = a + b(X_1) \tag{7.10}$$

and

$$\widehat{\ln(Y_2)} = a + b(X_2) \tag{7.11}$$

So, subtracting,

$$\widehat{\ln(Y_1)} - \widehat{\ln(Y_2)} = (a - a) + b(X_1 - X_2) = b \tag{7.12}$$

But we know from the properties of logs that

$$\ln(Y_1) - \ln(Y_2) = \ln(Y_1/Y_2) \tag{7.13}$$

So we have

$$\widehat{\ln(Y_1/Y_2)} = b \tag{7.14}$$

Then, exponentiating both sides (that is, making each an exponent of e), we have

$$\widehat{Y_1/Y_2} = e^b \tag{7.15}$$

Now let us look at the relationship of b to e^b for various values of b.

b	e^b	b	e^b
0.01	1.01	−0.01	0.99
0.05	1.05	−0.05	0.95
0.10	1.11	−0.10	0.90
0.15	1.16	−0.15	0.86
0.20	1.22	−0.20	0.82
0.30	1.35	−0.30	0.74
0.40	1.49	−0.40	0.67
0.50	1.65	−0.50	0.61

We see that for b less than about $|0.2|$, b is a good approximation to the expected proportional increase in Y for a one-unit increase in X. For larger values of b, b underestimates the proportional increase in Y.

To see how to interpret such results, consider the effect of education and hours worked on ln(income), by sex, using the 2004 GSS. We estimate a model of the form

$$\widehat{\ln(I)} = a + b(E) + c(H) + d(M) \tag{7.16}$$

where I = income in 2003, E = years of school completed, H = hours worked per week, and $M = 1$ for males and $= 0$ for females. (Note that although the present analysis is restricted to people with incomes, it is common to add a small constant, say 1, to the value of the dependent variable to ensure that zero values are not dropped; such transformed variables are known as "started logs" [Tukey 1977]. See the discussion of tobit analysis in Chapter 14 for an alternative way of dealing with zero values.) The estimated equation, based on 1,459 cases with complete data, is

$$\widehat{\ln(I)} = 7.41 + .125(E) + .0207(H) + .335(M); \quad R^2 = .257 \tag{7.17}$$

This equation tells us that each year of schooling would be expected to increase income by about 12 percent, within gender categories, among those working an equal number of hours per week. Correspondingly, each additional hour worked per week would be expected to increase income by 2.1 percent, within gender categories, among those with equal education. Finally, among those with the same education working an equal number of hours, men would be expected to earn about 40 percent more than women. Here the coefficient understates the male advantage because $e^{.335} = 1.398$. This reminds us that the b may only be directly interpreted as indicating the expected percentage increase for $b < |0.2|$. For larger bs, we should actually calculate the exponent.

Negative coefficients have the same interpretation. For example, a coefficient of -0.05 indicates that a one-unit increase in the independent variable would be expected to yield a 5 percent decrease in the dependent variable; that is, the expected value of the dependent variable would be 95 percent as large. Also, for $b < -0.2$, the percent loss will be smaller than implied by the coefficient. So, again, we should compute the exponentiated value.

Note that the equation expresses a linear relationship between the independent variables and the *natural log* of income, not income itself. This is evident from inspection of a graph of the relationship between education and ln(income), evaluated separately for males and females at the mean number of hours worked per week by all workers, males and females combined (42.67). The relationship is, of course, linear and the expected values for the two sexes differ only by a constant, as shown in Figure 7.2.

However, when we graph the expected relationship between income and education, the relationship is curvilinear and the lines are no longer parallel (Figure 7.3).

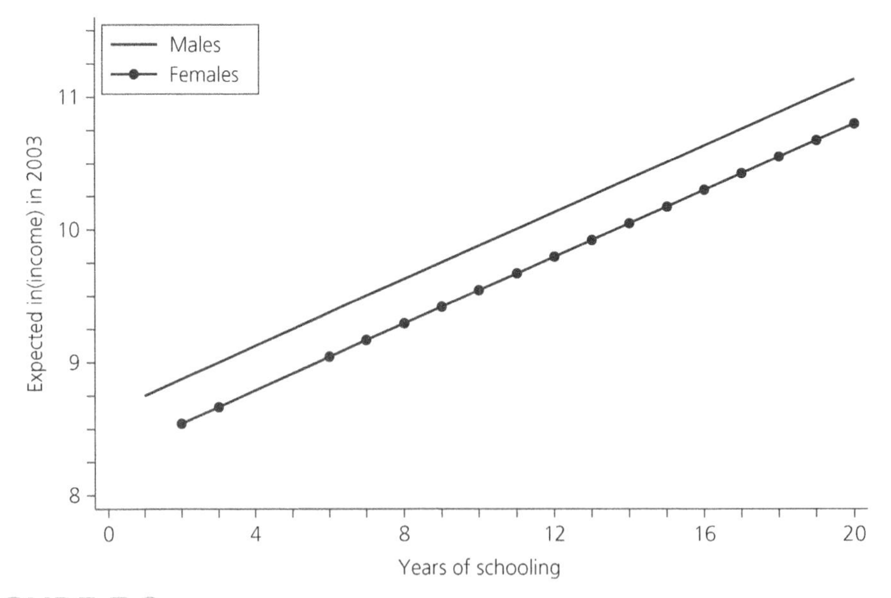

FIGURE 7.2. *Expected ln(Income) by Years of School Completed, U.S. Males and Females, 2004, with Hours Worked per Week Fixed at the Mean for Both Sexes Combined = 42.7 (N = 1,459).*

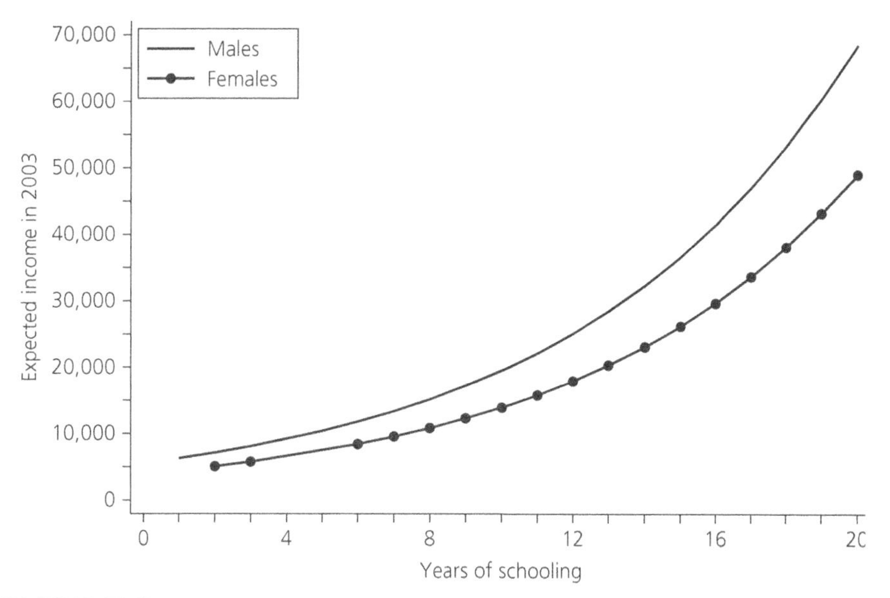

FIGURE 7.3. *Expected Income by Years of School Completed, U.S. Males and Females, 2004, with Hours Worked per Week Fixed at the Mean for Both Sexes Combined (42.7).*

Suppose we have an equation involving a logged dependent variable and squared terms for some of the independent variables. How can we interpret the coefficients?

Consider the equation

$$\widehat{\ln(Y)} = a + b(X) + c(X - k)^2 \tag{7.18}$$

One way to interpret this equation is to find the first derivative of $\widehat{\ln(Y)}$ with respect to X, and evaluate it for appropriate values of X, say the mean. Recall from first year calculus that

$$\frac{\partial(\widehat{\ln(Y)})}{\partial(\hat{X})} = b + 2cX - 2ck \tag{7.19}$$

A TRICK TO REDUCE COLLINEARITY A potential problem of excessive collinearity arises when a variable and some monotonic transformation of it are both in the same equation because such variables tend to be highly correlated. To reduce collinearity between a variable and its square, analysts sometimes subtract a constant before squaring. It can be shown that subtracting $b/2$, where b is the slope of the regression of X^2 on X, renders X and $(X - b/2)^2$ orthogonal (see Treiman and Roos 1983, 621).

The first derivative gives the slope of a straight line tangent to the curve at the point evaluated. Thus, evaluating the first derivative at the mean of X gives the partial slope relating ln(Y) to X at the mean.

However, because the function is not a straight line, the convenient interpretation of coefficients in semilog equations as indicating the proportional change in Y for a unit change in X is not available. The difficulty is that the slope does not take account of the curve in the function. However, we can derive an algorithm that will do this. Consider Equation 7.18 for two values of X, X_1 and X_2, where $X_2 = X_1 + 1$. Then we have

$$\widehat{\ln(Y_1)} = a + b(X_1) + c(X_1 - k)^2 \tag{7.20}$$

and

$$\widehat{\ln(Y_2)} = a + b(X_2) + c(X_2 - k)^2 \tag{7.21}$$

or

$$\widehat{\ln(Y_2)} = a + b(X_1 + 1) + c((X_1 + 1) - k)^2 \tag{7.22}$$

Then, subtracting Equation 7.20 from Equation 7.22, we have

$$\widehat{\ln(Y_2)} - \widehat{\ln(Y_1)} = b + 2cX_1 - 2ck + c \tag{7.23}$$

But because

$$\widehat{\ln(Y_2)} - \widehat{\ln(Y_1)} = \widehat{\ln(Y_2/Y_1)} \tag{7.24}$$

if we exponentiate both sides of Equation 7.23, we have

$$\widehat{Y_2/Y_1} = e^{(b+2cX_1-2ck+c)} \tag{7.25}$$

Thus, Equation 7.25 gives the expected proportional increase in Y for a one-unit increase in X, evaluated for any given X. So, if we set X equal to its mean, Equation 7.25 gives the proportional increase in Y for two individuals who are at the mean and one point above the mean on variable X.

Note that none of the preceding is affected by the presence of other variables in the equation. In the presence of other variables, these are partial effects, net of the effects of the other variables. For an application of this approach, see Treiman and Lee (1996).

Mobility Effects

Suppose we want to test the Durkheimian hypothesis that extreme social mobility, either upward or downward, leads to anomie. If we are willing to consider the effect of upward and downward mobility as symmetrical, we might estimate an equation of the form

$$\hat{A} = a + b(P_F) + c(P) + d(P - P_F)^2 \quad (7.26)$$

where A = the score on an anomie scale, P_F = the prestige of the respondent's father's occupation, and P = the prestige of the respondent's occupation. (Note that this specification of the hypothesis assumes that it applies to intergenerational mobility, that occupational mobility is a good indicator of social mobility and prestige a good measure of occupational status, and that extreme mobility should be most heavily weighted—by squaring the difference. In a substantive analysis, all of these assumptions need to be justified explicitly, not merely presented without justification.) A significantly positive coefficient d indicates that anomie increases as the discrepancy between respondent's and father's occupational prestige increases, controlling for the level of both respondent's and father's prestige. Thus d indicates the effect of mobility per se, controlling for the effect of status level. It is necessary to control for status level because anomie may be related to origin or destination status entirely apart from any effect of mobility.

Of course, many other transformations of variables can be used to represent different social processes. For some examples, see Goldberger (1968, Chapter. 8), Treiman (1970), and Stoltzenberg (1974, 1975).

TESTING THE EQUALITY OF COEFFICIENTS

Sometimes situations arise in which we want to determine whether two coefficients within the same equation are of equal size. You have already seen an example in the previous chapter, in the discussion of Equation 6.30. Here we consider an additional example. Suppose we are interested in assessing the effect of parental education on respondent's education and, in particular, in deciding whether the mother's or the father's education has a stronger effect. The hypothesis that educational transmission through the mother is stronger than through the father arises from the observation that mothers spend more time with their children than do fathers and hence are putatively more important socializing agents. The alternative hypothesis, that the father's education has a stronger effect, derives from the claim that the father's socioeconomic characteristics largely determine the family's socioeconomic status. Because education involves opportunity costs, it may well be that those whose fathers are poorly educated and hence typically have low incomes will be more likely to leave school early to switch from being a drain on the family financial resources to being an economic contributor to the family.

Armed with these two competing hypotheses, we might then estimate the regression of years of school completed on father's and mother's years of school completed. From the 1980 GSS I estimated an equation of the form

$$\hat{E} = a + b(E_F) + c(E_M) \quad (7.27)$$

where E = respondent's years of schooling, E_F = father's years of schooling, and E_M = mother's years of schooling. (I chose the 1980 GSS data to illustrate how to test the significance of an apparent true difference. The 2004 data yield virtually identical coefficients for mother's and father's education. Assessing cross-temporal trends in the relative effects of parental education and the reasons for such trends might yield an interesting paper.) Estimating Equation 7.27, with N = 985, yields

$$\hat{E} = 7.87 + .209(E_F) + .269(E_M); \qquad R^2 = .313 \tag{7.28}$$

This result appears to support the claim that mother's education has a somewhat stronger effect on educational attainment than does father's education. It is possible, however, that this result arises simply from sampling variability. How can we find out? The trick is to force the coefficients for mother's and father's education to be equal and then to assess whether the R^2 for the unconstrained equation (7.28) is significantly larger than the R^2 for the constrained equation. We constrain the coefficients to equality by estimating an equation of the form:

$$\hat{E} = a + b(E_P) \tag{7.29}$$

where $E_P = E_F + E_M$. Thus, we have

$$\begin{aligned} \hat{E} &= a + b(E_P) \\ &= a + b(E_F + E_M) \\ &= a + b(E_F) + b(E_M) \end{aligned} \tag{7.30}$$

Note that defining a variable as the *sum* of the years of schooling of the mother and father is equivalent, with respect to testing the hypothesis, to defining a variable as the *mean* of the years of schooling of the mother and father. If the mean were specified, the coefficient would simply double in size. In the present case the sum is more readily interpretable because it retains the metric of the separate measures for mother and father. Estimating the equation, we get

$$\hat{E} = 7.93 + .236(E_P); \quad R^2 = .312 \tag{7.31}$$

Next we compare the two models. First, we do an F-test of the equality of the coefficients, which is equivalent to testing the significance of the increment in R^2. This can be done very easily in Stata by using the `-test-` command. In fact, we don't even need to construct the constrained model. We simply issue the command `-test paeduc=maeduc-` after estimating the unconstrained model. This yields an F-value of 1.40, which is not significant (p = .236); note that because we have a two-tailed test (we have hypotheses expecting either mother or father to have greater in uence), the conventional significance level required to reject the null hypothesis is .025. An alternative way of comparing the models is, of course, to compare the *BIC*s for the two models. To do this we need to estimate the constrained model. The two *BIC*s, estimated in the usual way, are

Unconstrained: −355.4

Constrained: −360.9

In this case both the *BIC* and the *F*-test favor the constrained model of no difference. This general strategy can be applied to a wide variety of substantive problems.

TREND ANALYSIS: TESTING THE ASSUMPTION OF LINEARITY

As the GSS has matured, it has become an increasingly valuable resource for the study of cross-temporal trends. Because many questions have been asked in exactly the same way since the first GSS was conducted in 1972, it is possible to pool the data from all years to study a variety of trends. Moreover, if no cross-temporal variations are detected, the data for all years can be treated as a sample of the U.S. population in the late twentieth century to generate sufficient cases to study relatively small subsets of the population.

The simplest trend model (apart from the limiting case of no trend) is that there is a linear trend over time with respect to the outcome of interest. As a first step, it is useful to contrast such a model with a model that posits year-to-year variations in the outcome—what Sorokin many years ago (1927) described as "trendless uctuations." To do this, we estimate two models:

$$\hat{Y} = a + bT \tag{7.32}$$

and

$$\hat{Y} = a' + bT + \sum_{j=2}^{J} c_j T_j \tag{7.33}$$

where T is a linear representation of time (here, the year of the survey), and the T_j are dummy variables for each year the survey was conducted; note that two dummy variables must be omitted because the linear term uses up one degree of freedom. We then compare the two models in the usual way, via an *F*-test of the significance of the increment in R^2 and a comparison of *BIC* values. A convenient way to do the first in Stata is to estimate Equation 7.33 and then to test the hypothesis that all the c_j equal zero, via a Wald test using Stata's `-test-` command. (Note that Equation 7.33 is simply a different parameterization of an equation in which the linear term is omitted and only the dummy variables are included. The coefficients will, of course, differ. But the predicted values, R^2, and *BIC* will be identical.) If we conclude that no simple linear trend fits the data, we might then posit either a model with a smooth curve by including a squared term for T, or a model that tries to model particular historical events by grouping years into historically meaningful groups and identifying each group (less one) via a dummy variable, or a spline model (see the section "Linear Splines" later in the chapter). Because the variance explained by Equation 7.33 is the maximum possible from any representation of time (measured in years), the R^2 associated with Equation 7.33 serves as a standard against which to assess, in substantive rather than strictly statistical terms, how close various sociologically motivated constrained models come to fully explaining temporal variation in the dependent variable.

Although, to simplify the exposition, I have not included any variables in the model other than time, a model actually posited by a researcher typically would include a number of covariates (other independent variables) and also, perhaps, interactions between the covariates and the variables representing time. Exactly the same logic would apply to such an analysis as to the simpler analysis just described; the logic is also identical to the dummy variable approach to the assessment of group differences described in the previous chapter (although here the "groups" are years or, if warranted by the analysis, multiyear historical periods).

Predicting Variation in Gender Role Attitudes over Time: A Worked Example

Four items on attitudes regarding gender-role equality were asked in most years of the GSS between 1974 and 1998. The four variables are shown here with the percentage endorsing the pro-equality position, pooled over all years in which all four questions were asked:

- Do you agree or disagree with this statement? Women should take care of running their homes and leave running the country up to men (74 percent disagree).
- Do you approve or disapprove of a married woman earning money in business or industry if she has a husband capable of supporting her? (77 percent approve).
- If your party nominated a woman for President, would you vote for her if she were qualified for the job? (84 percent say yes).
- Tell me if you agree or disagree with this statement: Most men are better suited emotionally for politics than are most women (63 percent disagree).

To form a gender-equality scale, I simply summed the pro-equality responses for the four items, excluding all people to whom the questions were not asked and treating other nonresponses as negative values. The point of treating "don't know" and similar responses as negative values rather than excluding them is to save cases. But this would not be wise if there were not substantive grounds for doing so—in this case, it seemed reasonable to me to treat "don't know" as something other than a clear-cut endorsement of gender equality.

IN SOME YEARS OF THE GSS, ONLY A SUBSET OF RESPONDENTS WAS ASKED CERTAIN QUESTIONS Users of the GSS need to be aware that to increase the number of items that can be included in the GSS each year, some items are asked only of subsets of the sample. A convenient way to exclude people who were not asked the questions is to use the Stata `-rmiss-` option under the `-egen-` command to count the number of missing data responses and then to exclude people missing data on all items included in a scale. However, in the current analysis I excluded all those who lacked responses on any of the four items because in some years only some, but not all, of the questions were asked.

Estimating equations such as Equations 7.32 and 7.33 suggests significant nonlinearities in attitudes regarding gender inequality. The increment in R^2 implies $F = 3.54$ with 14 and 21,448 *d.f.*, which has a probability of less than 0.0001. However, the *BIC* for the linear trend model is more negative than the *BIC* for the annual variability model (the *BIC*s are, respectively, −959 and −871), suggesting that a linear trend is more likely given the data. Because *BIC* and classical inference yield contradictory results, a sensible next step is to graph annual variations in the mean level of support for gender equality, to see whether there is any obvious pattern to the nonlinearity. If substantively sensible deviations from linearity are observed, the annual variation model might be accepted, or a new model, aggregating years into historically meaningful periods, might be posited (keeping in mind the dangers of modifying your hypotheses based upon inspection of the data—see the discussion of this issue at the end of Chapter Six), or a smooth curve or spline function might be fitted to the data. Figure 7.4 shows both the linear trend line and annual variations in the mean. Inspecting the graph, it appears that deviations from linearity are neither large nor systematic. Given this, I am inclined to accept a linear trend model as the most parsimonious representation of the data, despite the *F*-test results. The linear trend is, in fact, quite substantial, implying an increase of 0.81 (= .0338*(1998 − 1974)) over the quarter of a century for which we have data; this is about 20 percent of the range of the scale and is about two-thirds of the standard deviation of the scale scores. Apparently, support for gender equality has been increasing modestly but steadily throughout the closing years of the twentieth century.

From a technical point of view, it may be helpful to compare the estimates implied by the two alternative ways of representing departures from linearity: Equation 7.33 and the

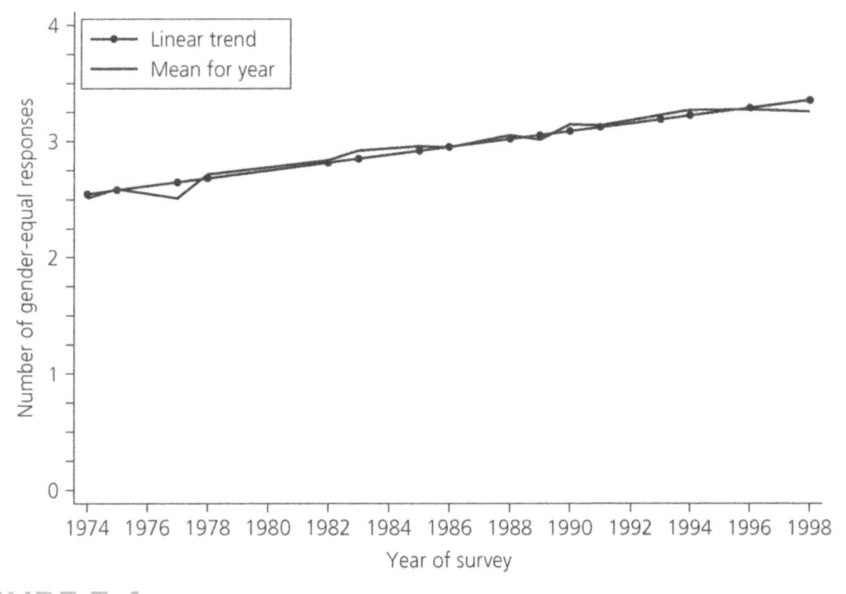

FIGURE 7.4. *Trend in Attitudes Regarding Gender Equality, U.S. Adults Surveyed in 1974 Through 1998 (Linear Trend and Annual Means; N = 21,464).*

alternative specification that does not include a linear term for year. When the linear term is included, two dummy variable categories are dropped, rather than one, because the linear term uses up one degree of freedom. However, the two procedures produce identical results, which is evident from inspection of Table 7.1.

Unfortunately, there is no simple correspondence between the coefficients in equations of the form of Equation 7.33 and deviations from the predictions of the linear equation. If you want to show annual departures from linearity, you need to construct a new variable, which is the difference between the predicted values for each year from Equation 7.32 and Equation 7.33. This is very easy to accomplish in Stata using the `-foreach-` or `-forvalues-` command.

LINEAR SPLINES

Sometimes we encounter situations in which we believe that the relationship between two variables changes abruptly at some point on the distribution of the independent variable, so that neither a linear nor a curvilinear representation of the relationship is adequate. For example, alcohol consumption may have no impact on health below some threshold, whereas above the threshold health declines in a linear way as alcohol consumption increases. Temporal trends also may abruptly change, as a result of policy changes, cataclysmic events such as depressions, wars, revolutions, and so on. In cases of this kind, it is useful to represent the relationships via a set of connected line segments, known as *linear splines*.

A Worked Example: Trends in Educational Attainment over Time in the United States

Consider changes in the average level of education over time. Figure 7.5 presents a scatter plot relating educational attainment to year of birth, estimated from the GSS. To create this graph, I combined data from all years between 1972 and 2004. However, I dropped those born prior to 1900 because the very small sample sizes produced unstable results. I also dropped all those less than age twenty-five at the time of the survey because many people do not complete their schooling until their mid-twenties. The graph is based on 5 percent of the cases, to make it readable, and is "jittered" to make clear where the density is greatest. Inspecting the figure, there appears to be a secular increase in education, but the nature of the increase is hard to discern—is it linear or is the trend better represented by some other functional form?

To discover how average educational attainment has increased over time, we can plot the mean years of school completed for each birth cohort. Figure 7.6 shows such a plot, made with the same specifications as the scatter plot. Inspecting the plot, you see that average education increased in a more or less linear way for those born between 1900 and 1947 but then leveled off. Because the plot bounces around a bit, probably because of the relatively small number of cases for each cohort, it might be better to plot a three-year moving average of mean years of schooling. Such a plot is shown as Figure 7.7. (Details on how these graphs were created are shown in the downloadable `-do-` and `-log-` files.) Inspecting this graph, we come to the same conclusion—there is a fairly abrupt change in the trend, with those born in the first half of the twentieth

TABLE 7.1. Demonstration That Inclusion of a Linear Term Does Not Affect Predicted Values.

Coefficient	$\hat{Y} = a + \sum c_j T_j$	$\hat{Y} = a' + bT + \sum c'_j T_j$
Year	—	$b = -0.0375872$
Constant	$a = 2.5114$	$a' = 71.68578$
R^2	.0464	.0464
BIC	−871	−871
Expected score on the gender equity scale, for selected years		
1974	$a = 2.5114$	$a' + b*1974 = -71.68578 + 0.0375872*1974 = 2.5114$
1975	$a + c_{1975} = 2.5114 + .0780 = 2.5894$	$a' + b*1975 + c'_{1975} = -71.68578 + 0.0375872*1975 + 0.0403799 = 2.5893$
1977	$a + c_{1977} = 2.5114 - 0.0009 = 2.5105$	$a' + b*1977 + c'_{1977} = -71.68578 + .0375872*1977 - 0.1136418 = 2.5105$
…	…	…
1998	$a + c_{1998} = 2.5114 + 0.7473 = 3.2587$	$a' + b*1998 + c'_{1998} = -71.68578 + 0.0375872*1998 - 0.1548326 = 3.2586$

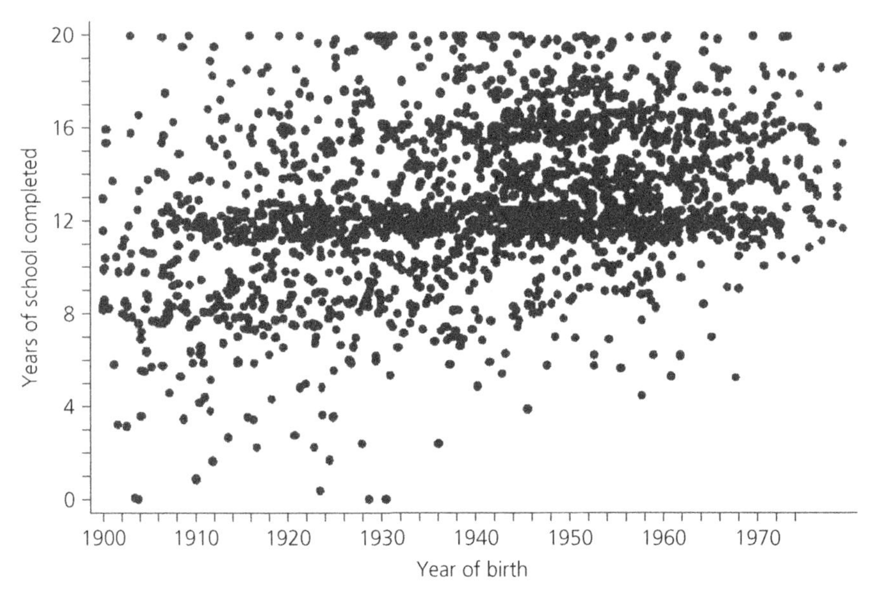

FIGURE 7.5. *Years of School Completed by Year of Birth, U.S. Adults (Pooled Samples from the 1972 Through 2004 GSS; N = 39,324; Scatter Plot Shown for 5 Percent Sample).*

century (precisely, until 1947) experiencing a fairly steady year-by-year increase in their schooling, but those born in 1947 or later experiencing no change at all. This suggests that the trend in educational attainment is appropriately represented by a linear spline with a knot at 1947, where "knot" refers to the point at which the slope changes.

This specification can be represented by an equation of the form:

$$\hat{E} = a + b_1(B_1) + b_2(B_2) \tag{7.34}$$

where B_1 = the year of birth for those born in 1947 or earlier and = 1947 otherwise, and B_2 = the year of birth – 1947 for those born after 1947 and = 0 otherwise. More generally, a spline function relating Y to X with segments $v_1 \ldots v_{n+1}$ and knots at $k_1, k_2, \ldots, k_n$ can be represented by

$$\hat{Y} = a + b_1(X_1) + b_2(X_2) + \cdots + b_{n+1}(X_{n+1}) \tag{7.35}$$

where $v_1 = \min(X, k_1)$, $v_2 = \max(\min(X - k_1, k_2 - k_1), 0), \ldots, v_{(n+1)} = \max(X - k_n, 0)$ (see Panis [1994]; the entry for Stata's `-mkspline-` command [StataCorp 2007]; and Greene [2008]). Each slope coefficient is then the slope of the specified line segment. We can see this concretely by going back to our example, Equation 7.34, and evaluating the equation separately for those born in 1947 or earlier and those born after 1947. For those born in 1947 or earlier, we have

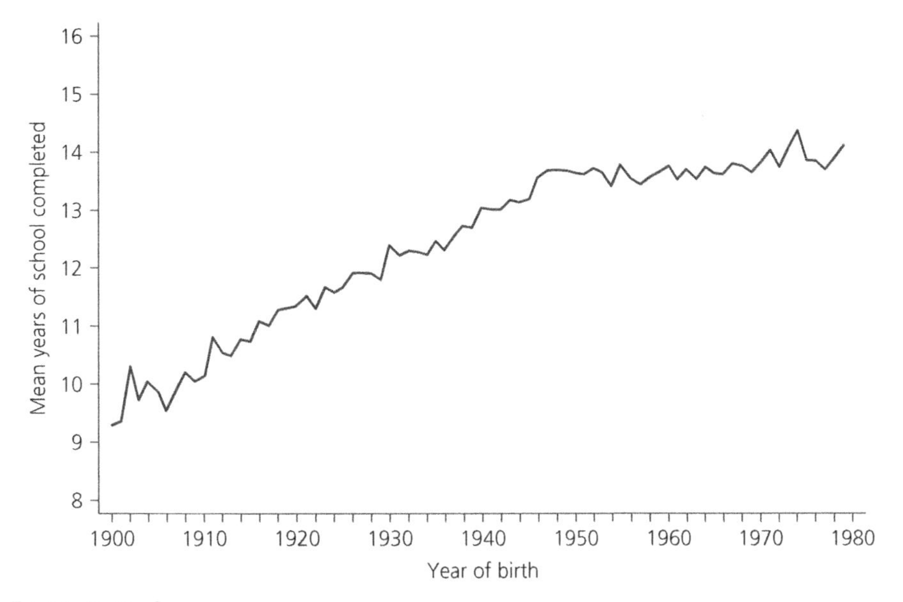

FIGURE 7.6. *Mean Years of Schooling by Year of Birth, U.S. Adults (Same Data as for Figure 7.5).*

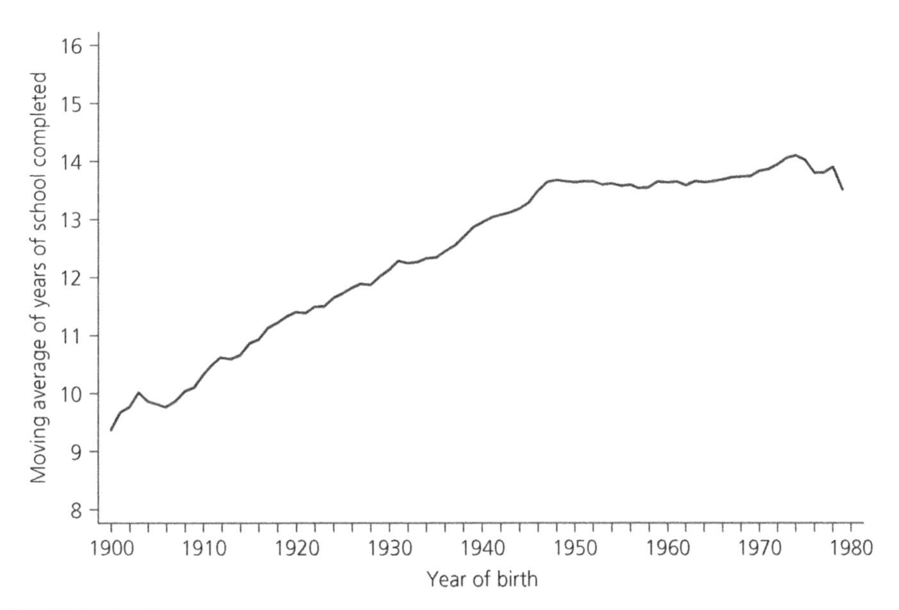

FIGURE 7.7. *Three-Year Moving Average of Years of Schooling by Year of Birth, U.S. Adults (Same Data as for Figure 7.5).*

$$\hat{E} = a + b_1(B) + b_2(0) = a + b_1(B) \tag{7.36}$$

and for those born later than 1947, we have

$$\begin{aligned}\hat{E} &= a + b_1(1947) + b_2(B-1947) \\ &= (a + 1947b_1) + b_2(B-1947)\end{aligned} \tag{7.37}$$

Notice that the intercept in Equation 7.37 is just the expected level of education for those born in 1947 and that b_2 gives the slope for those born after 1947. Thus, the expected level of education for those born in 1948 is just the expected level of education for those born in 1947 plus b_2; for those born in 1949 it is the expected level of education for those born in 1947 plus $2b_2$; and so on.

Estimating Equation 7.34 from the pooled 1972–2004 GSS data yields the coefficients in Table 7.2. By inspecting the *BIC*s for three models—the spline model, a linear trend model, and a model that allows the expected level of schooling to vary year-by-year—it is evident that the linear spline model is to be preferred. Note, however, that a comparison of R^2s indicates that by the criterion of classical inference, the model positing year-by-year variation in the level of schooling fits significantly better than the spline model. I am inclined to discount this result because it has no theoretical justification, is

AN ALTERNATIVE SPECIFICATION OF SPLINE FUNCTIONS An alternative specification represents the slope of each line segment as a deviation from the slope of the previous line segment. In this specification, a different set of new variables is constructed. Suppose there are *k* knots, that *X* is the original variable, and that $\nu_1, \ldots, \nu_{(n+1)}$ are the constructed variables. Then

$\nu_1 = X$

$\nu_2 = X - k_1$ if $X > k_1$; $= 0$ otherwise

. . .

$\nu_{(n+1)} = X - k_n$ if $X > k_n$; $= 0$ otherwise

To see this concretely, consider the present example specifying a knot at birth year 1947 in the trend in educational attainment. We would estimate the equation where ν_1 = birth year (*X*) and $\nu_2 = X - 1947$ if $X > 1947$ and = 0 otherwise. Then for those born in 1947 or earlier,

$$\hat{E} = a + b_1(X) + b_2(0) = a + b_1(X)$$

while for those born later than 1947

$$\hat{E} = a + b_1(X) + b_2(X - 47)$$

Thus, for those born in 1948, the expected level of education is given by $(a + 48b_1) + b_2$; for those born in 1949 it is $(a + 49b_1) + 2b_2$; and so on. From this, it is evident that b_2 gives the deviation of the slope for the previous line segment. For useful discussions of these methods, see Smith (1979) and Gould (1993).

TABLE 7.2. Coefficients for a Linear Spline Model of Trends in Years of School Completed by Year of Birth, U.S. Adults Age 25 and Older, and Comparisons with Other Models (Pooled Data for 1972–2004, N = 39,324).

	b	s.e.	t	p	
Slope (birth years 1900–1947)	.0856	.0014	63.26	.000	
Slope (birth years 1948–1979)	.0092	.0024	3.89	.000	
Intercept	−153.2	2.6	−58.55	.000	
Model Comparisons	*BIC*	R^2	F	d.f.	p
(1) Linear spline model	−5,401	.1288			
(2) Linear trend model	−4,870	.1167			
(3) Annual variability model	−4,685	.1310			
(1) vs. (2)	−531	.0121	545.2	1; 39321	.0000
(1) vs. (3)	−716	−.0022	1.29	77; 39244	.0462

clearly inferior by the *BIC* criterion, and occurs simply as a consequence of the large sample size. Thus, I accept the linear spline model as the preferred model.

The coefficients for the line segments indicate that for people born in 1947 or earlier, there is an expected increase of .086 years of schooling for each successive birth cohort. Thus, people born twelve years apart would be expected to differ on average by about a year of schooling. However, for people born in 1947 or later, there is no trend in educational attainment; the coefficient .0092 implies that it would take about a century for average schooling to increase by one year. This is a somewhat surprising result, especially because there have been substantial increases in average education among disadvantaged minorities, that is, Blacks and native-born persons of Latin American origin, and also among women. However, as Mare (1995, 163) notes, educationally disadvantaged proportions of the population have grown over time relative to the White majority. Disaggregation of the trend would be worthwhile but cannot be pursued here; it would make an interesting paper. The graph implied by the coefficients for the linear spline model is shown in Figure 7.8, together with a 2 percent random sample of observations for each cohort (reduced from 5 percent to 2 percent to make it easier to see the shape of the spline). In this figure the `-jitter-` feature in Stata is used to make it clear where in the graph there is the greatest density of points.

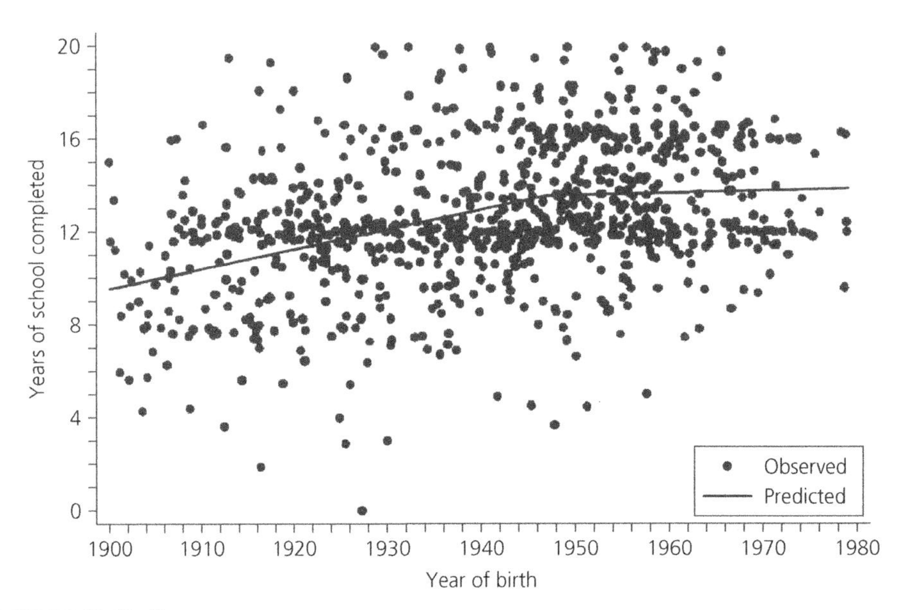

FIGURE 7.8. *Trend in Years of School Completed by Year of Birth, U.S. Adults (Same Data as for Figure 7.5; Scatter Plot Shown for 2 Percent Sample). Predicted Values from a Linear Spline with a Knot at 1947.*

A Second Worked Example, with a Discontinuity: Quality of Education in China Before, During, and After the Cultural Revolution

The typical use of spline functions is to estimate equations such as the one just discussed, in which all points are connected but the slope changes at specified points ("knots"). However, there are occasions in which we may want to posit *discontinuous* functions. The Chinese Cultural Revolution is such a case. It can be argued that the disruption of social order at the beginning of the Cultural Revolution in 1966 was so massive that it is inappropriate to assume any continuity in trends. Deng and Treiman (1997) make just such an argument with respect to trends in educational reproduction. They argue that there was then a gradual "return to normalcy" so that changes resulting from the end of the Cultural Revolution in 1977 were not nearly as sharp and were appropriately represented by a knot in a spline function rather than a break in the trend line.

Here we consider another consequence of the Cultural Revolution, the quality of education received (the example is adapted from Treiman [2007a]). Although primary schools remained open throughout the Cultural Revolution, higher level schools were shut down for varying periods: most secondary schools were closed for two years, from 1966 to 1968, and most universities and other tertiary-level institutions were closed for six years, from 1966 to 1972. Moreover, it was widely reported that even when the schools were open, little conventional instruction was offered; rather, school hours were taken up with political meetings and political indoctrination. Rigorous academic instruction was not fully reinstituted until 1977, after the death of Mao. Under the circumstances, we might well suspect that, quite apart from deficits in the *amount* of schooling acquired by those who were unfortunate enough to be of school age during the Cultural Revolution period, those cohorts also experienced deficits in the *quality* of schooling compared to those who obtained an equal amount of schooling before or after the Cultural Revolution.

To test this hypothesis, we can exploit the ten-item character recognition test administered to a national sample of Chinese adults that was also analyzed in Chapter Six (see Table 6.2). As before, I take the number of characters correctly identified as a measure of literacy and hypothesize that, net of years of school completed, people who turned age eleven during the Cultural Revolution would be able to recognize fewer characters than people who turned eleven before or after the Cultural Revolution period. Moreover, following Deng and Treiman (1997), I posit a discontinuity in the scores at the beginning but not at the end of the period. To do this, I estimate an equation of the form:

$$\hat{V} = a + b_1(B_1) + b_2(B_2) + c_2(D_2) + b_3(B_3) \tag{7.38}$$

where B_1 = year of birth (last two digits) if born prior to or in 1955 and = 55 if born subsequent to 1955; B_2 = 0 if born prior to 1956, = year of birth − 55 if born between 1956 and 1967, inclusive, and = 67 − 55 if born subsequent to 1967; B_3 = 0 if born prior to or in 1967 and = year of birth − 67 for those born after 1967; and D_2 = 0 for those born prior to or in 1955 and = 1 for those born after 1955. Note that the difference between this and Equation 7.35 is that I include a dummy variable to distinguish those born after 1955 from those born earlier; this is what permits the line segments to be discontinuous at 1955. If I were to have posited a discontinuity at 1967 as well, the equation would be the mathematical equivalent to estimating three separate equations, for the period before, during, and after the Cultural Revolution, in each case predicting the number of characters recognized from years of schooling and year of birth. The advantage of equations such as Equation 7.38 is that they permit the specification of alternative models within a coherent framework and by so doing permit us to select between models.

Estimating this equation yields the results shown for Model 4 in Tables 7.3 and 7.4. As in the previous example, I contrast my theory-driven specification with other possibilities: that there is a simple linear trend in the data; that there are year-by-year variations; that there are knots at both the beginning and the end of the Cultural Revolution, but no discontinuities; that there are discontinuities at both the beginning and the end of the Cultural Revolution; and, for the three spline functions, that there is a curvilinear relationship between year and knowledge of characters during the Cultural Revolution period.

TABLE 7.3. Goodness-of-Fit Statistics for Models of Knowledge of Chinese Characters by Year of Birth, Controlling for Years of Schooling, with Various Specifications of the Effect of the Cultural Revolution (Those Affected by the Cultural Revolution Are Defined as People Turning Age 11 During the Period 1966 through 1977), Chinese Adults Age 20 to 69 in 1996 (N = 6,086).

Model	*BIC*	R^2	F	df	p
(1) Linear trend model	−6638.0	.665	6037.06	2; 6083	.000
(2) Annual variability model	−6422.5	.676	251.78	50; 6035	.000
(3) Spline: knots at 1955 and 1967	−6680.4	.668	3062.36	4; 6081	.000
(4) Spline: knot at 1967, discontinuity at 1955	−6725.9	.671	2482.30	5; 6080	.000
(5) Spline: discontinuities at 1955 and 1967	−6723.9	.672	2071.62	6; 6079	.000
(6) Spline: (3) plus curve between 1955 and 1967	−6722.7	.671	2480.33	5; 6080	.000
(7) Spline: (4) plus curve between 1955 and 1967	−6724.1	.672	2071.72	6; 6079	.000
(8) Spline: (5) plus curve between 1955 and 1967	−6717.4	.672	1776.33	7; 6078	.000
Model Comparisons					
(2) vs. (1)	215.5	.011	4.26	48; 6035	.000
(3) vs. (1)	−42.4	.003	30.04	2; 6081	.000
(4) vs. (3)	−44.5	.003	54.43	1; 6080	.000
(5) vs. (4)	2.0	.001	6.65	1; 6079	.010
(5) vs. (3)	−43.5	.004	30.56	2; 6079	.000
(6) vs. (3)	−42.3	.003	51.17	1; 6080	.000
(7) vs. (4)	1.8	.001	6.86	1; 6079	.009
(8) vs. (5)	6.5	.000	2.19	1; 6078	.139

TABLE 7.4. Coefficients for Models 4, 5, and 7 Predicting Knowledge of Chinese Characters by Year of Birth, Controlling for Years of Schooling (*p*-Values in Parentheses).

	Model 4	Model 5	Model 7
Years of schooling	.443 (.000)	.443 (.000)	.444 (.000)
Born 1955 or earlier (age 11 1965 or earlier)	0.001 (.721)	0.001 (.734)	0.001 (.749)
Born 1956–1967 (age 11 1966–1977)	0.043 (0.000)	0.032 (0.000)	−0.047 (0.185)
Born 1968 or later (age 11 1978 or later)	0.041 (.000)	0.016 (.233)	0.028 (.012)
Discontinuity at 1955	−0.557 (.000)	−0.508 (.000)	−0.349 (.001)
Discontinuity at 1967		0.241 (.010)	
Curvilinear trend 1956–67			0.0066 (.009)
Intercept	0.770	0.770	0.771
R^2	0.671	0.672	0.672
s.e.e. (root mean square error)	1.29	1.29	1.29

A comparison of the *BIC*s suggests that three models—my hypothesized model, a model that in addition to a discontinuity at the beginning of the Cultural Revolution allows the trend during the Cultural Revolution period to be curvilinear, and a model positing discontinuities at both the beginning and the end of the Cultural Revolution—are about equally likely given the data, albeit with weak evidence favoring the single-knot model, and that all three are strongly to be preferred over all other models.

Again, *BIC* and classical inference yield contradictory results because the two alternative models fit significantly better (at the 0.01 level) than does the originally hypothesized model. Here I am in a bit of a quandary as to which model to prefer. I have already stated a basis for positing a single discontinuity, plus a knot at the end of the Cultural Revolution. However, another analyst might favor a two-discontinuity model, on the ground that the curricular reform in 1977 that restored the primacy of academic subjects was radical enough to posit a discontinuity at the end as well as at the beginning of the Cultural Revolution. A third analyst might argue that a linear specification of trends, especially in times of great social disruption, is too restrictive and that it makes more sense to posit a curvilinear effect of time during the Cultural Revolution period. In Treiman (2007a), I presented the model positing a discontinuity at 1955, a knot at 1967, and a curve between 1955 and 1967—see Figure 7.4 in that paper. However, the truth is that there is no clear basis for preferring any one of the three, except for the evidence provided by *BIC*, which suggests that the originally hypothesized model is slightly more likely than the others given the data. Again, my suggestion is, go with theory. If you have a theoretical basis for one specification over the others, that is the one to favor; but, at the same time, you must be honest about the fact that alternative specifications fit nearly equally well. In fact, the optimal approach is to present all three models and invite the reader to choose among them. A warning: if you do this, you probably will have to fight with journal editors, who are always trying to get authors to reduce the length of papers, and perhaps with reviewers, who sometimes seem to want definitive conclusions even when the evidence is ambiguous.

The estimated coefficients for all three models are shown in Table 7.4. In all three models each additional year of schooling results in nearly half a point improvement in the number of characters identified. However, the coefficients associated with trends over time are relatively difficult to interpret. Again, this is an instance in which graphing the relationship helps. Figure 7.9 shows, for each of the three preferred models, the predicted number of characters recognized by people with twelve years of schooling, that is, who have completed high school. Although the three graphs appear to be quite different, they all show a decline of about half a point in the number of characters identified for those who were age eleven during the early years of the Cultural Revolution period, relative to those with the same level of schooling who turned eleven before and after the Cultural Revolution. Thus, despite the difficulty in choosing among alternative specifications, together they strongly suggest that the quality of education declined during the Cultural Revolution. People who acquired their middle school (junior high school) education during the Cultural Revolution, in effect, lost a year of schooling—that is, displayed knowledge of vocabulary equivalent to those with one year less schooling who were educated before and after the Cultural Revolution.

Still, we should be cautious in our interpretation of Figure 7.9, where the Cultural Revolution effect appears to be quite large because of the way the data are graphed (with the y-axis ranging from 5.3 to 6.7 characters recognized). Indeed, Figure 7.10, in which the y-axis ranges from 0 to 10, suggests a rather different story—a very modest decline in the number of characters recognized. It is quite reasonable to report figures such as Figure 7.9 to make the differences among the models clear, but in such cases the

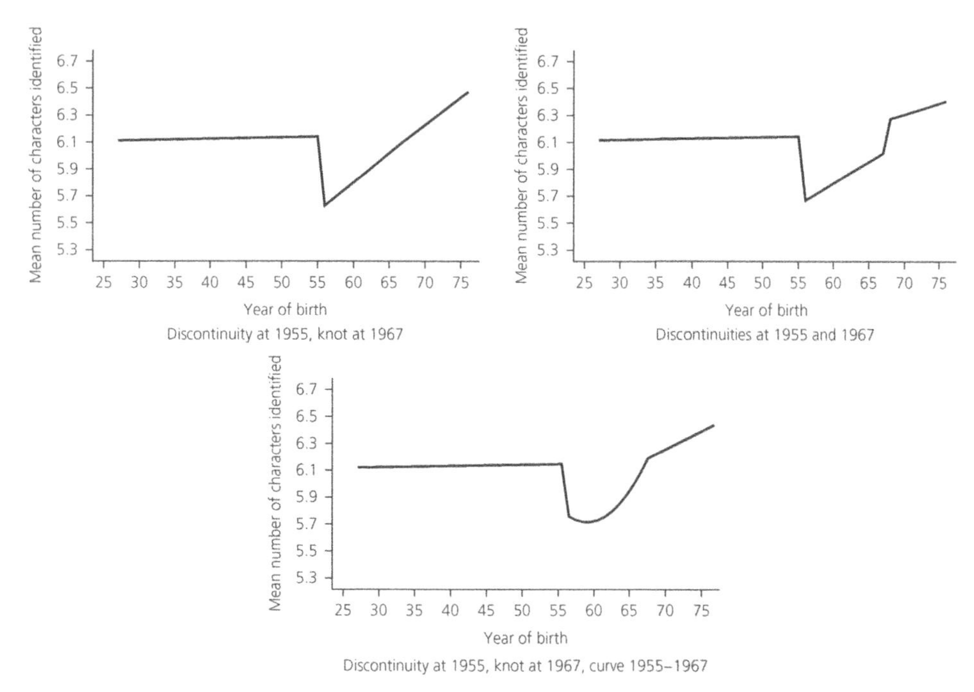

FIGURE 7.9. *Graphs of Three Models of the Effect of the Cultural Revolution on Vocabulary Knowledge, Holding Constant Education (at Twelve Years), Chinese Adults, 1996 (N = 6,086).*

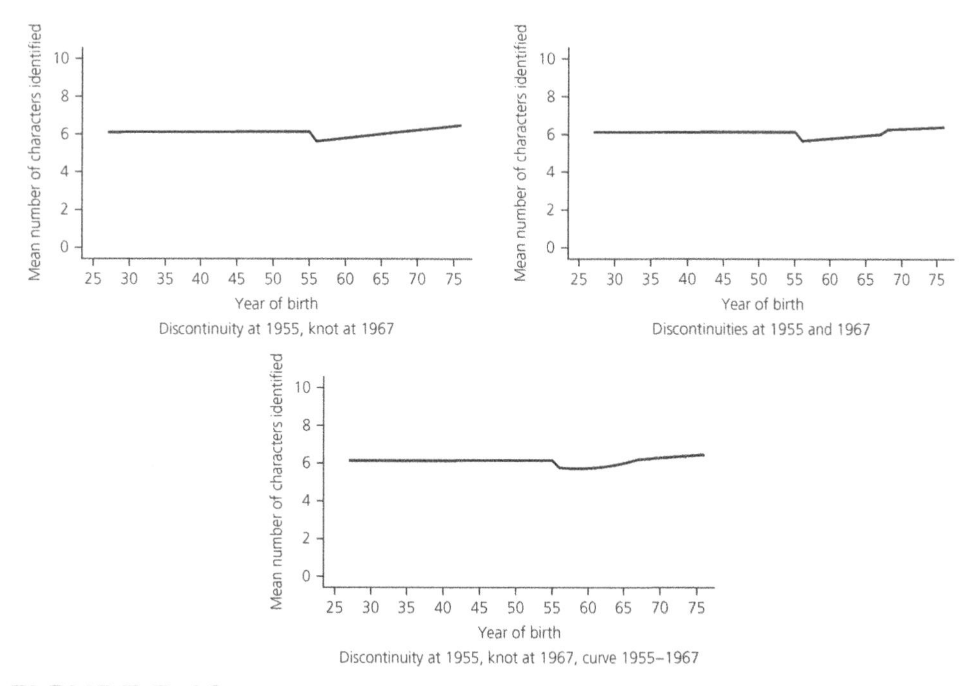

FIGURE 7.10. *Figure 7.9 Rescaled to Show the Entire Range of the Y-Axis.*

responsible analyst will call the reader's attention to the range of the y-axis to avoid misinterpretation.

EXPRESSING COEFFICIENTS AS DEVIATIONS FROM THE GRAND MEAN (MULTIPLE CLASSIFICATION ANALYSIS)

The conventional way of treating categorical independent variables is the approach presented in the previous chapter: omit one category and interpret the remaining coefficients as deviations from the expected value for the omitted category. Sometimes, especially when we have a large number of categories, it is preferable to express our coefficients as deviations from the mean of the dependent variable. We can do this by transforming the coefficients, making use of the following relationships:

$$a_{ij} = b_{ij} + Q_i; \quad Q_i = -\sum_j p_{ij} b_{ij} \tag{7.39}$$

where the a_{ij} are the coefficients for the *j*th category of the *i*th predictor, expressed as deviations from the mean of the dependent variable; the b_{ij} are the corresponding coefficients associated with the dummy variables; the Q_i are adjustment coefficients that constrain the weighted sum of the coefficients associated with the categories of each predictor to zero; and the p_{ij} are the proportion of total cases falling in the *j*th category of the *i*th predictor (Andrews et al. 1973, 45–47).

To see how these coefficients work, consider the relationship between religious denomination and tolerance. The analysis task includes two elements:

- To assess to what extent and in what way religious groups differ in their tolerance of the antireligious
- To assess to what extent the observed differences between religious groups can be attributed to the fact that they differ with respect to education and Southern residence because these variables are known to affect tolerance (with the more educated and non-Southern residents more tolerant than others)

I start by estimating two regression equations in the usual way—one with only the dummy variables for religious groups and one also including education and Southern residence—using pooled data from the 2000, 2002, and 2004 GSS; I pool three years of data to increase the sample size because some religious categories are quite small, and the tolerance questions were asked of only a subset of respondents in each year. The results are shown in the left-hand panel of Table 7.5. I then reexpress these coefficients as deviations from the mean of the dependent variable using Equation 7.39. The rightmost panel of the table shows the reexpressed coefficients.

Ordinarily you would not present both sets of coefficients, but would choose one form or the other—either a dummy variable representation or a multiple-classification representation. I present them both together here so that you can see the relationships between the coefficients.

TABLE 7.5. Coefficients of Models of Tolerance of Atheists, U.S. Adults, 2000 to 2004 (N = 3,446).

	Dummy Variable Coefficients (Deviations from Omitted Category)		MCA Coefficients (Deviations from Grand Mean)		N
	Model 1	Model 2	Model 1	Model 2	
Baptists	0.000	0.000	−0.422	−0.308	(644)
Methodists	0.395	0.298	−0.027	−0.010	(263)
Lutherans	0.647	0.487	0.224	0.179	(163)
Presbyterians and Episcopalians	0.502	0.262	0.080	−0.046	(182)
Other Protestants	0.340	0.242	−0.082	−0.066	(573)
Catholics	0.422	0.304	−0.000	−0.004	(835)
Other Christians	0.378	0.274	−0.044	−0.033	(96)
Jews	0.836	0.464	0.414	0.156	(71)
Other religions	0.643	0.412	0.221	0.104	(119)
No religion	0.869	0.709	0.447	0.401	(500)
Education		0.102		0.102	
Southern		−0.107		−0.068	(1,251)
Non-Southern		0.000		0.039	(2,195)
Intercept	1.685	0.476	2.107	2.107	
R^2	.061	0.136	0.061	0.136	(3,446)

Note: Since *p*-values are not readily computed for the MCA coefficients, and are not particularly meaningful for the dummy variable coefficients because they indicate the significance of the difference from the omitted category, they are not shown here.

Note that the *difference* between coefficients is the same in both versions. In Model 1, for example, the difference between the tolerance scores of Methodists and Baptists is $0.395 = 0.395 - 0 = -0.027 - (-0.422)$. Similarly, in Model 2 the difference is $0.298 = 0.298 - 0 = -0.010 - (-0.398)$.

What do the reexpressed coefficients tell us? I think they are easier to interpret. First consider Model 1. From this model we see that Baptists are considerably less tolerant than the average respondent whereas Jews and those without religion are considerably more tolerant than average, and Lutherans and "Others" are somewhat more tolerant than average. However, these differences, especially the high tolerance of Jews, are somewhat explained by religious differences in education and region of residence because, in general, the deviations from the mean decline when these two factors are controlled.

Southern residents are somewhat below average in tolerance, net of religious affiliation and education, whereas non-Southern residents are slightly above average in tolerance. Non-Southern residents will necessarily be closer to the overall mean because there are substantially more of them, and the weighted average of the coefficients (with weights corresponding to the proportion of the sample in the category) must sum to zero.

The coefficients for the religious groups and for Southerners versus non-Southerners are sometimes referred to as "adjusted group differences," where the adjustment refers to the fact that other variables in the model are controlled.

The slope coefficient for education does not change, but the scaling of the educational variable does. In the reexpressed ("MCA") representation, education is expressed as a deviation from its mean—in this case 13.4. Finally, the intercept in the reexpressed representation is just the mean of the dependent variable, the level of tolerance.

OTHER WAYS OF REPRESENTING DUMMY VARIABLES

Three other ways of representing the effects of categorical variables sometimes can facilitate interpretation. Two of them, *effect coding* and *contrast coding*, require representing the categories of the classification in a different way from conventional dummy variable coding (see Cohen and Cohen [1975, 172–210], Hardy [1993, 64–75], and Fox [1997, 206–11]). A third, which I label *sequential effects*, involves manipulating the output. None of these alternative ways of expressing the effects of a categorical variable alters the contribution of the categorical variable to the explained variance; that is, the R^2 is unaffected. All they do is reparameterize the effects, and so the only reason for using any of them is to make interpretation of the relationships in the data clearer.

To see how these alternatives work, let us consider a new problem—the effect of occupation and education on knowledge of vocabulary in the United States, using data from the 1994 GSS. The GSS includes a ten-item vocabulary test, a detailed classification of current occupation, and a measure of years of school completed. For the purpose of this example, I have collapsed the detailed occupational classification into four categories: upper nonmanual (managers and professionals), lower nonmanual (technicians, sales occupations, and administrative support occupations), upper manual (precision production, craft, and repair occupations), and lower manual (all other categories: service occupations; agricultural occupations; and operators, fabricators, and laborers). I expect that, net of current occupation, vocabulary scores will increase with years of schooling.

More interestingly, I expect vocabulary scores to increase with occupational status—that is, that lower manual workers, upper manual workers, lower nonmanual workers, and upper nonmanual workers will have increasingly high vocabulary scores, on average, net of years of schooling. The argument is that symbolic manipulation plays an increasingly large role in work as one moves up the occupational hierarchy, so that verbal skills are much more strongly reinforced and enhanced in high-status occupations than in low-status occupations. (Of course, in a serious analysis, I would need to consider the possibility that those with better verbal skills, relative to their education, would be more likely to end up in high-status occupations.)

The conventional approach to representing these effects is to estimate an equation of the form

$$\hat{V} = a + b(E) + \sum_{i=2}^{4} c_i O_i \tag{7.40}$$

where V is the vocabulary score, E is the years of school completed, and the O_i are the occupation categories with, say, $O_1 = 1$ for lower manual workers and $= 0$ otherwise, … , $O_4 = 1$ for upper nonmanual workers and $= 0$ otherwise. The top panel of Table 7.6 shows the "design matrix" (also known as the "model matrix") for the conventional coding of dummy variables to represent the separate effects of each occupation category; the resulting coefficients are shown in the top panel of Table 7.7.

As you can see, there are no surprises here. As expected, vocabulary knowledge increases with education and also increases monotonically with occupational status. However, let us now see how to represent the effects of occupational status in other, mathematically equivalent, ways.

Effect Coding

One possibility is to highlight the effect of each occupation category by contrasting it with the unweighted average of the effects of all categories. If we include a single categorical variable in the model, represented by a set of $k - 1$ trichotomous variables each coded -1 for the omitted category, 1 for the ith category, and 0 otherwise (refer to the second panel of Table 7.6), the resulting regression coefficients give the difference between the mean on the dependent variable for the specific category and the unweighted average of the means for all categories. That is,

$$e_i = \bar{Y}_i - \bar{Y} \tag{7.41}$$

where $\bar{Y}$ is the unweighted average of the sample means for each category, averaged over all categories of the classification; $\bar{Y}$ is also the intercept of the equation. The coefficient for the omitted category is just the negative sum of the coefficients for the $k - 1$ explicitly included categories, although in these days of high-speed computing, it usually is easier to simply change the omitted category. When other variables are included in the

TABLE 7.6. Design Matrices for Alternative Ways of Coding Categorical Variables (See Text for Details).

	Lower Manual	Upper Manual	Lower Nonmanual	Upper Nonmanual
Conventional Dummy Variable Coding				
D_2	0	1	0	0
D_3	0	0	1	0
D_4	0	0	0	1
Effect Coding				
E_2	−1	1	0	0
E_3	−1	0	1	0
E_4	−1	0	0	1
Contrast Coding				
C_1	.5	.5	−.5	−.5
C_2	1	−1	0	0
C_3	0	0	1	−1

regression equation, the same relationships hold except that now we have adjusted means rather than actual means.

As noted, the coding of categories that produces this outcome is shown in the second panel of Table 7.6. As always, we have $k - 1$ variables to represent k categories. The reference category is coded −1 on all the indicator variables (the set of variables used to represent the categorical classification). The remaining categories are successively coded 1 and 0 on the remaining indicator variables to create a contrast between the omitted category and each of the other categories while minimizing the in uence of the remaining categories.

Inspecting the coefficients in the second panel of Table 7.7, we see that the unweighted mean of the means for the four occupation categories is 2.48 and that lower manual workers have substantially lower than average vocabulary scores, net of education; upper manual

TABLE 7.7. Coefficients for a Model of the Determinants of Vocabulary Knowledge, U.S. Adults, 1994 (N = 1,757; R^2 = .2445; Wald Test That Categorical Variables All Equal Zero: $F_{(3,1752)}$ = 12.48; p < .0000).

	Coefficient	Standard Error	*p*-Value
Conventional dummy variable coding			
Education	0.277	0.018	0.000
D_1	0	0	–
D_2	0.226	0.154	0.142
D_3	0.520	0.108	0.000
D_4	0.764	0.134	0.000
Intercept	2.105	0.220	0.000
Effect coding			
Education	0.277	0.018	0.000
E_1	−0.377	0.076	0.000
E_2	−0.152	0.108	0.160
E_3	0.143	0.070	0.041
E_4	0.386	0.089	0.000
Intercept	2.482	0.239	0.000
Contrast coding			
Education	0.277	0.018	0.000
C_1	−0.529	0.106	0.000
C_2	−0.226	0.154	0.142

(Continued)

TABLE 7.7. **Coefficients for a Model of the Determinants of Vocabulary Knowledge, U.S. Adults, 1994 ($N = 1{,}757$; $R^2 = .2445$; Wald Test That Categorical Variables All Equal Zero: $F_{(3,1752)} = 12.48$; $p < .0000$).** *(Continued)*

	Coefficient	Standard Error	*p*-Value
C_3	−0.244	0.120	0.042
Intercept	2.482	0.239	0.000
Sequential coefficients			
Education	0.277	0.018	0.000
S_2	0.226	0.154	0.142
S_3	0.295	0.154	0.056
S_4	0.243	0.119	0.042
Intercept	2.105	0.220	0.000

workers have somewhat lower than average scores; lower nonmanual workers have somewhat higher than average scores; and upper nonmanual workers have substantially higher than average scores. Note that the differences between occupation categories are identical (within rounding error) in the two parameterizations and that neither the effect of education nor the R^2 is affected. This reparameterization is likely to be most useful when the categorical variable includes a large number of response categories, no one of which is a particularly useful reference category. Note also the contrast between this parameterization and that discussed in the previous section, which shows coefficients as deviations from the *weighted* average of the subgroup means. Here the coefficients are deviations from the *unweighted* average. Both are appropriate, and each may be useful under certain circumstances.

Contrast Coding

Sometimes we may want to compare the effects of subsets of variables. For example, we may want to contrast nonmanual and manual workers, and then to contrast the two nonmanual categories and the two manual categories. We can do this by constructing a set of contrasts of the subgroup means. That is, we form a set of contrasts of the form

$$C = \sum_{i=1}^{k} a_i \bar{Y}_i \tag{7.42}$$

subject to the constraints that the a_i sum to zero; that $k - 1$ contrasts be formulated to represent k categories; and that the codes for each pair of contrasts be linearly independent or, to put it differently, that the contrast codes be orthogonal—which condition is satisfied when, for each pair of contrasts, the sum of the products of the codes = 0.

A set of contrast codes is shown in the third panel of Table 7.6. Note that they satisfy all three constraints mentioned in the previous paragraph: three contrast variables are used to represent the four occupation categories; each row sums to zero; and the sum of the products of the codes = 0 in all cases (for example, for C_2 and C_3 we have $.5*1 + .5*(-1) + (-.5)*0 + (-.5)*0 = 0$; and similarly for C_2 and C_4 and for C_3 and C_4). This coding, plus a little further computation on the regression output, explained next, yields the coefficients shown in the third panel of Table 7.7.

Note that the intercept gives the unweighted mean of the category means, just as for effect coding, but the coefficients of the indicator variables have a somewhat different interpretation, which requires a little additional manipulation. Each contrast, j, gives the difference in the unweighted means of the means for the categories in the two groups being contrasted and is computed by

$$C_j = b_j \frac{n_{g1} + n_{g2}}{(n_{g1})(n_{g2})} \tag{7.43}$$

where n_{g1} is the number of categories included in the first group, n_{g2} is the number of categories included in the second group, and b_j is the regression coefficient for the contrast-coded dummy variable. Note that the standard errors also must be multiplied by the same factor as the regression coefficients.

Inspecting the contrast coefficients shown in the third panel of Table 7.7, we see that the manual groups average about a half point below the nonmanual groups in their mean vocabulary scores, which is highly significant; that upper manual workers average about a quarter point higher than lower manual workers in their vocabulary scores, but that this difference is significant only at the .14 level, which gives us little confidence that there is a true difference between these categories; and that upper nonmanual workers likewise average about a quarter point higher than lower nonmanual workers, and that this difference is significant at the 0.04 level, which means that we can have modest confidence that there is a true difference between these categories.

Sequential Coefficients

One additional way of presenting coefficients for categorical variables is sometimes helpful. When the underlying dimension is ordinal, or we want to treat it as ordinal, it may be useful to reexpress the coefficients as indicating the difference between each category and the preceding category. To do this is a simple matter of estimating the equation using conventional dummy variable coding but then subtracting each coefficient from the preceding coefficient. If we have k categories and omit the first one, then k_2 remains unchanged ($k'_2 = k_2 - 0$); $k'_3 = k_3 - k_2$; and so on. The appropriate standard errors for each coefficient are then the standard errors of the difference from the preceding coefficient. Again, the standard error for k_2 remains unchanged, and the standard errors of the remaining variables may either be computed by hand from the variance-covariance matrix of coefficients using the denominator of the formula shown in the boxed note, "How to Test the Significance of the Difference Between Two Coefficients," in Chapter Six, or computed simply by reestimating the regression equation, successively altering the omitted category.

Here we see (in the fourth panel of Table 7.7) that the differences between adjacent occupation categories are each about one-quarter of a point on the vocabulary scale and that all but the first of the differences is significant at conventional levels. Note also that the contrasts between the first and second categories (lower and upper manual workers) and between the third and fourth categories (lower and upper nonmanual workers) are identical (within limits of rounding error) to contrasts 2 and 3 of the previous panel—which, of course, must be so because the same categories are being contrasted in both cases.

DECOMPOSING THE DIFFERENCE BETWEEN TWO MEANS

A common problem in social research is to account for why two (or more) groups differ with respect to their average score or value on some variable. For example, we may observe that Blacks and non-Blacks differ with respect to their average earnings and may wonder how this comes about. In particular, we may wonder to what extent the difference arises from group differences in their "assets," the traits that enhance earnings, and to what extent they arise because the groups get different "rates of return" to their assets—that is, some groups gain a greater advantage from any level of assets than do other groups. Consider education, for example. To what extent do earnings differences between Blacks and non-Blacks arise because Blacks tend to be less well educated than non-Blacks, and to what extent do they arise because Blacks get a lower return to their education than do non-Blacks? A natural way to investigate the determinants of any outcome of interest is to regress the outcome on a set of suspected determinants and note the relative size of the coefficients associated with each independent variable. A natural extension of this approach for the purpose of comparing two groups is to compute parallel regressions for the two groups of interest, to subtract one equation from the other, and to note the size of the resulting differences.

Consider the following equations:

$$\hat{Y}_1 = a_1 + \sum_{i=1}^{k} b_{i1}X_{i1} \tag{7.44}$$

and

$$\hat{Y}_2 = a_2 + \sum_{i=1}^{k} b_{i2}X_{i2} \tag{7.45}$$

which represent some model with k independent variables estimated for groups 1 and 2. Because regression equations pass through the mean of each variable, it follows that

$$\bar{Y}_1 = a_1 + \sum_{i=1}^{k} b_{i1}\bar{X}_{i1} \tag{7.46}$$

and

$$\bar{Y}_2 = a_2 + \sum_{i=1}^{k} b_{i2}\bar{X}_{i2} \tag{7.47}$$

Then, taking the difference between (7.46) and (7.47), we have

$$\begin{aligned}\bar{Y}_1 - \bar{Y}_2 &= (a_1 + \sum b_{i1}\bar{X}_{i1}) - (a_2 + \sum_i b_{i2}\bar{X}_{i2}) \\ &= (a_1 - a_2) + \sum (b_{i1} - b_{i2})\bar{X}_{i2} + \sum b_{i2}(\bar{X}_{i1} - \bar{X}_{i2}) \\ &\quad + \sum (b_{i1} - b_{i2})(\bar{X}_{i1} - \bar{X}_{i2})\end{aligned} \tag{7.48}$$

(You can work out the equality for yourself. It is easier if you start with the expanded equation and derive the simple difference.)

Equation 7.48 also can be written as

$$\begin{aligned}\bar{Y}_1 - \bar{Y}_2 &= (a_1 - a_2) + \sum (b_{i1} - b_{i2})\bar{X}_{i1} + \sum b_{i1}(\bar{X}_{i1} - \bar{X}_{i2}) \\ &\quad + \sum (b_{i1} - b_{i2})(\bar{X}_{i2} - \bar{X}_{i1})\end{aligned} \tag{7.49}$$

(Again, you can convince yourself of this by working out the algebra.)

Equations 7.48 and 7.49 represent alternative decompositions of the difference between two means into the difference between the intercepts, the slopes, the means on the independent variables, and the interactions between the differences in slopes and differences in means. In Equation 7.48, Group 2 is used as a standard. Hence, the effect of the difference in slopes is evaluated at the mean for Group 2, and the effect of the difference in means is evaluated with respect to the slope for Group 2. In Equation 7.49, Group 1 is taken as the standard. These equations generally yield different answers, and there usually is no obvious way to decide between them. Hence, it is a good practice to present both sets of decompositions, as I do here. Differences in interpretation associated with use of the two standards will be discussed shortly.

In both these decompositions the coefficients representing the effect of the difference in means and the interaction are unchanged when a constant is added to or subtracted from the independent variables, but the coefficients representing the effect of the difference in intercepts and the difference in the rate of return to the independent variables do depend on the scaling of the variables (Jones and Kelley 1984). For this reason, it generally is advisable to combine these two terms. Doing so yields three components. From Equation 7.48 we have

$\bar{Y}_1 - \bar{Y}_2$	**Actual:** Observed group difference.
$\sum b_{i2}(\bar{X}_{i1} - \bar{X}_{i2})$	**Composition:** Portion due to differences in assets.
$(a_1 - a_2) + \sum (b_{i1} - b_{i2})\bar{X}_{i2}$	**Rate:** Portion due to differences in the rates of return to assets (that is, the difference remaining if assets were equalized).
$\sum (b_{i1} - b_{i2})(\bar{X}_{i1} - \bar{X}_{i2})$	**Interaction:** Portion due to valuing the differences in assets at the Group 2 rate of return rather than the Group 1 rate of return.

Equation 7.49 can, of course, be reorganized in the same way. Note that in Equations 7.48 and 7.49, the interaction terms have the same absolute value but opposite sign, which follows from the definition of the interaction term.

A Worked Example: Factors Affecting Racial Differences in Educational Attainment

Now let us consider a substantive problem to see what to do with these coefficients. Suppose we are interested in studying the factors affecting racial differences in the level of education completed. It is well known that on average Blacks have less education than do members of other races. Data from the GSS show that over the period 1990 to 2004, Blacks averaged about a year less in schooling than did others (12.8 years compared to 13.7 years for others). What factors might account for this difference?

To study this question, it is necessary to obtain a large enough sample of Blacks to yield reliable estimates. Although no one year of the GSS has enough Black respondents, pooling all years from 1990 to 2004, excluding 2002 (in which the race question was asked in a nonstandard way and thus the data are not comparable to those from other years), yields a sample of 2,105 Blacks with complete information on all variables. I thus pooled data from these years of the GSS, divided the sample into Blacks and non-Blacks, and estimated for each sample the regression of years of school completed on mother's years of schooling, number of siblings, and whether the respondent resided in the South at age sixteen. I chose these variables for study because they are known to affect educational attainment: mother's education is a measure of family cultural capital and is superior to father's education in the Black population due to the relatively large number of female-headed households—the higher the level of mother's education, the higher the expected level of respondent's education; the number of siblings is an indicator of the share of parental resources that can be devoted to any single child—the larger the number of siblings, the lower the expected level of educational attainment; and Southern residence at age sixteen is an indicator of inferior schooling—those who grew up in the South are expected to obtain less education than those who grew up in other parts of the country.

WHY BLACK VERSUS NON-BLACK IS BETTER THAN WHITE VERSUS NON-WHITE FOR SOCIAL ANALYSIS IN THE UNITED STATES When studying racial differences in the United States, it is reasonable to divide the population into Blacks and non-Blacks. Non-Blacks, of course, will be mainly White (in the GSS from 1990 to 2004 94 percent of non-Blacks are White). Including "Others" (those who are neither Black nor White—and, in fact, are mostly Asian) with Whites has little effect on the estimates but has the advantage of retaining the entire population rather than arbitrarily studying most but not all of the population. Including "Others" with Blacks makes less sense, both because "Others" are more similar to Whites with respect to most social characteristics and because they would constitute a larger fraction of the "non-White" population than of the "non-Black" population, thus making the category less homogeneous.

The question at issue is to what extent the observed nearly one-year difference in the average education of Blacks and non-Blacks is due to racial differences in the average level of mother's schooling, the average number of siblings, and the probability of living in the South, and to what extent the difference is due to the lower rates of return to Blacks from having educated mothers, coming from small families, and living outside the South. I start by estimating an equation of the form

$$\hat{E} = a + b(E_M) + c(S) + d(R) \tag{7.50}$$

separately for Blacks and non-Blacks, where E = years of school completed; E_M = years of school completed by the mother; S = number of siblings; and $R = 1$ if the respondent lived in the South at age sixteen and = 0 otherwise.

A COMMENT ON CREDIT IN SCIENCE In Stata the decomposition is carried out using an `-ado-` file, `-oaxaca-`, which can be downloaded from the Web: type "`-net search oaxaca-`", then click the entry for oaxaca. The name of the `-ado-` file is a telling reflection of the sociology of science. The decomposition technique was introduced by the demographer, Evelyn Kitagawa, in 1955 and was elaborated in a number of ways over the years by demographers and sociologists—see the section on "Additional Reading on Decomposing the Difference Between Means" later in the chapter. However, it was only when an economist, Ronald Oaxaca (1973), used the procedure that it gained general currency among economists. It is now known as the "Oaxaca decomposition" or the "Blinder-Oaxaca decomposition," due to a somewhat clearer exposition by another economist, Alan Blinder (1973).

Table 7.8 shows the means, standard deviations, and correlations among the variables included in the equation, separately for Blacks and non-Blacks. From the table we see that Blacks come from much larger families, are much more likely to have been raised in the South, and that both respondents and their mothers average nearly a year less schooling than non-Blacks. Table 7.9 shows the regression estimates, and Table 7.10 shows the decomposition.

For Blacks in the 1990 to 2004 GSS pooled sample, the estimated values for Equation 7.50 are

$$\hat{E} = 11.09 + .220(E_M) - .071(S) - .512(R) \quad (7.51)$$

whereas for non-Blacks the estimated values are

$$\hat{E} = 10.76 + .308(E_M) - .138(S) - .488(R) \quad (7.52)$$

Table 7.9 gives the regression coefficients for the two equations, together with standard errors. This table shows that the main differences between the determinants of education for Blacks and non-Blacks are, first, that the cost of coming from a large family is substantially greater for non-Blacks than for Blacks and, second, that the advantage of mother's education is greater for non-Blacks than for Blacks. Interestingly, the effect of

TABLE 7.8. Means, Standard Deviations, and Correlations for Variables Included in a Model of Educational Attainment for U.S. Adults, 1990 to 2004, by Race (Blacks Above the Diagonal, Non-Blacks Below).

	(E)	(E_M)	(S)	(R)	Mean	S.D.
(E) Years of school	–	0.350	−0.186	−0.168	12.8	2.68
(E_M) Mother's years of school	0.417	–	−0.274	−0.201	10.6	3.73
(S) Number of siblings	−0.232	−0.268	–	0.124	4.96	3.45
(R) Lived in South at 16	−0.102	−0.065	0.011	–	0.559	0.497
Mean	13.7	11.4	3.33	0.270	N = 2,105	
Standard deviation	2.83	3.46	2.67	0.444	N = 14,985	

TABLE 7.9. **Coefficients of a Model of Educational Attainment, for Blacks and Non-Blacks, U.S. Adults, 1990 to 2004.**

	Metric Coefficients (Standard Errors)	
	Blacks	Non-Blacks
Mother's years of school	0.220 (0.015)	0.308 (0.006)
Number of siblings	−0.071 (0.016)	−0.138 (0.008)
Lived in South at 16	−0.512 (0.112)	−0.488 (0.047)
Constant	11.09 (0.22)	10.76 (0.09)
R^2	0.140	0.195

growing up in the South differs little for the two races, which represents an important change from the past. Finally, less than a fifth of the variance in education is explained by the three variables in the model for non-Blacks and less than a sixth for Blacks, both substantially less than in the past.

The coefficients in Table 7.9 do not, however, permit a formal comparison of the determinants of educational attainment for Blacks and others. To see this we turn to Table 7.10, which gives a decomposition of the almost one-year difference in the average years of school completed by Blacks and non-Blacks. Decomposition 1, which takes Blacks as the standard, is constructed from Equation 7.48, whereas Decomposition 2, which takes non-Blacks as the standard, is constructed from Equation 7.49. In both cases non-Blacks are taken as group 1, and Blacks as group 2. So the decomposition is of the approximately one-year advantage in the average schooling of non-Blacks compared with Blacks. Both decompositions suggest that differences in assets—the fact that non-Blacks have better-educated mothers, fewer siblings, and are less likely to live in the South—are more important than differences in returns to assets. But the two decompositions differ in the contribution they assign to differences in maternal education and number of siblings, both of which are more important in Decomposition 2 than in Decomposition 1. The reason for this is straightforward: when Blacks are taken as the

TABLE 7.10. Decomposition of the Difference in the Mean Years of School Completed by Non-Blacks and Blacks, U.S. Adults, 1990 to 2004.

	Decomposition 1 (Black Standard)	Decomposition 2 (Non-Black Standard)
Total difference	0.89	
Differences in assets		
Mother's education	0.17	0.24
Number of siblings	0.11	0.22
Lived in South at 16	0.15	0.14
Total due to difference in assets	0.44	0.61
Differences in returns to assets		
Mother's education	0.93	1.00
Number of siblings	−0.34	−0.23
Lived in South at 16	0.01	0.01
Intercept	−0.33	−0.33
Total due to differences in returns	0.28	0.46
Interactions		
Mother's education	0.07	−0.07
Number of siblings	0.11	−0.11
Lived in South at 16	−0.01	0.01
Total due to interactions	0.17	−0.17

standard—that is, when the Black/non-Black differences in slopes is evaluated at the Black mean—the difference in expected values is smaller in both cases than when the difference in slopes is evaluated at the non-Black mean. (To convince yourself of this, sketch a graph of the two slopes for each of the two variables.) Finally, the interaction terms have relatively little importance in this decomposition because of the offsetting effects of number of siblings and Southern residence.

Additional Reading on Decomposing the Difference Between Means

For a good sense of how to carry out more complicated decompositions and what the interpretative issues are, read the papers by Duncan (1968), Winsborough and Dickinson (1971), Kaufman (1983), Treiman and Roos (1983), Jones and Kelley (1984), Kraus (1986), Treiman and Lee (1996), and Treiman, McKeever, and Fodor (1996).

WHAT THIS CHAPTER HAS SHOWN

In this chapter we have covered various elaborations of multiple regression procedures that give us improved ability to represent social processes and thus to test ideas about how the social world works. Specifically, we have considered nonlinear transformations of both dependent and independent variables; ways to test the equality of coefficients within an equation; how to assess the assumption of linearity in a relationship; how to construct and interpret linear splines to represent abrupt changes in slopes; alternative ways of expressing dummy variable coefficients; and a procedure for decomposing the difference between two means. Several of the worked examples have focused on trends over time, which gives us a model for how to use multiple regression procedures to study social change.

In the next chapter we return to perhaps the most vexing problem in nonexperimental social research—missing data on some but not all variables—and consider what is currently regarded as the gold standard for dealing with missing data: multiple imputation of missing values.

CHAPTER

MULTIPLE IMPUTATION OF MISSING DATA

WHAT THIS CHAPTER IS ABOUT

In this chapter we consider issues involved in the treatment of missing data, review various methods for handling missing data, and see how to use a state-of-the-art procedure for imputing missing data to create a complete data set, the method of *multiple imputation*. For a very useful overview of imputation methods, including multiple imputation, see Paul and others (2008), upon which this discussion draws heavily. Other useful reviews of the literature on missing data treatments include Anderson, Basilevsky, and Hum (1983), Little (1992), Brick and Kalton (1996), and Nordholt (1998).

INTRODUCTION

Missing data is a vexing problem in social research. It is both common and difficult to manage. Most survey items include nonresponse categories: respondents do not know the answers to some questions or refuse to answer; interviewers inadvertently skip questions or record invalid codes; errors are made in keying data; and so on. Administrative data, hospital records, and other sorts of data have similar problems—invalid or missing responses to particular items. Where information is missing because it is not applicable to particular respondents (for example, age at marriage for the never married), there is no problem; the analytic sample is simply defined as those "at risk" of the event. But in the remaining cases, in which in principle there could be a response, we need special procedures to cope with missing information.

In the statistics literature on missing data (Rubin 1987; Little and Rubin 2002), a distinction is made between three conditions: *missing completely at random* (MCAR), the condition in which missing responses to a particular variable are independent of the values of any other variable in the explanatory model and of the true value of the variable in question; *missing at random* (MAR), the condition in which missingness is independent of the true value of the variable in question but not of at least some of the other variables in the explanatory model; and *missing not at random* (MNAR) or, alternatively, *nonignorable* (NI), in which missingness depends on the true value of the variable in question and, possibly, on other variables as well.

Note that these distinctions refer to net effects. Thus, for example, if the probability that data are missing on the father's education is independent of the true value of the father's education after account is taken of the respondent's education but depends on the respondent's education, the data would satisfy the MAR condition. The fact that the typology refers to net rather than gross effects is very important because otherwise it is difficult to think of variables that satisfy the MAR condition. For example, it is likely that missingness on the father's education is *correlated* with the true value of the father's education simply because the father's and the respondent's education are correlated, and lack of knowledge of the father's education is greater among the poorly educated.

Unfortunately, at least in cross-sectional data, there is no way to empirically determine whether missingness is independent of the true value of the variable; this must be defended on theoretical grounds. Although it is likely that missingness is seldom completely independent of the true value of the variable, there are many cases in which it is plausible to assume that it is largely independent, net of the other variables in the explanatory model. These are the cases that concern us here.

The NI condition is often discussed under the rubric of *sample selection bias*, the situation where the sample is selected on the basis of variables correlated with the dependent variable. This topic is beyond what can be included in this book (but see Chapter Sixteen for a brief introduction). Accessible discussions of the issues involved in sample selection bias and possible corrections can be found in Berk and Ray (1982), Berk (1983), Breen (1996), and Stoltzenberg and Relles (1997).

Next we review a number of procedures for dealing with missing data, culminating in a discussion of *Bayesian multiple imputation*, the current gold standard, and presentation of a worked example using this method.

Casewise Deletion

The most commonly used method for dealing with missing data (which we have employed so far in this book) is simply to drop all cases with any missing data on the variables included in the analysis. If data mainly are missing completely at random, due to recording, keying or coding errors, or omission by design (the question is asked only of a random subset of the sample), the main cost is to reduce the sample size. This is bad enough because often the reduction in sample size is quite dramatic. For example, Clark and Altman (2003) reported a study of prognosis of ovarian cancer in which missing data on 10 covariates reduced the sample size by 56 percent, from 1,189 to 518.

WHY PAIRWISE DELETION SHOULD BE AVOIDED

Sometimes, to avoid substantial reductions in their sample size, analysts base their analysis on "pairwise-present" correlations—that is, correlations computed from all data available for each pair of variables. This is a bad idea because it can produce inconsistent, and often uninterpretable, results, especially when hierarchical models are contrasted, of the kind discussed in the section of Chapter Six on "A Strategy for Comparisons Across Groups."

However, the problem usually is much worse because data are not missing completely at random. Rather, the presence or absence of data on particular variables tends to depend on the value of other variables. For example, as noted previously, poorly educated people are less likely to know about their family histories, and hence their parents' characteristics, than are well-educated people; the refusal to answer certain kinds of questions, for example, those invoking political attitudes, may vary with political party affiliation; self-employed businessmen may refuse to divulge their income for fear that the information will wind up in the hands of the tax authorities; and so on. In such cases, coefficients estimated using casewise deletion are generally biased. Thus, to simply omit missing data is to risk seriously distorting our analysis.

Case deletion (also known as *listwise deletion*) is also appropriate when the model is perfectly specified, and the value of the dependent variable is not affected by the missingness of data on any of the independent variables (Paul and others 2008). But perfectly specified models are virtually unknown in the social sciences. The *mean imputation with dummy variables* method discussed a bit later provides a test of the dependence of the dependent variable on the missingness of the independent variable or variables; but we still are left with the problem of imperfect model specification. One circumstance in which case deletion *is* appropriate is when a question is asked only of a random subset of a sample because then the subset is still a probability sample of the population. But even here, there usually is a heavy cost to pay in terms of reduction in sample size.

Weighted Casewise Deletion

A similar approach, which is possible where the population distribution of some variables is known or can be accurately estimated (for example, from a census or high-quality survey), is to drop cases with any missing data but then to weight (or reweight) the sample so that it reflects the population distribution with respect to known variables, such as age, sex, ethnicity, education, and geographical distribution. The U.S. Census Bureau and a number of sample survey houses do this to correct their sample surveys for differential nonresponse, but the method also has been used to correct for item nonresponse. If the substantive model is perfectly specified, the method can result in unbiased estimates, albeit with inflated standard errors. In addition, weights that depart substantially from unity also inflate the standard errors. (Stata's `-pweight-` option provides correct standard errors when this method is used, but the standard errors typically will be inflated relative to standard errors for unweighted data.) However, because our models are virtually never perfectly specified, the validity of the procedure depends on how closely perfect specification is approximated, which requires a judgment on the part of the analyst.

Mean Substitution

Various methods for imputing missing data (rather than dropping cases) have been proposed. (The mean substitution methods provide a way to generate complete data with respect to the explanatory variables. In these methods, the dependent variable is not imputed; doing so would amount to artificially inflating the strength of the association by adding cases on the regression line.) Early studies often simply substituted the mean or mode of the nonmissing values, but this procedure is now regarded as entirely inadequate because doing so without further correction produces biased coefficients in regression models even under the MCAR condition (Little 1992) and also produces downwardly biased standard deviations of each distribution containing imputed data and hence downwardly biased standard errors and confidence intervals of calculated statistics.

Another approach, which has been widely used in the social sciences, is the *missing indicator* method: for each independent variable with substantial missing data, the mean (or some other constant) is substituted, and a dummy variable, scored 1 if a value has been substituted and scored 0 otherwise, is added to the regression equation. An advantage of the method is that it provides a test of the MCAR assumption: if any of the dummy variables has a (significantly) nonzero coefficient, the data are not MCAR. Cohen and Cohen (1975, 274), early proponents of this method, claim that it corrects for the nonrandomness of missing data. However, Jones (1996) has shown that it and related methods (for example, adding a category for missing data when a categorical variable has been converted to a set of dummy variables) produce biased estimates.

A final mean substitution method is *conditional mean imputation*, in which missing values are replaced by predicted values from the regression of the variable to be imputed (for the subset of cases with observations on that variable) on other variables in the data set; this is the method implemented by Stata 10.0 in its `-impute-` command. This method also results in (typically downwardly) biased coefficients and underestimated standard errors.

All the mean imputation methods suffer from the problem of over-fitting. Because missing data are replaced by a predicted value, the completed data set does not adequately represent the uncertainty in the process being studied—the error component in the values for each individual. This is manifest in standard errors that are too small, even in cases where the coefficients themselves are close to being unbiased. Because of this, another class of imputation methods builds in uncertainty. There are two main ways of doing this: *hotdeck imputation* and Bayesian multiple imputation. (A computationally less intensive modification of Bayesian multiple imputation, the *approximate Bayesian bootstrap* proposed by Rubin and Schenker [1986; Rubin 1987] is not considered here.)

Hotdeck Imputation

This is the method used by the U.S. Census Bureau to construct complete data public use samples. The sample is divided into strata (similar to the strata used in the weighted casewise deletion and conditional mean imputation methods). Then each missing value within a stratum is replaced with a value randomly drawn (with replacement) from the observed cases within the stratum. As a result, within each stratum the distribution of values for the imputed cases is (within the limits of sampling error) identical to the distribution of values for the observed cases. When the imputation model is correctly specified (that is, when all variables correlated with the missingness of values on a given variable are used to impute the missing values), this method produces unbiased coefficients but biased standard errors. It also tends to perform poorly when a substantial fraction of cases have at least one missing value (Royston 2004, 228).

Full Bayesian Multiple Imputation

This method, introduced by Rubin in 1987, is the current gold standard (but see Paul and others [2008] for a skeptical, or at least cautionary, view of the superiority of multiple-imputation in practice). Little and Rubin (2002) is the classical exposition of the method, but Schafer (1997, 1999) provides more accessible introductions, as does Allison (2001). For an early applications, see Treiman, Bielby, and Cheng (1988), and Schenker, Treiman, and Weidman (1993).

The essence of multiple imputation is this. First, equations are estimated that predict each variable with missing data from other variables in the data set. Then values drawn at random from the predicted distribution are substituted for the missing values. Because variables with missing data may be among the predictors for another variable with missing data, the process is repeated several times, cycling through the prediction equations and using updated values for the predictor variables. This procedure results in a complete data set, with the missing data imputed. Several such complete data sets are created—the convention is five, but there is some evidence that more may be helpful (Royston 2004, 236–240).

Each of these data sets is then analyzed in the usual way, and the resulting coefficients are averaged or otherwise combined, using what have become known as "Rubin's rules" (Rubin 1987, 76). This method produces unbiased coefficients and also, by taking account of the additional uncertainty created by the imputation process, unbiased standard errors. Specifically, the standard error of a coefficient based on M imputations is given by

$$S_b = \sqrt{\frac{1}{M}\sum_{m=1}^{M} s_m^2 + \left(1+\frac{1}{M}\right)\left(\frac{1}{M-1}\right)\sum_{m=1}^{M}(b_m - \bar{b})^2} \qquad (8.1)$$

That is, the standard error is estimated as the average of the standard errors based on each imputation (the leftmost term), which captures the uncertainty in the estimate within each imputation, plus a component for the variation in the estimated coefficients across imputations, which captures the uncertainty introduced by the imputation procedure.

For this procedure to produce proper imputations, two conditions must be satisfied: (1) that the analyst do a good job of predicting missingness, and (2) that if, in the substantive model, missingness is correlated with the outcome variable, the outcome variable be included in the imputation model.

Software that implements multiple-imputation procedures has been written for Stata by Royston (2004, 2005a, 2005b, 2007, building on earlier work by Van Buuren, Boshuizen, and Knook [1999]); to download the necessary `-ado-` files from within Stata (connected to the network), type `-lookup ice-` and click the fourth entry, "sj-7-4." (See also useful guides to using Royston's `-ado-` files written by Academic Technology Services of the University of California at Los Angeles [UCLA]; they will appear in response to the same `-lookup-` command.) Royston's software makes the process less tedious than it used to be. Nonetheless, implementing multiple imputation can add considerable complexity to your analysis. The difficult and time-consuming part of the work is to choose the predictor variables used to produce estimates of the missing values of each variable that has any missing values.

The essence of the procedure is to specify which variables to include, to create appropriate transformations (dummy variables and interactions), and to specify the relationships among the variables. These details are part of Royston's command `-ice-`, which should be built using the `-dryrun-` option so that the logic can be tested before beginning what can be a lengthy computation. The imputation is then carried out, and a data set is saved, consisting of multiple copies of the original data, each of which has complete data because the missing values are imputed. However, in each of the completed data sets, the imputed values generally will differ. This multiple-copy, or multiply imputed, data set can then be used to carry out any analysis, using the command `-micombine-`. This command carries out the specified estimation procedure, for example, multiple regression, using each of the imputed data sets and then combines the resulting coefficients to produce a single coefficient—usually the average of the coefficients estimated from each of the completed data sets—and a standard error that takes account of the additional uncertainty introduced by the imputation procedure (see Equation 8.1).

Typically, construction of the imputed data set is computationally intensive—in the worked example discussed next, it took about 3.5 minutes on my home computer (which has a 2.92 GHz processor)—but analysis using the imputed data set is nearly as fast as analysis using a single data set, typically a matter of seconds. As you increase the number of imputations, the time required to create the imputed data sets increases arithmetically. As you add variables to be imputed, the time required increases at a faster rate. For example, approximately doubling the number of variables to be imputed increased the time by a factor of four.

Perhaps the best way to convey what is entailed in using multiple imputation to create and analyze complete data sets is to carry out an example. This is what follows. The `-do-` and `-log-` files that produced the example are available for downloading. These files contain, just before the imputation step, a discussion of how to specify the `-ice-` command.

A WORKED EXAMPLE: THE EFFECT OF CULTURAL CAPITAL ON EDUCATIONAL ATTAINMENT IN RUSSIA

There is increasing evidence from many nations that the extent to which parents are engaged with the written word—measured by the number of books in the household when the respondent was growing up—is at least as important (and perhaps more important) a determinant of educational attainment as is the amount of formal schooling attained by parents (Evans and others 2005). The reasoning is straightforward: what matters about parental education is not the credentials it brings but the way it affects family life and child rearing. In households where reading is an important activity, children often learn to read at home, enjoy reading, and become good at it, all of which improves their ability to meet the demands of formal schooling. Thus, they tend to do well in school and in consequence to continue their education to advanced levels.

In this example I investigate whether books in the childhood home were important for educational attainment in Russia. I chose Russia for the worked example both because the number of books is probably a good indicator of family reading habits in Russia because the cost of books was very low during the Soviet period (my data pertain to adults surveyed in 1993, just after the collapse of the Soviet Union) and because—as a result of massive casualties during the Second World War—there is an unusually large amount of missing data on parental characteristics in the Russian data set, a national probability sample of 5,002 Russian adults age twenty and over (see Appendix A for details on the data and how to obtain them; see also Treiman and Szelényi 1993, Treiman 1994). The sample is restricted to those age twenty to sixty-nine to avoid understatement of educational attainment by those still in school (fewer than 3 percent of twenty-year-olds were still in school) and differential mortality and morbidity (rendering people unavailable to be interviewed) among those age seventy and older. This reduces the sample to 4,685. In the present analysis, the data are not weighted, although weights are permissible in `-ice-`, because weighting introduces an additional complication in comparing results obtained from a casewise-deleted data set and a multiply imputed data set.

Creating the Substantive Model

I first specify a conventional educational attainment model:

$$\hat{E} = a + b(E_P) + c(E_D) + \sum_{i=2}^{6} d_i(O_i) + e(C) + f(S) + g(M) + h(B) + i(E_pB) \qquad (8.2)$$

where E is the number of years of schooling of the respondent, E_p is the sum of the number of years of schooling of the father and the mother, and E_D is the difference in the number of years of schooling of the father and the mother; the O_i refers to the father's occupational category when the respondent was age fourteen; C is year of birth ("cohort"), which captures any effects of the secular increase in education in Russia over the course of the twentieth century; S is the number of siblings, which is known to negatively affect educational attainment (Maralani 2004; Lu 2005; Lu and Treiman 2008); M is scored 1 for males and 0 for females to test the possibility that the two sexes differ in their average education, something that is true in some places but not others; and, finally, B is an ordinal scale measuring the number of books in the household when the respondent was age fourteen (the categories are none, 1 or 2, around 10, around 20, around 50, around 100, around 200, around 500, and 1,000 or more), and E_pB is a product term that captures a possible interaction between parental education and the number of books—I would expect the size of the home library to be more important when parents are less educated, on the ground that well-educated parents are likely to provide school-relevant skills whether or not there is a family culture of reading but that this is less likely when the parents are poorly educated. That is, I would expect that parental education and parental involvement in reading are to some extent substitutes.

TECHNICAL DETAILS ON THE VARIABLES

- **Parental education.** I specify the sum and difference of the years of schooling of each parent rather than simply including the years of schooling of each parent as a separate variable. It can be shown that the two specifications are mathematically equivalent and that either can be derived from the other. But the specification I used is more readily interpretable because it gives the overall effect of parental education plus any additional effect resulting from a difference in the educational level of the parents.
- **Father's occupation.** The occupational categories are from the six-category version of the Erikson-Goldthorpe-Portocararo (EGP) occupational class scheme, modified by Ganzeboom and Treiman (1996).
- **Number of books.** I explored three specifications of this variable: the ordinal scale, midpoints of the number of books indicated by each category, and the natural log of the midpoint scale. Interestingly, the ordinal scale produced the best fit, probably because the log scale excessively diminished the effect of increases in large home libraries.

The problem for this analysis is that many of the variables in the model have substantial fractions of missing data. Table 8.1 shows the percentage of cases missing for each variable. If I were to simply drop all cases with missing data, I would be left with only 57 percent of the sample (2,661 cases). Moreover, because it is probable that missingness is correlated with other variables in the model, I would be analyzing a nonrandom subset of the original sample, thus completely undercutting the validity of any claim that the analysis characterizes the educational attainment process in late twentieth-century Russia. Evidence that missingness is not random is to be found in a comparison of the means and standard deviations based on the complete-data (casewise deletion) sample (N = 2,661) and the corresponding statistics computed over all observations available for each variable: in the complete-data subsample, the means of the socioeconomic status variables are generally higher, and the standard deviations are generally smaller than when computations are based on all observations available for each variable. Thus, I turn to multiple imputation of the missing data to create a valid complete data set.

TABLE 8.1. Descriptive Statistics for the Variables Used in the Analysis, Russian Adults Age Twenty to Sixty-Nine in 1993 (N = 4,685).

	Mean		SD			
Variable	All Observations	Casewise Deletion (N = 2661)	All Observations	Casewise Deletion (N = 2661)	Number of Nonmissing Observations	% Missing
Years of schooling	12.5	12.9	3.7	3.5	4,633	1.1
Father's years of schooling	8.4	8.5	4.0	3.9	3,880	17.2
Mother's years of schooling	7.4	7.9	4.4	4.3	4,469	4.6
Sum of parental years of schooling	16.0	16.4	7.6	7.5	3,807	18.7
Difference in parental years of schooling	0.8	0.7	3.4	3.2	3,807	18.7
Male	.41	.41	.49	.49	4,685	0.0
Number of siblings	2.2	2.1	2.0	1.9	4,219	10.0
Year of birth	1951	1953	13.7	13.1	4,685	0.0
Number of books in family home	150	172	245	259	4,305	8.1
Father's occupation	%[a]					
Professionals and managers	22.7	23.6				
Routine nonmanual	2.6	2.7				
Self-employed	1.7	1.6				
Skilled manual	31.8	32.5				
Semi- and unskilled manual	20.2	19.6				
Agricultural	21.0	20.0				
Total	**100.0**	**100.0**			3,265	30.3

[a]Distribution of cases for which responses are not missing.

Creating the Imputation Model

For each of the variables with any missing data, it is necessary to specify an imputation model—that is, a model predicting values on the variable from the cases for which observations are available. Van Buuren, Boshuizen, and Knook (1999, 687) suggest that although in principle the larger the number of variables in the imputation model the better, in practice (to avoid multicollinearity and computational problems), it is best to limit the predictor set to fifteen to twenty-five variables. They propose as criteria for inclusion:

1. Include (as predictors for each variable with missing data) all variables that will be included in the substantive (complete-data) model.
2. In addition, include (as predictors for a given variable with missing data) all variables thought to affect the missingness of that variable. Such variables can be identified by examining the association between missingness and candidate variables. If the association is not zero or close to zero, include the candidate variable.
3. In addition, include (as predictors for a given variable with missing data) all variables that are strong predictors of that variable. Such variables can be identified by examining the association between the given variable and candidate variables for cases in which the given variable is observed.
4. Remove from sets (2) and (3) those variables that themselves have substantial amounts of missing data.

An intermediate step that I skip in the present exposition is to confirm that the data are not MCAR by predicting missingness from other variables in the data set; if some of the coefficients are nonzero, we have evidence that the data are not MCAR. However, there is no way of deciding empirically whether they are MAR or NI. For each variable, missingness is dichotomous, so the appropriate estimation tool is binary logistic regression. However, because we will not cover this technique until Chapter Thirteen, this part of the worked example is omitted.

In the present case, we need to impute missing data for all variables in Table 8.1 except gender and year of birth (which have no missing data). Following the criteria of Van Buuren and his colleagues, my imputation model for the variables included in the substantive model is

$$
\begin{aligned}
E &= f(E_F, E_M, \Sigma O_i, C, S, M, B, E_p B) \\
E_F &= f(E, E_M, \Sigma O_i, C, S, M, B) \\
E_M &= f(E, E_F, \Sigma O_i, C, S, M, B) \\
O &= f(E, E_F, E_M, C, S, M, B) \\
S &= f(E, E_F, E_M, \Sigma Oi, C, M, B) \\
B &= f(E, E_F, E_M, \Sigma O_i, C, S, M)
\end{aligned}
\qquad (8.3)
$$

where the variables are those included in the substantive model defined in Equation 8.2. I did not have to restrict myself to variables included in the substantive model. Rather, following Van Buuren, Boshuizen, and Knook (1999), I might well have chosen additional variables, which predict the independent variables in the model or predict their missingness or do both; generally, this would be advisable. However, in the interest of keeping the worked example from becoming too complex, I settled for the prediction equations shown.

The `-ice-` command permits the specification of several different estimation models (OLS for continuous variables, and also binary, multinomial, and ordinal logistic regression for categorical variables). Because we have not yet covered logistic regression, I ask you here to take on faith that they are the appropriate techniques for dealing with these sorts of variables; these techniques will be exposited in Chapters Thirteen and Fourteen. As it happens, all variables to be imputed are continuous except for the father's occupation categories, which are imputed using a multinomial logistic regression model, and the number of books in the household, which is imputed using an ordinal regression model.

A very useful feature of `-ice-` is its ability to handle "passively imputed" variables, that is, variables such as interaction terms and sets of dummy variables that are mathematical transformations of other variables, some of which may include missing data. See Royston (2005a, 191–195) for a description of the procedures and the downloadable `-do-` and `-log-` files for the chapter for detailed discussion of how to specify the `-ice-` command.

Comparing Casewise Deletion and Multiple Imputation Results

Table 8.2 shows regression coefficients, standard errors, and t- and p-values for two models, one estimated using casewise deletion and the other estimated from a multiply imputed data set using Royston's command `-micombine-`. Although the results are not greatly different, they do lead to different substantive conclusions regarding three of the twelve variables in the model: if we accept the conventional .05 level of significance, we would conclude that all variables in the OLS model are significant, with the exception of the father's employment in a routine nonmanual job, which results in equal educational chances for offspring as does the father's employment in a managerial or professional job. In particular, we would conclude that it is beneficial when the mother is well educated relative to the father (because, net of the average level of parental education, the more the father's education exceeds the mother's, the lower the educational attainment of offspring). We also would conclude that, as expected, those from large families get less education, and also that males get less schooling than do females. However, none of these three coefficients exceeds the .05 level of significance in the imputed data.

Interestingly, the size of the home library is the most important variable in the model, as indicated by the standardized coefficients, shown in the rightmost column. But, as predicted, its importance diminishes as parental education increases, as is evident from the negative coefficient for the interaction term.

TABLE 8.2. Comparison of Coefficients for a Model of Educational Attainment Estimated from a Casewise-Deleted Data Set [C] (N = 2,661) and from a Multiply Imputed Data Set [M] (N=4,685; 10 imputations), Russian Adults Age Twenty to Sixty-Nine in 1993.

	b		Std. Error		t		p		ß
	C	M	C	M	C	M	C	M	C
Parents' years of school (sum)	.136	.146	.020	.015	6.76	9.43	.000	.000	.29
Parents' years of school (difference)	−.041	−.026	.020	.016	−2.04	−1.61	.042	.108	−.04
Father's occupation (professionals and managers omitted)									
Routine nonmanual	.02	−.17	.41	.40	.04	−.43	.968	.666	
Self-employed	−1.77	−1.90	.52	.51	−3.40	−3.71	.001	.000	
Skilled manual	−.86	−.99	.20	.19	−4.36	−5.33	.000	.000	
Semi- or unskilled manual	−1.10	−1.12	.22	.22	−4.91	−5.04	.000	.000	
Agricultural	−1.41	−1.42	.23	.25	−6.03	−5.72	.000	.000	

Year of birth	−.033	−.014	.006	.005	−5.45	−3.00	.000	.003	−.12
Number of siblings	−.082	−.052	.036	.028	−2.29	−1.86	.022	.063	−.04
Male	−.249	−.168	.125	.098	−2.00	−1.72	.046	.085	
Size of home library at fourteen	.541	.548	.068	.054	7.90	10.07	.000	.000	.35
Parents' years of school (sum) size of home library	−.011	−.012	.004	.003	−2.99	−4.11	.003	.000	−.20
Intercept	12.2	10.9	.376	.309	32.37	35.20	.000	.000	
R^2									.205

WHAT THIS CHAPTER HAS SHOWN

In this chapter we have considered different kinds of missing data, drawing upon a distinction between data that are "missing completely at random" (MCAR), data that are "missing at random" (MAR), and data that are "missing not at random" (MNAR or NI). We explored the properties of each of these missing data types and then considered a number of procedures for handling missing data, including listwise deletion and various forms of imputation of missing values. We determined that most of the procedures reviewed produce biased coefficients in prediction models. This circumstance motivated consideration of multiple imputation procedures, in which missing data are imputed several times and the results of each imputation are combined. Multiple imputation offers the best chance of producing results that are free from bias. We then considered, through a worked example (the role of cultural capital in educational attainment in Russia), how to implement multiple imputation procedures, using software written by the British medical statistician Royston.

Thus far we have carried out statistical inference on the assumption that our data were drawn from a simple random sample. However, most surveys, such as the GSS, are not based on simple random samples but rather on complex multistage probability samples. In the next chapter we consider various sample designs and see how to get correct standard errors when we are using data based on stratified or clustered multistage probability samples or both.

CHAPTER

SAMPLE DESIGN AND SURVEY ESTIMATION

WHAT THIS CHAPTER IS ABOUT

Thus far we have treated the issue of statistical inference as if we were analyzing simple random samples and our data conformed to the distributional properties assumed by ordinary least squares (OLS) regression. Neither condition is likely to hold in practice. Thus, now that we are comfortable with the manipulation and interpretation of regression models, it is time to expand our analytic toolkit to make correct inferences about data based on the kinds of complex samples typically used in national surveys. We also will want to consider how to identify and, if possible, correct for anomalous features of our data. As we will see, these two topics are fairly closely related.

I begin with a description of types of samples used in survey research. I then discuss the problem of statistical inference created by complex sample designs. Chapter Ten then considers various diagnostic procedures for OLS regression models and some ways of correcting problems revealed by the diagnostic procedures.

SURVEY SAMPLES

As we know from elementary statistics, to generalize from a sample to a population we need some sort of *probability sample*. For our purposes, there are three basic kinds of probability samples:

- **Simple random** samples, in which every individual in the population has an equal chance of being included in the sample. (The equal-probability-of-selection condition defines a sample as random.)
- **Multistage probability** samples. These are nothing more than complex random samples in which units are randomly sampled and then subunits of the sampled units are randomly sampled, and so on. Examples include area probability samples in which, say, cities and counties are randomly sampled, then blocks within areas, then households within blocks, then persons within households; and school samples in which, say, school districts are sampled, then schools within districts, then classrooms within schools, then pupils within classrooms.
- **Stratified probability** samples, which are also complex random samples. In stratified samples the population is divided into strata on the basis of certain characteristics (race, sex, place of residence, and so on). A probability sample is drawn within each stratum, with the strata often sampled at different rates—for example, with Blacks sampled at a higher rate than non-Blacks to ensure that there are enough Blacks for analysis.

Simple Random Samples

Let us first consider simple random samples. To draw a random sample requires a list of every individual in the population and a randomizing procedure to select a fraction of the individuals in the population. A typical way of drawing a random sample before the age of computers was to consult a list of random numbers. Table 9.1 shows a small portion of such a list.

Suppose we wanted to draw a random sample of 10 people out of a class of 40, using a table of random numbers such as Table 9.1. We would list the 40 people in the class

TABLE 9.1. **Portion of a Table of Random Numbers.**

10480	15011	01536	02011	81647	91646
22368	46573	25595	85393	30995	89198
24130	48360	22527	97265	76393	64809
42167	93093	06243	61680	07856	16376
...					

sequentially, from 1 to 40. Then we would devise a rule for moving through Table 9.1. We might decide to move through the table row by row and to take the first two and then the last two digits in each five-digit sequence in the table. (This decision rule is, of course, arbitrary. But this does not matter. Because the digits are arranged in random order, any decision rule is as good as any other, so long as it is consistent.) Following this rule, the first number we encounter is 10. Thus, we include the tenth person in our class list. The next number we encounter is 80. Because this is outside the range 1 through 40, we simply skip it. Then we encounter, successively, 15, 11, 1, 36, 2, and 11 again. Strictly speaking we should take the eleventh person twice. But this is very awkward from a practical standpoint. Thus, in carrying out practical work we almost always sample *without replacement*. This has certain statistical consequences, but they are of no importance for populations of the size we generally deal with in survey research. When sampling without replacement we simply skip the eleventh case when we encounter it the second time.

We complete our sample by taking the next four numbers falling into the range 1 through 40 and not previously chosen: 22, 25, 30, and 24. So we have a sample of 10 people, consisting of individuals 1, 2, 10, 11, 15, 22, 24, 25, 30, and 36 in our list of 40 people. This sample is random because by virtue of the way we drew it; each of the 40 people in the population had an equal chance of being included.

In current practice such samples are usually drawn with the assistance of a computer, by constructing a sequential list of all members of a population and randomly selecting from the list. But the principle is exactly the same. Clearly a list of all members of the population is necessary. Obtaining such a list is not always easy and is often impossible. For example, there is no list of the population of the United States.

Sometimes population lists are available, but not in machine-readable form. In such cases *systematic* sampling is sometimes substituted for random sampling. In systematic sampling a random starting point is chosen, and then every kth case is chosen from the list, where k is the ratio of the population size to the desired sample size. In the present example we would simply choose a random starting point, perhaps picking the first case by consulting a table of random numbers, and then we would take every fourth student in the class, wrapping from the end of the list to the beginning if necessary, to get a sample of 10 people from a class of 40.

The properties of systematic samples are, in practice, very similar to those of random samples (Sudman 1976, 56–57), and we generally treat them as if they were random samples. Indeed, in some respects they may be better than random samples because they effectively stratify the sample with respect to whatever characteristics are correlated with the ordering of cases on the list (for example, ethnicity when the population is listed alphabetically; see the discussion later in this chapter, in the section "Stratifying to Offset the Effect of Clustering"). When I discuss simple random sampling, it should be understood that what I say also applies to systematic sampling.

Multistage Probability Samples

Simple random sampling is practical only in limited circumstances—in particular, when a complete list of the population is available and when respondents can be accessed from a central location, by telephone, mail, or the internet. Neither condition applies in the case of national samples of the U.S. population in which face-to-face residential interviews are conducted—that is, interviews in which the interviewer goes to the home of the respondent. First, as just noted, there is no national register of the U.S. population, which makes it impossible to draw a simple random sample of the population. Second, even if it were possible to draw such a sample, it would be prohibitively expensive to travel to the homes of the selected respondents, who would almost certainly be scattered throughout the nation. Thus both sampling and fieldwork considerations lead us to devise multistage probability samples for national household surveys.

Such samples are created in stages (hence the name *multistage* probability samples). In the first stage, *primary sampling units* (cities, counties, and so on) are drawn at random, with probability proportional to size (PPS)—that is, in such a way that the chance of each city being selected is proportional to the size of its population. Suppose we want to draw a national sample of two thousand people and decide to do this by choosing one hundred primary sampling units (PSUs) and conducting twenty interviews in each place. Obviously, we could not simply list all the cities in the country and randomly choose some of them; this would give people in small towns a much higher chance of being included in the sample than people in large cities. For example, if we randomly chose Los Angeles

TELEPHONE SURVEYS Telephone surveys are increasingly widely used because they are less costly than face-to-face interviews. Currently in the United States, a one-hour face-to-face interview in a national probability sample conducted by one of the leading academic survey centers costs about $300 per respondent, compared with about $150 for a phone interview of the same length. Apart from cost, the major advantage of telephone surveys is that it is possible to access households that are very difficult to access in person, such as those in gated neighborhoods or security buildings, and also in high-crime neighborhoods where people are reluctant to answer the door (and interviewers are reluctant to work). On the other hand it is more difficult to establish rapport and to ask complex questions over the telephone, and interviews must be shorter than face-to-face interviews to minimize respondent fatigue. Moreover, in an era in which few people are skilled at reading aloud, it is difficult to find competent telephone interviewers.

A final difficulty with telephone surveys is that respondents are increasingly hostile toward them. Telemarketing has queered the field for legitimate survey research, especially since some telemarketing agencies claim to be doing a survey as a device to draw people into a phone conversation.

Sampling generally is easier with phone interviews than with in-person household interviews because, in principle, random sampling can be used, by means of random digit dialing. However, procedures must be devised to screen out business telephones and to adjust for multiple-telephone households. The proliferation of cell phones, fax machines, and caller-ID screening devices have created new difficulties. Still, since almost all households in the United States have telephones, sample bias is not much of a problem. In many other countries, of course, it would be a major problem.

or Santa Monica or Beverly Hills (the latter two are small cities in Los Angeles County) and then randomly selected twenty people in the chosen city (assuming we had a list of all residents), any person in Santa Monica or Beverly Hills would have a much higher chance of being included in the sample than would any person in Los Angeles.

So instead we group cities into *strata* on the basis of their size and randomly sample cities within strata, at a rate proportional to their size. For example, we might group the largest cities into a stratum, large cities into a second stratum, medium-sized cities into a third stratum, and so on. Suppose the population of the largest group averaged two million, the population of the second group one million, and the population of the third group five hundred thousand, and so on. We might then randomly choose every city in the first group, every other city in the second group, every fourth city in the third group, and so on. If we then interviewed the same number of people from each selected city, every person in the country would have an (approximately) equal chance of being included:

20/2 million = 1/100,000

20/(1 million/0.5) = 1/100,000

20/(500,000/0.25) = 1/100,000

and so on

MAIL SURVEYS Mail surveys are generally undesirable because they tend to produce low return rates, with a bias favoring the better educated. However, well-designed mail surveys can be effective, especially with extensive follow-up emphasizing the importance of the survey (through registered letters, telegrams, phone calls, and so on). Jonathan Kelley and Mariah Evans have achieved amazingly high response rates in mail surveys carried out in Australia—on the order of 65 percent—by doing extensive follow up. They also show that nonrespondents to their surveys are essentially no different from respondents (Evans and Kelley 2004, Chapter 20). Such surveys require a sampling frame that includes addresses. This is impossible in the United States but possible in countries that have registration systems, such as Australia, where voting registration is required. Noncitizens are excluded, but the sampling frame is good otherwise.

Another disadvantage of mail surveys is that one cannot ask complex questions or questions that are contingent on responses to previous questions; respondents have difficulty following the logic of complex contingencies, known as filters. On the other hand, one can ask questions with relatively long lists of alternatives because people can handle more alternatives when they can read and refer back to them than when the items are read to them. A final limitation of mail surveys is that they are vulnerable to being completed by committee—that is, by several members of the household consulting one another. For many topics, this poses little difficulty and may actually be advantageous, as, for example, when life histories are solicited; but where independent responses are required, this is a serious shortcoming.

WEB SURVEYS In recent years Web-based surveys have become increasingly widely used. In some respects Web surveys are like mail surveys in that they eliminate the interviewer and require a respondent to decide to participate and to complete the survey without the benefit of persuasion by a live person, which—when practiced by a skilled interviewer—can overcome trepidation, boredom, irritation, and other impediments to completing the interview. On the other hand, for the computer literate they are easier to complete than paper-and-pencil questionnaires, at least when they are well designed. They also have the advantage over all other modes in permitting complex filters, in which questions are included or omitted depending on responses to previous questions. In both face-to-face and telephone surveys, filters are used, but they are vulnerable to interviewer error. In paper-and-pencil surveys, using filters is difficult because respondent error is likely.

With respect to sample bias, Web surveys today face the same limitations as telephone surveys did in the United States in the first half of the twentieth century: a strong socioeconomic bias in computer access and computer literacy. In addition, there is no known sampling frame of Web addresses that corresponds to a population of people. Moreover, given the current flood of spam and concerted attempts to intercept it through spam filters, efforts to secure responses from a random sample of Web addresses will likely fail. Hence the usefulness of Web-based surveys is likely to be restricted to situations in which there is a well-specified sampling frame (such as a list of members of an organization) and the ability to address survey questionnaires to named individuals with suitable appeals and inducements to respond and assiduous follow-up efforts to convert nonresponses to responses.

The problem with this method is that the strata may be quite heterogeneous. For example, suppose all cities with populations of one million or more are included in the first stratum. Then, if cities were simply chosen at random, residents of Los Angeles would have only one-third the chance of being included in the sample as residents of San Diego, since the population of Los Angeles is about three times the population of San Diego.

To avoid this problem an alternative procedure is often used: within each stratum, units are sampled PPS. To accomplish this, all the units are arrayed in order according to their size, and the total population is cumulated. Then numbers are drawn at random, and units are chosen that include the randomly drawn numbers. For example, suppose we want to sample PPS five of the ten largest cities in California as PSUs, so we can interview one hundred people per PSU. (Here, because of the large variance in the size of the cities, it makes sense either to sample *with replacement* or to divide Los Angeles, and perhaps San Diego, into portions and treat each portion as a separate city. I have done the former.) Table 9.2 shows the population (here, according to the 1990 census), the cumulative population when cities are arrayed by size, and the percentage of the total population of the ten cities residing in each city.

TABLE 9.2. The Population Size, Cumulative Population Size, and Percentage of the Total Population Residing in Each of the Ten Largest Cities in California, 1990.

City	1990 Population	Cumulative Population	Percentage of Total Population of the 10 Cities
Los Angeles	3,485,398	3,485,398	43.2
San Diego	1,110,549	4,595,947	13.8
San Jose	782,225	5,378,172	9.7
San Francisco	723,959	6,102,131	9.0
Long Beach	429,433	6,531,564	5.3
Oakland	372,242	6,903,806	4.6
Sacramento	369,365	7,273,171	4.6
Fresno	354,202	7,627,373	4.4
Riverside	226,505	7,853,878	2.8
Stockton	210,943	8,064,821	2.6

Now we need to choose some random numbers. Going to a convenient random number table at the back of one of my statistics texts and arbitrarily deciding to take the third through ninth number in each row, I get the following:

9,732,533	Beyond the range (ignore)
4,204,805	Choose San Diego (since 4,204,805 falls within the range: 3,485,399 to 4,595,947)
2,268,953	Choose Los Angeles
1,902,529	Choose Los Angeles again
799,970	Choose Los Angeles still again
6,574,717	Choose Oakland

Note that Los Angeles is chosen three out of the five times. (Of course, since the population of Los Angeles is 43 percent of the total population of the ten largest cities in California, we would expect Los Angeles to be chosen about two out of five times on average if we repeated the sampling procedure many times.) We would thus divide Los Angeles into three equally sized sections and treat each of them as a primary sampling unit, together with San Diego and Oakland. By sampling in this way, and repeating the process for smaller units within each primary sampling unit, we ensure that every individual living in the ten cities has an approximately equal chance of being included in the sample, precisely because the chance of the city being included is exactly proportional to the size of the city.

Note that I say "approximately equal." This is because the multistage selection process introduces "lumpiness." Here, for example, each primary sampling unit represents exactly 20 percent of the population, but each city does not contain an exact multiple of 20 percent of the population. Although there always will be some lumpiness, the larger the number of sampling units at each stage, the smaller the problem becomes.

Typically a survey house will use the same primary sampling units repeatedly. For example, the National Opinion Research Center (NORC) changes its primary sampling units every ten years, when the new census data are available (these are needed to determine the population size). NORC does this because it maintains a staff of interviewers in each primary sampling unit and wants to avoid the expense of recruiting and training a new set of interviewers for each survey. The part of a sampling design that is fixed in advance and maintained over time is known as the *sampling frame*.

PHILIP M. HAUSER (1909–1994) was a demographer who spent his entire academic career at the University of Chicago, earning his BA in 1929, his MA in 1933, and his PhD in 1938, all in sociology. He made important organizational and academic contributions to the social sciences, working at the U.S. Bureau of the Census from 1939 to 1947, serving first as Assistant Chief Statistician of the Population Census and eventually as Assistant Director (and as Acting Director from 1949 to 1950). At the Bureau he played a major role in creating the 20 percent sample long form, used in the 1940 census for the first time, as well as methods to reduce the undercount, particularly of Blacks.

At Chicago he published on many topics. Perhaps most notable among his publications was a study of mortality differentials by race and class (Kitagawa and Hauser 1973). He also established the University of Chicago Population Research Center and served as its director for thirty years, training more than a hundred PhDs, many of them from developing nations. He is perhaps the only person to have served as president of three major professional associations in the social sciences: the American Sociological Association, the American Statistical Association, and the Population Association of America.

When sampling large, geographically diverse populations, the selection process typically is repeated several times, for successively smaller units. For example, in a 1996 national sample survey of China (Treiman 1998), we divided the country into urban and rural sectors. Then, within each sector, we sampled counties (or their urban equivalents), with probability proportional to size. Within each of the chosen counties, we sampled townships (or zip-code-sized districts of cities ["streets"]), with probability proportional to size. Then within each of the chosen townships we sampled villages (or neighborhoods of cities), with probability proportional to size.

Once small geographical units are identified—for example, villages in rural China or districts or neighborhoods of cities—there are four standard alternatives for choosing individuals to be interviewed:

- Random selection from a population register
- Random selection from a list of addresses (household samples) and further selection within households
- Random walk procedures (another way of selecting households)
- Quota selection

Population Register Samples In countries that maintain registers of the population (for example, Eastern Europe and China), it is common to randomly sample individuals meeting the study criteria (usually simply those falling within some age range) directly from the population register. This is a very good method because it allows strong control from the office—that is, it makes it very difficult for the interviewers to cheat by filling out the questionnaires themselves. A simple control procedure is to ask the respondent for the exact date of birth. This information typically is in the population register but will be unknown to the interviewer. Thus interviewers cannot make up an answer to this question from their kitchen table.

There are three (related) potential disadvantages to using population registers to draw samples. First, if the register is not kept up to date, it will miss those who tend to move around a lot. Second, often people are officially registered in one place (for example, their home village) but are away working somewhere else for an extended period. Thus they will be interviewed in neither place because it usually is extremely expensive to track them down. This is a major problem in China, where 25 percent of the population of Beijing, and comparable proportions in other cities, is "floating," working in the city but registered in a village. To obtain better records for official statistics (and also—indeed, mainly—to maintain tight social control of the population), the Chinese government began in 1994 to require that people residing in a place for more than three months register as "temporary residents"; nonetheless, many people fail to register. A third disadvantage to basing samples on population registers is that the registers are virtually always restricted to the *de jure* population rather than the *de facto* population. So a large resident alien population—like Germany's *Gastarbeiter* (guest workers)—will be excluded. This can result in rather odd samples. For example, German samples typically have far too few male unskilled workers because unskilled jobs are almost always done by *Gastarbeiter*.

Random Samples of Households and Further Selection within Households In the United States and other countries lacking population registers, the problem is to create a list of people to be sampled within each of the small geographic units chosen. This typically is done in three stages: by enumerating households, sampling them, and then, as part of the interviewing process, randomly choosing one person (or more) per household to be interviewed.

Households are enumerated (listed) by fieldwork staff who walk through the area, locating and recording every occupied dwelling unit. In suburban neighborhoods full of single-family houses, this is pretty easy—although one still has to be careful to include mother-in-law apartments and such. In places where people live in garages, rooms in the backs of shops, and other informal dwellings, it can be very difficult. (Contemporary urban China is such a case. For an account of challenges faced by those trying to conduct sample surveys in such environments, see Treiman, Mason, and others [2006].) It can also be difficult to get into security buildings and gated neighborhoods. This is a problem not only at the listing stage but at the interviewing stage as well.

Once the list is compiled, a random sample of dwelling units is drawn and interviewers are sent to conduct the interviews. The next problem is to randomly select one or more people within the household for interviewing. This is done by the interviewer, who lists all the residents of the household who meet the criteria and randomly selects one (or sometimes more, depending on the design of the study), using what is known as a *Kish table* (after the sampling statistician Leslie Kish) or similar methods (see Gaziano [2005] for a review of within-household respondent selection techniques). Suppose, for example, that the interviewer is instructed to interview a person aged eighteen to sixty-nine. The interviewer lists all household members between eighteen and sixty-nine and then chooses one by referring to a table of random numbers or using some other device, such as choosing the person whose birthday is closest to the interview day.

Household samples have the advantage of capturing the de facto population—the population actually living in a place. But they have three important disadvantages. First, it is fairly easy to cheat, interviewing whomever happens to be available rather than returning to complete an interview with a person chosen but not available. Interviewers are supposed to make a specified number of attempts to complete the interview (typically three) before abandoning the attempt. I picked up cheating in a survey I did in South Africa in the early 1990s by noting that 97 percent of Blacks were interviewed on the first try—a completely unbelievable proportion. (By contrast, about 80 percent of the White, Asian, and Coloured interviews were completed on the first try.) To make it possible to discover such problems, it is a very good idea to build information on the interviewing process into the data collection—for example, by having the interviewer record the date and time of each attempt to complete an interview and the outcome of the attempt, and also by collecting information on at least the age and sex of each household member and including this information in the analytic data set, which permits the analyst to compare the distribution of completed cases to the distribution of household members. In my South African study I used such information to determine that men had been undersampled and was able to get the survey house to collect a supplementary sample of men.

A second disadvantage of household samples is that they are not true probability samples of the population, because people in large households have a smaller probability of being selected than do people in small households. For example, in 2000, 34 percent of U.S. households included only one adult, 54 percent had two adults, and the remaining 12 percent had three or more adults (data from the GSS). Obviously the chance of an adult in a single-adult household being included in a sample is twice as great as the chance of an adult in a two-adult household being included.

We typically convert household samples to person samples by weighting the data by the number of eligible people in the household, normalized to retain the original sample size. This is very easy to do in Stata. For example, suppose that the target population is adults, that we wish to correct for the number of adults in the household, and that we have a count of the number of adults in each household. In Stata we simply specify `[pweight = adults]` (or, depending on the command, `[aweight = adults]`). Now suppose the average number of adults per household was 2.0. Then a household with four adults would get a weight of 2, whereas a household with one adult would get a weight of 0.5. Also, the mean weight would be 1 and the sum of the weights would be N, the number of cases in the sample.

As it happens, weighting the GSS to take account of differential household size makes little difference for most variables, which means that the analysis we have done in previous chapters using the GSS is for the most part not far off the mark. Still, it is important to get it right. Moreover, sometimes correcting for differential household size does matter—for example, when we consider family income. Correctly weighting for differential household size increases the estimate of family income by about 10 percent in the 2002 U.S. GSS (for the evidence, see Part 1 of downloadable file "ch09.do"):

unweighted mean = \$50,102

mean weighted by household size = \$54,880

A third disadvantage of household samples is the increasing difficulty in securing high response rates. In Eastern Europe during the communist period it was common to complete more than 90 percent of the attempted interviews, but response rates dropped with the fall of communism. The same is true in China, where response rates once exceeded 95 percent but have been falling steadily, especially in urban areas where people increasingly live in high-rise apartment buildings with restricted access. The GSS typically gets about a 75 percent response rate, and other U.S. surveys do much worse, which creates a strong possibility that respondents will be a nonrandom subset of the target population. In the GSS, for example, men are usually undersampled relative to women because of differential nonresponse (Smith 1979). Any population estimate in which men and women differ—which is often the case with attitude items—will be biased.

A SUPERIOR SAMPLING PROCEDURE A superior alternative to listing and then sampling households and, within selected households, listing and then sampling eligible individuals, is to conduct a minicensus in each small area, visiting each household and recording the age, sex, and other identifying characteristics of each resident, and then sampling directly from the list of eligible individuals. This approach radically increases costs but is far more accurate than simply listing addresses, because household sizes tend to vary substantially and because, especially in crowded neighborhoods, there often are "doors behind doors"—that is, separate households that would be missed without interviewing local residents and inquiring about the presence of such households.

Two ways of improving coverage are typically used: drawing a sample somewhat larger than the target number of completed interviews, to offset nonresponses; and substitution by survey interviewers of a new case, typically from the same small area, when an interview cannot be completed. Both methods increase the number of completed interviews but do nothing to overcome biases that are due to the differential availability of potential respondents.

Random Walk Samples Random walk samples are a variant of household samples. Within each small area, the interviewer is instructed to start at a particular location (a particular street intersection) and to proceed in a specified way, taking every *n*th address (or even varying the interval according to a schedule of random numbers) and turning in a specified direction at each intersection. This amounts to doing the address listing on the fly. This is not a desirable method because, in addition to the other weaknesses of household samples, it results in difficult-to-find dwelling units being overlooked, even by honest interviewers. Also, cheating is even easier than with conventional household samples because enumeration, household selection, and interviewing are all done by the same person: and typically there is little or no documentation of the potential sample, only those actually interviewed. It is used because it is less expensive than population-register sampling and conventional household sampling. In the first two years of the GSS (1972 and 1974), a random walk procedure combined with a quota sample was used.

Quota Samples A quota sample is a sample in which the interviewer is instructed to obtain information on a given number of people with specified characteristics—females under forty, females forty and over, working women, and so on. Often quota procedures are combined with multistage probability sampling: small areas are selected using multistage probability sampling methods, and then, within each small area, the interviewer is instructed to obtain interviews to fulfill specified quotas.

In general, quota samples are not a good idea, for two reasons: first, they do not meet the conditions permitting valid statistical inference—they are not a probability sample of any population. Second, they typically produce a biased sample of the population they purport to represent, overrepresenting the kind of people who tend to be available when interviewing is carried out. Still, carefully controlled quota sampling can be useful under conditions in which probability sampling is prohibitively difficult, because in such circumstances coverage of the population might actually be better.

Stratified Probability Samples

Multistage probability samples are sometimes *stratified*, that is, designed to treat various segments of the population as if they are separate populations. For example, an initial distinction might be made between urban and rural areas, with separate samples drawn from the urban portion and the rural portion. The main reason for stratifying a sample is to ensure that a sufficient number of cases are drawn from each stratum to permit analysis. For example, to get estimates of some phenomenon for each state in the United States it would be necessary to stratify a national sample by state because otherwise only a small number of respondents, or perhaps none at all, would be likely to be chosen from small states. A second reason for using a stratified sample design is to minimize the effect of clustering, a point discussed in more detail later in the chapter.

SOURCES OF NONRESPONSE The main reason for nonresponse is failure to start the interview, because the interviewer cannot contact the target household (as sometimes happens in gated communities and high-rise apartments), because no one is home, or because the householder refuses to answer the door. For this reason high-quality survey operations often attempt to contact targeted households by mail to explain the survey and pave the way for the interviewer. Once contacted, relatively few people refuse to be interviewed (although refusals are increasing, especially in urban areas), and almost no one terminates an interview after it starts.

DESIGN EFFECTS

The fact that national sample surveys generally are based on multistage area probability samples creates a problem—standard statistical packages, which assume random sampling, tend to understate the true extent of sampling error in the data. The reason for this is that when observations are *clustered* (drawn from a few selected sampling points), for many variables the within-cluster variance tends to be smaller than the variance across the population as a whole. This in turn implies that the between-cluster variance—the variance of the cluster means, which gives the standard error for clustered samples—is inflated relative to the variance of the same variable computed from a simple random sample drawn from the same population. Reduced within-cluster variance, especially with respect to sociodemographic variables, is typical within the small areas that make up the third stage of multistage probability samples: areas of a few blocks tend to be more homogeneous with respect to education, age, race, and so on than the population of the entire country. The result is that when we use statistical procedures based on the assumption of simple random sampling, our computed standard errors typically are too small. What we need to do is to take account not only of the variance among individuals within a cluster but of the variance between clusters. This is what *survey estimation* procedures do. (For a useful introduction to such procedures, especially as implemented in Stata, see Eltinge and Sribney [1996]. However, note that Stata's survey estimation procedures have greatly expanded since that paper was published: they are now capable of handling multistage designs with more than two levels, and survey versions of many more estimation procedures are available.)

To illustrate what can happen to our standard errors when we take account of *design effects*—the fact that we have a clustered sample—I draw upon some sampling experiments conducted in the course of designing my 1996 national sample survey of China (Treiman and others 1998). Because this survey was to be conducted by sending interviewers from Beijing to each sampling point, cost was a strong incentive to minimize the number of sampling points. However, since China is a very heterogeneous country, it was possible that a highly clustered sample would produce an unacceptably high level of sampling error. To estimate the potential damage that could result from clustering, we conducted some analysis using a 1:100 sample of the 1990 Census of China.

Although we carried out several experiments, I draw upon only a subset to illustrate the potential problem of clustering—a three-stage design for a rural sample. The first stage consisted of fifty counties, chosen randomly with probability proportional to size. In the second stage two villages within each county were chosen randomly with probability proportional to size. In the third stage thirty people between ages twenty and sixty-nine were chosen at random within each village. Altogether, this design created a sample of three thousand people. We also drew a corresponding sample from the urban population. To assess whether the clustered samples produce larger sampling variability than would corresponding random samples of the same population, we computed several statistics summarizing features of the Chinese population and estimated the design effect (*deff*) for each statistic. *Deff* is the ratio of the variance calculated taking the clustered sample design into account to the estimated sampling variance from a hypothetical survey of the same size with observations collected from a simple random sample. It also can be thought of as a factor for the sample size; thus a

LESLIE KISH (1910–2000) was one of the leading survey statisticians of the twentieth century, publishing the pioneering monograph *Survey Sampling* (1965), which became the standard for the field. He made major contributions to the development of inference procedures for complex samples and other applications. (Kish invented the *deff* and *meff* statistics.) He also helped to found the Institute for Survey Research at the University of Michigan and to design its sample.

Born in Poprad, now in Slovakia, of Hungarian parentage, he came with his family to the United States in 1925. His father died shortly thereafter, so he completed his BA in mathematics in night school at City College in New York, studying while helping support his mother and siblings. He also took two years off to fight the fascists in Spain as a member of the International Brigade. After completing his BA he moved to Washington, D.C., where he worked first at the Census Bureau and then at the Department of Agriculture. He then again volunteered for military service, this time in the U.S. Army. In 1947 he moved to the University of Michigan, where, in addition to helping found the Institute for Social Research, and teaching, he completed his MS and PhD. He remained at Michigan for the rest of his life.

design effect of two indicates that we would need twice as large a sample with our clustered design to achieve the same standard error we would obtain from a simple random sample (Kish 1965, 259).

Note that design effects must be estimated separately for each statistic since they may vary substantially from variable to variable. In some cases—especially when the sample is stratified—design effects may even drop below one (which indicates that the sampling variability of the design-based sample is actually smaller than we would expect from a corresponding random sample).

The leftmost panel of Table 9.3 (under the heading "Without Stratification") shows design effects for several statistics computed from the 1990 Chinese census using the design just described. The figures in the second column show that the design effects are unacceptably large, especially in the urban sample. For example, the design effect of mean years of schooling is 13.43, which implies that a 3,000-person sample clustered in the way we drew the Chinese census sample would have a true standard error as large as a random sample of 223 people. Although most of the design effects in Table 9.3 are not nearly so large, they are large enough to suggest that the design we studied is inadequate. How can we do better?

Stratifying to Offset the Effect of Clustering

An interesting and powerful feature of the statistics of sampling is that under certain conditions we can more or less completely offset the effect of clustering by appropriately *stratifying* our sample—that is, by dividing the population into several subpopulations based on specified criteria and sampling each of the subpopulations. By stratifying the sample on a variable (or variables) substantially correlated with variables of analytic interest, we often can virtually eliminate the design effect. The reason for this is that a stratified sample is the equivalent of a set of independent samples, one for each stratum. But well-chosen strata will tend to be more homogeneous than the sample as a whole, not only on the stratifying variable or variables—which is a necessary consequence of the way the strata are constructed—but on other variables as well. This homogeneity can be exploited to produce smaller estimates of standard errors.

The key to successful sample stratification is to stratify the sample with respect to characteristics that are strongly correlated with the variables of central analytic interest. In the Chinese sampling experiments, we stratified counties with respect to the proportion of the adult population with at least a lower middle school (junior high school) education. The effect of doing this was dramatic—we reduced the design effect to *below unity* for every comparison in the experiment, as can be seen in the rightmost panel of Table 9.3 (under the heading titled "With Stratification"). These results imply that for analysis of variables of the sort shown in the table, in the Chinese case a sample stratified by education is actually superior to a simple random sample.

Later in this chapter I present a worked example illustrating survey estimation procedures with data from the 1996 Chinese survey that was designed on the basis of the results of the sampling experiments I just reviewed. First, however, we need to consider some additional features of sample designs.

TABLE 9.3. **Design Effects for Selected Statistics, Samples of 3,000 with Clustering (50 Counties as Primary Sampling Units, 2 Villages or Neighborhoods per County, and 30 Adults Age 20 to 69 per Village or Neighborhood), With and Without Stratification, by Level of Education.**

		Without Stratification		With Stratification	
Statistic	**Sample**	**Coefficient**	**Deff**	**Coefficient**	**Deff**
Mean years of schooling	Urban	8.45	13.43	8.41	0.78
	Rural	5.49	8.22	5.61	0.87
Mean age	Urban	38.06	2.69	38.21	0.96
	Rural	38.55	1.73	38.71	0.99
Mean ISEI score	Urban	33.35	5.68	33.44	0.87
	Rural	24.02	2.44	24.61	0.91
Percent with local registration	Urban	95.37	6.08	94.67	0.87
	Rural	99.43	2.73	99.30	0.96
Percent employed	Urban	81.07	4.19	81.13	0.93
	Rural	88.97	2.99	87.47	0.93
Percent with at least middle school	Urban	65.70	10.73	63.77	0.82

		Rural	32.07	3.36	32.10	0.92
Regression of ISEI on years of schooling	Int.	Urban	12.90	4.70	10.67	0.92
		Rural	18.91	2.81	17.45	0.96
	$b_{Educ.}$	Urban	2.42	6.36	2.71	0.90
		Rural	0.93	1.80	1.28	0.94
Regression of ISEI on years of schooling, age, and sex	Int.	Urban	16.23	1.83	13.57	0.97
		Rural	27.85	1.67	22.88	0.95
	$b_{Educ.}$	Urban	2.23	3.31	2.54	0.94
		Rural	0.49	1.70	0.89	0.94
	b_{Age}	Urban	−0.08	2.75	−0.08	0.95
		Rural	−0.20	1.50	−0.14	0.97
	b_{Sex}	Urban	2.70	1.43	2.81	0.99
		Rural	2.31	1.43	4.14	0.98

Note: This is the 1:100 sample of the 1990 Census of China, first version.

HOW THE CHINESE STRATIFIED SAMPLE USED IN THE DESIGN EXPERIMENTS WAS CONSTRUCTED Because stratified samples are just multistage probability samples, with each stratum treated as a separate sample, the procedure for creating such samples is similar to that described earlier, using as an example California cities. To create the Chinese sample, we first divided all county-level units (counties, county-level cities, and districts of large cities) into an urban and a rural sector, using data from the 1990 Chinese census. We treated these sectors as two separate populations—the population of urban China and the population of rural China. Consider the population of rural China first, which consisted of about 2,400 counties. We arrayed these counties in the order of the proportion of the adult population with at least a lower middle school education. We then divided the counties into twenty-five strata of approximately equal size so that counties totaling approximately 4 percent of the population were included in each stratum. We then chose two counties from each stratum, with probability proportional to size, picking the first one at random and the second one systematically by adding half the population of the stratum to the original number and picking the county within which the sum fell, wrapping as necessary. The remaining stages were sampled PPS in the usual way. We then created the urban sample in the same way.

As noted earlier in the chapter, a second reason for using stratified samples is to sample different subpopulations at different rates. We did this in the Chinese sample. Although for convenience I have presented the Chinese data as if they consisted of two separate samples (an urban sample and a rural sample), the urban-rural distinction may be thought of simply as a second stratification variable. However, because China was about 75 percent rural at the time the survey was conducted, we sampled the urban population at three times the rate at which we sampled the rural population in order to achieve urban and rural samples of the same size, which we wanted for our analysis. The same strategy was used in the 1982 and 1987 GSS to achieve a sample of the Black population of the United States large enough to sustain a separate analysis of Blacks and non-Blacks.

Weighting When portions of the population are sampled at different rates, the sample is, of course, no longer representative of the entire population. Thus any statistics computed over the entire sample will be biased. For example, if we naively computed the mean level of education in the Chinese sample, we would overstate the true level of education in the Chinese population since the urban population, which was oversampled relative to the rural population, is much better educated. A similar naive computation using the 1982 or 1987 GSS, which oversampled Blacks, would understate the level of education in the population given the lower level of education of Blacks compared with non-Blacks. To correct for such distortions, we *weight* the data proportionally to the inverse of the sampling rate.

For example, in the 1996 Chinese survey, which includes (approximately) three thousand rural cases and three thousand urban cases, to correct for the fact that the urban population was sampled at three times the rate of the rural population we would assign a weight, w_u, to the urban population and a weight, w_r, to the rural population, where $w_r = 3w_u$. Note that we would not want to simply assign a weight of 0.33 to the urban population and a weight of 1.0 to the rural population since this would result in a weighted sample size of 4,000, whereas the true sample size is 6,000. Rather, we would adjust the data back to the original sample size by dividing the initial weight by the mean weight, 0.67. (This is, of course, just what we have done to convert household samples to person samples). Thus we would create a new variable (weight) that has the value 0.5 for urban cases and the value 1.5 for rural cases. This yields a weighted sample size of 6,000 (which is identical to the unweighted sample size) and a weighted sample size of 1,500 urban cases and 4,500 rural cases, which corresponds to their relative population sizes. Then we can compute unbiased summary statistics for the entire population. Note, however, that this procedure overstates the reliability of rural responses (there are actually only 3,000 rural respondents, but we are treating the data as if there were 4,500) and similarly understates the reliability of urban responses.

WEIGHTING DATA IN STATA Weights can be included in Stata computations by using the notation `[<type of weight>=<name of weight variable>]` in the Stata command, before any options. For example, to get the regression of *Y* on *X* for a sample with a weight variable named *WT*, we would issue the following Stata command: `reg y x [pweight=wt]`. Stata permits several kinds of weights; see the *User's Guide* (StataCorp 2007) for details. In general, probability weights (pweights) are the appropriate choice for stratified probability samples, and these weights are used in Stata's survey estimation commands. However, Stata does not permit probability weights for all commands, and it requires that frequency weights be integers. I thus recommend that, in the relatively rare situations in which it is appropriate to weight data but not to do survey estimation (survey estimation procedures are discussed later in the chapter), analytic weights (aweights) be used whenever probability weights are not permitted. Stata automatically normalizes pweights and aweights to the unweighted total number of cases included in the analysis, which makes it unnecessary for the analyst to carry out this step.

Sometimes more complex weights are devised. For example, in the Chinese case we first corrected for differential household size by using the number of adults in the household as our first weight. Then we devised a weight to correct for oversampling the urban population. We then multiplied the two weights together to achieve an overall weight, which is appropriate since each weight is normed to a mean of 1.0—which is another way of saying that the sum of the weighted data is identical to the sum of the unweighted data.

As noted in the previous chapter, some survey houses construct a complex set of weights to take account of differential nonresponse. That is, they weight the data so that the distribution of key variables (geographic location, sex, age, education, and so on) in the sample conforms to the distribution in a standard population, such as the census. (This procedure is implemented in Stata 10.0 using the `-poststrata( )-` and `-postweight( )-` options in the `-svyset-` command.) This can be useful when nonresponse rates differ substantially across population groups of interest, but it also is potentially misleading since it assumes that nonrespondents are identical to respondents within the groups formed by the *n-way* cross-tabulation of the variables used to create the weights.

The use of weights is somewhat controversial. Some argue that you should never weight your data but rather should include in your analysis all the variables used to devise the weights. The claim is that weights sweep problems under the rug, masking effects that should be explicitly modeled. There is much to be said for this position. Certainly, urban-rural distinctions are crucial in China, and racial distinctions are crucial in the United States. Thus, generally it will be far more informative to explicitly describe the urban-rural distinction in China or race in the United States, plus appropriate interactions with other variables, in one's analysis than to weight the data and ignore these distinctions. However, from a practical point of view weighting is sometimes unavoidable, particularly in the computation of descriptive statistics. If we want to accurately estimate educational attainment in China, we do need to weight the data to reflect the oversample of the better-educated urban population; and so on. In addition, it is sometimes tedious to model *nuisance effects*—factors that might affect the outcome but that are not central to the substantive analysis. The effect of differential household size is such an example. Thus a case can be made for weighting the data to correct for such effects without focusing on them. Of course, the counter is that either they are unimportant, in which case weights are unnecessary, or they are important, in which case they should be modeled explicitly.

Perhaps the most important point to make about weights is that it is imperative that the analyst fully understand the weighting scheme used in the data being analyzed. Often weights are quite complex, and just as often weighting schemes are badly documented. Although it often takes a good deal of effort, full understanding of weighting schemes can save a great deal of trouble—and considerable embarrassment arising from errors in the analysis—down the road. In general, whenever you begin to use a new data set, you should try to obtain as much documentation as possible about the sample design and execution—and then, of course, read it.

Survey Estimation Using Stata To get correct estimates of standard errors from multistage samples, we need to use estimation procedures specifically designed for such samples. Stata provides a set of *survey estimation* commands to estimate standard errors for many common statistics, including means, proportions, OLS regression coefficients, and logistic regression coefficients. These commands make it possible to take account of both clustering and stratification at each level of a multistage sample, albeit with some restrictions.

LIMITATIONS OF THE STATA 10.0 SURVEY ESTIMATION PROCEDURE Although Stata 10.0 is much improved over previous versions in its ability to correctly estimate standard errors and design effects for multistage samples, one important limitation remains: because of the way Stata estimates standard errors, the default for strata with only one sampling unit is to report missing standard errors. Stata 10.0 provides three alternatives, which are helpful if only an occasional stratum contains just one sampling unit—although in this case Stata recommends that the offending units be combined with others (*Survey Data*, 154 [StataCorp 2007]). The alternatives are inappropriate when, by design, each stratum contains only one sampling unit. (Note that in Stata's implementation "the sampling units from a given stage pose as strata for the next sampling stage" [*Survey Data*, 154 (StataCorp 2007)].) This is the design used by the GSS, which has one PSU per stratum and is the design used in the 1996 Chinese survey analyzed here, in which one township was sampled per county.

The solution I adopt is to ignore the stage containing only one unit per stratum; but this understates the degree of clustering. For example, in the Chinese case two villages per county were sampled, but both were drawn from a single township; ignoring the township level results in this aspect of clustering not being taken into account. Although not optimal, this solution strikes me as better than ignoring subsidiary stages altogether, which is what Stata did before release 9.0.

To show the effect of using survey estimation procedures, I first repeat the analysis, presented in Chapter Seven, of the determinants of knowledge of Chinese characters, using survey estimation procedures. I then follow with an analysis of race differences in income among U.S. women to show how to do survey estimation for subsamples. I conclude with an analysis of race differences in education in the United States (the same example I used to discuss the decomposition of differences in means in the previous chapter) to show how to do survey estimation when combining several years of the GSS (or, by extension, other data sets). See Appendix A for descriptions of both the Chinese data and the GSS, and see Appendix B for a discussion of how to do survey estimation using the GSS.

A Worked Example: Literacy in China

What follows is a comparison of the regression estimates and standard errors derived two days: by using survey estimation procedures, and by assuming that the data were from a simple random sample, as we did in Chapter Seven (see Table 9.4). The 1996 Chinese survey analyzed here used a design similar to the design of the sampling experiments described earlier in the chapter, except that in the sample survey we sampled one township per county and two villages per township. (See Appendix A for details on how to access the documentation for the survey, including information on the sample design [Appendix D of the documentation] and how to obtain the data.)

TABLE 9.4. Determinants of the Number of Chinese Characters Correctly Identified on a 10-Item Test, Employed Chinese Adults Age 20–69, 1996 (N = 4,802).

	Unweighted		Weighted		Design Based				
Variable[a]	b	s.e.	b	s.e.	b	s.e.	Deff	Meff	$Meff_w$
E: Years of school	0.378	0.006	0.393	0.006	0.393	0.010	2.99	2.93	1.52
E_F: Father's years of school[a]	0.002	0.006	0.009	0.007	0.009	0.007	1.27	1.45	1.01
N: Nonmanual occupation	0.206	0.046	0.211	0.057	0.216	0.055	1.01	1.42	0.97
U: Urban origin	0.281	0.045	0.177	0.054	0.172	0.050	0.88	1.21	0.92
M: Male	0.366	0.037	0.385	0.044	0.385	0.049	1.70	1.80	1.12
C: Cultural capital	0.759	0.101	0.866	0.118	0.872	0.129	1.53	1.64	1.10
Intercept	0.706	0.040	0.546	0.039	0.544	0.060	3.25	2.31	1.53
R^2	0.681		0.687			0.688			
s.e.e.	1.24		1.24						

[a]All variables are significant at or beyond the .001 level except for father's education. For the unweighted data, $p = .690$. For the weighted data, $p = .195$, and for the design-based analysis, $p = .186$.

Stata requires that information regarding the properties of the data be set before specifying estimation commands. Once this is done, using the `-svyset-` command, estimation is carried out in the usual way, except that the survey version of the estimation command is substituted for the nonsurvey version. The specific commands used to do survey estimation for the Chinese literacy example are shown in downloadable file "ch09.do" (Part 2). See also the `-log-` file "ch09.log" for the output.

The Stata 10.0 survey estimation commands provide four design effect statistics: *meff*, the misspecification effect; *deff*, the classic design-effect statistic developed by Kish (1965) and discussed earlier in the chapter; and *meft* and *deft*, which are the approximate square roots of *meff* and *deff*. Of these I find the first two the most useful. These coefficients are reported in Table 9.4, which also includes three estimates of the determinants of literacy in China: regression coefficients assuming simple random sampling without weights; simple random sampling with weights; and survey estimation procedures (shown in the rightmost panel, "Design Based"). Finally, the table shows another heuristic design statistic, which I call $meff_W$.

Meff

Meff is the ratio of the sampling variance (the square of the standard error) computed using the design-based estimation command to the sampling variance computed on the assumption of *unweighted simple random sampling*. *Meff* thus informs us just how badly we would err in our estimates of sampling variance were we to naively compute statistics without taking account either of clustering or of differential sampling rates—as we have done in previous chapters; for the current example, these are the computations shown in the first two columns of Table 9.4. Note that as specified by the definition of *meff*, in the first row $meff = 2.93 = 0.010^2/0.006^2$ (or, precisely [from the downloadable `-log-` file] $0.0095421^2/0.0055767^2$), where the ratio is formed by the squared standard error estimated using design-based estimation divided by the squared standard error assuming simple unweighted random sampling. Sometimes, as in this example, the underestimate of the sampling variability can be substantial; thus it would be completely inappropriate to use naive estimating procedures for the Chinese data.

It also is insufficient to weight your data but to ignore clustering and stratification, as the computations in the rightmost column of Table 9.4 demonstrates. This coefficient, which I have labeled $meff_W$, gives the ratio of the design-based sampling variance to the sampling variance estimated by weighting the data but not taking account of clustering or stratification. (Because this coefficient is not among the Stata options—I created it for heuristic purposes—it must be computed by hand, unless you want to program Stata to do it for you. See downloadable file "ch09.do" [Part 2] to see how I used Stata to do the computations.) As is evident, the variance estimates can be quite different. Thus once again we see the importance of taking account of the sampling design to get correct estimates of the standard errors of our coefficients.

Deff

As noted earlier in the chapter, *deff* is the ratio of the design-based estimate of the sampling variance of a statistic that has been collected under a complex survey design to the estimated sampling variance from a hypothetical survey of the same size with observations collected through simple random sampling. Thus *meff* is different from *deff* in that it gives the ratio of the sampling variances obtained from our actual data under two conditions: (1) when we use design-based estimation to account for clustering and sample weights and (2) when we ignore clustering and weights and estimate statistics appropriate for simple unweighted random samples; *deff*, by contrast, gives the ratio of the design-based sampling variance to the sampling variance that we would expect if we had actually carried out a survey using a simple random sample. In this sense *meff* is mainly of didactic value because it reveals the consequences of naive estimation.

Deff can be thought of as a variance inflator, indicating the extent to which the sampling variance is inflated because of the clustering of the observations in the sample. Because the standard error is a function of the square root of the sample size, *deff* also can be thought of as indicating how much larger than a simple random sample a sample based on the clustered design would have to be for both samples to yield standard errors of the same size.

In the present case, despite our best efforts to stratify the sample, we still have a relatively large *deff* for years of school completed: 2.99. This implies that with respect to the measurement of years of schooling, our clustered sample of about 6,000 cases has the precision of a simple random sample of about 2,000 cases. Although this is a great improvement over the design effects of 8.22 and 13.43 that we obtained for the rural and urban samples in our design experiment based on the 1990 Chinese census, it still is quite large. Fortunately, none of the remaining variables in the model has design effects nearly as large (although the intercept does).

In the course of carrying out analysis of an existing survey there is, in fact, little reason to compute *deff* because *deff* provides information that is useful primarily in designing a new survey (as in the Chinese census analysis discussed earlier). Rather, for samples for which we have adequate information on the design, we simply carry out our analysis in the standard way, but using the survey estimation commands rather than commands that assume simple random samples. Unfortunately, such information often—indeed, usually—is not included in survey documentation, especially for older surveys.

In such cases there is a next-best approach. You can approximate design effects by treating your sample as somewhat smaller than it actually is, by weighting your data by 0.75 or 0.67 or 0.50 (this is easily accomplished, either by creating a weight variable = 0.75, 0.67, or 0.50, or whatever you judge the reciprocal of the design effect to be; or by multiplying any existing weight variable by your judgment of the reciprocal of the design effect). Weighting by 0.75 is tantamount to assuming that each statistic in your analysis has a design effect of 1.33 (= 1/0.75). Because, as we have seen, design effects can vary substantially, this is hardly an optimal solution, but it is superior to blithely assuming that the multistage probability samples upon which almost all survey data are based are as precise as simple random samples, which is what we do when we make no correction for design effects.

In the GSS the design effect is typically about 1.5 for attitude items and about 1.75 for sociodemographic items, which tend to vary more across clusters (Davis and Smith 1992). So we could get an approximation to the correct standard errors by weighting our sample by the reciprocal of the design effect, for example, weighting the GSS by 0.57 (= 1/1.75), to be conservative. However in recent years the GSS has included the variable SAMPCODE, which permits the use of survey estimation procedures. The GSS uses a complex design, which, moreover, has changed somewhat every decade, with the shift to a new sampling frame. Hence the correct use of survey estimation procedures for trend analysis is a somewhat cumbersome business even when, as is reasonable, years are treated as strata. However, for the analysis of a single year of the GSS, the task is somewhat easier. (See Appendix B for a discussion of how the sample design of the GSS has changed over time and the implications of the changes for how to do survey estimation using the GSS.)

AN ALTERNATIVE TO SURVEY ESTIMATION If there is clustering at only one level and no stratification information is available, an appropriate alternative to survey estimation is to use the conventional regression command, `-regress-`, with the `-robust-` and `-cluster-` options. This will produce standard errors that are (within rounding error) identical to the estimates produced by the survey estimation command when no strata are specified. That is, `-robust-` and `-cluster-` options take account of clustering but not of stratification. In general, but not always, failure to take account of stratification will produce larger standard errors.

This approach may make it possible to provide a partial correction for clustering even when information on the sample design is not available in the survey documentation. Because almost all large population surveys are clustered on the basis of geography, you may be able to use geographic place identifiers as a cluster variable. In addition, you may have a data set that includes information on households and also on several individuals within a household. In such cases you can treat the household identifier as a cluster variable (in addition to any geographic identifier in the data set).

HOW TO DOWNWEIGHT SAMPLE SIZE IN STATA When it is known or suspected that a complex sampling scheme was used but no other information is available, it would be prudent to downweight the data by a factor of two or three to reflect an approximate design effect by inflating the standard errors. To accomplish this in Stata, use the `[iweight]` specification, which creates weights that are *not* renormed to the sample size. Note that using the `[iweight]` specification is the equivalent of using `[aweights]` rather than `[pweights]`, but using `[aweights]` when doing model estimation is in general incorrect and typically will produce smaller standard errors than when `[pweights]` are used. So this clearly is a suboptimal solution. It thus is generally well worth the (often considerable) effort to determine the actual sample design and to obtain the variables necessary to implement survey estimation procedures that correctly reflect complex sampling designs—even if it means imposing upon original investigators who have gone on to other research and do not want to be bothered trying to document something they paid little attention to in the first place.

Analysis of Subpopulations: Effect of Education and Race on Income Among Women

One special feature of the design-based estimation procedure in Stata needs to be highlighted: when analysis is restricted to a subset of the data, it is inappropriate simply to exclude cases not meeting the selection criterion. The reason for this is that the sample design features pertain to the entire sample, not the subset selected for analysis. Stata correctly handles analysis of subsamples through the `-subpop-` option in the estimation command (see Part 3 of the downloadable files "ch09.do" and "ch09.log"). To illustrate the use of this option and to further illustrate how the use of survey estimation can change substantive conclusions, I here carry out a simple analysis, using the 1994 GSS, of the effect of education and race on income among women.

The 1994 GSS is a stratified multistage sample. The units for the first stage were 2,489 U.S. metropolitan areas and nonmetropolitan counties divided into 100 strata, with one PSU per stratum selected at random with probability proportional to size; then, within PSUs, 384 second-stage units (groups of blocks) were chosen PPS, and in some instances a third-stage selection was made as well. However, the documentation for the GSS identified only the PSUs, using the variable SAMPCODE. Because, as noted, Stata does not permit the specification of one PSU per stratum, I set the PSU but not the strata, and I treated the analysis as including only one sampling stage. This procedure probably underestimates the true standard errors but is the best option given the documentation available.

Table 9.5 shows the coefficients and standard errors for three models, each estimated three ways: treating the sample as if it were a simple random sample of the population; weighting the sample to correct for differential household size; and taking account of the clustering created by the first stage of the multistage design used in the GSS. In these models education is expressed as a deviation from the mean years of schooling of women in the sample.

Considering first the contrasts among models, it is evident that adjusting for differential probabilities of selection resulting from differential household sizes has a nontrivial effect on the results. The estimation that treats the data as an unweighted simple random sample (panel I) yields a significant increment in R^2 for Model 3 in contrast to Model 1, leading us to conclude that the determinants of income differ for Black and non-Black women. By contrast, neither the weighted (panel II) nor survey (panel III) estimation procedure yields the same conclusion. From the weighted and design-based estimates, we would be led to accept the null hypothesis of *no* racial difference in returns to education for women. Here is a case where taking account of the imprecision caused by treating household samples as if they are person samples changes a substantive conclusion in an important way.

An alternative way to do this without weighting the data would be to introduce a set of dummy variables for the number of adults in the household, plus interactions between the set of dummy variables and, respectively, race and education, and perhaps three-way interactions as well. Unless the focus of the analysis is on how race and education differences vary by the number of adults in the household, this alternative strikes me as excessively complex and tedious. I think the example makes it clear why,

TABLE 9.5. Coefficients for Models of the Determinants of Income, U.S. Adult Women, 1994, Under Various Design Assumptions (N = 1,015).

	Education			Black			Interaction			Intercept	R^2
	b	s.e.	p	b	s.e.	p	b	s.e.	p	b	
I. Assuming simple random sample											
Model 1	2,736	193	.000							18,644	.1650
Model 2	2,739	194	.000	260	1,421	.855				18,607	.1650
Model 3	2,548	205	.000	−10	1,419	.994	1,755	621	.005	18,747	.1716
II. Assuming weighted random sample											
Model 1	2,643	276	.000							18,173	.1559
Model 2	2,656	282	.000	1,257	1,907	.510				18,001	.1565
Model 3	2,419	247	.000	892	1,710	.602	2,225	1,434	.121	18,156	.1670
III. Assuming weighted and clustered sample											
Model 1	2,643	326	.000							18,173	.1559
Model 2	2,656	338	.000	1,257	1,772	.480				18,001	.1565
Model 3	2,419	269	.000	892	1,639	.588	2,225	1,117	.049	18,156	.1670

Contrasts (Model 3 versus Model 1)

	d.f.(1)	*d.f.*(2)	*F*	*p*
I.	2	1,011	4.01	.0184
II.	2	1,011	1.22	.2951
III.	2	98	2.14	.1212

Notes: *b* is the net regression coefficient; *s.e.* is the standard error of the coefficient; *p* is the associated probability; *d.f.*(1) and *d.f.*(2) are the numerator and denominator degrees of freedom; and *F* is the value of the *F*-statistic for the contrast between models.

in general, we want to do survey estimation when we have the information to be able to do so.

Note also that the R^2s for corresponding models in panels II and III are identical even though the standard errors differ. This follows from the definition of R^2 as a function of the ratio of variance around the regression surface to regression around the mean of the dependent variable. Because the point estimates are the same for panels II and III, the R^2s are also the same, even though in panel III the point estimates have wider confidence intervals.

Note also that I have not shown *BIC* estimates. Although it is legitimate to compute *BIC* for simple random samples, as we did in Chapters Six and Seven, *BIC* is not appropriate for weighted or clustered samples. For such designs, pseudolikelihood functions are estimated, which may be substantially different from true likelihoods and may even vary in a non-monotonic way across nested models. Thus neither likelihood ratio tests nor *BIC*, of which the log likelihood is a component, should be used to compare models for weighted or clustered data. Rather, Wald statistics, implemented in Stata as the `-test-` and `-svytest-` commands, should be used. (For a discussion of maximum likelihood estimation, which is used by most of the procedures we will explore in Chapters Twelve through Fifteen, see Appendix 12.B.)

Combining GSS *Data Sets for Multiple Years*

Previously I suggested that under some circumstances it is useful to merge several samples drawn from the same population into a single data set. In particular, if it can be assumed that a social process is consistent over time, it would be reasonable to combine GSS samples drawn in different years to increase the number of cases. I did this in the worked example in Chapter Seven, decomposing the difference between two means. Here I use the same data in a slightly modified way to study racial (non-Black versus Black) differences in educational attainment over the period 1990 through 2004. The point of the present exercise is to illustrate what is entailed in combining several data sets (see downloadable file "ch09.do," Part 4, for the Stata code). In carrying out this analysis, I treat year as the stratum variable, on the ground that the sample for each year is fixed. I then manipulate the data a bit to create a weight variable that is consistent across years. (See the downloadable file for details on this process.) Having appropriately weighted the data, I carry out survey estimation in the usual way. The results are shown in Table 9.6.

For our present purposes both the *deff* and *meff* coefficients are instructive. The largest *deff* tells us that in our estimation of the coefficient for Southern origins, we have the same efficiency as a random sample of 8,754 (= 15,932/1.82). Of course, since our sample is so large (because we have combined data from eight GSS samples), we still have the equivalent of a very large sample. The *meff* coefficients also are large, especially for mother's years of schooling. This once again suggests that naive analysis that takes no account of weighting or of clustering can be misleading, although again the very large size of the sample protects us. Although the results are substantively interesting, I forgo further commentary on them since it would largely repeat the discussion in Chapter Seven.

TABLE 9.6. Coefficients of a Model of Educational Attainment, U.S. Adults, 1990 to 2004 (N = 15,932).

Predictor Variable	Coef.	s.e.	p	Deff	Meff
Mother's years of school	0.288	.010	.000	1.64	2.27
Number of siblings	−0.133	.010	.000	1.36	1.59
Southern residence, age 16	−0.531	.065	.000	1.82	1.82
Black	0.434	.331	.190	1.31	1.75
Black*mother's years of school	−0.093	.023	.000	1.32	1.78
Black*siblings	0.057	.021	.007	1.13	1.20
Black*Southern residence	−0.006	.154	.971	1.54	1.47
Intercept	10.961	.135	.000	1.63	2.26
R^2	0.182				

CONCLUSION

The lesson to be drawn from each of the analyses in this chapter is that we are likely to badly underestimate the degree of sampling variability if we fail to take account of and correct for the fact that large sample surveys typically use multistage designs that result in substantial clustering of observations. Note that this is true not only of area probability samples but also of organizationally based samples, such as samples of students (often sampled by first selecting schools, then classrooms, then individuals within classrooms), hospital or clinic patients, and so on. Survey estimation procedures should be used for such surveys as well.

Even when complete information on the sampling design is unavailable—which is unfortunately all too common—it sometimes is possible to approximate the design by using information on the place where the interview was conducted, since almost all surveys are clustered by place. Analysts are well advised to explore their data sets for information that will enable them to approximate the sampling design, and then to use the design-based procedures available in Stata, to avoid overstating the reliability of their findings. Understating the sampling error, and thus increasing the chance of Type I error (rejecting the null hypothesis when it is true), is the usual consequence of treating multistage samples as if they were simple random samples.

Most of the standard statistical procedures we deal with in this book have survey-based versions, and these should be used whenever they are available. For procedures not yet included in the package of survey-based estimation commands, it may be possible to approximate survey-based estimation by using `[pweights]` and the `-cluster-` option, along the lines suggested in this chapter. Also, where there is only one sampling stage and no information on any stratum variables, the `-cluster-` option used with nonsurvey estimation procedures yields results identical to the survey estimation procedures discussed in this chapter (except when a subpopulation is analyzed, in which case survey estimation procedures should be used).

WHAT THIS CHAPTER HAS SHOWN

This chapter has taken us from "textbook" analysis, in which simple random sampling is assumed, to the kinds of samples actually used in social research and the implications of such sample designs for statistical analysis. We reviewed the main types of samples, with particular focus on multistage probability samples; the idea of stratifying samples, both to reduce sampling error and to gain sufficient cases for small subgroups of the population to permit analysis; and the conditions under which it is necessary or desirable to compute weighted estimates. We then turned to survey estimation, a set of procedures for correctly estimating standard errors that take account of features of the sample design, particularly clustering of cases. Finally, we considered how to interpret two statistics, *deff* and *meff*, that quantify the effect of departures from random samples for sampling error.

CHAPTER

REGRESSION DIAGNOSTICS

WHAT THIS CHAPTER IS ABOUT

In this chapter we consider ways of identifying, and under some circumstances correcting, troublesome features of our data that might lead to incorrect inferences. We do this by reanalyzing one of my published papers as a way of seeing how to apply and interpret various regression diagnostic tools.

Apart from not adjusting our standard error estimates to take account of complex sample designs of the kind we considered in the previous chapter, there are other ways we can be led to incorrect inferences. Even if we are completely attentive to the complexity of our sample, we may err either because we have specified the wrong model or because our sample includes anomalous observations—topics we briefly touched on in Chapter Five. Here I give these topics extended treatment.

I treat these possibilities together both because, in at least some cases, the same set of observations may be thought of either as anomalous with respect to some posited process or as conforming to some other process that can be captured by introducing additional variables or changing the functional form of one or more predictors, and also because the same methods can be used to detect and correct both sets of problems. First we consider regression diagnostics, a set of procedures for detecting troubles. Then we consider robust regression, an approach to correcting a subset of these troubles. This discussion relies heavily on Fox (1997, Chapters 11–12) and the discussion of various regression diagnostics in the *Stata 10.0* section "Regress Postestimation" (StataCorp 2007). I recommend these sources for further study.

INTRODUCTION

To illustrate the kinds of troubles that can befall the naive analyst who is inattentive to the properties of his or her data, consider the four scatter plots shown in Figure 10.1. These plots were contrived to produce the same regression estimate (slope and intercept), the same correlation between the variables, and the same standard error of the regression coefficients. However, only plot (a) is reasonably summarized by a linear regression line. Plot (b) shows a curvilinear relationship. Plot (c) shows a linear relationship with one value that distorts what is otherwise a perfect linear relationship. Plot (d) shows a data set with variance in X and the slope relating Y to X created entirely by a single point (where X is the variable on the horizontal axis and Y is the variable on the vertical axis). Clearly there is a cautionary lesson in this—it is a very good idea to visually inspect the relationships among variables to ensure that your specified model adequately captures the true relationships in your data.

Apart from these examples we need to be sensitive to still other ways our regression models may fail to adequately capture relationships observed in our data. In particular, important variables may be omitted from our model, as illustrated in Figure 10.2. Here it is obvious that the regression of Y on X is misleading, because the three middle observations have expected values of Y three points higher than those to the left and right. It should be evident that an equation of the form

$$\hat{Y} = a + b(X) + c(Z) \quad (10.1)$$

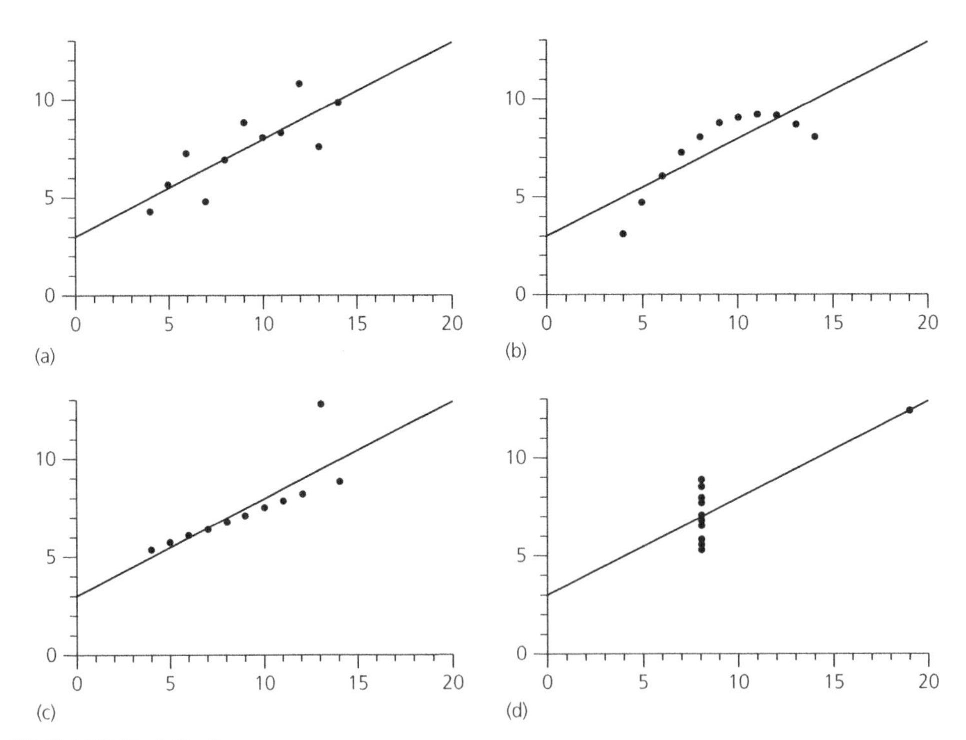

FIGURE 10.1. *Four Scatter Plots with Identical Lines.*

Source: Anscombe 1973, 19–20.

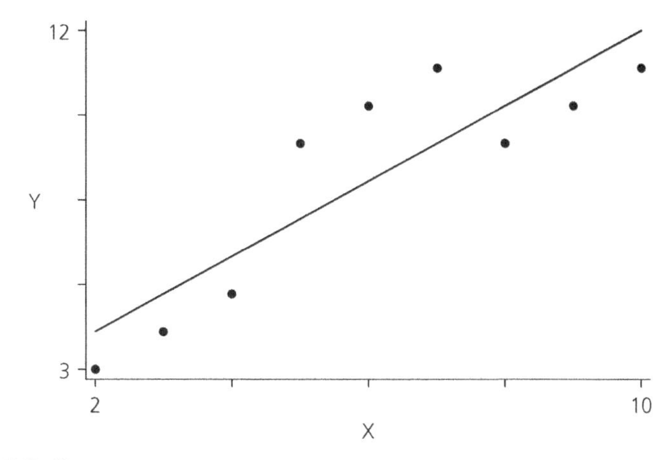

FIGURE 10.2. *Scatter Plot of the Relationship Between X and Y and Also the Regression Line from a Model That Incorrectly Assumes a Linear Relationship Between X and Y (Hypothetical Data).*

where Z is scored 1 for the three middle observations and is scored 0 otherwise perfectly predicts Y. Visual inspection of a scatter plot such as in Figure 10.2, or a component-plus-residual plot, discussed later in this chapter, can sometimes reveal the need to consider additional variables—although empirical examples are not usually so clear cut.

Still another potential problem is *heteroscedasticity,* unequal error variance around the regression surface at different predicted values, which results in inaccurate standard errors of regression coefficients. Heteroscedasticity is fairly common because in many cases the variance of observations increases with the mean. Fortunately, modest violations—the largest error variances less than ten times as large as the smallest—have little effect on the standard errors. Still, we need to check for large violations.

To detect anomalous relationships in our data in the case of simple two-variable models, we can simply plot the relationship between X and Y as in Figures 10.1 and 10.2. However, for multiple regression equations, zero-order scatter plots between each of the independent variables and the dependent variable are likely to miss important nonlinearities and anomalies. Thus we need to exploit a set of additional procedures known collectively as *regression diagnostics.*

Nonetheless, it is useful to start with a very simple example, if only to illustrate how insidious the problem can be in actual research situations. Suppose an analyst fails to notice that missing data are represented by very large codes (recall the boxed comment in Chapter Four, "Treating Missing Values as if They Were Not"). Consider the relationship between number of siblings and years of school completed in the 1994 GSS. For both *SIBS* and *EDUC,* code 98 = "Don't know" and code 99 = "No answer." If we naively assumed that the data were complete and correlated the two variables, we would conclude that the amount of education obtained is unrelated to the number of siblings, because $r = .006$. Excluding the missing data from both variables yields a more plausible estimate: $r = -.246$.

What can we do, apart from simply being alert and careful, to protect ourselves against such an error? The first and most obvious step is simply to make and inspect a

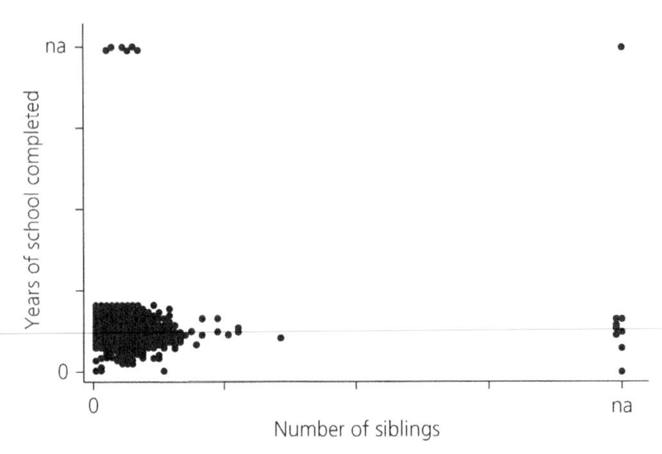

FIGURE 10.3. *Years of School Completed by Number of Siblings, U.S. Adults, 1994 (N = 2,992).*

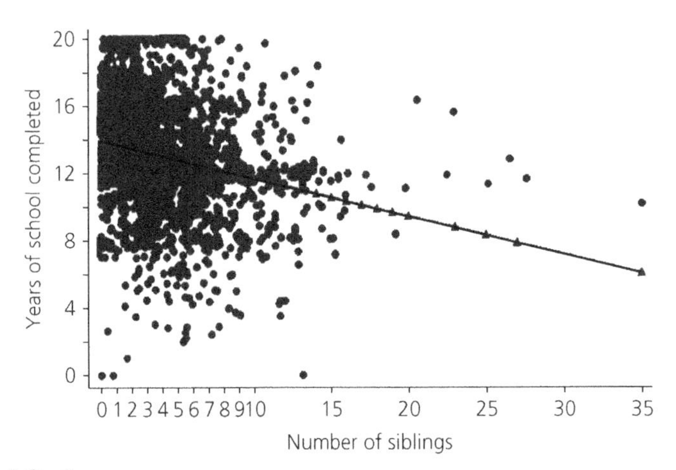

FIGURE 10.4. *Years of School Completed by Number of Siblings, U.S. Adults, 1994.*

scatter plot of the relationship between the two variables. As we have seen in Figure 10.1, such scatter plots are enormously instructive, not only in revealing gross errors such as the inclusion of missing value codes but also in indicating other anomalies in the data: curvilinear relationships, discontinuities, patterns that suggest the possibility of omitted variables, and heteroscedasticity. Figure 10.3 shows a plot of the relationship between number of siblings and years of school completed, in the 1994 GSS.

This plot immediately reveals trouble and would do so just as clearly if the numerical value of the "NA" category, 99, were shown. Inspecting the plot, we see that the missing values must be omitted to obtain meaningful results. Doing so results in the plot shown in Figure 10.4, with the regression line included. Note that the new plot is based on 2,975 cases, a reduction of only 17 cases, but the regression estimate differs substantially. Because so few cases are missing, we need not be concerned about imputing the missing data (recall the discussion in Chapter Eight). However, even after omitting missing cases, we still need to be concerned about the possibility that the regression estimate is unduly influenced by the relatively small number of respondents with many siblings. The following section addresses how to assess such a possibility.

A WORKED EXAMPLE: SOCIETAL DIFFERENCES IN STATUS ATTAINMENT

The failure to omit missing cases in the preceding example is a particularly blatant error, easy to note and easy to fix. Sometimes, however, errors are more subtle. Thus we need a set of procedures for detecting anomalies in our data. Regression diagnostic procedures are at present not very well systematized. There are many graphical methods and tests, often doing more or less the same thing, and considerable confusion about nomenclature (the same procedures are called by different names, and the same names are used for different procedures). I have illustrated a subset of these procedures that seem useful, favoring those that are or easily can be implemented in Stata. (For a useful exposition of regression diagnostic procedures, see Bollen and Jackman [1990].)

As a concrete example of how to carry out regression diagnostic procedures, I reanalyze an article I completed with a former graduate student, Kam-Bor Yip (Treiman and Yip 1989). In this multilevel analysis we were interested in how macrosocial characteristics affect the process of status attainment. For a very simple model predicting men's offspring's occupational status from their fathers' occupational status and their own education in eighteen nations, we hypothesized that the effect of the son's education should be stronger and the effect of the father's occupational status weaker in more-industrialized countries and in countries with less income inequality and less educational inequality in the father's generation.

The first step, after converting all of our data to a common metric, was to estimate the micromodel separately for each nation. The second step was to predict the size of the regression coefficients resulting from the first step, using measures of industrialization and inequality. This sort of two-step estimation procedure, although statistically suboptimal, is conceptually clear. (For statistically optimal multilevel procedures, see Raudenbush and Bryk [2002], and for a brief introduction to multilevel analysis, see the discussion in Chapter Sixteen.)

Here I reanalyze the results shown in Equation 7 of Treiman and Yip:

$$\hat{b}_E = -.39(EI) - .19(II) + .31(D); \quad R^2 = .55; \quad Adj.\ R^2 = .46 \tag{10.2}$$

where b_E is the metric regression coefficient relating occupational status to education in each microequation, EI is a measure of educational inequality, II is a measure of income inequality, and D is a measure of economic development. The coefficients estimated for Equation 10.2 are expressed in standard form. However, regression diagnostic procedures operate on metric coefficients. Here is the corresponding equation, with coefficients expressed in metric form.

$$\hat{b}_E = 2.02 - .35(EI) - .32(II) + .30(D) \tag{10.3}$$

Although regression diagnostic procedures are helpful for samples of all sizes, they are particularly useful for analyses based on small numbers of observations because such samples are particularly vulnerable to the undue influence of one or a few extreme observations.

Downloadable files "ch10.do" and "ch10.log" show the Stata `-do-` and `-log-` files for my reanalysis; you should study these along with the text because many details are provided only in the commentary in these files.

Preliminaries

As I would do in any reanalysis of published results, I start by trying to replicate the published figures. The Stata `-log-` file shows a listing of the data set, various summary statistics, and estimates for the regression equation reported in the published article. All agree with the corresponding figures in the published article. This is not always the case. A surprisingly large number of published articles contain errors—coefficients that do not correspond to estimates derived from data sets that are listed by the authors or are available from archives. Sometimes this is because the authors have dropped cases or transformed variables without informing the reader, but sometimes the authors have simply made mistakes. Given the ease of email communication, it often is possible to clear up such problems relatively painlessly and is certainly worth the effort.

The first time I tried to replicate the published equation I got an absurd minimum value for the educational inequality measure and regression estimates that disagreed with the published estimates. It turned out that the explanation was simple—the scanning operation that input the data from the published article (which was written many years ago, before I started systematically keeping logs of my work) recorded −69 rather than −0.69 for Britain, and I failed to notice this when I proofread the file. It is not worth the space, or your time, to detail my effort to detect the source of and correct the problem, but the lesson is clear: devise as many checks as possible and study them carefully at each step before proceeding.

There also is a lesson here regarding good professional practice—in your publications always describe your procedures in sufficient detail to make it possible for a competent analyst to exactly replicate your coefficients given only your paper and your original data set. Doing so is not only a matter of courtesy; it will help you discover your own errors before they are published and become vulnerable to snide correction by some graduate student looking for a quick publication. Whenever you produce a paper for publication (or even a semipublication such as a deposit on a Web page or a submission as a term paper or dissertation chapter), your last step before submission should be to rerun your complete `-do-` file and check every coefficient in your paper against the `-do-` file. You will be surprised how many errors you find!

Leverage

Having established that I can replicate the published results, I now consider whether they provide a reasonable representation of the relationships in the data. I start by considering whether any observations have particularly high *leverage,* where leverage refers to the difference between the value or values of the independent variable or variables for a particular observation, and the mean or centroid of the values for all observations. Plot (d) in Figure 10.1 illustrates this case. The observation with a score of nineteen on the horizontal axis has high leverage. Such observations are troublesome because they may exert undue influence on the regression slope; obviously, in (d) of Figure 10.1 the slope would be infinite except for the single high leverage point.

A conventional measure of leverage is the diagonal elements of the *hat-matrix,* which provides a scale-free measure of the distance of individual observations from the centroid. Computing the hat-matrix for the eighteen nations in our data set (search for "hat" in the downloadable file "ch10.do"), we note that India has an unusually large value, nearly four times the mean hat-value. This suggests the possibility that India is unduly influencing the regression estimates.

Outliers

Before acting on this possibility, however, we need to further explore the data. Our next step is to discover whether there are any extreme *outliers,* observations far from the regression surface. To do this we need to adjust for the fact that observations with high leverage tend to have small residuals precisely because the least-squares property pulls the regression surface toward such observations. The *studentized residual* (E_i^*) provides such an adjustment by basing each residual on a regression equation estimated with the observation omitted. The studentized residual is attractive because it follows a t distribution with $N - k - 2$ degrees of freedom (where N is the number of observations and k is the number of independent variables), which makes it possible to assess the statistical significance of specific residuals.

However, because we usually do not have a priori hypotheses regarding particular observations, we need to adjust our tests of significance for simultaneous inference. A simple way to do this is to make a *Bonferroni adjustment* by dividing our desired probability cutoff value (conventionally .025 for a two-tailed test) by the number of possible comparisons, which in this case is the number of observations. Thus the procedure for the analysis here is to compute studentized residuals and to identify outliers as unlikely to have arisen by chance if the p-value is less than $.025/18 = .00139$. As it happens, none of the outliers is statistically significant, because the largest studentized residual, for Denmark, is 3.349, with $18 - 3 - 2 = 13$ *d.f.,* which implies a t-value of .00523 (search for "estu" in the downloadable file "ch10.do"). It probably would be unwise to take such tests of significance too seriously, especially given the very small sample. Fox (1997, 280) argues that studentized residuals greater than two in absolute value are worthy of concern. This suggests that we need to further consider Denmark ($E_i^* = 3.35$) and perhaps India ($E_i^* = 1.91$).

Influence

Measures that take simultaneous account of both leverage and outliers are known as *influence statistics.* Several relatively similar measures are available; here we focus on *Cook's Distance measure* (Cook's D), which is a scale-free summary measure of how a set of regression coefficients changes when each observation is omitted. Taking $4/N$ as a cutoff point for Cook's D, we note that only India is exceptionally influential, with the United States marginally so (search for "cooksd" in the downloadable file "ch10.do").

Plots for Assessing Influence

Despite our focus thus far on numerical summary measures, a generally more useful approach to diagnosing regression ills is to plot the relationships among various indicators.

Two useful plots that combine measures of leverage and residuals are the *leverage-versus-residual-squared plot* (the `-lvr2plot-` command in Stata) and the *studentized-residual-versus-hat plot weighted by Cook's D* proposed by Fox (1997, 285) and easily implemented in Stata (to see how I did this, search for "what=hat" in the downloadable file "ch10.do"). Figures 10.5 and 10.6 show these plots. The point of squaring the residual in Figure 10.5 is to indicate the influence of the outliers because the regression procedure minimizes the sum of *squared* errors. Still, Figure 10.6 seems to do a better job of revealing the overall influence of specific observations. Clearly India stands out from the remaining observations. Denmark, however, has the largest outlier.

Added-Variable Plots

Our next task is to try to discover any systematic relationships among the variables that might account for either the large residuals or the highly influential observations. A good way to do this is to construct *added-variable plots,* also known as *partial-regression leverage plots* or simply *partial-regression plots.* Such plots provide a two-dimensional analog to the kind of scatter plot with a regression line through it that we can construct in the simple regression case. Added-variable plots do this by showing a plot of the relationship between two residual values: (1) the residual values from a regression of the dependent variable on all variables except one and (2) the residual values from the regression of the variable omitted from (1) on the remaining independent variables.

Graph (a) in Figure 10.7, assessing the effect of educational inequality (*EI*), suggests that India is highly influential; it has very high educational inequality relative to its income inequality and level of industrialization, and it also displays a stronger effect of education on occupation than would be expected from its levels of income inequality and industrialization. Interestingly, the plot reveals that if India were removed or downweighted, the slope relating educational inequality to the level of occupational returns to education would

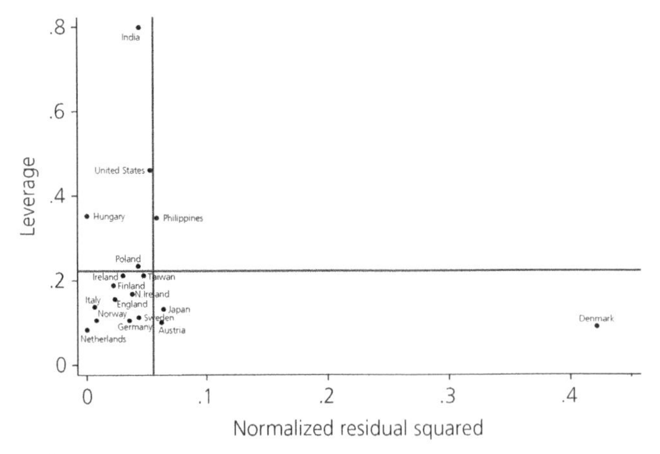

FIGURE 10.5. *A Plot of Leverage Versus Squared Normalized Residuals for Equation 7 in Treiman and Yip (1989).*

Note: The horizontal and vertical lines are the means of the two variables being plotted.

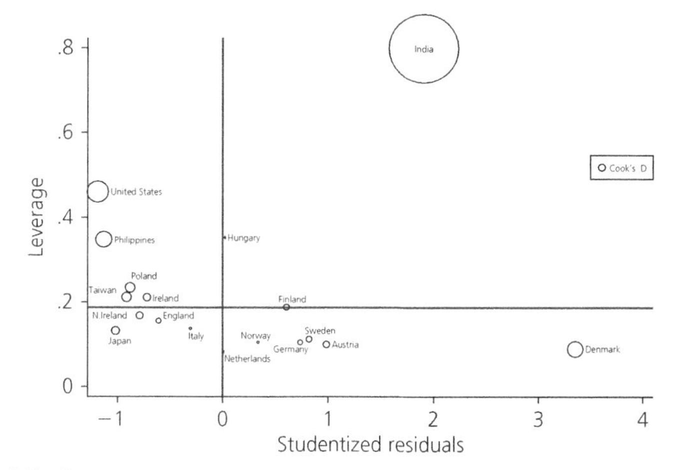

FIGURE 10.6. *A Plot of Leverage Versus Studentized Residuals for Treiman and Yip's Equation 7, with Circles Proportional to the Size of Cook's D.*

Note: The horizontal line is at the mean hat-value, and the vertical line is at zero.

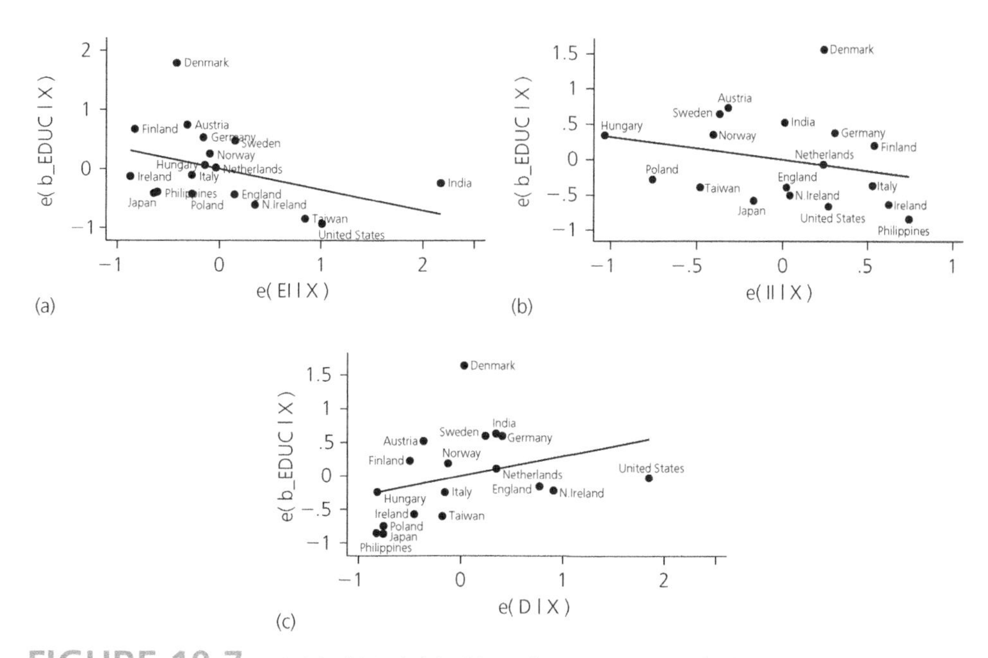

FIGURE 10.7. *Added-Variable Plots for Treiman and Yip's Equation 7.*

become *increasingly negative.* Denmark, by contrast, has unusually low educational inequality relative to its levels of income inequality and industrialization, but it has a much stronger education-occupation connection than would be expected from its position on the other two variables, so the omission or downweighting of Denmark would decrease the effect of educational inequality. Graph (b), assessing the effect of income inequality

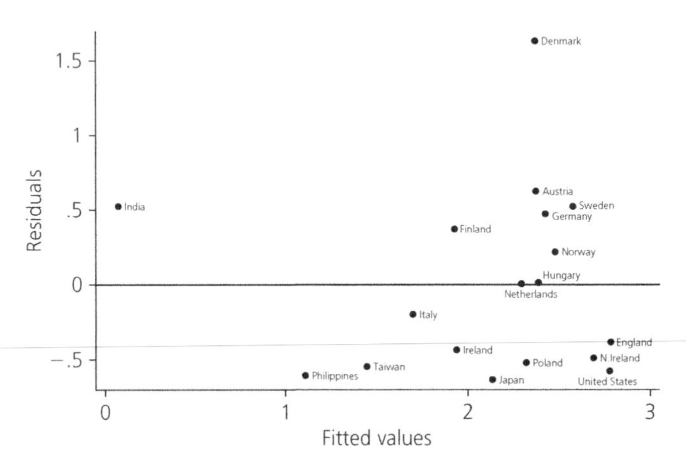

FIGURE 10.8. *Residual-Versus-Fitted Plot for Treiman and Yip's Equation 7.*

(*II*), reveals that only Denmark is a large outlier. Otherwise, the plot is fairly unremarkable. Graph (c), assessing the effect of industrialization (D), shows the United States to be a high-leverage observation, with a very high level of industrialization relative to its level of educational and income inequality. Because the United States is below the regression line, its omission would increase the slope.

Residual-Versus-Fitted Plots and Formal Tests for Patterns in the Data

A graphic procedure, **residual-versus-fitted plots**, and two formal tests, for heteroscedasticity and for omitted variables, are useful to determine whether the pattern of residuals is systematic. Residual-vs-fitted plots are just what they are called—they plot residuals against predicted values. Fig. 10.8 shows such a plot for our data. This plot suggests considerable heteroscedasticity in the data, even ignoring India and Denmark. A formal test for excessive heteroscedasticity is provided by Stata's `-hettest-` command, which tests whether the squared errors are a linear function of the values of the independent variables. Interestingly, applying this test shows no basis at all for rejecting the null hypothesis of homoscedasticity, perhaps because the residuals increase and then decrease. Here again the graph seems more informative.

A second test, Stata's `-ovtest-` command, assesses the possibility of omitted variables by testing whether the fit of the model is improved when the second through fourth powers of the fitted values are added to the equation. Given the small sample size, I take the *p*-value of .08 resulting from this test as suggesting the possibility of omitted variables. *Component-plus-residual plots* (also known as **partial-residual plots**) are useful in revealing the functional form of relationships and, by extension, the possibility of omitted variables. Such plots differ from added-variable plots because they add back the linear component of the partial relationship between Y and X_j to the least-squares residuals, which may include an unmodeled nonlinear component. Figure 10.9 shows such plots for our data, using the "augmented" version available in Stata (search for "acprplot" in the downloadable file "ch10.do").

The plots in Figure 10.9 continue to show Denmark as a large outlier. But otherwise they do not appear orderly; and—with one exception—I can think of no omitted variables. The exception derives from work by Müller and Shavit (1998) that suggests

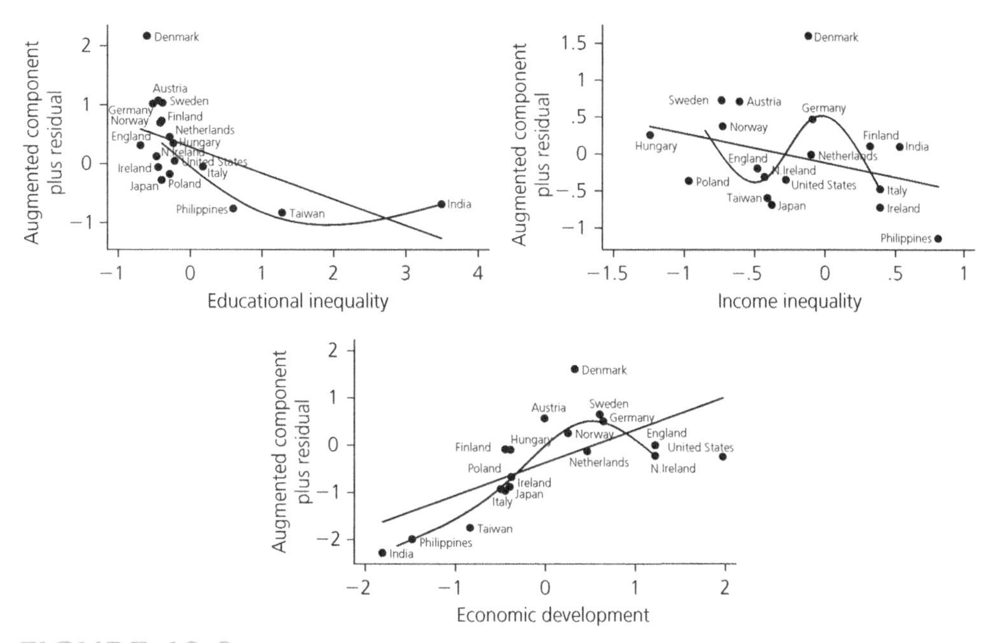

FIGURE 10.9. *Augmented Component-Plus-Residual Plots for Treiman and Yip's Equation 7.*

that the education-occupation connection is especially strong in nations with well-developed vocational education systems and especially weak in nations with poorly developed vocational education systems. In our data Denmark, Germany, Austria, and the Netherlands have especially strong vocational education systems, and the United States, Japan, and Ireland have very weak vocational education systems. The relationship found by Müller and Shavit seems to hold in our data, with the nations with strong vocational education systems above the regression line and the nations with weak vocational education systems below the regression line. This result suggests adding the strength of the vocational education system as a predictor. To do this I add two dummy variables to distinguish the three sets of nations (strong, weak, and neither especially strong nor especially weak vocational education systems). I then reestimate Equation 7, which yields the coefficients shown in the second column of Table 10.1 (for convenience, Column 1 shows the metric coefficients from Treiman and Yip's original Equation 7, that is, those shown in Equation 10.3 of this chapter); the remaining columns show various additional estimates discussed in the following paragraphs.

The specification shown in Column 2 produces a better representation of the determinants of the strength of the education-occupation connection in the eighteen nations studied here than does the original specification. The adjusted R^2 increases substantially and, as expected from the pattern of residuals, the coefficients for strong and weak vocational educational systems both have the expected signs. (I discuss the standard errors later in this chapter.)

However, the question remains as to whether the results are still substantially driven by India and Denmark. To determine this I repeated all the diagnostic procedures discussed previously with the new equation. The Stata log contains the commands I used, but in the interest of saving space and avoiding tedium, I have not shown the resulting

TABLE 10.1. Coefficients for Models of the Determinants of the Strength of the Occupation-Education Connection in Eighteen Nations.

	18 Observations		17 Observations (Expanded Model)	
	Original Model (Metric Coefficients)	Expanded Model (OLS Estimates)	OLS	Robust Regression
Educational Inequality	−0.354 (0.532)	−0.292 (0.565)	−0.821 (2.268)	−0.769 (1.292)
Income Inequality	−0.320 (0.299)	−0.342 (0.324)	−0.321 (1.594)	−0.349 (0.572)
Industrialization	0.299 (0.279)	0.287 (0.275)	0.208 (1.049)	0.215 (0.558)
Strong Voc. Ed. System		0.836 (0.410)	0.707 (0.644)	0.584 (0.590)
Weak Voc. Ed. System		−0.403 (0.414)	−0.476 (2.518)	−0.452 (0.740)
Intercept	2.021 (0.222)	1.899 (0.251)	1.814 (0.631)	1.804 (0.509)
R^2	.553	.762	.792	–
Adjusted R^2	.457	.662	.698	–
s.e.e.	0.672	0.529	0.471	–

Note: Bootstrapped standard errors, derived from 2,000 repetitions, are shown in parentheses. The bootstrapped standard errors are often considerably larger than the conventional standard errors, which are 0.220, 0.332, 0.231, and 0.176 for Column 1; 0.179, 0.275, 0.186, 0.333, 0.372, and 0.190 for Column 2; 0.304, 0.245, 0.170, 0.303, 0.333, and 0.174 for Column 3; and 0.331, 0.267, 0.185, 0.330, 0.362, and 0.189 for Column 4.

plots and will not discuss the results except to note that India continues to be a high leverage point and Denmark continues to be a large outlier, although the diagnostic indicators for both are somewhat less extreme than the corresponding indicators just reviewed.

ROBUST REGRESSION

So, what to do? Because we have no clear basis for modifying or omitting particular observations, nor for transforming our variables to a different functional form, we need an alternative way of handling outliers and high leverage points. One alternative is *robust estimation,* which does not in general discard observations but rather downweights them, giving less influence to highly idiosyncratic observations. Robust estimators are attractive because they are nearly as efficient as ordinary least-squares estimators when the error distribution is normal and are much more efficient when the errors are heavy-tailed, as is typical with high leverage points and outliers. There are, however, several robust estimators, and there are no clear-cut rules for knowing which to apply in what circumstances. The best advice is to explore your data as thoroughly as time and energy permit. (For further details on robust estimation, consult Fox [1997, 405–414; 2002], Berk [1990], and Hamilton [1992a; 1992b, 207–211].)

One class of robust estimators, known as *M estimators,* works by downweighting observations with large residuals. It does this by performing successive regressions, each time (after the first) downweighting each observation according to the absolute size of the residual from the previous iteration. Different M estimators are defined by how much weight they give to residuals of various sizes, which can be shown graphically as *objective functions.* The objective functions of three well-known M estimators are shown in (a), (b), and (c) of Figure 10.10 The OLS objective function ([a] of Figure 10.10) increases exponentially, as it must given that OLS regression minimizes the sum of *squared* residuals. The Huber function ([b] of Figure 10.10) gives small weight to small residuals but weights larger residuals as a *linear* function of their size. The bi-square objective function ([c] of Figure 10.10) gives sharply increasing weight to medium-sized residuals but then flattens out so that all large residuals have equal weight. Because Huber weights deal poorly with severe outliers (whereas bi-weights sometimes fail to converge or produce multiple solutions), Stata's implementation of robust regression first omits any observations with very large influence (Cook's $D > 1$), uses Huber weights until the solutions converge, and then uses bi-weights until the solutions again converge. Because of the way it is defined, robust regression takes account only of outliers but not of high-leverage observations with small residuals. For some problems this can be a major limitation.

Panel 2 of Table 10.1 shows (in Column 4) robust regression estimates for the elaborated model of the education-occupation connection we have been studying. There is no robust regression estimate in Panel 1 because the procedure dropped India at the outset because of its large Cook's D. Column 3 shows the corresponding OLS estimates with India omitted. Interestingly, the OLS and robust regression estimates differ very little in Panel 2, with the exception of the effect of strong vocational education, which is reduced in the robust estimate because Denmark, with its large residual, is downweighted. The agreement between different estimators does not always hold and should not be taken as an indication that robust estimation is unnecessary. However, the stability of the estimates under different estimation procedures gives us added confidence in them.

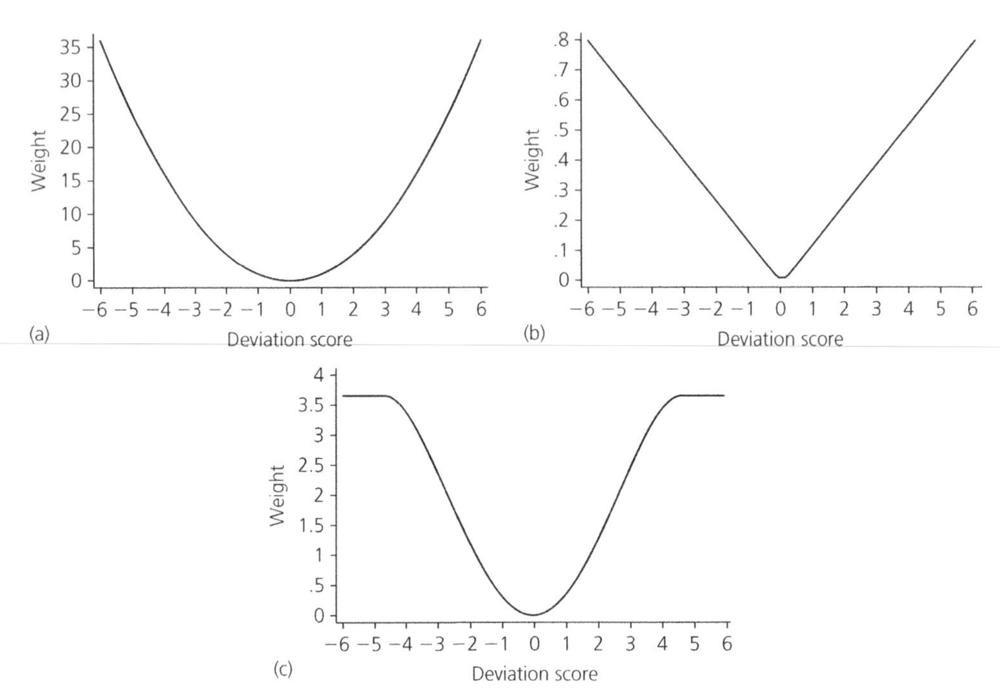

FIGURE 10.10. *Objective Functions for Three M Estimators: (a) OLS objective function, (b) Huber objective function, and (c) bi-square objective function.*

By contrast, the omission of India strongly affects the educational inequality coefficient, increasing it by more than a factor of two. The coefficient for strong vocational education is modestly reduced, and the coefficient for industrialization is even more modestly reduced. A reasonable conclusion is that the education-occupation connection is governed by a different process in India than in other nations and that a summary of the general relationship between industrialization, inequality, and status attainment should properly set India aside for separate consideration.

BOOTSTRAPPING AND STANDARD ERRORS

A powerful property of many statistics, including both ordinary least squares and robust regression, is that even if the observed values are not normally distributed, the distribution of errors is asymptotically normal—that is, as the sample size increases the distribution of errors approaches normality. However, for small samples ($N - k < 30$, where N is the number of observations and k is the number of independent variables) the approximation tends to break down badly, which means that analytically derived standard errors are often incorrect.

One way around this problem is to *bootstrap* the standard errors. In principle, if we had access to the population from which a sample is drawn, we could take repeated samples from that population (sampling with replacement); then, for each sample, estimate any statistic of interest, for example, a regression coefficient; and estimate the standard error of the statistic as the standard deviation of the distribution of estimates from the repeated samples. But of course we almost never are able to resample a large number of times from a given population; moreover, the "population" from which our "sample" of eighteen nations is drawn does not actually exist since we took all nations for which data were available. Thus we need to resort to an approximation.

Bootstrapping approximates resampling by taking the observed sample as a proxy for the population and repeatedly sampling, with replacement, observations from the observed sample. Thus, in our current example, we would randomly draw (with replacement) a first sample of eighteen cases from our eighteen observations, say Norway, Netherlands, India, Ireland, Austria, United States, Finland, Philippines, Denmark, Italy, Taiwan, Sweden, India, Ireland, Finland, Denmark, Denmark, Taiwan. Note that England, Germany, Hungary, Japan, Northern Ireland, and Poland do not fall into the sample; Austria, Italy, the Netherlands, Norway, the Philippines, Sweden, and the United States are included once; Finland, Ireland, India, and Taiwan are included twice; and Denmark is included three times. From this sample we would estimate our regression equation and record the coefficients. Then we would draw a second sample with replacement, a third, and so on. The result is, for each coefficient, a distribution of values equal in size to the number of samples we have drawn. From this distribution we estimate the standard error as the standard deviation of the distribution. (For further discussion of bootstrapping, see Fox [1997, 493–514], Stine [1990], Hamilton [1992a; 1992b, 313–325], and the entry for `-bootstrap-` in the Stata 10.0 manual.)

This method provides a good estimate of the standard error of a statistic, provided that the sample in fact represents the population from which it is drawn and that the resulting distributions are approximately normal. With very small samples with outliers and high leverage points, as in our case, there tends to be high variability from sample to sample. Thus it is wise to draw many samples to get stable estimates of the sampling distribution. In the present case I drew 2,000 samples to estimate the standard errors in each of the columns of Table 10.1 (see the section on "Bootstrapped Standard Errors" in the downloadable file "ch10.do"). I experimented with smaller numbers of samples but got unsatisfactory variability across trials in my estimates of the standard errors. With 2,000 replications the standard errors seem to be reasonably stable but hardly normally distributed—see Figure 10.11. The outliers in these distributions derive from the random omission or multiple presence of high-leverage observations. (With seventeen observations sampled with replacement, the probability of a given country being excluded from a particular sample is 36 percent—more precisely, $0.357 = [1 - 1/N]^N = [1 - 1/17]^{17}$.)

Note that the standard errors are sometimes much larger than the corresponding asymptotic standard errors reported in the note to the table, especially those for the educational inequality measure. This result alerts us to the danger of naively accepting computed standard errors from general purpose statistical programs, especially when working with small samples. On the other hand, as noted previously, it is unclear in the present case that much should be made of the standard errors, given that our "sample" is both very small and hardly a probability sample of any population. It is reasonable to tentatively accept the estimated model, specifically the robust regression estimates for seventeen societies reported in Column 4 of Table 10.1, which have far smaller standard errors than do the corresponding OLS estimates. Nonetheless, we must note that the results are only suggestive and require confirmation with more and better data before being regarded as definitive.

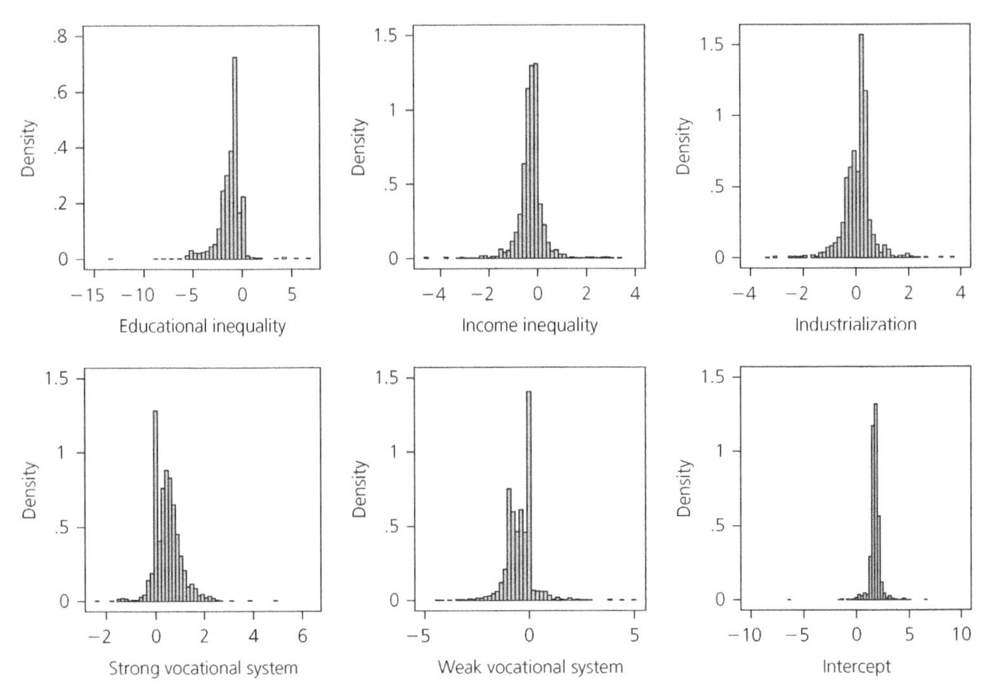

FIGURE 10.11. *Sampling Distributions of Bootstrapped Coefficients (2,000 Repetitions) for the Expanded Model, Estimated by Robust Regression on Seventeen Countries.*

Note: These are the bootstrapped coefficients for column 4 of Table 10.1.

However, when we have genuine probability samples of larger populations, the standard errors and the confidence intervals they imply assume much greater importance. The calculation of appropriate confidence intervals for bootstrapped statistics is an unsettled and ongoing area of statistical research. Stata provides four different 95 percent confidence intervals based on different assumptions. There is considerable controversy as to which of these estimates provides the best coverage of the true standard error. But the weight of the evidence to date seems to support bias-corrected estimates, and this is the default in Stata.

WHAT THIS CHAPTER HAS SHOWN

In this chapter we have seen how to check our data for anomalous observations and violations of the assumptions underlying OLS regression, how to use the information obtained to generate new hypotheses, how to use robust regression procedures to achieve estimates with smaller standard errors, and how to use bootstrap procedures to calculate standard errors in situations in which the assumption that sample statistics are normally distributed cannot be sustained. The main lesson of this chapter is that much can be learned by graphing relationships in the data. Indeed, often the best way to understand your data is to graph what you think you are observing. The results are often surprising, and usually informative.

CHAPTER

SCALE CONSTRUCTION

WHAT THIS CHAPTER IS ABOUT

In this chapter we see how to improve both the validity and reliability of measurement by constructing multiple-item scales. We consider three ways to construct scales: additive scaling, factor-based scaling, and effect-proportional scaling. We also consider two variants of regression analysis: *errors-in-variables regression*, which corrects for unreliability of measurement, and *seemingly unrelated regression*, which is used to compare regression equations with (some or all of) the same independent variables but different dependent variables.

INTRODUCTION

In social research we often wish to study the relationship among concepts for which we have no direct and exact measures. Examples include, from social stratification, class, status, and power; from attitude research, anomie, alienation, and authoritarianism; and from political sociology, liberalism versus conservatism. For concepts of this kind, it is difficult to imagine that any single measure of what people believe or how they behave would adequately reflect the concept. Suppose, for example, that we are interested in differentiating among members of Congress on the basis of how liberal their voting record is. We would hardly be content to measure "liberal voting" by choosing a single vote—say, support for foreign aid—and declaring those who voted for it to be liberals and those who voted against it to be conservatives. For any particular vote, factors besides "liberalism" or "conservatism" may come into play—objection to the particular language of the legislation, the need to discharge a political debt, the judgment that in tight times funds are better spent on social welfare at home, and so on. Although extraneous factors may affect any particular vote, we would still expect that on average "liberals" would be more likely to support foreign aid, domestic welfare, civil liberties, voting rights, affirmative action, and such, than would "conservatives." (Of course, we might want to refine our concept, differentiating domains of liberalism or conservatism—for example, social values, fiscal policy, and internationalism versus isolationism. But the same basic point holds: any one item tends to be a poor measure of an underlying concept because extraneous factors may affect the response to a single item.) Therefore a useful strategy for constructing *operational indicators* of underlying concepts is to create *multiple-item scales.* That is, we take the average of the responses to each of a set of items thought to reflect an underlying concept as *indicating* or *measuring* the strength of the concept. Multiple-item scales should satisfy two criteria: they should be *valid,* and they should be *reliable.*

VALIDITY

An indicator is valid if it measures what it is supposed to measure—that is, if it adequately measures the underlying concept. Unfortunately, there is in general no technical way to evaluate the validity of a scale, although, as we will see later in the chapter when we discuss factor-based scaling, we can gain confidence in the validity of a scale by determining that it has the relationships to other variables that we would expect on theoretical grounds. Assessment of validity is mainly a matter of constructing an appropriate theoretical link for the relationship between the concept and its indicator or indicators, and between the concept and other variables.

Many of the most important arguments in science are about the validity of measures. This is as true in the physical and biological sciences as in the social sciences. Indeed, I recommend a fascinating account of a search for life on Mars (Burgess 1978) that includes a vivid portrayal of the ongoing dispute between the "pro-life" camp and the "anti-life" camp as to whether particular indicators being sent back by the Mars Lander could validly be interpreted as evidence of the presence of life on Mars.

The first requirement for devising a valid measure is to be clear about what you are trying to measure. This is not as obvious as it sounds. More often than not our concepts are formulated rather vaguely. Just what do we mean by "social class," for example? If we take a Marxist approach and define class by the "relationship to the means of production," we have merely shifted the problem, because we then have to say exactly what we mean by the relationship to the means of production. If we take a Weberian approach and define class by "market position," we have exactly the same problem.

Anyone who thinks I am constructing a straw man is advised to look at the writings of Erik Olin Wright and his followers, who are to be commended for trying to do serious quantitative research within a Marxist framework (see, for example, Wright and others 1982, Wright 1985, and Wright and Martin 1987). A good portion of the writings of Wright and his group is preoccupied with the validity of alternative indicators.

Even seemingly straightforward variables often have the same difficulties. Just what underlying concept are we trying to measure when we devise a scale of educational attainment: skill, knowledge, credentials, values, conformity to external demands, or still something else? In principle our theory, as represented in the specification of the concept of interest, should dictate our choice of indicators. For example, if we are interested in the gatekeeping function of education in channeling access to particular kinds of jobs, we may want to measure educational attainment by the highest degree one obtains. If we regard schooling as enhancing cognitive skills, we may wish simply to count the number of years of schooling a person obtains.

Sometimes, of course, we are restricted to extant data and must work in the other direction, constructing an argument about what underlying concept is represented by the measure we have at hand. In either event clarity is crucial in your own mind, and on the written page as well, regarding what concepts your indicators measure. (For a brief introduction to different types of validity, see Carmines and Zeller [1979, 17–26].)

RELIABILITY

Reliability refers to consistency in measurement. Different measures of the same concept, or the same measurements repeated over time, should produce the same results. For example, if one individual scores high and another individual scores low on a measure of interracial tolerance, we would like to get the same difference between the two individuals if we used a different (but equally valid) measure of interracial tolerance; to the extent the measures yield similar results, we say that both measurements are reliable. Also, if the same respondent were asked his attitude at two points in time, we would like to get the same result (assuming he has not changed his attitude).

From this definition it is easy to see why multiple-item scales are generally more reliable than single-item scales. When responses are averaged over a set of items, each of which measures the same underlying dimension, the idiosyncratic reasons that individuals respond in particular ways to particular questions tend to get "averaged out." Of course, this is true only to the extent that each item in a scale reflects the same underlying *dimension* (the conceptual variable). If an item is capturing some other underlying dimension instead of, or in addition to, the one of interest to the analyst, it will undercut the reliability (and validity) of a scale. For example, suppose responses to a question about willingness to have people of a different race as neighbors reflected differences in economic anxiety, with some people rejecting potential neighbors not because of racial intolerance but for fear (rightly or wrongly) of a reduction in property values. We would not want to include such an item in a scale of racial tolerance because it would tend to make the scale less reliable—with scale scores determined to some extent by whether we happened to include a lot of people with economic anxieties or only a few such people.

An important reason for creating reliable scales is that, all else equal, unreliable scales tend to have lower correlations with other variables. This follows from the fact that unreliable scales contain a lot of "noise." We might think of scales as having a "true" component and an "error" component. The "true" component is represented by the correlation of the observed measurement with the true underlying dimension; the size of this correlation gives the reliability of the scale. The "error" component—the portion uncorrelated with the underlying dimension—reflects idiosyncratic determinants of the observed measurement. From this definition of reliability, it follows that the lower the reliability of each of two measures, the lower the correlation between their observed values relative to the true correlation between the underlying dimensions. Formally, we can estimate the "true" correlation between variables by knowing their observed correlation and the reliability of each variable. The true correlation is given by

$$\rho_{X_T Y_T} = \frac{r_{XY}}{\sqrt{r_{XX_T} r_{YY_T}}} \tag{11.1}$$

where $\rho_{X_T Y_T}$ is the correlation between the true scores, r_{XY} is the observed correlation between X and Y, and r_{XX_T} and r_{YY_T} are the reliability coefficients for X and Y, respectively. Equation 11.1 is also referred to as a formula for correcting for attenuation caused by unreliability; $\rho_{X_T Y_T}$ is the correlation between X and Y corrected for attenuation. For example, if two scales each have a reliability of .7 and an observed correlation between them of .3, the correlation corrected for attenuation will be $.3/\sqrt{(.7)(.7)} = .43$. Clearly, correlations can be strongly affected by the reliability of the component variables.

WAYS TO ASSESS RELIABILITY There are several ways to measure the reliability of a scale:

- *Test-retest reliability* is the correlation between scores of a scale administered at two points in time.
- *Alternate-forms reliability* is the correlation between two different scales thought to measure the same underlying dimension.
- *Internal-consistency reliability* is a function of the correlation among the items in a scale. Cronbach's alpha, discussed in the following paragraphs, is an internal-consistency measure.

Sometimes analysts will correct correlations for attenuation before undertaking a multiple regression analysis and will then estimate the regression model from the matrix of corrected correlation coefficients (see Jencks and others [1972, 1979] for examples of the extensive use of such corrections). This is, of course, only possible when the analyst knows or is able to estimate the reliability coefficients. The Stata command `-eivreg-` (errors-in-variables regression) provides a convenient way to use information about the reliability of variables in regression estimation. Later in this chapter a worked example illustrates this concept.

In general the internal-consistency reliability of scales depends on two factors: the number of items and the average intercorrelation among items. As each component increases, the reliability increases. There are several measures of internal-consistency reliability, of which Cronbach's alpha is the most widely used:

$$\alpha = \frac{N\bar{r}}{1 + \bar{r}(N - 1)} \tag{11.2}$$

where N is the number of items and $\bar{r}$ is the average correlation among items. In Table 11.1 we can see what this implies about the reliability of scales with an average interitem correlation of, respectively, .09, .25, and .49 (which correspond to average factor loadings of .3, .5, and .7).

It is clear that even for items with as low an average correlation as .25, scales composed of at least seven or eight items produce reasonably reliable results. Nonetheless, it is important to keep in mind that "reasonably reliable" is not the same as "highly reliable." One reason the prediction of attitudes is generally weak is that correlations involving attitude scales typically are substantially attenuated because of unreliability, especially when only a few items are included in the scale.

WHY THE SAT AND GRE TESTS INCLUDE SEVERAL HUNDRED ITEMS The effect of the number of items in a scale on reliability makes clear why examinations such as the SAT and GRE comprise several hundred items. Because college or graduate school admissions and financial aid are heavily affected by test scores, one wants to have an extremely reliable test, one that would in principle produce essentially the same rank order of students if the same students repeated the test two or three months later (without having prepared for the test in the interim). Interestingly, even with several hundred items, a person's SAT or GRE scores often vary greatly from trial to trial, presumably because performance is affected by idiosyncratic factors such as the level of anxiety and fatigue at the time the test is taken, and also, of course, the degree of preparation.

TABLE 11.1. Values of Cronbach's Alpha for Multiple-Item Scales with Various Combinations of the Number of Items and the Average Correlation Among Items.

N	.09	.25	.49
2	.17	.40	.66
3	.23	.50	.74
4	.28	.57	.79
5	.33	.62	.83
6	.37	.67	.85
7	.41	.70	.87
8	.44	.73	.88
9	.47	.75	.90
10	.50	.77	.91
20	.66	.87	.95
50	.83	.94	.98
100	.91	.97	.99
200	.95	.99	.99

SCALE CONSTRUCTION

In this chapter we consider three strategies to create multiple-item scales: additive scaling, factor-based scaling, and effect-proportional scaling. (For a brief general introduction to scaling, see McIver and Carmines [1981]. For a recent extended treatment, see Netemeyer, Beardon, and Sharma [2003]. A classic but still useful treatment is Nunnally and Bernstein [1984].)

Additive Scaling

The simplest way to create a multiple-item scale is simply to sum or average the scores of each of the component items—which is what we have been doing up to now. Where items are dichotomous, this amounts to counting the number of positive responses. Where the items themselves constitute scales—for example, continuous variables for education or income or attitude items ranging from "strongly agree" to "strongly disagree"—we ordinarily standardize the variables before averaging (by subtracting the mean and dividing by the standard deviation). If we fail to do this, the item that has the largest variance will have the greatest weight in the resulting scale. The effect of the variance on the weight is easy to see by considering what would happen if a researcher decided to make a socioeconomic status (SES) scale by combining education and income, and did so simply by adding, for each respondent, the number of years of school completed and the annual income. He may think he has an SES scale, but what he actually has is an income scale with a very slight amount of noise, because education typically ranges from zero to twenty years (and, in the United States, effectively from eight to twenty years) whereas income ranges in the tens of thousands of dollars. By dividing each variable by its standard deviation, the analyst gives each variable equal weight in determining the overall scale score. (I first came to appreciate this point many years ago in graduate school when a professor told me that he and another member of the faculty effectively controlled who passed and who failed the collectively graded PhD exams. They did so simply by using all one hundred points in the scale the faculty had devised for scoring exams, while most of their colleagues gave failing examinations a score of fifty or so.)

The trouble with simple additive scales of the sort just described is that the items included may or may not re ect a single underlying dimension. A scale with a heterogeneous set of items runs the risk of being both invalid, because in addition to what the analyst thinks the scale is measuring, it is also measuring something else, and unreliable, because at least some of the items are weakly or even negatively correlated.

Factor-Based Scaling

How can we determine whether the items we propose to include in a scale re ect a single dimension? First we identify a set of candidate items that we believe measure a single underlying concept. Then we empirically investigate two questions: (1) Do the items all "hang together" as a whole, or do one or more items turn out to be empirically distinct from (in the sense of having low correlations with) the remaining items, even though we thought they re ected the same conceptual domain? If so, we must reject the offending items. (2) Does each item have approximately the same net relationship to the dependent variable of interest? If not, the deviant items should not be used because this is again evidence that they do not measure the same concept (or that they also measure other concepts besides the one of interest). Assessing the second question is a simple matter of regressing the dependent variable on the set of tentatively selected components of the scale, plus additional control variables where appropriate. When the scale will be used as a dependent variable, the coefficients relating the component items to the independent variables should be inspected. In both situations, what we are looking for is evidence that the candidate items all have the same sign and approximately the same magnitude.

Education, occupational status, and income are good examples of items that are positively correlated but that tend to have quite different net effects on various dependent variables. For example, fertility is known to be negatively related to education net of income but positively related to income net of education. Similarly, various measures of tolerance tend to be positively related to education net of income but unrelated or negatively related to income net of education. For this reason the common practice of constructing scales of socioeconomic status should be avoided, and each of the component variables should be included as a separate predictor of the dependent variable of interest.

A useful procedure for deciding whether items "hang together" is to submit them to a *factor analysis.* Factor analysis (or more precisely, *exploratory factor analysis*) is a procedure for empirically determining whether a set of observed correlations can, with reasonable accuracy, be thought of as re ecting, or as generated by, a small number of hypothetical underlying factors. Factor analysis is a well-developed set of techniques, with many variations. However, this chapter is concerned not with the intricacies of factor analysis but with its use as a tool in scale construction. For our present purposes, the optimal procedure is to use *principal factor analysis with iterations* and a *varimax rotation* and then to inspect the *rotated factor matrix.* The varimax rotation rotates the factor matrix in such a way as to maximize the contrast between factors, which is what we want when we are trying to determine whether we can find distinctive subsets of items within a larger set of candidate items. We then choose the items that have high *loadings* on one factor and low loadings on the remaining factor or factors. A rule of thumb for "high" is loadings of .5 or more (which are consistent with correlations of about $.5^2 = .25$ or higher).

TRANSFORMING VARIABLES SO THAT "HIGH" HAS A CONSISTENT MEANING In the context of factor analysis, "high" refers to the absolute value of a factor loading. We would thus regard a loading less than or equal to $-.5$ or greater than or equal to .5 as high. It is important to appreciate, however, that a high negative loading implies that a variable is *negatively* related to the underlying concept. For this reason, it is desirable to transform all variables so they conceptually run in the same direction—that is, so that a high value on the variable indicates a high level of the underlying dimension (from which it then follows that all the indicators should be positively correlated). For example, consider the GSS items *SPKCOM* (*"Suppose this admitted Communist wanted to make a speech in your community. Should he be allowed to speak, or not?"*) and *COLCOM* (*"Suppose he is teaching in a college. Should he be fired, or not?"*). Clearly, a positive response to the first item and a negative response to the second item both indicate support for civil liberties. So to make the interpretation of the factor analysis less confusing, it would be desirable to reverse the scaling of the second item. This can be accomplished easily by transforming the original variable, X, into a reverse-scaled variable, X', using the relation $X' = (k + 1) - X$, where there are k response categories. Similar transformations are helpful in any kind of multivariate analysis.

We then choose those items that meet both criteria—high loadings on the factor and similar relationships to the dependent variable—and combine them into a single scale by first standardizing them (subtracting the mean and dividing by the standard deviation) and then averaging them. These procedures typically produce scales with a mean near zero and a range from some messy negative number around –2.*x* or –3.*x* to some equally messy positive number. For convenience of exposition, it is useful to convert the scale into a range extending from zero to one because then the coefficient associated with the scale gives the expected (net) difference on the dependent variable between cases with the lowest and highest scores on the scale. Such a conversion is easy to accomplish, by solving two equations in two unknowns as you did in school algebra:

$$\begin{aligned} 1 &= a + b(\max) \\ 0 &= a + b(\min) \end{aligned} \qquad (11.3)$$

where "max" is the maximum value of a scale, S, in the data, and "min" is the minimum value of S in the data. This yields a and b, which you then use to transform S into a new variable, S', as follows:

$$S' = a + b(S) \qquad (11.4)$$

CONSTRUCTING SCALES FROM INCOMPLETE INFORMATION When you construct multiple-item scales, it often is useful to compute scale scores even when information on some items is missing. This reduces the number of missing cases. For example, if I am constructing a five-item scale, I might compute the average if data are present for at least three of the five items. This is easy to accomplish in Stata by using the `-rowmean-` command to compute the mean and the `-rowmiss-` command to count the number of missing items, replacing the scale score with the missing value code if the number of missing items exceeds your chosen limit—in the present example, if more than two of the five items have missing values.

If several factors emerge from the factor analysis, we can, of course, construct several scales. Here the problem of validity looms again. Because we ordinarily start with a set of candidate items that a priori we think measure a single underlying concept, we are on the firmest ground if only one factor emerges. If more than one factor emerges, we are forced to consider what concept each factor is measuring. Working from indicators to concepts poses the very real danger that our sociological imagination will get the better of us and that we will invent a concept to explain a set of correlations that reflect sampling error more than some underlying reality. The danger is compounded if we forget that we have invented the concept to explain the data and start treating it as if it has an independent reality—that is, if we *reify* our concept. To be sure that we have actually discovered some underlying reality, we should replicate the items and the scale in some independent data set (perhaps by using a random half of our sample to develop our scales and fit our models and then using the other half of the sample to verify the adequacy of both scales and models). Unfortunately, this seldom is done, because we usually want larger samples no matter how large our sample is. However, the GSS provides such opportunities because an analysis developed using data from one year often can be replicated using data from the preceding or following year. I strongly encourage this kind of independent validation.

Readers familiar with factor analysis may wonder why I suggest choosing a set of candidate items, weighting them equally, and averaging them, in contrast to constructing a scale by using the *factor scores* as weights. The reason is that using the factor scores maximizes the association between the hypothetical underlying concept and the constructed scale *in the sample*. That is, it capitalizes on sampling variability. The result is that the correlations between a scale constructed in this way and other variables are likely to be substantially smaller if the same analysis is replicated using a different data set. By contrast, the *factor-based scaling procedure,* in which the items are equally weighted, is much less subject to cross-sample shrinkage. In this sense factor-based scales are more reliable than are scales constructed using factor scores as weights.

A Worked Example: Religiosity and Abortion Attitudes (Again)

Abortion has become an increasingly salient and emotionally charged issue in recent years. Fundamentalist religious groups (and others) oppose abortion as "murder" while feminists (and others) defend the right of women to control their own bodies. Despite the sharp polarization of opinion regarding abortion, most Americans evidently support the availability of legal abortion under at least some circumstances. Many people find abortion acceptable for medical or therapeutic reasons but not for reasons of personal preference or convenience. Considering the theological underpinning of the "right to life" movement—that a fetus is a person and hence that abortion is tantamount to murder—we might expect strongly religious people to adamantly oppose abortion for personal preference reasons but to be less opposed to abortion for therapeutic reasons, when the "rights" of the fetus must be weighed against the health and safety of the mother. Those who are less religious, by contrast, might be expected to make less of a distinction between the acceptability of abortion for personal preference and therapeutic reasons. If these suppositions are correct, we would expect religiosity to have a weaker effect on attitudes regarding therapeutic abortion than on attitudes regarding abortion for personal preference reasons.

To test this hypothesis, I use data from the 1984 GSS, a representative sample of 1,473 adult Americans. (See downloadable files "ch11.do" and "ch11.log" for estimation details.) I use the 1984 survey because it contains items suitable for constructing a scale of religiosity (discussed later). Specifically, I compare the coefficients in two regression equations:

$$\hat{T} = a + b(F) + c(E) \tag{11.5}$$

and

$$\hat{P} = a' + b'(F) + c'(E) \quad (11.6)$$

where *T, P,* and *F* are, respectively, scales of the acceptability of abortion for therapeutic reasons, the acceptability of abortion for personal preference reasons, and religiosity (faith). *E* is years of school completed, introduced as a control variable because it is known that acceptance of abortion increases with education and that religious fundamentalism is negatively correlated with education in the United States.

The three scales were constructed by factor analyzing items thought to represent the dimension being measured, eliminating items with low factor loadings, converting each item to standard score form, and averaging items. To facilitate interpretation of the regression coefficients, the resulting scales were then transformed so that each had a range of zero (for the lowest level of religiosity and the lowest acceptance of abortion) to one (for the highest level of religiosity and the highest acceptance of abortion).

Candidate items for the scale of religious fundamentalism included the following:

1. *ATTEND: How often do you attend religious services?* (Range: never . . . several times a week).
2. *POSTLIFE: Do you believe there is a life after death?* (no, yes).
3. *PRAY: About how often do you pray?* (Range: never . . . several times a day).
4. *RELITEN: Would you call yourself a strong* [religion named by respondent in response to question on religious preference] *or not a strong* [preference]? (not very strong; somewhat strong [volunteered]; strong).
5. *BIB:* Alternative versions of this question were asked of two-thirds and one-third of the sample, respectively:

   I. *Which of these statements comes closest to describing your feelings about the Bible?*

      a. *The Bible is the actual word of God and is to be taken literally, word for word.*
      b. *The Bible is the inspired word of God but not everything in it should be taken literally, word for word.*
      c. *The Bible is an ancient book of fables, legends, history, and moral precepts recorded by men.*

   II. *Here are four statements about the Bible, and I'd like you to tell me which is closest to your own view:*

      a. *The Bible is God's word and all it says is true.*
      b. *The Bible was written by men inspired by God, but it contains some human errors.*

c. *The Bible is a good book because it was written by wise men, but God had nothing to do with it.*

d. *The Bible was written by men who lived so long ago that it is worth very little today.*

Corresponding categories from versions I and II were combined, except that categories (c) and (d) from version II were combined with category (c) from version I to create a new variable, *NEWBIB.* Before these five items were factor analyzed, *ATTEND* was reversed so that it positively correlated with the remaining items. Also, "don't know" and "no answer" responses were omitted. The sample size was weighted to correct for bias caused by differential household size. After cases with missing values were eliminated, the number of cases available for the factor analysis was 1,292. The items were factor analyzed using iterated principal factoring and varimax rotation.

A single dominant factor emerged, which explained 86 percent of the total variance, with loadings after rotation:

ATTEND	**.787**
POSTLIFE	.196
PRAY	**.573**
RELITEN	**.664**
NEWBIB	.260

Given the pattern of factor loadings, it appears from simple inspection that a three-item scale formed from the three items with the highest loadings is more reliable than a scale that includes either *NEWBIB* alone or both *NEWBIB* and *POSTLIFE.* However, we need not rely on simple inspection. Applying Equation 11.2, we have $\alpha_3 = .665$, $\alpha_{(3+BIB)} = .656$, and $\alpha_5 = .644$. (It turns out that there is an important lesson here: "inspection" can easily lead to error—I thought that the three-item scale would be substantially more reliable than the scales that included items with much lower factor loadings; but in fact they are about equally reliable. In such a case I still would choose the scale composed only of items with high loadings, because it would provide the "purest" measurement of the concept.) Doing this, I then standardized and averaged the three items *ATTEND, PRAY,* and *RELITEN* to form a scale of *FAITH.* To minimize the loss of cases because of missing data, I averaged scores for all individuals who had valid responses to at least two of the three items. This scale was then transformed to a range from zero to one, with one indicating the greatest faith or religiosity.

To create scales measuring attitudes regarding abortion, I factor analyzed the following seven items, using the methods just described.

Please tell me whether or not ***you*** *think it should be possible for a pregnant woman to obtain a* ***legal*** *abortion . . .*

1. *ABDEFECT:* If there is a strong chance of serious defect in the baby?
2. *ABNOMORE:* If she is married and does not want any more children?

3. *ABHLTH:* If the woman's own health is seriously endangered by the pregnancy?
4. *ABPOOR:* If the family has a very low income and cannot afford any more children?
5. *ABRAPE:* If she became pregnant as a result of rape?
6. *ABSINGLE:* If she is not married and does not want to marry the man?
7. *ABANY:* If the woman wants it for any reason?

In each case the possible responses were "Yes," "No," "Don't know," and "No answer." "Don't know" and "No answer" responses were coded as intermediate between "Yes" and "No." Although, as indicated previously, I hypothesized that there are distinctive responses to abortion for therapeutic and personal preference reasons, I nonetheless factor analyzed both subsets of items together to confirm empirically that the two sets of items do in fact behave distinctively.

Two nontrivial factors were extracted, which together explained 96 percent of the variance in the items. Table 11.2 shows the loadings before rotation.

As is evident, all seven items load strongly on Factor 1. But some are positive and some are negative on Factor 2. The pattern of positive and negative loadings on Factor 2 suggests that these items can be subdivided into two distinct factors. Table 11.3 shows the result of executing a varimax rotation, a rotation of the axes of the factor matrix that maximizes the distinction between factors.

TABLE 11.2. Factor Loadings for Abortion Acceptance Items Before Rotation.

	Factor 1	Factor 2
ABDEFECT	.622	.533
ABNOMORE	.864	−.263
ABHLTH	.483	.542
ABPOOR	.831	−.183
ABRAPE	.612	.412
ABSINGLE	.869	−.249
ABANY	.826	−.302

TABLE 11.3. Abortion Factor Loadings After Varimax Rotation.

	Factor 1	Factor 2
ABDEFECT	.274	**.777**
ABNOMORE	**.880**	.208
ABHLTH	.149	**.708**
ABPOOR	**.805**	.247
ABRAPE	.321	**.657**
ABSINGLE	**.876**	.217
ABANY	**.870**	.152

Inspecting these loadings we see that, as hypothesized, two factors underlie abortion attitudes. *ABNOMORE, ABPOOR, ABSINGLE,* and *ABANY* all load strongly on Factor 1 (shown in bold) and weakly on Factor 2, whereas the remaining three items load strongly on Factor 2 (shown in bold) and weakly on Factor 1. These two sets of items correspond to the a priori distinction I made between abortion for personal preference reasons (Factor 1) and abortion for therapeutic reasons (Factor 2).

Figure 11.1 demonstrates that the unrotated and rotated factor structures are simply mathematical transformations of one another and do nothing to change the relationships among the variables. The rotation merely presents the results in a form that makes them more readily interpretable. As noted previously, in the unrotated matrix (solid axes), all items load positively on Factor 1 but some items have positive loadings on Factor 2 and some items have negative loadings. After I rotate the axes 30 degrees counterclockwise (to the dashed lines), all items have positive loadings on both factors, but four (the personal preference reasons) load strongly on the first factor and weakly on the second factor while three (the "therapeutic" reasons) load weakly on the first factor and strongly on the second factor.

Given these results, two separate scales are warranted. I therefore constructed a scale of acceptance of abortion for personal preference reasons, using the four items loading strongly on Factor 1, and a scale of items for therapeutic reasons, using the three items loading strongly on Factor 2. In each case the items were converted to standard form and averaged. I computed averages if valid responses were available for at least three of the four personal preference items and at least two of the three therapeutic items. Again the

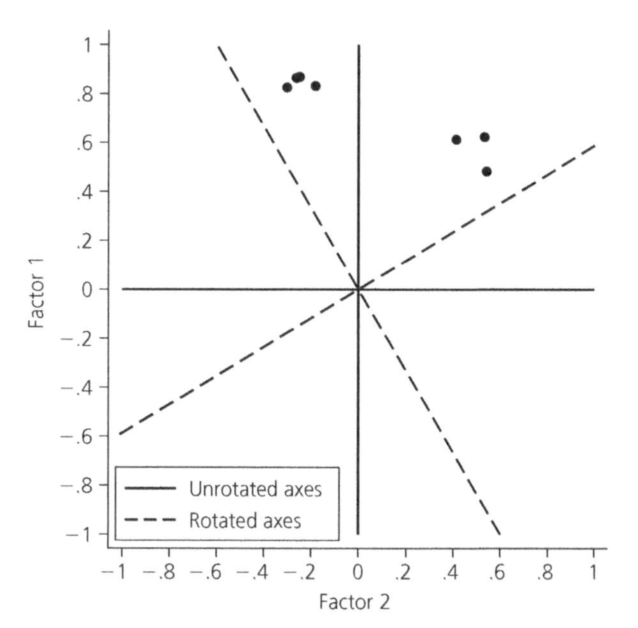

FIGURE 11.1. *Loadings of the Seven Abortion-Acceptance Items on the First Two Factors, Unrotated and Rotated 30 Degrees Counterclockwise.*

scales were transformed to range from zero to one, with one indicating high acceptance of abortion.

The second criterion for scale validity is whether the component items all bear approximately the same relationship to the other variables in the analysis. Ideally, one should assess both the zero-order and net relationships between the component items and the dependent variables. Here, however, the dependent variables are the two abortion attitudes scales. Thus I assess the consistency of the relationships simply by inspecting the correlations among each of the components of all three scales plus the remaining independent variable, education. These correlations are shown in downloadable file "ch11.log." All of the components of each scale show consistency with respect to sign and gross similarity with respect to magnitude in their correlations with the remaining variables. Thus I conclude that combining these items into scales as I have done is appropriate.

Table 11.4 shows the means, standard deviations, and correlations among the three scales and years of school completed, and Table 11.5 shows the coefficients estimated for Equations 11.5 and 11.6. Not surprisingly, the mean for the scale of acceptance of therapeutic abortion is much higher than the mean for the scale of acceptance of abortion for personal preference reasons. (Because each scale is calibrated by converting the lowest score in the sample to zero and the highest score in the sample to one, comparison of the means across scales is not, strictly speaking, legitimate. However, they do indicate where the typical respondent falls relative to the respondents most accepting and least accepting of each category of abortion, and hence can be used to compare the relative acceptance of the two types of abortion.)

TABLE 11.4. Means, Standard Deviations, and Correlations Among Variables Included in Models of the Acceptance of Legal Abortion, U.S. Adults, 1984 (N = 1,459).

Variable	F	T	P	Mean	SD
E: Education	−.007	.164	.252	12.4	3.09
F: Religiosity		−.331	−.347	.675	.223
T: Acceptance of abortion for therapeutic reasons			.474	.823	.307
P: Acceptance of abortion for personal preference reasons				.425	.441

TABLE 11.5. Coefficients of Two Models Predicting Acceptance of Abortion, U.S. Adults, 1984.

	T: Therapeutic		P: Personal Preference	
Metric coefficients	**b**	**s.e.**[a]	**b**	**s.e.**[a]
E: Education	.0161	.0033	.0356	.0038
F: Religiosity	−.454	.040	−.682	.052
Intercept	.930	.050	.444	.066
R^2	.136		.182	
Standardized coefficients				
E: Education	.161		.249	
F: Religiosity	−.330		−.345	

[a]Standard errors were computed using the Stata 10.0 survey estimation procedure, to account for clustering resulting from the fact that the GSS is a multistage sample. All coefficients are larger than twice their standard errors.

As predicted, acceptance of abortion for reasons of personal preference is somewhat more strongly socially structured than is acceptance of abortion for therapeutic reasons. The R^2 for the former is .182, compared with .136 for the latter. Moreover, both of the metric coefficients are substantially larger for the personal preference equation than for the therapeutic equation, indicating that both education and religiosity have a greater impact on attitudes regarding abortion for personal preference reasons than regarding abortion for therapeutic reasons. However, the standardized effect of religiosity is about equally strong for both sets of abortion reasons, whereas the standardized effect of education is much stronger for personal preference abortion.

Seemingly Unrelated Regression

A formal test of whether corresponding coefficients differ significantly in the two equations is available through Zellner's *seemingly unrelated regression* procedure, implemented in Stata as `-sureg-`. This procedure simultaneously estimates models containing some or all of the same independent variables but different dependent variables. When the independent variables are identical across models, the coefficients and standard errors are identical to those from separately estimated equations, but `-sureg-` provides two additional kinds of information—an estimate of the correlation between residuals from each equation and a test of the significance of the difference between corresponding coefficients. In the present case, the correlation between residuals is .38, which tells us that whatever factors other than education and religiosity lead to acceptance of abortion for therapeutic reasons also tend (modestly) to lead to acceptance of abortion for personal preference reasons. The tests of the equality of corresponding coefficients reveal that, as hypothesized, the coefficients for education and religiosity are significantly larger in the personal preference equation than in the therapeutic equation. (See downloadable file "ch11.do" for details on how to implement `-sureg-`.)

Effect-Proportional Scaling

A special kind of scaling problem arises when we have an independent variable that has a nonlinear relationship to the dependent variable of interest. In Chapter Seven I discussed procedures for assessing whether relationships are nonlinear and for representing nonlinear relationships by changing the functional form of equations. One possibility I discussed was to represent nonlinear relationships by converting variables into sets of categories and studying the relationship between category membership and the outcome variable. In this section I describe an extension of categorical representations of variables: *effect-proportional scaling,* which is available in situations in which the dependent variable has a clear metric. (For an example of a research use of effect-proportional scaling, see Treiman and Terrell [1975].)

Suppose, for example, that we are interested in the relationship between educational attainment and occupational status in a nation with a multitrack school system. We might well expect that in such systems occupational attainment depends not only on the *amount* of schooling but on the *type* of schooling completed. How to represent the effect of schooling in a succinct way becomes a difficult problem in such situations. We could, of course, create and report the coefficients for a typology of type-by-extent of schooling, but this is likely to require the presentation of many coefficients. An alternative would be to go one step further and scale educational categories in terms of their *effect* on occupational status. From a technical point of view, this is very simple. We estimate the relationship between occupational status (measured, say, by the International Socioeconomic Index of Occupations [ISEI] [Ganzeboom, de Graaf, and Treiman 1992; Ganzeboom and Treiman 1996]) and a set of dummy variables corresponding to our typology of type-by-extent of schooling, and then we form a new education variable in which each category in the typology is assigned its predicted occupational status.

Doing this maximizes the correlation between educational attainment and occupational status—no other scaling of education would produce a higher correlation (given the same set of categories), and, of course, the correlation is identical to the correlation ratio. Thus the interpretation of the education variable becomes "the highest level of education achieved, calibrated in terms of its average occupational status return." So long as the analyst is candid with the reader that this is what has been done, there can be no objection. The clear advantage of the procedure is that it allows educational attainment to be included succinctly in subsequent analysis and thus permits assessment of how the relationship between educational attainment and occupational status is affected by other factors, and how the relationship differs across subpopulations, for example, by gender or ethnicity.

Here is an example of the construction and use of such a scale. (No log file is shown for this worked example because no new computing techniques are introduced.) In the 1996 Chinese survey analyzed earlier (in Chapters Six, Seven, and Nine; see Appendix A for details on the data set and how to obtain it), education was solicited with a question that included the categories shown in Table 11.6. Although, with the exception of the last two categories, the classification appears to form an ordinal scale of increasing education, it is not evident whether the scale has a monotonic relationship to occupational status. In fact it does not, as can be seen from the means on the ISEI shown in Table 11.6. In particular, vocational and technical middle school graduates tend to achieve substantially higher occupational status than do academic upper middle school graduates who do not go on to university.

I thus created a new education variable in which each category was assigned the mean ISEI score shown in Table 11.6. (A convenient way to do this in Stata is to regress ISEI on the education categories and get predicted values from the regression. The R^2 associated with this regression, .372, is, of course, just the square of the correlation ratio that we encountered in Chapter Five, η^2.) This scale can then be used in other analyses. For example, we might wish to assess the dependence of occupational status on education and father's occupational status for several nations, including China, to assess national similarities and differences in the relative importance of achievement and ascription in occupational status attainment.

ERRORS-IN-VARIABLES REGRESSION

As noted previously, unreliable measurement generally produces weaker measured effects. Thus when variables are measured with differential reliability, the multivariate structure of relationships can be substantially distorted. Because attitude variables often have poor

TABLE 11.6. Mean Score on the ISEI by Level of Education, Chinese Males Age Twenty to Sixty-Nine, 1996.

Level of Education	Mean ISEI	N
Illiterate	18.2	173
Can read	16.0	2
Sishu (before liberation)	20.1	23
Lower primary (*chu xiao*)	21.0	181
Upper primary (*gao xiao*)	22.6	503
Lower middle	28.5	918
Vocational and technical middle	49.7	159
Upper middle (also specialized)	35.5	272
Specialized, including "five big"	61.0	111
University (four years or more)	65.1	65
Graduate school	78.0	1
Imperial degree holder (*xiucai, juren*)	30.5	4
Other	39.0	1
Total	**28.5**	**2,413**

reliability, analyses including such variables often can be misleading. A way of correcting this problem, when measures of reliability are available, is to correct correlations for attenuation caused by unreliability. The Stata command `-eivreg-` (errors-in-variables regression) does this conveniently. The analyst supplies an estimate of the reliability of each variable, and the command makes the adjustment and carries out the regression estimation. (If no estimate is supplied, the variable is assumed to be measured with perfect reliability.)

To show how this procedure works and what consequences it can have, I here present an analysis of the effect of abortion attitudes and religiosity (the three scales created previously) plus race, region of residence, an interaction between race and region, and the natural log of income on political conservatism.

From the previous analysis we have the reliability of the three scales. I take the reliability of the income measure, .8, from Jencks and others (1979, Table A2.13) and assume that race and region of residence are measured without error. Table 11.7 shows the results of OLS estimation without correcting for unreliability of measurement, and errors-in-variables estimation that does correct for unreliability. Because `-eivreg-` does not permit correction of the standard errors for clustering and requires aweights (or fweights), I carried out both conventional and errors-in-variables regression with these specifications.

TABLE 11.7. Coefficients of a Model of the Determinants of Political Conservatism Estimated by Conventional OLS and Errors-in-Variables Regression, U.S. Adults, 1984 (N = 1,294).

	Conventional OLS			Errors-in-Variables		
	b	s.e.	p	b	s.e.	p
Religiosity	0.692	0.170	.000	1.066	0.297	.000
Personal preference abortion	−0.282	0.091	.002	−0.220	0.113	.051
Therapeutic abortion	−0.388	0.134	.004	−0.443	0.200	.027
Black	−0.036	0.173	.836	−0.040	0.174	.816
South	0.144	0.080	.073	0.141	0.080	.079
Black*South	−0.498	0.235	.034	−0.506	0.235	.031
Log income	0.125	0.041	.002	0.148	0.053	.005
Intercept	2.924	0.425	.000	2.467	0.541	.000
R^2	0.068			0.079		
s.e.e.	1.24			1.23		

The effect of adjusting for differential reliability is dramatic—the coefficient associated with religiosity increases by 54 percent. In addition, the coefficients associated with acceptance of therapeutic abortion and with income increase slightly, and the coefficient associated with acceptance of abortion for personal preference reasons decreases slightly. These results indicate clearly how the relative effects of variables in a multiple regression can be distorted if the variables are measured with differential reliability, as these are. (Recall that the reliabilities of the religiosity, therapeutic abortion, income, and personal preference abortion measures are, respectively, .66, .78, .80, and .93.)

Note that with one exception, all the coefficients are about what we would expect: political conservatism increases with religiosity, with income, and, for non-Blacks, with Southern residence (although this last effect is only marginally significant) and decreases as acceptance of both kinds of abortion increase. The unexpected result is that acceptance of therapeutic abortion is a much stronger predictor of political conservatism than is acceptance of abortion for personal preference reasons. From the analysis I presented earlier showing that acceptance of personal preference abortion is more strongly predicted by education and religiosity than is acceptance of therapeutic abortion, I would have expected the opposite result. However, upon re ection, it is evident that the effect of attitudes regarding personal preference abortion is muted by the inclusion of the powerful religiosity variable in the model.

WHAT THIS CHAPTER HAS SHOWN

In this chapter we have seen why multiple-item scales are advantageous: they improve reliability of measurement. We have considered two ways to construct such scales that go beyond simple counts of the kind used in previous chapters. In this chapter we focused primarily on factor-based scaling, which provides a means of purging a scale of items that do not re ect the same underlying dimension as the remaining items or that re ect other dimensions in addition to the main dimension. We also considered effect-proportional scaling, which is useful in establishing a metric for a set of categories by scaling the items according to their effect on some criterion variable. Finally, we considered two extensions of OLS regression: errors-in-variables regression, which corrects regression coefficients for attenuation caused by unreliability—something that can alter our substantive conclusions when variables in a model are measured with differential reliability; and seemingly unrelated regression, which provides a means for comparing models with different dependent variables but (at least some of) the same independent variables.

CHAPTER

LOG-LINEAR ANALYSIS

WHAT THIS CHAPTER IS ABOUT

Log-linear analysis is a technique for making inferences about the presence of particular relationships in cross-classification tables. The first three chapters of this book were devoted to percentage tables. In those chapters we spent considerable time on cross-tabulations, developing rules of thumb for deciding how large a difference between two percentages had to be before we were willing to take it seriously, how to detect interactions among variables, and so on. Log-linear analysis provides a way of formalizing the analysis of cross-tabulations, permitting an assessment of whether relationships observed in a cross-tabulation constructed from sample data are likely to exist in the population from which the sample is drawn and also providing a way of describing the relationships. In this chapter we first consider how to fit a log-linear model to multiway tables, to get the mechanics straight. We then move on to more parsimonious models for two-way tables, using the study of intergenerational occupational mobility as our main substantive example, although these techniques can be applied in many other contexts as well. Additional expositions of log-linear analysis can be found in Knoke and Burke (1980) and in Powers and Xie (2000, chap. 4). This chapter draws heavily on Powers and Xie.

INTRODUCTION

In one sense the model-fitting aspect of log-linear analysis is nothing more than a generalization of the X^2 (chi-square) test for the independence of two variables. Recall that in the usual (Pearson) X^2 test, the observed frequencies in each cell are contrasted with a model of perfect independence, in which the expected frequencies in each cell are simply the product of the marginal frequencies divided by the total number of cases in the table. The size of X^2 then depends on the extent to which the observed frequencies depart from the frequencies expected from the model of independence.

This approach can be generalized to more complex relationships, albeit with a change in the formula. For a bivariate frequency distribution we can write a general formula for expected cell frequencies:

$$F_{ij} = \eta\tau_i^X\tau_j^Y\tau_{ij}^{XY} \tag{12.1}$$

where η (eta) is the geometric mean of the cell frequencies (the geometric mean of k values is the kth root of their product); τ_i^X is the "effect parameter" for the ith category of the X variable (τ is pronounced "tau"); τ_j^Y is similarly defined; and τ_{ij}^{XY} is the effect parameter for the "interaction" of the ith category of X and the jth category of Y.

In Log-Linear Analysis "Interaction" Simply Means "Association"

Note that in the log-linear literature "interaction" is the term for what is called "association" in the older literature on cross-tabulation tables. It is important to recognize that it is not the same as what is called an interaction in both the older tabular literature and in the literature on multiple regression. In those literatures "interaction" refers to the situation in which the relationship between two variables depends on the value of one or more other variables.

The relationship expressed by Equation 12.1 can be shown to hold when the τ's are defined as functions of *odds ratios* (see Appendix 12.A). The *odds* of an observation being in a given category of a variable is just the ratio of the frequency of observations in the category to the frequency of observations not in it. Thus in a class of 20 men and 10 women, the odds of a student in the class being a man are 20/10 = 2:1 ("two to one").

Analyzing the data in Table 12.1, we see that the ratio of the odds of being a letters and science (LS) student, given that one is male, to the odds of being an LS student, given that one is female, is (9/11)/(9/1) = 1:11. So men are one-eleventh as likely to be LS students as are women (and, of course, women are eleven times as likely to be LS students as are men). Odds ratios vary around unity; if the odds of being an LS student were the same for males and females, the odds ratio would be 1.0. An odds ratio of less than 1.0 indicates, in this case, that the odds of being an LS student are smaller for males than for females whereas an odds ratio of greater than 1.0 indicates that the odds of being an LS student are greater for males than for females.

TABLE 12.1. Frequency Distribution of Program by Sex in a Graduate Course.

	Male	Female	Total
Letters and science	9	9	**18**
Management	11	1	**12**
Total	**20**	**10**	**30**

Now suppose we take the natural log of both sides of Equation 12.1. This gives us

$$\begin{aligned}\ln(F_{ij}) &= \ln(\eta\tau_i^X\tau_j^Y\tau_{ij}^{XY}) \\ &= \ln(\eta) + \ln(\tau_i^X) + \ln(\tau_j^Y) + \ln(\tau_{ij}^{XY})\end{aligned} \quad (12.2)$$

which has a *log-linear* form—that is, the left side of the equation is a linear function of the logs of the quantities on the right side of the equation—hence the term, *log-linear analysis*.

Equation 12.2 is sometimes (for example by Leo Goodman [1972, 1043], one of the creators of the method) expressed as

$$G_{ij} = \theta + \lambda_i^X + \lambda_j^Y + \lambda_{ij}^{XY} \quad (12.3)$$

where the λs (lambdas) are (natural) logs of the τ's, θ (theta) is the log of η, and G_{ij} is the log of F_{ij}. An alternative notation, used by Powers and Xie (2000, 107), is

$$\begin{aligned}\ln F_{ij} &= \ln(\tau) + \ln(\tau_i^R) + \ln(\tau_j^C) + \ln(\tau_{ij}^{RC}) \\ &= \mu + \mu_i^R + \mu_j^C + \mu_{ij}^{RC}\end{aligned} \quad (12.4)$$

where μ (mu) = $\ln(\tau)$, and so on. An even more convenient notation, which we also will use, is [XY], which implies that the model of interest includes the explicitly specified interaction and all of the lower order effects. This is sometimes called the *fitted marginals* notation. Equations 12.1 through 12.4 can easily be generalized to more than two variables, as we will see in the following section.

CHOOSING A PREFERRED MODEL

In Equation 12.1 the observed cell frequencies, f_{ij}, are exactly equal to the predicted cell frequencies, F_{ij}, and thus are perfectly predicted because all possible effect parameters are present in the model. Hence Equation 12.1 is known as a *saturated model*.

Such models are, however, generally of little interest. Ordinarily we would search for a simpler representation of the data, in which some of the parameters are assumed to be 1.0 (or in log form, 0)—that is, in which some of the possible effects are absent. In such cases each F_{ij} represents the frequency in the *ij*th cell that *we would expect if the model were true* (and of course each G_{ij} represents the natural log of the expected frequency).

There are two basic approaches to the task of choosing a preferred model. The more common is to engage in relatively atheoretical "data dredging," positing successively more complex models until we get a reasonably close fit to the data. (I use the term "*relatively* atheoretical" because virtually all analysis is informed by some idea, no matter how implicit or how vague.) The second approach, which I prefer, is explicitly to test hypotheses by estimating and comparing models that represent competing hypotheses. It is the latter approach that constitutes a formalization of our previously encountered procedures for interpreting percentage tables.

Model Selection Based on Goodness of Fit

Let us first work through the data-dredging approach, using the data from Table 12.1. The simplest model for a two-variable cross-tabulation posits an equal number of cases in every cell, in which case $F_{ij} = \eta$ for all *ij* (because the odds ratios, and hence the τ's, would all equal 1.0). Although simplest, this model is not very interesting, because equal cell frequencies are rarely substantively plausible. Nonetheless we can test whether this model fits the data (within the limits of sampling error) by estimating L^2, known as the *likelihood ratio* X^2 or, in Stata, as the *deviance*:

$$L^2 = 2\sum_{i=1}^{I}\sum_{j=1}^{J}\left(\frac{f_{ij}}{F_{ij}}\right) \tag{12.5}$$

where F_{ij} is, as noted previously, the expected frequency of the *ij*th cell under a given null hypothesis, f_{ij} is the observed frequency of the *ij*th cell, and the sum is over all cells in the table. (The usual X^2 statistic also can be used to estimate the goodness of fit of alternative models, and L^2 and X^2 are asymptotically equivalent. Advantages of L^2 are discussed in Bishop, Fienberg, and Holland 1975, 57–62.) Under the null hypothesis, L^2 is distributed approximately as X^2 with degrees of freedom equal to the number of nonredundant parameters set to 1.0. Four independent parameters can be estimated in a two-by-two table (because for dichotomous categories the odds of being in category one rather than category two are necessarily the reciprocal of the odds of being in category two rather than category one, and the same is true of the conditional odds in two-by-two tables):

$$\begin{aligned} &\eta \\ &\tau_1^X = 1/\tau_2^X \\ &\tau_1^Y = 1/\tau_2^Y \\ &\tau_{11}^{XY} = \tau_{22}^{XY} = 1/\tau_{12}^{XY} = 1/\tau_{21}^{XY} \end{aligned} \tag{12.6}$$

Because for the simplest model we estimate only η, we have three remaining (residual) degrees of freedom. As expected, the fit to Table 12.1 is poor: $L^2 = 10.96$, which implies that the observed distribution would occur by chance only about 1 percent of the time if the cells sizes were equal in the population (precisely, $p = .012$). So we conclude that this model does not fit the data; that is, we reject the null hypothesis that the cell sizes are equal. (For details on how to estimate such models, see the worked example on anticommunist sentiment later in the chapter and also the downloadable files "ch12_1.do" and "ch12_1.log.")

To be sure, the "population" in this example is problematic because we are presumably studying the characteristics of all individuals enrolled in a particular course and hence might think of these individuals as constituting the population rather than a sample. However, we might regard the individuals enrolled in the course at any given time as a sample of all possible sets of individuals ever enrolled in the course, and hence generalize from the particular set of observations to what we might expect "for this course over the long run" or for "courses like this." This sort of use of statistical inference is in fact quite common in research practice (see the discussion of the concept of *superpopulation* in Chapter Sixteen).

L^2 Defined It can be shown that L^2 is minus twice the difference in the log likelihoods for restricted and unrestricted models, where the unrestricted model is, unless otherwise indicated, the saturated model. (See Appendix 12.B for a brief introduction to maximum likelihood estimation and a definition of the likelihood.)

Next we might test the possibility that the two variables X and Y are independent, so that the cell frequencies are simply a function of the marginal distributions. We would write this: [X][Y]. In this case we are estimating three parameters: η, τ^X, and τ^Y. Only τ^{XY} is set to 1.0, so we have 1 degree of freedom. In this case $L^2 = 6.35$, so once again the fit is poor ($p = .012$, only coincidentally the same as for the previous model), indicating that there is an association between X and Y—we cannot predict the cell frequencies simply from the marginals.

To obtain a good fit in this example, it is necessary to estimate all four parameters, which uses up all degrees of freedom (and hence, as we have noted, ensures a perfect fit). We write this: [XY]. Note that in this exposition we are dealing with hierarchical models, which means that every higher-order relationship implicitly contains all lower-order relationships. Hence [XY] $\Rightarrow$[η][X][Y][XY]. We will return to this point later in the chapter.

So far we have done nothing that could not be done with the usual χ^2 test for independence. However, the same procedures apply to cross-tabulations containing more than two variables, and also to polytomies as well as dichotomies. Consider Table 12.2,

TABLE 12.2. Frequency Distribution of Level of Stratification by Level of Political Integration and Level of Technology, in Ninety-Two Societies.

	No Metalworking		Metalworking at Least	
	Stateless	State	Stateless	State
Egalitarian	24	5	4	1
Status distinctions only	8	4	4	7
Two or more classes	1	4	2	28

Source: Computed from Murdock and Provost (1973).

a cross-tabulation of level of stratification by level of political integration and level of technology in ninety-two societies.

In the data-dredging approach to log-linear analysis, it is common to posit an initial, or *baseline*, model of complete independence among the variables in the table—in the present case the model of no association between technology [T], political integration [P], and stratification [S]. We do this by fitting the model [T][P][S]. For this model, $L^2 = 84.68$, with 7 degrees of freedom; the goodness-of-fit statistics for this model, and several others, are shown in Table 12.3. Clearly, this model does not fit the data ($p < .0000$), but we will nonetheless make use of it momentarily.

We might next posit an association, or interaction, between level of political integration and degree of stratification and assume that neither of these variables is related to the level of technology. That is, we fit [T][PS] (Model 2 in Table 12.3). This model posits that the observed cell frequencies can be accounted for (within the limits of sampling error) by the univariate distribution of level of technology and the bivariate distribution of the degree of stratification by the level of political integration. Estimating this model yields $L^2 = 41.54$, with 5 *d.f.*

Although the large L^2 tells us that the model does not provide an accurate fit to the data ($p < .0000$), we might still want to know whether the prediction is improved relative to the baseline model of complete independence. To see this, we subtract one L^2 from the other and similarly subtract the degrees of freedom, and then we get the *p*-value associated with the new L^2 and new *d.f.* It also is common to show the L^2 for each of the subsequent models as a proportion of the L^2 for the baseline model, L^2_B, and to show the index of dissimilarity, Δ, between the observed frequencies and the frequencies expected under the model, and also *BIC*. The differences in these measures can be computed as well; the differences between Models 1 and 2 are shown in the penultimate row of Table 12.3. All of these computations provide information on the goodness of fit of the models and the improvement in goodness of fit realized by positing successive elaborations of a model.

TABLE 12.3. Models of the Relationship Between Technology, Political Integration, and Level of Stratification in Ninety-Two Societies.

Model	L^2	d.f.	p	*BIC*	L^2/L^2_B	Δ
(1) [T][P][S]	84.68	7	.000	53.0	1.00	41.6
(2) [T][PS]	41.54	5	.000	18.9	.49	30.4
(3) [P][TS]	46.08	5	.000	23.5	.54	31.5
(4) [S][TP]	60.48	6	.000	33.3	.71	31.4
(5) [TP][TS]	21.88	4	.000	3.8	.26	16.8
(6) [TP][PS]	17.34	4	.002	−0.8	.20	15.2
(7) [TS][PS]	2.94	3	.401	−10.6	.03	5.3
(8) [TP][TS][PS]	0.60	2	.739	−8.4	.01	2.5
(2) versus (1)	43.14	2	.000[a]	−34.1[b]	.51	11.2
(8) versus (7)	2.34	1	.126	2.2	0.02	2.8

[a]This is the probability for L^2 = 43.14 (= 84.68 – 41.54), with 2 (= 7–5) *d.f.* It can be obtained by using Stata's `-chi2tail-` command.

[b]Here a negative *BIC* indicates an improvement in fit.

Because the probability of L^2 = 43.2, with 2 *d.f.*, is less than .000, we conclude that positing an association between political integration and stratification significantly improves the fit of the model. Similarly, the difference in *BIC* tells us that the second model is much more likely than the first, given the data (although neither is as likely as the saturated model because both *BIC*s are positive).

We can get a quantitative estimate of the extent of improvement in the fit of a model from the two remaining sets of coefficients. From the ratio of the L^2s, we see that positing an association between the degree of stratification and the level of political integration reduces the lack of fit of the model to the data by about half relative to the baseline model of complete independence among the three variables.

Finally, from the rightmost column of Table 12.3 we note that the model of complete independence misclassifies about 42 percent of the cases in the table (that is, 42 percent of the cases would have to shift categories for the expected distribution to be identical to the observed distribution—recall the discussion of the Index of Dissimilarity, Δ, in Chapter Three), whereas the second model misclassifies only 30 percent of the cases.

Because model [T][SP] does not fit the data well, we would evaluate still other models, searching for the most parsimonious model that does fit well. Table 12.3 shows goodness of fit statistics for eight models (all the logically possible models except the saturated model and the model that assumes all cells have the same frequency). Continuing to examine the coefficients in Table 12.3, we see that Model 7, [TS][PS], fits the data quite well. This model posits that both the level of technology and the level of political integration are associated with the degree of stratification but that the level of technology and the level of political integration are unrelated, net of their association with stratification. It misclassifies only about 5 percent of the cases in the table and also reduces the baseline L^2 by 97 percent (= 100*(1.0 − 0.03)).

Although Model 8, which posits that each pair of variables is associated, provides an even better fit, it might be argued that it overfits the data. The penultimate model, [TS][PS], fits nearly as well and would be my choice as the final model on the ground of parsimony, especially because the improvement in fit between Model 7 and Model 8 is not significant (2.94 − 0.60 = 2.34; $p = .126$).

Note that the use of tests of significance in this context is the opposite of their usual role as a decision rule for *rejecting* a null hypothesis; here we want to decide whether to *accept* a null hypothesis; that is, a model. Accordingly, we would like to minimize Type II (β) error (the probability of accepting a null hypothesis when it is false) rather than Type I (α) error (the probability of rejecting a null hypothesis when it is true). Unfortunately, there is no direct way to do this, and so we must settle for a computation of Type I error. A useful rule of thumb is to accept a model if α is greater than 0.2. However, the larger the sample size, the smaller α will tend to be, so for very large samples we may wish to accept a model even when α is quite small. As we will see momentarily, *BIC* offers an alternative and more satisfactory method of model selection.

One additional coefficient is shown in Table 12.3, *BIC*, the Bayesian Information Criterion (Raftery 1986, 1995a, 1995b), which we first encountered in Chapter Six. Recall *BIC*'s definition:

$$BIC = -2[\ln(B)] \tag{12.7}$$

where B is the ratio of the (unknown) probability of some model, M, being true to the (unknown) probability of the saturated model being true, given the data. For log-linear models, *BIC* is estimated by

$$L^2 - (d.f.)[\ln(N)] \tag{12.8}$$

where L^2 is the likelihood ratio χ^2 for Model M; $d.f.$ is the residual degrees of freedom for Model M; and N is the number of cases in the table. When *BIC* is negative, Model M is preferred to the saturated model. When several models are compared, the model with the most negative *BIC* is most preferred because it has the greatest likelihood of being true given the data. Here, Model 7 is more likely than Model 8 given the data. Combining the information obtained from the L^2 and *BIC* contrasts of Models 7 and 8, we see that Model 7 is to be preferred.

The real value of *BIC* is in the comparison of models for very large samples because when the sample is large, often no model (except, of course, the saturated model) fits the data by conventional standards. When that happens, *BIC* is of great use in helping us choose among models. For this reason *BIC* has become the conventional method for assessing alternative models in log-linear analysis. An additional advantage of *BIC*, noted in Chapter Six, is that it can be used to compare nonnested models.

Theory-Based Model Selection

The second approach to model selection is to contrast models that represent alternative hypotheses about relationships among variables—that is, to do theory-driven rather than data-dredging model selection. For example, we might ask whether the association between the degree of stratification and the level of political integration can be explained by their mutual dependence on the level of technology. If the answer is yes, we would expect [TP][TS] to fit the data, because this model implies that the observed frequencies in the table are generated by an association between technology and political integration and an association between technology and stratification in the absence of an association between political integration and stratification. As we see in Table 12.3 (Model 5), this model does not fit the data, because $L^2 = 21.88$ with 4 *d.f.* ($p < .000$). Hence we reject the hypothesis.

Effect Parameters

As is shown in Appendix 12.A, the parameters associated with the interaction terms in log-linear models (for example, τ_{ij}^{XY} in Equation 12.1) can be interpreted to indicate the direction and strength of associations in cross-tabulation tables. Note, however, that the parameters for two-way interactions involving dichotomies are shown relative to geometric means of the expected frequencies. When more than two-way interactions or more than two categories are involved, the interpretation becomes more complex. Moreover, by default Stata uses a "dummy-variable" parameterization. When a dummy-variable parameterization is used, the parameters for two-way interactions give the odds ratios (or log odds) for the specified categories relative to the reference categories.

Because the effect parameters are not very straightforward, most analysts use log-linear analysis to test hypotheses about the presence or absence of particular associations (interactions) in the table but then discuss the table in terms of percentage differences, which are much more familiar to ordinary readers. This is particularly so when the software used to estimate the models shows the parameters in dummy variable form—that is, as deviations from an omitted category that is set to 0 in the log form or to 1.0 in the multiplicative form—because coefficients expressed in dummy variable form are difficult to interpret in the log-linear context.

My recommendation is that you use log-linear modeling when you want to test specific hypotheses about relationships in cross-tabulation tables, because it is an extremely powerful tool for doing this job. However, once you settle on a preferred

TABLE 12.4. Percentage Distribution of Expected Level of Stratification by Level of Political Integration and Level of Technology, in Ninety-Two Societies (Expected Frequencies from Model 7 Are Percentaged).

	No Metalworking		Metalworking at Least	
	Stateless	State	Stateless	State
Egalitarian	78.1	33.2	33.1	2.6
Status distinctions only	20.5	37.2	46.2	15.7
Two or more classes	1.4	29.6	20.7	81.7
Total	**100.0**	**100.0**	**100.0**	**100.0**
N	(30.6)	(15.4)	(12.4)	(33.6)

model, I suggest you interpret either the observed distribution or the expected distribution implied by the model. The point of percentaging the expected rather than the observed frequencies is that unsystematic variability is removed; however, you should be sensitive to the possibility that deviations of observed from expected cell frequencies may reveal relationships not adequately captured by the model.

Table 12.4 shows the percentage distribution of level of stratification by level of political integration and level of technology implied by Model 7, which posits an association between the level of technology and the degree of stratification and between the level of political integration and the degree of stratification but not between the level of technology and the level of political integration. Because the model fits well, the distribution of expected percentages closely parallels what we would have found had we percentaged Table 12.2. As we see, within levels of technology, state societies tend to have more complex stratification systems than stateless societies and, within levels of political integration, societies with metalworking technology tend to have more complex stratification systems than societies lacking metalworking technology. (One limitation of this approach is that the marginal frequencies in the expected table generally do not match those in the corresponding table of observations. For a method that recovers the marginal distributions, see Kaufman and Schervish [1986].)

Another Worked Example: Anticommunist Sentiment

The optimal way to carry out log-linear analysis using Stata is to use the `-glm-` (generalized linear model) command, which permits the estimation of a wide variety of linear models. Indeed, as should be evident from Equation 12.2, log-linear analysis is just a specific case of the familiar linear model, in which the dependent variable is the natural log of the number of cases in a cell of a multiway cross-tabulation and the independent

TABLE 12.5. Frequency Distribution of Whether "A Communist Should Be Allowed to Speak in Your Community" by Schooling, Region, and Age, U.S. Adults, 1977 (N = 1,478).

Age (A)	Region (R)	Schooling (S)	Communist Speaker (C)	
			Allow	Not Allow
39 or younger	South	No college	72	71
		College	55	22
	Non-South	No college	161	92
		College	157	25
40 or older	South	No college	65	162
		College	23	23
	Non-South	No college	197	214
		College	107	32

variables are dummy variables for the categories that make up the variables included in the cross-tabulation. Although a user-written Stata `-ado-` file (Judson 1992, 1993) can be used to do hierarchical log-linear analysis, the advantage of using `-glm-` is twofold: it retains the linear model framework, and all of the Stata post-estimation commands are available. To show how to carry out log-linear analysis using the `-glm-` command, I analyze Table 10 from Knoke and Burke (1980); a comparison of my results with theirs may provide additional insight.

Suppose we are interested in the relationship between age (thirty-nine and younger versus forty and older), region of residence (South versus non-South), schooling (some college versus high school or less), and tolerance for civil liberties, as measured by a question on whether a communist should be allowed to give a speech in your community. A multiway frequency distribution of these variables, based on data from the 1977 GSS, is shown in Table 12.5.

Analysis Strategy The first step in carrying out a log-linear analysis of Table 12.5 is to estimate a baseline model. Because my interest is in the effect of age, region, and schooling on tolerance of communists, a reasonable baseline model is [C][ARS]. That is, I fit the three-variable relationship among age, region, and schooling exactly, but I assume that none of these variables is related to tolerance of communist speakers. As a second step I posit [CA][CR][CS][ARS]. That is, I continue to fit the three-variable relationship exactly and in addition posit effects of each of the independent variables on tolerance of communist speakers ("interactions" between each of age, region, and schooling, respectively, and tolerance of communist speakers). If my second model yields a good fit, I then try to simplify the model by omitting specific two-variable interactions. If my second model does not yield a good fit, I explore more complicated models by fitting various three-variable interactions involving tolerance of communists plus pairs of the independent variables.

Implementation To carry out the analysis using `-glm-` in Stata, I first read in the contents of Table 12.5 as a data set, where each cell is an observation and the variables are the response categories for each variable plus an additional variable that gives the count in each cell. I thus create a data set, call it "knoke.raw":

```
1   1   1   1    72
1   1   1   2    71
1   1   2   1    55
1   1   2   2    22
1   2   1   1   161
1   2   1   2    92
1   2   2   1   157
1   2   2   2    25
2   1   1   1    65
2   1   1   2   162
2   1   2   1    23
2   1   2   2    23
2   2   1   1   197
2   2   1   2   214
2   2   2   1   107
2   2   2   2    32
```

and then read the data into Stata with the following command:

```
infile a r s c count using knoke.raw, clear
```

Recall that the baseline model [C][ARS] is a shorthand way of representing the model [C][A][R][S][AR][AS][RS][ARS]. Thus, I need to specify each of the terms in the model. Because the Stata command to create product terms for categorical variables, `-xi-`, does not permit more than two-way products, I take advantage of a user-written `-ado-` command, `-desmat-` (Hendrickx 1999, 2000, 2001a, 2001b), to specify the required variables. (See the downloadable files "ch12_1.do" and "ch12_1.log" for details.) Also, because `-glm-` does not provide all the coefficients shown in Table 12.3 and produces an incorrect estimate of *BIC* (given the way I have specified the problem, `-glm-` counts as cases the number of cells in the table rather than the number of people in the sample), I have written a small -do- file, -gof.do- (for "goodness of fit"), and a terse version, -gof2.do-, to provide these coefficients; these -do- files also are available in the packet of downloadable files for this chapter.

The -glm- command works just like every other Stata estimation command, with one exception: because it can handle many kinds of linear models you must specify which sort of model you want, using the -family- option (to indicate the distributional form). Here we want a "Poisson" model because the Poisson distribution is appropriate for count variables such as our dependent variable, "count." Specifying the Poisson family is what makes this a log-linear model.

Invoking -gof.do- (or -gof2.do-) after the -glm- command generates the coefficients shown in the first line of Table 12.6. I then repeat the process for a model that allows A, R, and S each to be related to C but not to interact (in the conventional sense of interaction) in their relationship to C, that is, [ARS][AC][RC][SC]. These commands yield the coefficients shown on the bottom line of Table 12.6 (Model 8). Clearly, this model fits the data well by all criteria, indeed so well as to suggest that a simpler model might also fit the data well. To determine this, I estimate all intermediate models. Doing so yields the remaining coefficients in the table.

Inspecting these statistics, we see that none of these models fits the data adequately. Hence, I settle on [ARS][AC][RC][SC] as my preferred model. Evidently, age, region,

TABLE 12.6. Goodness-of-Fit Statistics for Log-Linear Models of the Associations Among Whether "a Communist Should Be Allowed to Speak in Your Community," Age, Region, and Education, U.S. Adults, 1977.

Model	L^2	d.f.	p	*BIC*	L^2/L^2_B	Δ
(1) [ARS][C]	200.48	7	.000	149.4	1.00	15.1
(2) [ARS][AC]	138.48	6	.000	94.7	.69	10.7
(3) [ARS][RC]	149.57	6	.000	105.8	.75	13.1
(4) [ARS][SC]	87.75	6	.000	44.0	.44	8.2
(5) [ARS][AC][RC]	84.72	5	.000	48.2	.42	9.9
(6) [ARS][AC][SC]	48.69	5	.000	12.2	.24	7.8
(7) [ARS][RC][SC]	44.74	5	.000	8.2	.22	7.7
(8) [ARS][AC][RC][SC]	2.92	4	.571	−26.3	.01	1.5

TABLE 12.7. Expected Percentage (from Model 8) Agreeing That "A Communist Should Be Allowed to Speak in Your Community" by Education, Age, and Region, U.S. Adults, 1977.

Age	Region	No College	College
39 or younger	South	46.3 (143)	73.2 (77)
	Non-South	65.7 (253)	85.8 (182)
40 or older	South	29.2 (227)	56.7 (46)
	Non-South	47.8 (411)	74.3 (139)

Note: Cell frequencies are shown in parentheses.

and education all affect attitudes regarding communist speakers. To see what these effects are, I percentage the table of frequencies predicted by this model. (To see how to get these, consult the downloadable files, "ch12_1.do" and "ch12_1.log.") These percentages are shown in Table 12.7. The table clearly shows that, controlling for each of the other factors, those who are better educated, younger, and non-Southern are more likely to support the right of a communist to give a speech. In each comparison, the percentage differences always go in the same direction and are quite substantial.

The attitudes reported here are from thirty years ago, during the height of the Cold War. It would be of interest to determine whether the same pattern holds today. To do this within a log-linear framework you would need to construct a second data set, based on recent data (for example, the 2006 GSS), to append the second data set to the first, with an additional variable (*T*, for "time"), and then to assess whether it is necessary to posit an effect of time (or of interactions between time and any of the two-variable associations) to adequately represent the data. That is, you would estimate [ARS][AC][RC][SC], [ARS][AC][RC][SC][T], and [ARS][ACT][RCT][SCT], and perhaps some intermediate models, and compare their goodness of fit. If none of the more elaborate models produced a better fit than [ARS][AC][RC][SC], which is just Model 8 replicated for the pooled data, you would conclude that attitudes regarding the rights of communists have not changed between 1977 and 2006. If [ARS][AC][RC][SC][T] emerged as the preferred model, you would conclude that there had been an across-the-board change (presumably an increase) in support for the civil liberties of communists. If [ARS][ACT][RCT][SCT] emerged as the preferred model, you would conclude that the structure of the relationships between age, region, and education, respectively, and support for the civil rights of communists changed between 1977 and 2006. If an intermediate model were preferred, you would conclude that the structure of some, but not all, of the relationships had changed.

Doing Log-Linear Analysis with Polytomous Variables

Although the example we have worked through, regarding the civil rights of communists, happened to involve only dichotomous variables, log-linear analysis is equally appropriate for the analysis of polytomous variables. However, in this case we need to form product terms for each polytomous variable involving each of the $k - 1$ dummy variables required to represent the k categories of the variable. To see how to do this, consider Table 12.8, which shows the frequency distribution for a four-way cross-tabulation of voting [V] by race [R], education [S], and voluntary association membership [M] (this is Table 3 in Knoke and Burke 1980). Voting, race, and membership are dichotomous variables, but education is a three-category variable and thus requires that we create two dummy variables: S2 (= 1 for high school graduates and = 0 otherwise) and S3 (= 1 for those with at least some college and = 0 otherwise), with those lacking high school graduation as the omitted category.

Suppose we are interested in estimating a model in which race, education, and membership all affect voting, but—as in the previous example—we do not care about the relationships among race, education, and membership and hence posit a three-way interaction among them, which allows this part of the model to fit perfectly. Our model is then [VR][VS][VM][RSM] (note the similarity to Model 8 in Table 12.6). We would specify this model with the `-glm-` command:

```
glm count r s2 s3 m rs2 rs3 rm s2m s3m rs2m rs3m v vr vs2
   vs3 vm,family(poisson)
```

where each of the compound variables is a product term—for example, `rs2=r*s2`, and so on. (See the downloadable files "ch12_1.do" and "ch12_1.log" for the specification of additional models and for the output. Note especially how I use Stata "macros" to reduce the tedium of specifying very long commands such as the one just shown. Using the macros also provides the great advantage of minimizing the chance of error.)

Doing Log-Linear Analysis with Individual-Level Data

So far we have seen how to carry out log-linear analysis from existing tabular data. However, we more commonly are in the situation of working with a data set consisting of individual records from a sample survey or census. Thus we need a way to carry out our analysis starting from unit-record data. This is easy to do in Stata, by using the `-collapse-` command. (Downloadable file "ch12_1.do" shows the details; see also "ch12_1.log.")

PARSIMONIOUS MODELS

Thus far we have dealt with models that posit a global association, or absence of global association, between combinations of variables. Often, however, we would like to test particular hypotheses regarding the *structure* of cross-classification tables—that is, whether tables can be described by relatively simple models that generate the observed

TABLE 12.8. Frequency Distribution of Voting by Race, Education, and Voluntary Association Membership.

Race	Education	Membership	Voted	Didn't Vote
White	<High School	None	114	122
		One or more	150	67
	High School Graduate	None	88	72
		One or more	208	83
	College+	None	58	18
		One or more	264	60
Black	<High School	None	23	31
		One or more	22	7
	High School Graduate	None	12	7
		One or more	21	5
	College+	None	3	4
		One or more	24	10

Source: Adapted from Knoke and Burke (1980, Table 3).

pattern of frequencies in the table. The development of such models to describe patterns of intergenerational occupational mobility has been a lively enterprise over the past thirty years or so, but the formal models developed in this context have applications far beyond the study of social mobility (for example, Radelet and Pierce 1985; Schwartz and Mare 2005; Roberts and Chick 2007; Domanski 2008). Still, it is convenient to illustrate these models in the context of mobility analysis. (See downloadable files "ch12_2.do" and "ch12_2.log" for details on the Stata procedures used to estimate the models in the remainder of the chapter.)

It is helpful to begin by deriving a general expression for log odds ratios. Recall Equation 12.4, which gives the natural log of expected frequencies for a two-variable table as a function of a set of μ parameters. From Equation 12.4 we can write an expression for the log odds ratio of the expected frequencies for cells formed from any pair of rows (i and i') and columns (j and j') in a two-variable table:

$$\log\theta = \log\frac{F_{ij}/F_{ij'}}{F_{i'j}/F_{i'j'}} = \log\frac{F_{ij}F_{i'j'}}{F_{ij'}F_{i'j}} = \log F_{ij} + \log F_{i'j'} - \log F_{ij'} - \log F_{i'j}$$

$$= (\mu + \mu_i^R + \mu_j^C + \mu_{ij}^{RC}) + (\mu + \mu_{i'}^R + \mu_{j'}^C + \mu_{i'j'}^{RC}) \tag{12.9}$$

$$-(\mu + \mu_i^R + \mu_{j'}^C + \mu_{ij'}^{RC}) - (\mu + \mu_{i'}^R + \mu_j^C + \mu_{i'j}^{RC})$$

$$= \mu_{ij}^{RC} + \mu_{i'j'}^{RC} - \mu_{ij'}^{RC} - \mu_{i'j}^{RC}$$

When dummy-variable coding is used, as in Stata's `-glm-` command, and i' and j' are the reference categories, the right side of Equation 12.9 simplifies to μ_{ij}^{RC}, which makes clear that the interaction parameters represent the log odds ratios for each cell relative to the omitted categories (ordinarily the first row and first column).

(Note that to uniquely identify the coefficients, it is necessary to impose constraints. Two different constraints, or "normalizations," typically are used. One is effect coding [used in Equation 12.6 and Appendix 12.A], which expresses coefficients as deviations from the grand total by requiring that the log-form coefficients for each variable sum to zero. The other constraint, dummy-variable coding, codes one category [in Stata, the first category] of each variable as zero.)

In the fully saturated model there is a unique coefficient for each cell of the table except, with dummy-variable coding, the cells in the first row and first column. This model can be represented by the following *design matrix* (for a seven-by-seven table):

1	1	1	1	1	1	1	
1	2	3	4	5	6	7	
1	8	9	10	11	12	13	
1	14	15	16	17	18	19	= full_dm
1	20	21	22	23	24	25	
1	26	27	28	29	30	31	
1	32	33	34	35	36	37	

Note that a design matrix is simply a variable, with one value per cell, that imposes equality constraints on some subset of cells—all cells with the same value are constrained to have equal coefficients. This design matrix specifies that all the coefficients for the first row and first column are equal; in fact, they are (implicitly) zero by virtue of the dummy-variable coding. None of the remaining coefficients is constrained to be equal. This model uses all the available information, and the observed counts in each cell are fit exactly.

Note that in Stata's `-glm-` command, the specification

```
xi:glm count i.X i.Y i.full_dm,family(poisson)
```

TABLE 12.9. Frequency Distribution of Occupation by Father's Occupation, Chinese Adults, 1996.

Father's Occupation When R Age 14	Respondent's Occupation in 1996							
	Prof.	Cadre	Cler.	Sales	Ser.	Man.	Agric.	Total
1. Professional	57	11	8	30	9	61	69	**245**
2. Cadre	36	13	17	16	6	35	31	**154**
3. Clerical worker	14	4	8	8	0	22	16	**72**
4. Sales worker	9	9	2	21	7	28	20	**96**
5. Service worker	10	5	4	11	10	12	10	**62**
6. Manual worker	46	22	23	50	26	236	124	**527**
7. Agricultural worker	131	61	43	140	59	416	2,495	**3,345**
Total	**303**	**125**	**105**	**276**	**117**	**810**	**2,765**	**4,501**

produces results identical to the usual way of specifying the saturated model:

```
xi:glm count i.X*i.Y,family(poisson)
```

That is, `-glm-` creates a design matrix like that of "full_dm" when the interaction is specified.

For the sake of concreteness, let us consider some real data. Table 12.9 shows the bivariate frequency distribution of respondent's occupation by father's occupation, for my 1996 Chinese sample (used in several earlier chapters and documented in Appendix A). Although it is customary to analyze the mobility of men and women separately, I have pooled men and women to increase the sample size. My justification for doing this is that the data more or less satisfy Powers and Xie's two conditions for *collapsibility* of a three-way table into a two-way table: that the three-way interaction is zero and that one of the two-way interactions involving the variable to be collapsed is zero (Powers and Xie 2000, 133–135).

To test the first condition, I contrast a model (call this Model A) that posits that, net of the marginals, the father-respondent mobility pattern is the same for men and women (that is, [SF][SR][FR], where S = sex, F = father's occupation when the respondent was age fourteen, and R = respondent's occupation in 1996) against the saturated model. Because the saturated model fits perfectly, this amounts to assessing the fit of my posited model. It fits fairly well: $L^2 = 52.0$ with 36 *d.f.*, which yields $BIC = -251$; $\Delta = 2.4$; and $p = .041$, which is only marginally significant. Given the relatively large size of the sample, I am inclined to focus on the *BIC* rather than the *p*-value and conclude that the first condition is satisfied.

To test the second condition, I contrast a model (call this Model B) that omits the interaction between sex and father's occupation—that is, [SR][FR]—against Model A. The substantive argument for this is that in China, where almost all women are in the labor force, we should expect no difference in the distribution of father's occupation for employed men and women. To contrast the two models, I take the difference in the L^2 and the difference in the degrees of freedom to get the *p*-value for the improvement of goodness of fit resulting from the addition of [SF] and also get the difference in *BIC* values. Although the fit of Model A is significantly better by classical standards ($p = .019$ [$L^2_B - L^2_A = 67.18 - 52.03 = 15.15$; $d.f._B - d.f._A = 42 - 36 = 6$]), Model B is more likely given the data ($BIC_B - BIC_A = -285.9 - [-250.6] = -35.3$). Again, I am inclined to put more weight on the *BIC* difference and conclude that the second condition is satisfied. Thus I am willing to pool men and women for the subsequent analysis, which effectively doubles the sample size.

Table 12.10 shows the coefficients for the saturated model (see "ch10_2.do" to see how these coefficients were computed using Stata). As we have seen, these coefficients are not readily interpreted directly. However, in the present case it may be of interest to contrast particular cells in the table. For example, we might ask about the relative chances of the child of an agricultural worker becoming an agricultural worker instead of a manual worker compared to the corresponding odds for the child of a manual worker. From Equation 12.9 it is evident that the log odds ratio can be computed as

$$\begin{aligned}\log\theta &= \mu^{RC}_{77} + \mu^{RC}_{66} - \mu^{RC}_{76} - \mu^{RC}_{67} \\ &= 2.756 + 1.567 - 1.088 - .801 \\ &= 2.434 \qquad (12.10)\end{aligned}$$

which implies that the relative odds are 11.40 ($= e^{2.434}$); that is, the children of agricultural workers are more than eleven times more likely to become agricultural workers themselves, rather than becoming manual workers, than are the children of manual workers. Similarly, the odds that the child of a professional will become a professional instead of becoming a cadre, compared to the corresponding odds for a child of a cadre, are

$$\begin{aligned}\log\theta &= \mu^{RC}_{11} + \mu^{RC}_{22} - \mu^{RC}_{12} - \mu^{R}_{21} \\ &= 0 + .627 - 0 - 0 \\ &= .627 \qquad (12.11)\end{aligned}$$

which implies that the relative odds are 1.87 ($= e^{0.627}$). Clearly, in China (as elsewhere) the "inheritance" of farm occupations relative to inflow from the children of manual workers is much stronger than the inheritance of professional occupations relative to inflow from the children of cadres.

Topological, or Levels, Models

Having shown how to interpret the interaction coefficients, I next address whether the table can be simplified. In particular, given the lack of differentiation between sales and

TABLE 12.10. **Interaction Parameters for the Saturated Model Applied to Table 12.9.**

	Respondent's Occupation in 1996						
Father's Occupation When R Age 14	**Prof.**	**Cadre**	**Cler.**	**Sales**	**Ser.**	**Man.**	**Agric.**
1. Professionals	0	0	0	0	0	0	0
2. Cadres	0	0.627	1.213	−0.169	0.054	−0.100	−0.341
3. Clerical workers	0	0.392	1.404	0.082	−15.591	0.384	−0.058
4. Sales workers	0	1.645	0.459	1.489	1.595	1.067	0.607
5. Service workers	0	0.952	1.047	0.737	1.846	0.114	−0.191
6. Manual workers	0	0.908	1.270	0.725	1.275	1.567	0.801
7. Agricultural workers	0	0.881	0.850	0.708	1.048	1.088	2.756

service workers in the Chinese economy, I suspect that these two categories might reasonably be collapsed into one. To determine this, I specify a design matrix that forces the cells involving the fourth and fifth row and fourth and fifth column to be identical:

1	1	1	1	1	1	1	
1	2	3	4	4	5	6	
1	7	8	9	9	10	11	
1	12	13	14	14	15	16	= ss_dm
1	12	13	14	14	15	16	
1	17	18	19	19	20	21	
1	22	23	24	24	25	26	

Because this model fits well ($L^2 = 16.06$, with 11 *d.f.*—because only twenty-five of the thirty-six nonredundant coefficients are estimated—which yields $p = .139$; $BIC = -76.5$; and $\Delta = .53$), I conclude that a six-by-six table captures the intergenerational mobility pattern about as well as a seven-by-seven table. I thus carry out the subsequent analysis for the six-by-six table.

If you exploit log-linear models for no other purpose, you should keep in mind this particular application and use it whenever you are trying to decide whether to collapse categories of a table. The procedure just described provides a rigorous basis for deciding whether it is possible to combine categories of a table without significant loss of information.

More generally, the specification of particular cells of a table as having identical coefficients (that is, as forming "levels") permits exploration of a wide variety of models (for examples, see Hauser 1978, 1980; and Erikson and Goldthorpe 1992b).

Quasi-Independence Models

Sometimes the claim is made that if people are able to free themselves from the social class defined by their father's occupation, their social origins are no longer connected to their destinations—that is, the "off-diagonal" cells of the table are independent. To test this hypothesis (on the collapsed six-by-six table), I specify a design matrix that exactly fits the diagonal cells of the table but otherwise forces all interaction parameters to be identical:

2	1	1	1	1	1	
1	3	1	1	1	1	
1	1	4	1	1	1	
1	1	1	5	1	1	= diag_dm
1	1	1	1	6	1	
1	1	1	1	1	7	

As we can see from the first row of the second panel of Table 12.11, this model is a huge improvement over the independence model, which is the baseline model in Table 12.11. Although it does not fit by classical standards, it is more likely than the saturated model and only misclassifies about 2 percent of the cases. Still, other models might fit even better.

TABLE 12.11. Goodness-of-Fit Statistics for Alternative Models of Intergenerational Occupational Mobility in China (Six-by-Six Table).

Model	L^2	d.f.	p	*BIC*	L^2/L^2_B	Δ
Independence	1080	25	.000	869	1.000	19.62
Crossings	58.8	20	.000	−109	.054	2.83
Uniform association	634	24	.000	432	.587	13.87
Linear-by-linear, ISEI	451	24	.000	249	.418	11.16
Linear-by-linear, urban *hukou*	157	24	.000	−45.2	.145	5.28
Linear-by-linear, ISEI + urban *hukou*	150	23	.000	−43.8	.139	4.99
Row_and_column effects I	324	16	.000	190	.300	8.40
Row_and_column effects II (RC)	106	16	.000	−28.6	.098	3.35
Diagonal cells fitted exactly						
Quasi-independence	53.5	19	.000	−106	.050	2.16
Quasi-symmetry	21.8	10	.016	−62.4	.020	1.24
Crossings	29.3	16	.022	−105	.027	1.37
Uniform association	34.5	18	.011	−117	.032	1.51
Linear-by-linear, ISEI	33.7	18	.014	−118	.031	1.61
Linear-by-linear, urban *hukou*	37.2	18	.005	−114	.034	1.72
Linear-by-linear, ISEI + urban *hukou*	33.7	17	.009	−109	.031	1.61
Row_and_column effects I	10.3	10	.415	−73.8	.010	0.66
Row_and_column effects II (RC)	9.0	10	.533	−75.1	.008	0.01

Quasi-Symmetry Models

An important issue in social mobility research is whether, net of any shift in the marginals, the relative odds of upward and downward mobility between corresponding categories are symmetrical. The following design matrix specifies this model for the six-by-six table:

$$\begin{matrix} 2 & 1 & 1 & 1 & 1 & 1 \\ 1 & 3 & 8 & 9 & 10 & 11 \\ 1 & 8 & 4 & 12 & 13 & 14 \\ 1 & 9 & 12 & 5 & 15 & 16 \\ 1 & 10 & 13 & 15 & 6 & 17 \\ 1 & 11 & 14 & 16 & 17 & 7 \end{matrix} \quad = \text{qi\_dm}$$

As we see in Table 12.11, this model fits slightly better than the quasi-independence model by the likelihood ratio standard but not nearly so well by the *BIC* standard.

Crossings Models

Suppose we were to take the occupational categories in our six-by-six table as representing social classes, with boundaries that constitute barriers to mobility. Suppose further that, in an analogy to movement across physical space, it is necessary to "cross" each barrier between adjacent classes to achieve mobility between nonadjacent classes. We can represent this model (following Powers and Xie 2000, 117) as

$$F_{ij} = \eta \tau_i^R \tau_j^C \nu_{ij}^{RC} \tag{12.12}$$

where

$$\nu_{ij}^{RC} = \begin{Bmatrix} \prod_{u=j}^{i-1} \nu_u & for\, i > j \\ \prod_{u=i}^{j-1} \nu_u & for\, i < j \\ \xi_i & for\, i = j \end{Bmatrix}$$

This specification implies the following interaction parameters for the cells of the six-by-six table (with the diagonal cells fitted exactly):

$$\begin{matrix} \xi_1 & \nu_1 & \nu_1\nu_2 & \nu_1\nu_2\nu_3 & \nu_1\nu_2\nu_3\nu_4 & \nu_1\nu_2\nu_3\nu_4\nu_5 \\ \nu_1 & \xi_2 & \nu_2 & \nu_2\nu_3 & \nu_2\nu_3\nu_4 & \nu_2\nu_3\nu_4\nu_5 \\ \nu_1\nu_2 & \nu_2 & \xi_3 & \nu_3 & \nu_3\nu_4 & \nu_3\nu_4\nu_5 \\ \nu_1\nu_2\nu_3 & \nu_2\nu_3 & \nu_3 & \xi_4 & \nu_4 & \nu_4\nu_5 \\ \nu_1\nu_2\nu_3\nu_4 & \nu_2\nu_3\nu_4 & \nu_3\nu_4 & \nu_4 & \xi_5 & \nu_5 \\ \nu_1\nu_2\nu_3\nu_4\nu_5 & \nu_2\nu_3\nu_4\nu_5 & \nu_3\nu_4\nu_5 & \nu_4\nu_5 & \nu_5 & \xi_6 \end{matrix}$$

These parameters can be estimated by summing six design matrices, one for each crossings parameter plus one for the diagonal design matrix (diag_dm), and taking antilogs. The corresponding model that does not fit the diagonal exactly is estimated in the same way except that the diagonal design matrix is omitted. Here are the five design matrices for the crossings parameters:

0 1 1 1 1 1	0 0 1 1 1 1	0 0 0 1 1 1	0 0 0 0 1 1	0 0 0 0 0 1
1 0 0 0 0 0	0 0 1 1 1 1	0 0 0 1 1 1	0 0 0 0 1 1	0 0 0 0 0 1
1 0 0 0 0 0	1 1 0 0 0 0	0 0 0 1 1 1	0 0 0 0 1 1	0 0 0 0 0 1
1 0 0 0 0 0	1 1 0 0 0 0	1 1 1 0 0 0	0 0 0 0 1 1	0 0 0 0 0 1
1 0 0 0 0 0	1 1 0 0 0 0	1 1 1 0 0 0	1 1 1 1 0 0	0 0 0 0 0 1
1 0 0 0 0 0	1 1 0 0 0 0	1 1 1 0 0 0	1 1 1 1 0 0	1 1 1 1 1 0
cr1_dm	cr2__dm	cr3_dm	cr4_dm	cr5_dm

As we see in Table 12.11, the crossings model fits better than any of the other models we have reviewed so far. Interestingly, fitting the diagonal cells exactly degrades the fit slightly by the *BIC* standard, presumably because moving between the diagonal and off-diagonal cells is captured well by the crossings parameters, and the additional degrees of freedom used by fitting the diagonal exactly are penalized by *BIC*.

The crossings parameters for the simpler crossings model are

$$\nu_1 = -0.138$$
$$\nu_2 = 0.002$$
$$\nu_3 = -0.203$$
$$\nu_4 = -0.228$$
$$\nu_5 = -1.033$$

Clearly, by far the most difficult transition (crossing) is between farm and nonfarm occupations (specifically, manual occupations); this is true everywhere, and China is no exception. Interestingly, the least difficult transition is between cadre and clerical occupations. Again, this is no particular surprise, because in China no sharp distinction is made between clerical and administrative tasks and the best and the brightest of the clerical staff are often tapped to become cadres. The known intragenerational mobility pattern may well carry over to intergenerational mobility, with clerical positions seen as reasonable starting points for the children of administrative cadres and cadre positions as attainable upward mobility goals for the children of clerical workers. Finally, this result could be due to the fact that the analysis here combines males and females. It could well be that the daughters of cadres disproportionately tend to become clerical workers.

Uniform Association Models

When the categories of a table are ordered, it is possible to estimate more parsimonious models than are available for nominal categories. The simplest such model assumes that the differences between each pair of adjacent categories are equal, so that the scale for each variable can be represented by consecutive integers. That is, the model is

$$\log F_{ij} = \mu + \mu_i^R + \mu_j^C + \beta ij \tag{12.13}$$

where the strength of the association between the row level and the column level is indexed by β. From this it follows that the log odds ratio between row categories i and i' and column categories j and j' is just

$$\log\theta = \beta(i - i')(j - j') \tag{12.14}$$

Table 12.11 shows goodness-of-fit statistics for the uniform association model with and without the main diagonal fitted exactly. As you see, when the diagonal cells are not fit exactly, the uniform association model fits very badly. The reason for this is simple: people disproportionately tend to remain in the same occupation category as their fathers. This tendency is captured by fitting the diagonal cells exactly but is ignored otherwise. For this reason none of the association models shown in Table 12.11 fits well unless the diagonal cells are estimated exactly.

However, when the diagonal cells are estimated exactly, the uniform association model fits quite well. It yields $\beta = .046$. From Equation 12.14 we can see that this implies, for example, that the odds that the child of a professional will become a professional rather than a farmer are more than three times the corresponding odds for the child of a farmer: $.046(1 - 6)(1 - 6) = 1.150$; $e^{1.150} = 3.158$. This is actually a rather low odds ratio, which is consistent with the general sense that intergenerational mobility in China is easier than in most other nations (but see Wu and Treiman [2004, 2007] for a counterargument).

Linear-by-Linear Association Models

Now suppose we have more information than simply a rank order of categories—for example, socioeconomic status scores. We can then estimate a linear-by-linear association model, where the scale scores are substituted for the category indexes. That is, instead of Equation 12.13, we have

$$\log F_{ij} = \mu + \mu_i^R + \mu_j^C + \beta x_i y_j \tag{12.15}$$

with the log odds ratio given by

$$\log \theta = \beta(x_i - x_{i'})(y_j - y_{j'}) \tag{12.16}$$

Estimating this model for the Chinese data, with occupation categories scored by their mean occupational status (ISEI; see Ganzeboom, de Graaf, and Treiman 1992), we achieve a model that fits marginally better than the uniform association model, by the *BIC* criterion. For this model, $\beta = .000483$. Thus for the same categories as in the uniform association example, we have $.000483(16.2 - 63.7)(16.2 - 63.7) = 1.090$; $e^{1.090} = 2.974$. We are hereby led to a qualitatively similar conclusion: the odds that the child of a professional will become a professional rather than a farmer are about three times as high as the corresponding odds for the child of a farmer.

Note that it is possible to include more than one scaling of the categories of a table, to represent different concepts. Table 12.11 shows goodness-of-fit statistics for two additional linear-by-linear models, one of which scales occupations by the proportion of incumbents who have permanent urban registration (urban *hukou*) and the other of which uses both the ISEI and urban registration measures. As it happens, neither fits as well as the ISEI and uniform association models. However, if we wished to assess the log odds ratio using, say, the model that includes both measures, we would simply apply Equation 12.16 to both variables and compute the sum. (For a well-known application of this kind of model, see Hout 1984.)

Row-Effects (and Column-Effects) Models

Sometimes we are confident that one variable can be scored with an integer scale—that is, that the difference between each pair of adjacent categories is the same—but we are uncertain about how to order the other variable. In such cases we can estimate the unknown scores. In this model the expected frequencies are given by

$$\log F_{ij} = \mu + \mu_i^R + \mu_j^C + j\phi_i \qquad (12.17)$$

where the j index the categories of one variable and the ϕ_i are the estimated scale scores for the other variable. The log odds ratio is given by

$$\log \theta = (\phi_i - \phi_{i'})(j - j') \qquad (12.18)$$

As an example of a situation in which these conditions might hold, consider the relationship between size of place of origin and educational attainment, for the 1996 Chinese survey we have been using. Table 12.12 shows the bivariate frequency distribution for adults not currently attending school. In constructing this table, I have collapsed education so that the categories represent approximate three-year intervals in median schooling. The size-of-place categories are from the official administrative hierarchy of China, which strongly affects the flow of resources to places. Thus, in addition to the general advantage of urban residence for educational attainment (greater exposure to the written word and such), we would expect educational attainment to be greater for places higher in the administrative hierarchy because such places are the beneficiaries of more resources from the central government.

The row-effects model fits well ($BIC = -135$, $\Delta = 2.96$) although not by classical inference ($p < .000$). But contrary to my expectation, the estimated scores

TABLE 12.12. Frequency Distribution of Educational Attainment by Size of Place of Residence at Age Fourteen, Chinese Adults Not Enrolled in School, 1996.

	Level of Schooling						
	None	Lower Primary	Upper Primary	Lower Middle	Upper Middle	Tertiary	Total
Village	1,100	668	1,192	1,385	390	69	**4,804**
Town	33	30	50	129	41	15	**298**
County seat	18	14	25	67	72	28	**224**
County-level city	8	8	15	52	51	21	**155**
Prefecture-level city	14	14	47	125	90	27	**317**
Provincial capital	6	7	9	64	68	24	**178**
Province-level city	3	3	4	37	26	13	**86**
Total	**1,182**	**744**	**1,342**	**1,859**	**738**	**197**	**6,062**

for size of place of origin suggest a non-monotonic relationship to education. The scores are

Village	0.00
Town	0.36
County seat	0.74
County-level city	0.86
Prefecture-level city	0.73
Provincial capital	1.01
Province-level city	0.98

According to this model, people from county-level cities (medium cities) get somewhat more education than do people from prefecture-level cities, although it would be unwise to make too much of this because the confidence intervals overlap (the 95 percent confidence interval is 0.71 to 1.01 for county-level cities and 0.63 to 0.84 for prefecture-level cities).

Column-effects models are formally identical to row-effects models, but with the role of rows and columns reversed. A column-effects model of the relationship between size of place at age fourteen and educational attainment does not fit as well as the corresponding row-effects model ($BIC = -108$, $\Delta = 2.98$, and $p < .000$), which suggests that the assumption of equal scale differences between adjacent size-of-place categories is probably incorrect. This is hardly surprising given the deviation from equal differences in the estimated coefficients for size-of-place categories in the row-effects model and, especially, the non-monotonicity of the scores relative to my a priori ordering.

Row-and-Column-Effects Model I Another analytic possibility is to treat both the row and column effects scores as unknown quantities to be estimated. However, in this case it is important to have the correct ordering of both the row and column categories because the results are not invariant under different orderings. For the Chinese example we have been exploring—the relationship between the size of the place of origin and educational attainment—this creates a bit of a dilemma. Is it better to reorder the size-of-place categories according to the scale scores derived from the row effects model or to retain the a priori ordering derived from the Chinese administrative hierarchy? One possibility is to estimate the model both ways and to compare goodness-of-fit statistics. Doing this reveals that the model with the reordered categories is clearly more likely given the data ($BIC = -152$, $p = .304$, and $\Delta = 1.20$, compared with $BIC = -136$, $p = .009$, and $\Delta = 1.65$ using the a priori categories). For the row-and-column effects model with the reordered categories, the scale scores are as follows:

Village	0.00	No schooling	0.00
Town	−0.60	Lower primary	0.90
Prefecture-level city	−1.23	Upper primary	1.67
County seat	−2.22	Lower middle	2.78
County-level city	−3.10	Upper middle	3.84
Province-level city	−4.00	Tertiary	4.80
Provincial capital	−4.95		

Formally, the row-and-column effects model (often called Row-and-Column-Effects Model I to distinguish it from a log-multiplicative model also proposed by Goodman [1979] and known as Row-and-Column-Effects Model II, which we will discuss in the next section), is given by

$$\log F_{ij} = \mu + \mu_i^R + \mu_j^C + j\phi_i + i\phi_j \tag{12.19}$$

with the log odds ratio given by

$$\log\theta = (\phi_i - \phi_{i'})(j - j') + (\varphi_j - \varphi_{j'})(i - i') \tag{12.20}$$

Thus, for example, from Equation 12.20 we can calculate the log odds ratio of a tertiary versus an upper primary education for a person raised in a provincial capital compared with a person raised in a village as log θ = (−4.95 − 0)(6 − 3) + (4.80 − 1.67)(7 − 1) = 3.93, which implies that the odds ratio is 50.9 (= $e^{3.93}$). That is, the odds of people obtaining a tertiary education rather than a primary education are more than fifty times as great for those living in provincial capitals as for those living in rural villages. When people from Chinese rural villages make it to university, they are overcoming stupendous odds.

Row-and-Column-Effects Model II (the RC or Log-Multiplicative Model)

As I noted in the previous section, a serious limitation of Row-and-Column-Effects Model I is that correct estimation of the scale scores depends on correctly ordering the categories. For this reason an alternative model proposed by Goodman (1979), Row-and-Column Effects Model II (also called the RC model or the Log-Multiplicative model), which is invariant under any ordering of categories, and which estimates scale scores from the data, has become much more widely used. In this model the expected frequencies are calculated as

$$\log F_{ij} = \mu + \mu_i^R + \mu_j^C + \phi_i \varphi_j \qquad (12.21)$$

and the log odds ratios as

$$\log\theta = (\phi_i - \phi_{i'})(\varphi_j - \varphi_{j'}) \qquad (12.22)$$

An alternative parameterization of Equation 12.21, which includes a term for the overall strength of association in the table (particularly useful for comparisons between groups, which I do not cover here) is

$$\log F_{ij} = \mu + \mu_i^R + \mu_j^C + \beta\phi_i \varphi_j \qquad (12.23)$$

with the odds ratios given as

$$\log\theta = \beta(\phi_i - \phi_{i'})(\varphi_j - \varphi_{j'}) \qquad (12.24)$$

For the data shown in Table 12.12, estimation of Equation 12.23 yields a very good fit: $p = .140$ and $BIC = -147.3$. Interestingly, the estimated scale scores preserve the rank order of the row-and-column effect scores reported earlier:

Village	0.00	No schooling	0.00
Town	0.42	Lower primary	0.14
Prefecture-level city	0.76	Upper primary	0.17
County seat	0.82	Lower middle	0.50
County-level city	0.91	Upper middle	0.80
Province-level city	1.00	Tertiary	1.00
Provincial capital	1.04		

In China, size of place of origin appears to be very strongly associated with educational attainment, reflected in the association parameter $\beta = 4.17$. Moreover, the greatest gap is between rural villages and any urban place, with the next largest gap between towns and prefecture-level cities. Making the same comparison as for the row-and-column effects model, from Equation 12.24 we can calculate the log odds ratio of a tertiary versus an upper primary education for a person raised in a provincial capital compared to a person raised in a village as $\log \theta = 4.17(1.04 - 0)(1.00 - 0.17) = 3.60$, which implies that the odds ratio is 36.6 ($= e^{3.60}$). That is, the RC model implies that the odds of people obtaining a tertiary education versus a primary education are about thirty-seven times as great for those living in provincial capitals as for those living in rural villages. Although the odds ratio implied by this model is not as large as that implied by the row-and-column effects model (which yields an odds ratio of fifty-one), it is still extremely large.

Although in this example the scaling of size-of-place categories was reasonably close to my a priori assumptions, and the rank ordering of the education categories was exactly what I anticipated, there is nothing in the method that guarantees such a close correspondence. Because the scale scores are computed to maximize the association between the row and column variables, they provide a test of the correctness of a priori assumptions. We can see this clearly by estimating an RC model for the Chinese intergenerational occupational mobility table analyzed earlier. In contrast to the typical outcome in Western nations (Ganzeboom, Luijkx, and Treiman 1989), the resulting scale scores for China deviate very substantially from my a priori ordering of occupation categories based on their socioeconomic position (perhaps because our data include both males and females whereas most research on occupational mobility for other nations, including that carried out by Ganzeboom, Luijkx, and Treiman [1989] and also Wu and Treiman's 2007 analysis of these data, is restricted to males). The following coefficients are from a model with the diagonal blocked.

	Father's Occupation	Respondent's Occupation
Professionals	0.00	0.00
Cadres	−27.68	−0.27
Clerical workers	−13.76	−0.18
Sales and service workers	−12.97	0.77
Manual workers	−2.33	0.87
Agricultural workers	1.00	1.00

Clearly, the children of cadres are much more likely than other offspring to move into high-status positions. By contrast, the children of professionals are hardly protected at all from downward mobility, which may reflect the rather heterogeneous character of this category; it includes village accountants and school teachers and many technical positions that do not require tertiary education. The scale scores for respondents' occupations are somewhat more orderly, revealing a sharp manual-nonmanual divide, although mobility into the professions from all sorts of origins appears easier than mobility into clerical or cadre positions. Wu and Treiman (2007) also obtain distinctive results, albeit not as extreme as these, in their male-only analysis and argue that their results reflect a distinctive Chinese institution, the residential registration system, which makes the children of rural nonagricultural workers vulnerable to downward mobility into agriculture but also creates an extreme upward mobility route into the professions for the bright children of peasants.

The contrast between these results and results from the corresponding Row-and-Column-Effects Model I is instructive:

	Father's Occupation	**Respondent's Occupation**
Professionals	0.00	0.00
Cadres	−0.66	0.25
Clerical workers	−0.86	0.51
Sales and service workers	−1.15	0.92
Manual workers	−1.33	1.29
Agricultural workers	−1.66	1.53

The row-and-column effects model gives orderly results, consistent with my a priori ordering of categories. Thus, an analyst might be tempted to settle for this model because by the likelihood-ratio criterion it has by far the best fit among all the models estimated in this chapter except for the RC model (see Table 12.11)—although it does not have the most negative *BIC*. However, the row-and-column effects model is clearly incorrect, even though it is nearly as likely as the RC model by the *BIC* standard.

From the RC model we can calculate the relative odds that the child of a professional will become a professional rather than a farmer, compared with the corresponding odds for the child of a farmer. Because the association coefficient, β, for the Chinese mobility table is 0.0455 (another indicator of the lack of association in the table), from Equation 12.24 we have $\log \theta = .0455(0 - 1)(0 - 1) = 0.0455$, which implies that the odds ratio is 1.047 ($= e^{0.0455}$). Apparently, the odds that the children of professionals will follow their fathers' footsteps rather than going into the fields are hardly larger than the odds that the children of farmers will become professionals rather than following their fathers into the fields.

This is a very different result from what we calculated from the uniform and linear-by-linear association models, and it brings home in a dramatic way the importance of finding the right model before making inferences. (It also is quite different from the corresponding result from Row-and-Column-Effects Model I, which implies that the children of professionals are about twice as likely [because $e^{0.65} = 1.92$] as the children of peasants to become professionals rather than peasants.) Nonetheless, here we might be well advised to settle for the linear-by-linear association in which mobility is a function of differences in status between occupation categories, on the ground that it has the most negative *BIC*.

Extensions

It is quite possible to extend the parsimonious models presented here to more than two variables. The most common application is to compare two-variable tables across contexts (time periods, nations, ethnic groups, and so on), but more general extensions are also possible. Many of these procedures are discussed in the literature briefly reviewed in the following section.

A BIBLIOGRAPHIC NOTE

A number of treatments of log-linear analysis are available, ranging from those intended for social scientists with limited mathematical backgrounds to full-blown treatises in mathematical statistics. The most accessible treatments include those by Davis (1974), Knoke and Burke (1980), Gilbert (1981), and Powers and Xie (2000), of which Power and Xie is probably the best. The Knoke and Burke book is very clear and provides an excellent introduction, but it does not include many types of models that have appeared in the research literature over the past twenty-five years. Somewhat more mathematical, but still accessible to social scientists, are the books by Bishop, Fienberg, and Holland (1975), Upton (1978), Fienberg (1980), Agresti (2002), and some papers and books by Goodman (for example, 1972, 1978, 1984).

Also available is a highly developed literature motivated mainly by the analysis of intergenerational occupational mobility, but it is appropriate for many other applications. In addition to the works by Goodman, important papers and monographs include those by Hauser (1978, 1980), Duncan (1979), Clogg (1982), Grusky and Hauser (1984), Hout

OTHER SOFTWARE FOR ESTIMATING LOG-LINEAR MODELS GLIM is available for purchase from http://www.nag.co.uk/stats/GDGE_soft.asp. The worked examples appearing in Goodman and Hout (1998) can be downloaded as two Microsoft Office Excel 97 workbook files from the Carnegie Mellon University Statistics Department's StatLib: http://lib.stat.cmu.edu/DOS/general. These files, "mobility.xls" and "voting.xls," include the raw data, GLIM results, and graphical displays from the examples presented in the article.

Vermunt's (1997) software, *l*em, and the accompanying documentation (Vermunt 1997) can be downloaded free of charge. The easiest way to find the download site is to query a search engine for "homepage jeroen vermunt." The documentation is very cryptic, but the software comes with many worked examples that can easily be adapted.

Pisati's Stata `-ado-` file to estimate uniform layer-effect models can be downloaded from within Stata (connected to the internet) by typing "net search pisati" and then clicking "sg142 from http://www.stata.com/stb/stb55."

John Hendrickx has written an `-ado-` file, `-rc2-`, that estimates the RC model (Equation 12.23). To download `-rc2-` from within Stata, type "net search rc2." Then click rc2 from http://fmwww.bc.edu/RePEc/bocode/r and follow the instructions. I thank Maarten Bois for pointing me to Hendrickx's program.)

(1983, 1984), Sobel, Hout, and Duncan (1985), Yamaguchi (1987), Becker and Clogg (1989), Mare (1991), Xie (1992), Erikson and Goldthorpe (1992a, 1992b), Hout and Hauser (1992), Goodman and Hout (1998), Fu (2001), Pisati (2001), and Park and Smits (2005).

The 1998 paper by Goodman and Hout is particularly valuable for analysts who wish to compare log-linear models across contexts. Goodman and Hout estimated their models using GLIM, a powerful British competitor to Stata. These models, and virtually any other log-linear or log-multiplicative model, can be estimated using *l*em developed by Jeroen Vermunt at Tilburg University, the Netherlands. A subset of the models discussed by Goodman and Hout can be estimated using a Stata `-ado-` file by Pisati (2001); see also Yamaguchi (1987) and Xie (1992), who originally proposed versions of these models.

Applications of log-linear modeling to substantive problems other than social mobility can be located by searching *Sociological Abstracts* or other bibliographic databases. (I got 810 hits searching *Sociological Abstracts* for "log linear" [and variants "loglinear" and "log-linear"] as a key word on 24 November 2007.)

WHAT THIS CHAPTER HAS SHOWN

In this chapter we have seen how to use log-linear analysis to test hypotheses regarding the presence or absence of associations among variables in multiway tables. These tools give us a powerful way of testing hypotheses pertaining to percentage tables. In addition, we have seen how to apply various models to parsimoniously summarize patterns of association in two-way tables, and to determine which of several alternative models fits best. Although the substantive examples for discussing parsimonious models were drawn mainly from studies of social mobility, the topic that has driven most model development, these models can be applied to a wide variety of substantive problems.

APPENDIX 12.A DERIVATION OF THE EFFECT PARAMETERS

To see what the η and τs are, consider the saturated model for a two-by-two table. Recall from Equation 12.1 that the expected frequencies in each cell of a table can be expressed in terms of τs:

$F_{11} = \eta\tau_1^X\tau_1^Y\tau_{11}^{XY}$	$F_{12} = \eta\tau_1^X\tau_2^Y\tau_{12}^{XY}$
$F_{21} = \eta\tau_2^X\tau_1^Y\tau_{21}^{XY}$	$F_{22} = \eta\tau_2^X\tau_2^Y\tau_{22}^{XY}$

(12.A.1)

First, multiplying one of the equations by the other three and simplifying (recalling the relationships among the τs shown in Equation 12.6), we have

$$\eta = (F_{11}F_{12}F_{21}F_{22})^{1/4} \tag{12.A.2}$$

Thus η is just the geometric mean of the expected cell frequencies (the geometric mean of a set of n numbers is the nth root of their product). In this sense, η is a scale factor; all it does is take account of the fact that tables have different average numbers of cases per cell.

Next we express the row effect as a function of the cell frequencies. We do this by writing the product of the two conditional odds as a function of ηs and τs and simplifying:

$$\left[\frac{F_{11}}{F_{21}}\right]\left[\frac{F_{12}}{F_{22}}\right] = \left[\frac{\eta\tau_1^X\tau_1^Y\tau_{11}^{XY}}{\eta\tau_2^X\tau_1^Y\tau_{21}^{XY}}\right]\left[\frac{\eta\tau_1^X\tau_2^Y\tau_{12}^{XY}}{\eta\tau_2^X\tau_2^Y\tau_{22}^{XY}}\right] = \frac{\left[\tau_1^X\right]^2}{\left[\tau_2^X\right]^2} = \left[\tau_1^X\right]^4 \quad (12.A.3)$$

And so

$$\begin{aligned}\tau_1^X &= [(F_{11}/F_{21})(F_{12}/F_{22})]^{1/4} \\ &= [(F_{11}F_{12})/(F_{21}F_{22})]^{1/4}\end{aligned} \quad (12.A.4)$$

That is, we see from Equation 12.A.4 that τ is a function of the product of the two conditional odds. But we can get a more readily interpretable alternative expression by multiplying both the numerator and the denominator of the second line of 12.A.4 by $(F_{11}F_{12})^{1/4}$ and simplifying. This yields

$$\tau_1^X = \frac{(F_{11}F_{12})^{1/2}}{(F_{11}F_{12}F_{21}F_{22})^{1/4}} \quad (12.A.5)$$

From Equation 12.A.5 we see that τ_1^X, the effect parameter for the first row, is just the ratio of the average size of the cells in the first row to the average size of all cells in the table (where by averages I mean geometric means). Thus τs larger than one indicate that a disproportionately large share of all the cases in the table is in the first row, and τs smaller than one indicate that a disproportionately small share of all the cases in the table is in the first row. In a similar way we can derive corresponding expressions for the effect parameter associated with the second row and also the effect parameters for columns.

Finally, we can derive interpretable expressions for the interaction effect parameters. To see this, we write the expected odds ratio, $(F_{11}/F_{21})/(F_{12}/F_{22})$, as a function of ηs and τs and simplify as we did earlier:

$$\frac{F_{11}/F_{21}}{F_{12}/F_{22}} = \frac{(\eta\tau_1^X\tau_1^Y\tau_{11}^{XY})/(\eta\tau_2^X\tau_1^Y\tau_{21}^{XY})}{(\eta\tau_1^X\tau_2^Y\tau_{12}^{XY})/(\eta\tau_2^X\tau_2^Y\tau_{22}^{XY})} = \frac{\tau_{11}^{XY}\tau_{22}^{XY}}{\tau_{21}^{XY}\tau_{12}^{XY}} \quad (12.A.6)$$

This yields

$$\left[\tau_{11}^{XY}\right]^4 = (F_{11}/F_{21})/(F_{12}/F_{22}) \quad (12.A.7)$$

and so

$$\tau_{11}^{XY} = \left[(F_{11}/F_{21})/(F_{12}/F_{22})\right]^{1/4} \tag{12.A.8}$$

That is τ_{11}^{XY} is a function of the ratio of the two conditional odds. Once again we can get a more readily interpretable expression by multiplying both the numerator and the denominator of the right side of Equation 12.A.8 by the geometric mean of the expected frequencies, $(F_{11}F_{12}F_{21}F_{22})^{1/4}$, and simplifying:

$$\tau_{11}^{XY} = \frac{(F_{11}F_{22})^{1/2}}{(F_{11}F_{12}F_{21}F_{22})^{1/4}} \tag{12.A.9}$$

From Equation 12.A.9 we see that the interaction effect parameter, τ_{11}^{XY}, is just the ratio of the average size of the two diagonal cells to the average size of all cells. If this τ is greater than one, there is a positive association (or *interaction,* in log-linear terms) between X and Y. If τ is smaller than one, there is a negative association between X and Y (assuming that Category 1 is the "positive" value in each case). In a similar way we can derive expressions for the other interaction effect parameters.

These relationships can be generalized beyond the two-by-two case, but that is beyond the scope of the present discussion. Those wishing to pursue this topic should consult the sources listed in the Bibliographic Note section of this chapter.

APPENDIX 12.B INTRODUCTION TO MAXIMUM LIKELIHOOD ESTIMATION

Maximum likelihood estimation is one of several methods used to obtain parameter estimates for the models presented in this and the following chapters. The principle is straightforward, although both the underlying mathematics and the computational procedures are often quite complex and go beyond what is dealt with in this book. For good introductions to the topic, see King (1989), Eliason (1993), Long (1997, 25–33, 52–61), and Powers and Xie (2000, Appendix B).

Suppose we observe a random sample of values on some variable, $x_1, x_2, \ldots, x_n$, drawn independently from a population distribution $f(x_1, x_2, \ldots, x_n | \theta)$ governed by an unknown parameter θ. We may then ask what is the probability of obtaining the observed sample for any given value of θ. This is the *likelihood* of the sample. What we want to do is to find the value for θ that maximizes the likelihood of the sample; this is the *maximum likelihood estimate* of θ. More generally, maximum likelihood estimation consists of procedures for finding estimates of unknown parameters that maximize the likelihood of the observed data; the resulting parameter estimates are called maximum likelihood estimates.

Maximum likelihood estimation involves two steps: determining the *likelihood function,* which expresses the probability of the observed data as a function of the unknown parameters; and maximizing the likelihood function. We can write a general expression for the likelihood function:

$$\Lambda(\boldsymbol{\theta}) = \prod_{i=1}^{N} f(x_i; \boldsymbol{\theta}) \tag{12.B.1}$$

where $\boldsymbol{\theta}$ is a column vector of unknown parameters; note that there may be only one unknown parameter, in which case $\boldsymbol{\theta}$ is a scalar. Equation 12.B.1 holds because the observations are assumed to be independent, which means that their joint distribution may be written as a product of their individual, marginal distributions. However, because product terms are mathematically intractable, we convert Equation 12.B.1 into log form (which is permissible because the relationship between a variable and its log is monotonic). Thus we have

$$\lambda(\boldsymbol{\theta}) = \ln\left[\prod_{i=1}^{N} f(x_i; \boldsymbol{\theta})\right] = \sum_{i=1}^{N} \ln[f(x_i; \boldsymbol{\theta})] \tag{12.B.2}$$

We then find the values of $\boldsymbol{\theta}$ (denoted $\hat{\boldsymbol{\theta}}$) that maximize the log likelihood; because of the monotonic relationship, these also maximize the likelihood.

MEAN OF A NORMAL DISTRIBUTION

Consider a simple case. Suppose we want to find the maximum likelihood estimate of the mean, μ, for a random sample of observations from a normally distributed population with variance σ^2. Because the likelihood for a single observation is

$$L(\mu, \sigma^2) = \frac{1}{\sqrt{2\pi\sigma^2}} \exp\left(-\frac{(x_i - \mu)^2}{2\sigma^2}\right) \tag{12.B.3}$$

it follows (from Equations 12.B.1 and 12.B.2) that the log likelihood of the sample is

$$\begin{aligned} \lambda(\mu) &= \sum_{i=1}^{N} \ln\left[\frac{1}{\sqrt{2\pi\sigma^2}} \exp\left(-\frac{(x_i - \mu)^2}{2\sigma^2}\right)\right] \\ &= -N\left(\ln\sqrt{2\pi\sigma^2}\right) - \frac{1}{2\sigma^2}\sum_{i=1}^{N}(x_i - \mu)^2 \end{aligned} \tag{12.B.4}$$

However, we can disregard the leftmost term on the right side of the equation because it does not depend on the x_i. We also can discard the $\frac{1}{2\sigma^2}$ term because σ^2 is assumed known. This leaves us with the *kernel* of the log likelihood:

$$-\sum_{i=1}^{N}(x_i - \mu)^2 \tag{12.B.5}$$

Thus our task is reduced to maximizing the kernel of the log likelihood, Equation 12.B.5. We do this by taking the first derivative with respect to μ and setting it to zero. This yields either a maximum or minimum value for the kernel of the likelihood. We then take the second derivative to determine whether we have a maximum or a minimum: a second derivative less than zero indicates a maximum because the curve in the region of the extreme value is concave. The first derivative of Equation 12.B.5 with respect to μ is

$$\frac{\partial}{\partial \mu}\left(-\sum_{i=1}^{N}(x_i - \mu)^2\right) = 2\sum_{i=1}^{N} x_i - 2N\mu \tag{12.B.6}$$

Setting Equation 12.B.6 to zero and solving for μ yields the maximum likelihood estimate for μ:

$$\hat{\mu} = \frac{\sum_{i=1}^{N} x_i}{N} \tag{12.B.7}$$

We know this is the maximum because the second derivative of Equation 12.B.5 with respect to μ is $-N/\sigma^2$, which is always less than zero. Note that the maximum likelihood estimate of the population mean is the sample mean. Similarly, it can be shown that for a normal distribution the maximum likelihood estimate of the population variance is the sample variance.

LOG-LINEAR PARAMETERS

Now let us consider a more difficult estimation problem: finding maximum likelihood estimates of the parameters of log-linear models. To keep the example as simple as possible, consider the independence model for a two-by-two table. The usual representation of expected frequencies for such a model is

$$\ln(F_{ij}) = \mu + \mu_i^R + \mu_j^C \tag{12.B.8}$$

However, it is convenient to convert the usual representation of such models from tabular form to column form, with the scale factor $\mu = \beta_0$, the coefficient for the row variable $\mu_i^R = \beta_1$, and the coefficient for the column variable $\mu_j^C = \beta_2$. Further, we represent the values associated with each in dummy-variable form. Thus we have

x_0	x_1	x_2	y_j
1	0	0	y_1
1	1	0	y_2
1	0	1	y_3
1	1	1	y_4

The independence model can then be written as a model for counts:

$$m_i = \exp(\beta_0 + \beta_1 x_{1i} + \beta_2 x_{2i}) \quad (12.B.9)$$

where m_i is the expected frequency in the ith cell. Under Poisson sampling, the kernel of the log likelihood is

$$\lambda(\boldsymbol{\beta}) = \sum_{i=1}^{k} (y_i \log m_i - m_i) \quad (12.B.10)$$

and so we need to maximize Equation 12.B.10. Because the model is nonlinear, an iterative solution is required in which we repeatedly update the estimates of the βs using the first and second derivatives of Equation 12.B.10 with respect to β. For our purposes, it is not necessary to consider further how the estimation is actually carried out. For additional detail, see Eliason (1993); Gould and Scribney (1999); and Powers and Xie (2000, Appendix B).

CHAPTER

BINOMIAL LOGISTIC REGRESSION

WHAT THIS CHAPTER IS ABOUT

This chapter introduces binomial logistic regression, a technique for estimating models with a dichotomous dependent variable. We start by considering the relationship of binomial logistic regression to log-linear analysis and then see, by studying a worked example, how to estimate and interpret logistic regression models. We then consider three additional worked examples to expand the applications of binary logistic regression to educational progression and similar models, to discrete-time hazard-rate models, and to case-control designs.

INTRODUCTION

Often social scientists are confronted with the need to analyze categorical *dependent* variables—whether people vote, for whom they vote, their degree of agreement with a particular attitude, their choice of occupation, and so on. Although as we have seen, OLS regression procedures can easily handle categorical *independent* variables, they are not appropriate for categorical *dependent* variables, even dichotomies. In the case of dichotomous dependent variables, the assumptions of multiple regression, including in particular that errors of prediction are normally distributed, break down badly, often yielding seriously misleading results; moreover, predicted values often lie outside the logically possible range (zero to one). For these reasons a variety of procedures have been developed for dealing with dichotomous dependent variables, of which one of the most powerful is *logit analysis* or (synonymously) *logistic regression,* which uses maximum likelihood estimation. Logistic regression can be readily extended to handle dependent variables with more than two categories (multinomial logistic regression) and ordered sets of categories (ordered logistic regression). We will consider these two extensions in the next chapter. But we start with binomial logistic regression.

MAXIMUM LIKELIHOOD ESTIMATION *Maximum likelihood estimation* refers to a framework for estimating parameters of statistical models. The principle, which underlies estimation of log-linear and logistic regression models, is to find the value of the parameter that maximizes the likelihood of observing the sample data. (See Appendix 12.B for a brief overview of maximum likelihood estimation; King [1989], Eliason [1993], Long [1997, 25–33, 52–61], and Powers and Xie [2000, Appendix B] for accessible introductions to the topic; and Gould and Sribney [1999] for a technical discussion of how to do likelihood estimation in Stata.)

PROBIT ANALYSIS An alternative to logistic regression, more widely used in economics than in sociology, is probit analysis. The two procedures generally yield similar results, and the choice between them is largely a matter of professional convention. (See Appendix 13.B for a brief introduction to probit analysis.)

Binomial logistic regression is a procedure for predicting, from a set of independent variables, the log odds that individuals will be in each of two categories of a dichotomous dependent variable. The formula for logistic regression is

$$\ln\left(F_{1|X_{1...K}}/F_{2|X_{1...K}}\right)=a+\sum_{k=1}^{K}b_kX_k \qquad (13.1)$$

for K independent variables, X_k, where the a and b_k are coefficients analogous to OLS regression coefficients, and the dependent variable is the natural log of the expected odds of being in category 1 of the dependent variable rather than in category 2, conditional on the values of the independent variables. Thus, logistic regression is another specific case of the general linear model.

It is also true (and can be easily shown by dividing through by N) that the log odds of the expected conditional *frequency* distribution of the dichotomous dependent variable equals the log of the ratio of the expected *probabilities* of being in each of the two categories:

$$\ln\left(\frac{F_{1|X_{1...k}}}{N}\Big/\frac{F_{2|X_{1...k}}}{N}\right)=\ln((P=1)/(P=2))=\ln((P=1)/(1-(P=1))) \qquad (13.2)$$

The dependent variable (the log odds) is known as the *logit*. As we have just seen, logits may be expressed either in terms of frequencies or in terms of probabilities.

RELATION TO LOG-LINEAR ANALYSIS

The relationship of the logit specification to log-linear analysis is straightforward, as can be seen with the help of a little algebra. Consider a log-linear analysis in which there are three variables, Y (a dichotomous variable that we wish to regard as the dependent variable), A, and B. Now consider the saturated model relating the three variables: [ABY]. Under this model, expected cell frequencies are estimated as

$$\ln(F_{ijk}^{ABY})=\theta+\lambda_i^A+\lambda_j^B+\lambda_k^Y+\lambda_{ij}^{AB}+\lambda_{ik}^{AY}+\lambda_{jk}^{BY}+\lambda_{ijk}^{ABY} \qquad (13.3)$$

Now, because the dependent variable, Y, is dichotomous, we can easily derive the log odds of being in category 1 (rather than in category 2) of Y from Equation 13.3 (for those rusty on school algebra, some algebraic relationships involving logs and exponents are shown in Appendix 13.A):

$$\begin{aligned}\ln\left(F_{ij1}^{ABY}/F_{ij2}^{ABY}\right)&=\ln\left(F_{ij1}^{ABY}\right)-\ln\left(F_{ij2}^{ABY}\right)\\&=\left(\theta+\lambda_i^A+\lambda_j^B+\lambda_1^Y+\lambda_{ij}^{AB}+\lambda_{i1}^{AY}+\lambda_{j1}^{BY}+\lambda_{ij1}^{ABY}\right)\\&\quad-\left(\theta+\lambda_i^A+\lambda_j^B+\lambda_2^Y+\lambda_{ij}^{AB}+\lambda_{i2}^{AY}+\lambda_{j2}^{BY}+\lambda_{ij2}^{ABY}\right)\\&=\left(\lambda_1^Y-\lambda_2^Y\right)+\left(\lambda_{i1}^{AY}-\lambda_{i2}^{AY}\right)+\left(\lambda_{j1}^{BY}-\lambda_{j2}^{BY}\right)+\left(\lambda_{ij1}^{ABY}-\lambda_{ij2}^{ABY}\right)\end{aligned} \qquad (13.4)$$

But because the λs must sum to zero across each dimension, $\lambda_1^Y = -\lambda_2^Y$, and so on. So we have

$$\ln\left(F_{ij1}^{ABY} / F_{ij2}^{ABY}\right) = 2\lambda_1^Y + 2\lambda_{i1}^{AY} + 2\lambda_{j1}^{BY} + 2\lambda_{ij1}^{ABY} \tag{13.5}$$

In short, the log odds of being in one category of the dependent variable rather than the other are given by the sum of twice the usual log-linear coefficients relating the dependent variable to each of the independent variables, alone or in combination; this relationship also holds for nonsaturated models. Note that the coefficient λ_{ij}^{AB}, expressing the association between the independent variables, drops out of the equation because $\lambda_{ij}^{AB} - \lambda_{ij}^{AB} = 0$. Thus we can carry out binomial logistic regressions on the data in published tables simply by doing a log-linear analysis and multiplying the resulting coefficients by two. However, the relationship holds only when the log-linear coefficients are expressed in dummy-variable format—that is, as deviations from a reference category with an implicit coefficient of zero.

Although *logit analysis* and *logistic regression* are mathematically identical, they have separate origins. Logit analysis was developed as a special case of log-linear analysis, in which one (dichotomous) variable is regarded as dependent on a set of other categorical variables. Logistic regression was developed by statisticians and econometricians to deal with the problems that dichotomous dependent variables create for ordinary least-squares regression. Therefore it was developed to handle continuous independent variables. (For a good introduction to the statistical theory underlying logistic regression, see Hosmer and Lemeshow [2000]. For treatments with sociological examples, see Long [1997] and Powers and Xie [2000]. For a Stata-oriented text, see Long and Freese [2006].)

A WORKED LOGISTIC REGRESSION EXAMPLE: PREDICTING PREVALENCE OF ARMED THREATS

Suppose we are interested in what affects the likelihood that a person has ever been threatened with a gun. Moreover, suppose we are interested in ascertaining whether the prevalence of armed threats has changed over time. (Investigating this latter question provides another occasion for demonstrating how to make cross-temporal comparisons using the GSS.) We might suspect that males are more likely to have experienced such threats than are females. Not only has some fraction of the male population been in combat, unlike women (until very recently), but men tend to be more likely to be involved in gang activity, bar fights, and other confrontational situations than are women. Second, exposure to armed threats should be negatively correlated with socioeconomic status, given patterns of residential segregation and status differences in leisure activities. For convenience, I take education as an indicator of socioeconomic status (SES): unlike occupational status and income, it is essentially fixed over the adult life course and is interpretable equivalently for men and women. Third, it is likely that Blacks are subject to more armed threats than are members of other racial groups, net of SES, given patterns of residential discrimination that force even middle-class Blacks to live in high-crime neighborhoods. Finally, claims about the breakdown of civility in America would suggest that the prevalence of armed threats has been increasing over time.

Data to assess these possibilities are available in the GSS. In most years from 1973 through 1994 respondents were asked, "Have you ever been threatened with a gun, or shot at?" In addition, the sex, race (White, Black, or Other), and education (years of school completed, ranging from 0 to 20) of each respondent was ascertained. I omitted, in sequence, 5,031 cases in which the gun question was not answered (mostly because in several years the question was asked only of a subsample of respondents), an additional 52 cases in which information on education was missing, and an additional 16 cases in which information on the number of adults in the household (used to construct the weight variable) was missing. These procedures yielded an effective sample of 19,260 people covering the years 1973 through 1994. I carried out the analysis using survey estimation procedures and treating each year as a stratum. (For estimation details, see Appendix B and the downloadable files "ch13_1.do" and "ch13_1.log.")

Table 13.1 confirms that a substantially higher percentage of males than of females and a moderately higher percentage of Blacks than members of other races have ever been threatened by a gun. It is difficult to see a consistent pattern with respect to either educational attainment or year, but it is possible that each variable suppresses the effect of the other because education has been increasing over time.

TECHNICAL POINT ON TABLE 13.1

Note that in Table 13.1 the percentages are based on weighted frequencies but the unweighted percentage bases are shown. I weighted the data to take account of differential household size, to adjust for an oversample of Blacks in 1987, and to equalize the contribution of each year (see downloadable file "ch13_1.do" for details). For descriptive statistics, it is necessary to use the weighted data to get correct estimates for the population. But it is desirable to show the unweighted N's to reveal to the reader the actual number of cases on which each computation is based.

My first task is to choose a preferred model. Table 13.2 shows goodness-of-fit statistics for five models. Model 1 is a baseline model, positing that sex, race, and education significantly affect the odds of being threatened by a gun. Model 2 in addition posits a linear trend in the (log) odds of being threatened, net of the effects of sex, race, and education. If the likelihood of being threatened has been increasing over time, the coefficient associated with year should be positive. Model 3 posits year-to-year variation around any linear trend in the (log) odds of being threatened. Models 1, 2, and 3 stand in a hierarchical relation to each other. Model 4 posits that the log odds of being threatened depend on sex, race, and education; that the log odds increase over time in a linear fashion; and that sex and race interact—the hypothesis being that gender differences in the likelihood of being threatened will be smaller for Blacks than for others because, owing to

TABLE 13.1. Percentage Ever Threatened by a Gun, by Selected Variables, U.S. Adults, 1973 to 1994 (N = 19,260).

	Percent Threatened[a]	Percentage Base[b]
Sex		
Male	31.0	(8,332)
Female	10.0	(10,928)
Race		
White	18.8	(16,342)
Black	25.0	(2,470)
Other	17.3	(448)
Education		
Less than high school	21.8	(5,545)
High school grad	17.3	(6,346)
Some college	21.2	(3,921)
College grad+	18.0	(3,448)
Year		
1973	16.8	(1,497)
1975	18.0	(1,483)
1976	17.0	(1,489)
1978	19.8	(1,518)
1980	21.0	(1,452)

1983	20.3	(1,594)
1984	18.9	(1,467)
1986	19.5	(1,466)
1987	20.4	(1,806)
1988	22.0	(994)
1989	19.3	(1,030)
1990	19.5	(895)
1991	20.6	(1,007)
1993	20.1	(1,053)
1994	18.7	(509)
Total	**19.5**	**(19,260)**

[a]Based on weighted frequencies—see the box "Technical Point on Table 13.1."
[b]Unweighted frequencies.

residential discrimination, Blacks are more likely than others to live in dangerous neighborhoods, and hence Black women are particularly vulnerable to being threatened. Model 5 extends the same argument to include an interaction between race and education, positing a smaller effect of education on the odds of gun threat for Blacks than for others because of the residential vulnerability of even well-educated Blacks.

If the GSS were a simple random sample of the U.S. population, it would be possible to compare nested models by using the likelihood ratio χ^2s, or L^2s (reported in Stata output from the `-logistic-` command as LR chi2); L^2 is defined as twice the difference between the model log likelihood and the log likelihood for a model with no independent variables. If we were analyzing a simple random sample, the L^2s would be distributed approximately as χ^2, as would the difference between any pair of L^2s. In such cases we could assess whether one model fits significantly better than another by assessing the significance of the difference between two L^2s, with degrees of freedom calculated as the difference in the degrees of freedom associated with the two models. However, when we use weighted data, clustered data, or design-based estimation procedures, what Stata estimates is actually

TABLE 13.2. Goodness-of-Fit Statistics for Various Models Predicting the Prevalence of Armed Threat to U.S. Adults, 1973 to 1994.

	F	d.f.(1)	d.f.(2)	p
Models				
(1): Sex, education, race	326.99	3	1,387	.0000
(2): (1) + Year of survey	246.23	4	1,386	.0000
(3): (2) + Year of survey (categorical)	59.56	17	1,373	.0000
(4): (2) + Sex by race	194.10	5	1,385	.0000
(5): (4) + Education by race	162.44	6	1,384	.0000
Contrasts				
(2) − (1)	7.09	1	1,389	.0078
(3) − (2)	1.07	13	1,377	.3821
(4) − (2)	2.85	1	1,389	.0916
(5) − (2)	2.65	2	1,388	.0711
(5) − (4)	2.56	1	1,389	.1097

a *pseudo–log likelihood* for which the relation to χ^2 does not hold. Thus instead we use an adjusted Wald test to choose a preferred model. Recall from Chapter Six that the Wald test assesses whether a set of coefficients is collectively equal to zero. Thus we can contrast two (hierarchical) models by subjecting the variables that distinguish them to a Wald test. A significant result leads us to prefer the more complex model—that is, to reject the null hypothesis that the simpler model adequately represents the data in favor of the more complex model. A nonsignificant result, of course, leads us to prefer the simple model because it tells us that none of the coefficients that differentiate the two models is significantly different from zero.

For the same reason another statistic sometimes reported in conjunction with logistic regression models, *pseudo-*R^2, is not appropriate for weighted or clustered data.

Pseudo-$R^2 = 1 - L_1/L_0$, where L_0 is the log likelihood for a *constant-only* model (that is, a model with no independent variables), and L_1 is the log likelihood for the estimated model. Obviously, if the dependent variable is perfectly explained by the set of independent variables, $L_1 = 0$ and *pseudo-*$R^2 = 1$, and if the independent variables explain nothing, *pseudo-*$R^2 = 0$. Thus the *pseudo-*R^2 gives a sense of how well a model does. However, if, as in the case of weighted or clustered data, pseudo–log likelihoods are estimated, the pseudo–log likelihoods can increase rather than decrease for more complete models, and hence the *pseudo-*R^2s can decrease, which makes little sense. More generally, when pseudo–log likelihoods are estimated, there is no simple relationship between changes in the pseudo–log likelihoods and improvements in the goodness of fit, so the *pseudo-*R^2s become uninterpretable. For the same reason, *BIC* is inappropriate for designed-based estimation because it also is based on a comparison of log likelihoods. (For random samples *BIC* for logistic regression is estimated by $-L^2 + (d.f.)[\ln(N)]$. The signs are opposite those for Equation 12.8 because here the comparison of interest is not with a saturated model but with a baseline model in which predictions are based on the intercept alone.) When we have data based on complex samples as in the current case, survey estimation (invoked by the Stata command `-svy:logistic-`) is the best available tool, with models compared through adjusted Wald tests.

LIMITATIONS OF WALD TESTS The appropriate way to do statistical inference for complex samples is at present an unsettled issue. As we saw in Chapter Nine, when the clustering of observations typical of multistage probability samples is ignored the standard errors of statistics may be substantially biased—they typically are underestimated but in some instances may be overestimated. But the proposed corrections have their own limitations, both theoretically and practically. In particular, Wald tests are known to have poor properties, which may produce misleading results (Gould and Sribney 1999, 7–8); and as noted, *BIC* is not available for weighted or clustered samples. The optimal solution may be to treat clustered samples in a multilevel context, estimating either fixed- or random-effects models (Mason 2001), which can be done in Stata using the `-xt-` or `-gee-` command; these procedures go beyond what can be covered in this book, but see Chapter Sixteen for a brief introduction to multilevel analysis. Although even now much that is published, even in leading journals, simply ignores complex sample designs and treats data as if they were generated by random sampling procedures, this is generally inappropriate and can lead to incorrect inferences. For the present, I suggest for logistic regression in its various forms that when you have data that are weighted or clustered you carry out your estimation using Stata's survey estimation commands and rely on adjusted Wald tests for model selection. Be cautious, however, in your interpretation and explore alternative specifications. Only where you have a true, unweighted, random sample should you use the `-logistic-` command and likelihood ratio test (`-lrtest-`). Further, whenever possible, eschew weighting in favor of including the variables used to create the weights in the model.

Inspecting the Wald-test statistics in the bottom panel of Table 13.2, we see that Model 2 fits better than Model 1, but no model fits significantly better than Model 2. We thus conclude that the likelihood of armed assault depends on gender, race, and education and also changes over time in a linear way. To see the nature of these relationships, we examine the coefficients in Table 13.3. I also have included the coefficients for Model 4 in Table 13.3, even though Model 4 is only a marginally significant improvement over Model 2 (p = .092). I do this to illustrate how to deal with interaction terms in the context of logistic regression.

TABLE 13.3. Effect Parameters for Models 2 and 4 of Table 13.2.

Independent Variable	b	Standard Error	p	e^b
Model 2				
Male	1.4235	0.0466	.000	4.1514
Education	–0.0185	0.0066	.005	0.9816
Year	0.0101	0.0038	.008	1.0102
Black[a]	0.4463	0.0687	.000	1.5625
Intercept	–2.8941	0.3178	.000	-
Model 4				
Male	1.4543	0.0507	.000	4.2817
Education	–0.0191	0.0065	.004	0.9811
Year	0.0101	0.0038	.008	1.0101
Black[a]	0.5690	0.1007	.000	1.7665
Black*Male	–0.2125	0.1259	.092	0.8085
Intercept	–2.9037	0.3178	.000	-

[a]Black versus non-Black. Non-Blacks consist of "Whites" and "Others."

There are two alternative (but equivalent) ways to discuss effects estimated from logistic regression models: to consider the *additive* effects of each independent variable on the *log odds* of the dependent variable, and to consider the *multiplicative* effects of each independent variable on the *odds* of the dependent variable. First consider how to interpret the log-odds effects, the effects on the *logits*. As is evident from Equation 13.1, we can interpret the contributions to log odds, the bs, just as we would the coefficients in an OLS regression equation: a one-unit difference in the independent variable results in b units difference in the log odds of being threatened by a gun, net of all other variables. Thus for example, in Model 2 of Table 13.3, the difference in the log odds of males and females being threatened by a gun is 1.42, holding constant race, education, and the year of the survey. Also, in each successive year the expected log odds of having been threatened increase by 0.0101, net of sex, race, and education, so that the *expected log odds* of having been threatened in 1994 are about 20 percent larger than in 1973 (precisely, 0.2121 = 0.0101*[1994–1973]), all else equal. And so on.

Although the interpretation is straightforward, log odds are not very intuitively meaningful. Hence, a more appealing possibility is to interpret the antilogs of the bs, the e^bs. Here a one-unit difference in the independent variable results in an increase (or decrease) of e^b units in the relative *odds* of being threatened by a gun, net of all other variables. This follows directly from Equation 13.6 (which is the same as Equation 13.1):

$$\ln\left(F_{1|X_{1\ldots K}}/F_{2|X_{1\ldots K}}\right) = a + \sum_{k=1}^{K} b_k X_k \tag{13.6}$$

Exponentiating both sides of the equation, we get

$$\begin{aligned} F_{1|X_{1\ldots K}}/F_{2|X_{1\ldots K}} &= e^{(a+\sum_{k=1}^{K} b_k X_k)} \\ &= e^a \prod_{k=1}^{K} e^{b_k X_k} \end{aligned} \tag{13.7}$$

That is, the odds of being in category 1 rather than category 2 of a dichotomous variable is given by the product of the antilogs of the coefficients. The antilogs of the bs are interpreted as contributions to *odds ratios,* that is, the ratios of the odds for people who differ by one unit with respect to a given independent variable, holding constant all other independent variables. Thus for example, in Model 2 the expected odds of males being threatened by a gun are 4.15 ($= e^{1.4235}$) times greater than the odds of females being threatened, holding constant race, education, and the year of the survey. Also, the odds of having been threatened increase by 1.0102 each year, net of sex, race, and education, so that the expected net odds of having been threatened in 1994 are about 25 percent larger than in 1973 (precisely, 1.2363 [$= e^{0.0101(1994-1973)} = (e^{0.0101})^{(1994-1973)}$]). And so on.

So what can we conclude substantively? Net of other factors, the expected odds of males ever having been threatened are four times greater than for females; the expected odds of having been threatened decline slightly with increasing education (the odds of being threatened for those with at least a BA are about 14 percent less than for those with only an eighth grade education—precisely, $0.8624 = e^{(-0.0185(16-8))}$), but increase modestly over time, as we have seen; and the odds of Blacks having been threatened in any given year are more than 1.5 times as great as for non-Blacks of the same sex with the same amount of education (precisely, $1.56 = e^{0.4463}$).

Now let us consider Model 4. Note that the coefficients for education and year hardly change. Thus we can restrict ourselves to the interpretation of the coefficients for *MALE* and *BLACK* and their interaction. A convenient way to see how to interpret these coefficients is to evaluate the equation for fixed values of *EDUCATION* and *YEAR*. Let us take 1994 and twenty years of education as our values for these variables, to assess the effect of race and sex on the probability of having ever been threatened by a gun. We thus compute a new intercept: $a' = a + b_E{*}20 + b_Y{*}94 = -2.9037 - 0.0191{*}20 + 0.0101{*}94 = -2.3363$ (where b_E is the coefficient for education and b_Y is the coefficient for year of survey). Then we write out the expected log odds of having experienced a gun threat (for convenience, call this G) by race and sex (where b_M is the coefficient for *MALE*, b_B is the coefficient for *BLACK*, and b_{BM} is the coefficient for the interaction term).

For non-Black females we have

$$\begin{aligned} G &= a' \\ &= -2.3363 \end{aligned} \tag{13.8}$$

For Black females we have

$$\begin{aligned} G &= a' + b_B \\ &= -2.3363 + 0.5690 = -1.7673 \end{aligned} \tag{13.9}$$

For non-Black males we have

$$\begin{aligned} G &= a' + b_M \\ &= -2.3363 + 1.4543 = -0.8820 \end{aligned} \tag{13.10}$$

For Black males we have

$$\begin{aligned} G &= a' + b_B + b_M + b_{BM} \\ &= -2.3363 + 0.5690 + 1.4543 - 0.2125 = -0.5255 \end{aligned} \tag{13.11}$$

We see, both from the coefficients in Table 13.3 and from the sums just shown, that the expected log odds of being threatened are 1.45 larger for non-Black males than for non-Black females; that the expected log odds of being threatened are 0.57 larger for Black females than for non-Black females; but that Black males do not face the full double jeopardy of being male and Black, because their expected log odds are 0.21 less than the sum of the *MALE* coefficient and the *BLACK* coefficient. Or, to put it differently, the gender difference is greater for non-Blacks than for Blacks, and the race difference is greater for females than for males. These results are as hypothesized except that the interaction is too weak for us to have much confidence in it.

Again, the interpretation is easier if we consider the odds ratios rather than the logits. One way to do this is simply to take the antilog of the logits we just computed (the Gs). Doing this, we see that the expected odds of ever having been threatened by a gun among people with twenty years of schooling in 1994 are 0.10 for non-Black females ($= e^{-2.3363}$), 0.17 for Black females, 0.41 for non-Black males, and 0.59 for Black males. Note that the odds ratios are just what are shown in the rightmost column of Table 13.3 (within the limits of rounding error): the odds of non-Black males having been threatened are 4.3 times as large as the odds of non-Black females having been threatened ($0.4140/0.0967 = 4.2813 \cong 4.2817$); the odds of Black males having been threatened are about 3.5 times as large as the odds of Black females having been threatened ($0.5913/0.1708 = 3.4619 \cong 3.4618 = 4.2817*0.8085$); and so on. We can see this most clearly by writing out the expected odds just as we did for the logits.

For non-Black females we have

$$\begin{aligned} e^G &= e^{a'} \\ &= .0967 \end{aligned} \tag{13.12}$$

For Black females we have

$$\begin{aligned} e^G &= e^{a'} e^{b_B} \\ &= (0.0967)(1.7665) = 0.1708 \end{aligned} \tag{13.13}$$

For non-Black males we have

$$\begin{aligned} e^G &= e^{a'} e^{b_M} \\ &= (0.0967)(4.2817) = 0.4140 \end{aligned} \tag{13.14}$$

For Black males we have

$$\begin{aligned} e^G &= e^{a'} e^{b_B} e^{b_M} e^{b_{BM}} \\ &= (0.0967)(1.7665)(4.2817)(0.8085) = 0.5913 \end{aligned} \tag{13.15}$$

One other coefficient is sometimes useful—the percentage change in the odds, given by $100(e^b - 1)$. For example, from Model 2 in Table 13.3 we would conclude that all else equal, the odds of Blacks having ever been threatened or shot at are 56 percent greater than the corresponding odds for non-Blacks because $100(1.56 - 1) = 56$.

However, even though odds ratios are readily interpreted, expected odds are still not particularly intuitive. Thus it would be useful to convert expected odds into percentages. For example, in the present case it would be helpful to get the expected percentage of individuals in each race-by-sex group who have ever been threatened, net of education and survey year. That is, we would like to get the *adjusted percentages* implied by the model so that we can assess percentage differences between race and sex groups, controlling for education and year of survey. We can do this by making use of the relationship

$$Pct(Y) = 100\left(\frac{x}{x+1}\right) \qquad (13.16)$$

where x is the odds of Y for specified values of the independent variables. Note that because the relationship between the odds and the percentages is nonlinear, we need to choose specific values of the independent variables at which we wish to make the conversion. Here I use the same values at which we evaluated Model 4; that is, I get expected percentages by race and sex among people with twenty years of education in 1994. For example, for non-Black women, we have $Pct(Y) = 100*[0.0967/(0.0967 + 1)] = 8.8$. The corresponding percentages are, respectively, for Black women, 14.6; for non-Black men, 29.3; and for Black men, 37.2. If we wished, we could, of course, construct an entire table of such expected percentages for various values of education and year of survey. Doing this requires a fair amount of hand calculation. However, in conjunction with their text on Stata procedures for handling limited dependent variables, Long and Freese (2006) developed a set of Stata `-ado-` files that automate the computation of these and other statistics for interpreting logistic regression coefficients. (Those wishing to explore these files should start with Long's web page: http://www.indiana.edu/~jslsoc. Follow the links to Long and Freese's book.)

A SECOND WORKED EXAMPLE: SCHOOLING PROGRESSION RATIOS IN JAPAN

In the educational stratification literature an important hypothesis is that the dependence of educational attainment on the social status of one's parents decreases as education increases. This hypothesis has been operationally specified in terms of *progression ratios* from one level of schooling to the next (Mare 1980, 1981). That is, we can ask what affects the odds that those at any given level of education go on to the next level: that primary school graduates enter secondary school, that those who enter secondary school graduate, that secondary school graduates go on to college or university, and so on. Once we specify the problem in this way, it is evident that it is a logistic regression problem, but one of a particular kind. The distinctive feature of this sort of problem is that any individual may make several transitions. It also should be evident that the formal structure of the problem is identical to that of many nonreversible transitions; for example, in criminology, from arrest to arraignment to trial to conviction to sentencing; in medical research, the transition through various stages of a disease; and so on. We tackle problems with this sort of formal structure by pooling data for all transitions into a single data set and then analyzing not a sample of *people* but rather a sample of *transitions*.

To see how this is done, consider an analysis of trends in educational attainment in Japan, carried out by Treiman and Yamaguchi (1993). Here, to illustrate the method,

I present only the portion of our analysis concerned with the transition from middle school to higher secondary school and from higher secondary school to college or university in postwar Japan. The data set included 1,320 men who completed their education during the postwar period. Because education up to the middle school level is compulsory in Japan, our base is 1,320 middle school graduates; that is, 1,320 men were "at risk" of obtaining at least some higher secondary education. Of these, 1,056 did so and hence were "at risk" of continuing on to college or university. Pooling those at risk of making the first and second transitions, we have 2,376 (= 1,320 + 1,056) *transition possibilities* to study. For each of these cases, we create a dummy variable, *SUCCESS* (*S*), scored 1 if the transition was made and 0 otherwise. We distinguish the two transitions by a dummy variable, *TRANSITION* (*T*), scored 1 for the transition from higher secondary to tertiary education, and 0 otherwise. We then estimate a series of logistic regression equations in which the dependent variable is the natural log of the odds of successfully making a transition and the independent variables are the transition variable, measures of the status of parents, year of birth (to study trends), and various interactions among the variables. Table 13.4

TABLE 13.4. Goodness-of-Fit Statistics for Various Models of the Process of Educational Transition in Japan (Preferred Model Shown in Boldface).

	Model L^2	d.f.	*BIC*
Model[a]			
(1): Social origins	265	2	−251
(2): (1) + Transition + (Social origins)*(Transition)	675	5	−639
(3): (2) + Year + (Year)*(Transition)	**703**	**7**	**−653**
(4): (3) + (Social origins)*(Year)	706	9	−641
Contrasts			
(2) − (1)	410	3	−388
(3) − (2)	28	2	−14
(4) − (3)	3	2	12

Source: Adapted from Treiman and Yamaguchi (1993, Table 10.4).

[a]All models and contrasts are significant at the .000 level except the (4) − (3) contrast, which is significant at the .223 level.

TABLE 13.5. Effect Parameters for Model 3 of Table 13.4.

	b	e^b
Intercept	−6.16	0.002
E: Parents' education	0.348	1.416
P: Prestige of father's occupation	0.0569	1.059
T: Transition	1.23	3.421
*T***E*	−0.0503	0.951
*T***P*	−0.0180	0.982
Y: Year of birth	0.0519	1.053
*T***Y*	−0.0439	0.957

Source: Adapted from Treiman and Yamaguchi (1993, Table 10.5); Treiman and Yamaguchi did not report the standard errors.

shows goodness-of-fit statistics for various models of the educational transition process, and Table 13.5 shows effect parameters for the preferred model. (This analysis was carried out before designed-based estimation was generally available. Thus no account was taken of the clustering of the sample. In addition to the usual clustering by sampling points typical of national surveys, transition-ratio models are clustered by person because the transitions made by any one person are hardly independent. Thus in addition to any other adjustments for clustered samples, the observations for each individual should be treated as nonindependent.)

From Table 13.4 we see that Model 3 fits best according to both the likelihood ratio and *BIC*. The model posits that the effect of social origins varies across transitions and also that the odds of making the two transitions change over time (but the effect of social origins does not change over time). From the point of view of our a priori hypothesis—that the effect of social origins declines with successive transitions—the contrast between Models 2 and 1 is particularly noteworthy. Model 1 posits that the odds of moving to the next higher level of education are affected by the social status of one's parents (specifically, parents' education and father's occupational status, measured by prestige) but that the relationship is the same regardless of which transition is considered. Model 2, by contrast, posits both that the odds of making the transition depend on which transition is considered and also that the relationship between social origins and the odds of making a transition depend on which transition is considered. Model 2 represents our a priori hypothesis.

As we see, Model 2 is far more likely than Model 1 given the data, but Model 3, which also posits a temporal shift in the odds of making each transition, is still more likely. Thus we have preliminary support for our hypothesis, but we also have evidence that the transition process had changed over time. (This point is further explored in the paper but need not concern us here.)

In retrospect, the contrasts presented by Treiman and Yamaguchi are not wholly satisfactory. It would have been better to include a model intermediate between Model 1 and Model 2; that is, a model that posits a difference in the odds of making successive transitions but with the effect of social origins constrained to equality across transitions. The difficulty is that we do not know whether Model 2 is superior to Model 1 because the odds of making the transition vary across transitions, or because the effect of social origins varies across transitions, or both. The same point can be made with respect to the effect of birth year—a model intermediate between Model 2 and Model 3 would have been desirable.

Actually, all that the coefficients in Table 13.4 tell us is that a model that posits *different* effects of social origins for different transitions is more likely than a model that posits the same effect. To pin down our claim, we need to inspect the effect parameters, reported in Table 13.5, to be sure that they have the predicted sign.

Table 13.5 shows the effect parameters associated with the preferred model. Note that I do not report the standard errors or *p*-values for individual coefficients. Because all of the "main effects" in the model also appear in "interaction terms," the appropriate way to assess the effect of a single dimension is to contrast models with and without the variables representing that dimension. I have done this in Table 13.4, but only for selected contrasts rather than every possible pair of models. (Raftery discusses *S-Plus* software that makes it possible to choose the most-likely-model-given-the-data from among all possible models involving a given set of variables. Interested readers should consult Raftery [1995a].)

Note that the treatment of standard errors in Table 13.5 contrasts with Table 13.3, where I do show the standard errors and *p*-values. The difference is that Table 13.3 shows only one interaction, so the *p*-value associated with the interaction term indicates the significance of the difference in the fit of models including and not including the interaction term. Where a model includes both variables for which individual significance tests are meaningful and variables for which they are not because they are confounded by interactions (or other transformations such as squared terms), the usual practice is to report all significance tests and *p*-values. It might be preferable, however, only to report significance statistics when they are meaningful, in order to preclude incorrect interpretation.

This model suggests that the process of moving from one level of education to the next in Japan is about as we would expect it to be: the odds of making each transition vary positively with parents' education and with the status of the father's occupation. Of greater interest are the coefficients of the interaction terms $T*E$ and $T*P$. These are both negative, which indicates that, as hypothesized, in these data the effect of social origins on going on to the next level of education is weaker for the transition from higher secondary school to university than for the transition from middle school to higher secondary school. Each year of average parental education increases the odds of making the first transition, from middle school to higher secondary school, by about 40 percent (because $e^{0.3480} = 1.416$), but increases the odds of the second transition, from secondary school to university, by only about 35 percent (because $e^{(0.3480-0.0503)} = 1.347$). Thus, for example, all else equal, the odds that a son of a university graduate will go on to higher secondary school are more than 11 times as great as the corresponding odds for the son of a middle school graduate (because $1.416^{(16-9)} = 11.414$). By contrast, among those who managed to get into higher secondary school, the odds that the son of a university graduate will go on to university are only eight times as great as the corresponding odds for the son of a middle school graduate (because $[(1.416)(0.951)]^{(16-9)} = 8.030$). Similarly, the net effect of each unit increment in the prestige of the father's occupation is to increase the odds of the first transition by about six percent (because $e^{0.0569} = 1.059$) but to increase the odds of the second transition by only 4 percent (because $e^{(0.0569-0.0180)} = 1.040$). Thus, for example, the net odds of the son of a shopkeeper (prestige score = 42) making the transition from middle school to higher secondary school are more than twice as great as the net odds of the son of a factory worker (prestige score = 29) making the transition (because $1.059^{(42-29)} = 2.107$). But the net odds of a shopkeeper's son making the transition from secondary to tertiary education are only about 66 percent greater than those for a factory worker's son (because $1.040^{(42-29)} = 1.665$). The effects of year of birth and of the interaction between transition and year of birth can be interpreted in a similar way.

As a reminder, the interpretation of contributions to log odds in models involving interaction terms is exactly the same as in ordinary least-squares regression (see Chapter Six): the appropriate coefficients are added. However, as we saw in the first worked example, exponentiated coefficients (*contributions to odds ratios*) are not added but rather multiplied. Thus, for example, the coefficient for parental education is 0.3480 for the first transition and 0.2977 (= 0.3480 − 0.0503) for the second transition. The corresponding exponentiated coefficients are 1.4162 for the first transition and 1.3468 (= 1.4162*0.9509) for the second. Of course, $1.3468 = e^{0.2977}$.

A THIRD WORKED EXAMPLE (DISCRETE-TIME HAZARD-RATE MODELS): AGE AT FIRST MARRIAGE

One of the most powerful uses of binary logistic regression procedures is to estimate *discrete-time hazard-rate models,* sometimes called *event history models*. Hazard-rate models are those for the rate at which events occur or the likelihood that an event will occur at a specified time. There is a well-developed statistical technology for estimating such models, most of which is beyond the scope of this book. However, for a particular class of these models, in which time is treated as a set of discrete values and the interest is in estimating the likelihood that an event occurs in each period of time, conventional binomial logistic regression procedures can be used once the data are appropriately arranged. Indeed, as we will see, discrete-time hazard-rate models are formally equivalent to the educational transition model we just discussed.

The basic procedure is to create a person-period data set by *stacking* replicates of the original data set for each period for which each individual is "at risk" of the event occurring.

For example, suppose we are interested in estimating the likelihood that individuals marry at specified ages—say, at each year of age from 15 to 36. We can do this by creating a new data set consisting of one observation for each person for each year of age at which the person has not yet married, plus the age at which the person married if he or she did, up to and including the individual's current age. The dependent variable is a dichotomy, scored 1 if the person married at that age and scored 0 otherwise. For ever-married individuals, the dependent variable takes on the value 0 for each age, from age 15 until the year before they married, and is scored 1 for the age at which they marry. Observations representing subsequent ages are dropped from the data set, because once they marry, people are no longer "at risk" of (first) marriage. For never-married individuals, the dependent variable is scored 0 for all years, from age 15 up to their current age. Ages greater than their current age are dropped from the data set because they obviously are not at risk of marriage for ages they have not yet reached. We then analyze this data set in the usual way, estimating a binomial logistic regression equation.

At this point you may be wondering why we go to all this fuss when it would be easy to treat age at first marriage as a ratio variable and simply carry out an OLS regression with age at first marriage as the dependent variable. This might be a reasonable procedure if we had a sample of persons old enough to no longer be at risk of marriage. However, this typically is not the case, because we usually analyze representative samples of a population and thus include adults of all ages, some of whom have not yet married but will do so in the future. These cases are *censored* because we have stopped observing them while they are still at risk for the event. Under these circumstances OLS regression gives misleading results whereas discrete-time hazard-rate models give correct estimates of the likelihood of marrying at each age for those who are still at risk because they have reached that age without ever having married.

To illustrate the practical procedures for carrying out such an analysis, I use the 1994 GSS to estimate the likelihood of marrying for the first time as a function of age, mother's education, sex, and race (Blacks versus non-Blacks). Given marital norms in the late twentieth-century United States, we would expect the likelihood of marrying to increase with age up to the mid-twenties but then to decline, and we also would expect males to marry later than females. We would expect those from well-educated families (measured by mother's education) to marry later, in part because they themselves tend to become well educated and to delay marriage until they complete their education (although for some people marriage affects the likelihood of continuing in school). Finally, we would expect Blacks to be less likely to marry than non-Blacks, both because of the socioeconomic position of Blacks and because of racial differences in norms regarding childbearing outside of marriage; Blacks are less likely to be coerced into marriage by their families in the case of unanticipated pregnancies.

The downloadable files "ch13_2.do" and "ch13_2.log" show the specific commands I used to carry out the analysis, together with comments. Because I have extensively documented these files, further comment on the specific Stata commands is for the most part unnecessary. The only novel feature of the Stata setup is the use of the `-reshape-` command to create a suitable data set, shown with results in the Stata log file. This command converts data from wide to long form; that is, in this case from a file of people to a file of person-years, where there is one observation for each person for each year at which he or she remains unmarried, plus the year of marriage. This is a very efficient way to create a suitable data set for a discrete-time hazard-rate analysis, which requires only a few lines of code. Study my setup plus the relevant section of the Stata manual to be sure you understand the logic of this command.

I began by defining the risk set as including ages 15 through 56, because (after excluding a small number of cases with missing values on the independent variables) no one in the sample married for the first time before age 15 or after age 56. I then estimated an equation of the form

$$\ln\left(\frac{W}{1-W}\right) = a + \sum_{i=16}^{56} b_i A_i \tag{13.17}$$

where W is the probability of first marriage, conditional on a respondent's age, and the A_i are dummy variables for age at risk, with 15 the omitted category. This regression produced the expected probabilities shown in the Stata log (converted from the logits estimated in Equation 13.17) and graphed in Figure 13.1. Inspecting the figure, we see that the right tail is not very orderly.

In particular, the probability of first marriage appears to increase among those in their 40s and 50s. Inspection of the downloadable Stata `-log-` file makes it clear why this is so: by the time people are in their late 30s, almost everyone has married. Thus one or two marriages constitute a nonnegligible proportion of all those at risk. The graph is somewhat misleading in another way as well because all ages for which prediction is perfect because

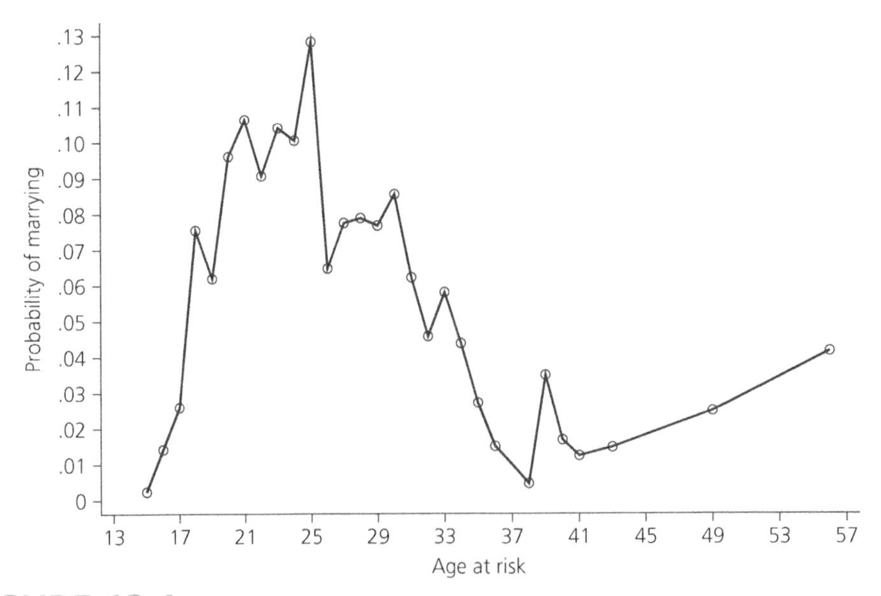

FIGURE 13.1. *Expected Probability of Marrying for the First Time by Age at Risk, U.S. Adults, 1994 (N = 1,556).*

no one at risk married at that age (37, 42, 44 through 48, 50 through 55) are dropped from the equation and hence from the graph. The lesson here is that very sparse data may give misleading results. Before continuing with the analysis, I dropped all ages greater than 36.

I next reestimated the model, predicting the probability of marriage at discrete ages (using an equation similar to Equation 13.17) and then substituted a fourth-degree polynomial for discrete years to fit a smooth curve to age at risk. (I decided that a fourth-degree polynomial, estimated by

$$\ln\left(\frac{W}{1-W}\right) = a + b(A) + c(A^2) + d(A^3) + e(A^4) \tag{13.18}$$

was required by testing the significance of successive powers of risk age.) The two curves are shown as Figures 13.2 and 13.3. Visual comparison suggests that they are quite similar, although a formal test of significance reveals significant discrepancies at specific ages. When I determined this, I had to consider whether to continue with a discrete or smooth representation of age at risk. I opted for a discrete representation, which more faithfully represents the data, although a smooth representation of age at risk, which is far more parsimonious, also would have been reasonable.

I then proceeded to estimate two additional models: including first the other variables I hypothesized to affect age at marriage (sex, race, and mother's education) and then interactions between these three variables and age at risk. Wald tests, for all interactions and for interactions with each of the main effects (shown in the Stata log), made it clear that the model including interactions is the preferred model; all tests are significant at beyond the .000 level. Thus, the likelihood of marrying at each age varies by sex, race, and mother's education. Table 13.6 shows contributions to odds ratios, which are the antilogs of the coefficients estimated from an equation of the form

$$\begin{aligned}\ln\left(\frac{W}{1-W}\right) &= a + b(E) + c(M) + d(B) + \sum_{i=15\,to\,24}^{26\,to\,36} e_i A_i \\ &\quad + \sum f_i A_i E + \sum g_i A_i M + \sum h_i A_i B\end{aligned} \tag{13.19}$$

where W is the probability of marrying given that one is at risk; E is the number of years of school completed by the respondent's mother, expressed as a deviation from the sample mean; M is scored 1 for males and 0 for females; B is scored 1 for Blacks and 0 for non-Blacks; and the A_i are dummy variables for age at risk, with age 25 the reference category.

In Table 13.6 the first column, labeled "Main Effect," shows the expected odds of marrying for non-Black females whose mothers' years of schooling are at the sample mean, expressed in ratio to the effect for those age 25. The remaining three columns show the interactions of age at risk with mother's education, sex, and race, except that the coefficients for these variables at age 25 are the main effects. These odds ratios can be used to make any comparison of interest. For example, among women who have never married by age 21, the odds of Blacks marrying at that age are about three-fifths the odds for

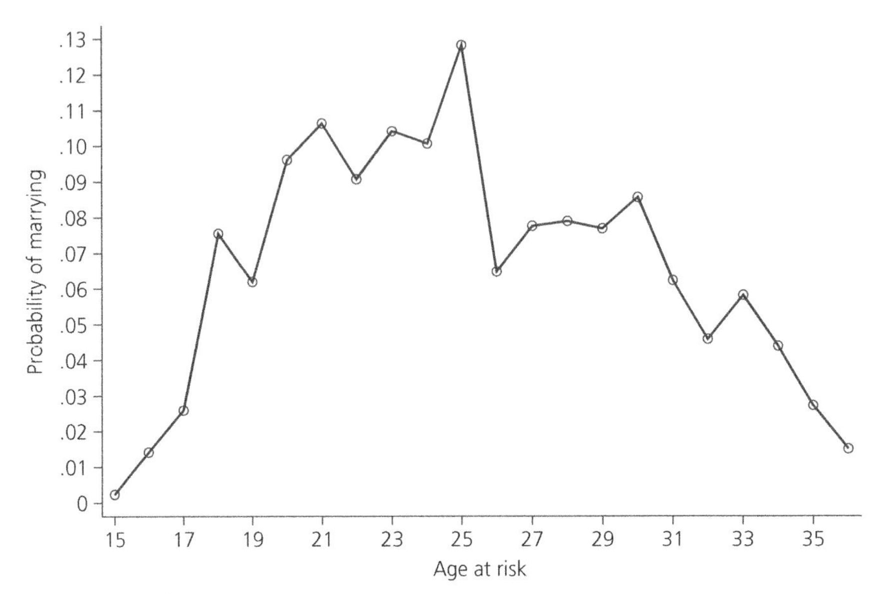

FIGURE 13.2. *Expected Probability of Marrying for the First Time by Age at Risk (Range: Fifteen to Thirty-Six), Discrete-Time Model, U.S. Adults, 1994.*

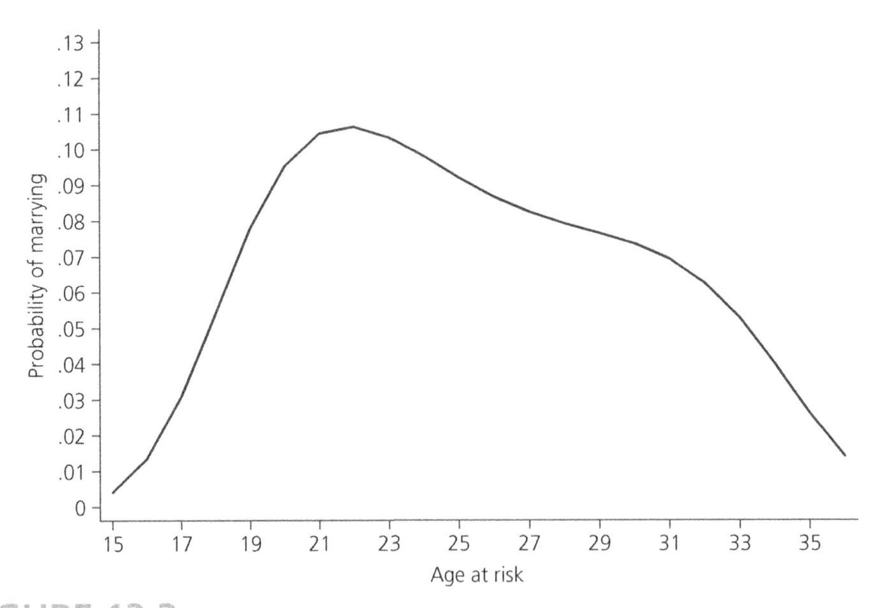

FIGURE 13.3. *Expected Probability of Marrying for the First Time by Age at Risk (Range: Fifteen to Thirty-Six), Polynomial Model, U.S. Adults, 1994.*

TABLE 13.6. Odds Ratios for a Model Predicting the Likelihood of Marriage from Age at Risk, Sex, Race, and Mother's Education, with Interactions Between Age at Risk and the Other Variables.

Age at Risk	Main Effect	Interaction with		
		Mother's Education	Sex (Male)	Race (Black)
15	0.021	0.879	[a]	[a]
16	0.121	0.986	0.064	9.847
17	0.204	0.913	0.272	3.022
18	0.801	1.026	0.243	2.850
19	0.558	0.999	0.381	4.426
20	0.843	1.033	0.575	3.156
21	0.715	0.993	1.021	3.108
22	0.786	0.988	0.669	1.361
23	0.765	1.018	0.826	4.516
24	0.918	1.067	0.588	2.814
25 (reference. category)[b]	1.000	0.918	0.930	0.190
26	0.498	1.081	0.887	1.628

(*Continued*)

TABLE 13.6. Odds Ratios for a Model Predicting the Likelihood of Marriage from Age at Risk, Sex, Race, and Mother's Education, with Interactions Between Age at Risk and the Other Variables. *(Continued)*

Age at Risk	Main Effect	Interaction with Mother's Education	Interaction with Sex (Male)	Interaction with Race (Black)
27	0.350	0.989	1.689	4.904
28	0.452	1.008	1.679	[a]
29	0.671	1.032	0.724	1.710
30	0.617	1.114	1.008	2.270
31	0.440	1.092	0.884	3.942
32	0.093	1.221	3.541	11.409
33	0.249	1.089	1.785	6.528
34	0.271	1.056	0.933	5.694
35	0.069	0.997	3.577	5.097
36	0.058	1.119	2.458	[a]

[a]No persons at risk married. Thus the logits are undefined.

[b]Because this is the reference category, the odds ratio for age is, implicitly, one. The three remaining coefficients are the main effects of mother's education, sex (male), and race (Black).

SMOOTHING DISTRIBUTIONS Smoothing refers to a class of techniques for making the general shape of a distribution clear by removing "noise"—deviations from the underlying trend that result from sampling error or idiosyncratic factors. Perhaps the simplest smoother is a *moving average*. A moving average is the average value of several consecutive data points. Consider the worked example in this section. A three-year moving average of the expected probability of marriage at each age would be constructed by first taking the average of the expected probabilities for ages fifteen, sixteen, and seventeen; then the average of the expected probabilities for ages sixteen, seventeen, and eighteen; and so on. At the time the age-at-first-marriage example was created, the Stata subcommand `-ma-` ("moving average") was available within the `-egen-` command. However, this subcommand is no longer documented in Stata 10 (although it still works), and has been replaced by `-smooth-`, which generates medians of the included points rather than means. Another smoother available in Stata is `-lowess-`.

non-Blacks (precisely, $0.591 = 0.190*3.108$). Among 30-year-old never-married people, the odds of marrying in that year among those whose mothers are college graduates are nearly 10 percent higher than the odds for those of the same race and sex whose mothers are high school graduates (precisely, $1.094 = (0.918*1.114)^4$).

Despite the usefulness of Table 13.6 for making specific contrasts, the overall pattern implied by the coefficients is difficult to discern. Again, graphs help. Figures 13.4 and 13.5 show three-year moving averages of the expected probability of first marriage by age at risk, separately for Blacks and non-Blacks. In each graph, separate lines are shown for males and females whose mothers had twelve and sixteen years of schooling (as a convenient way of visually representing the effect of mother's education). Moving averages are shown because there is a great deal of " oat and bounce" for individual years, which is evident from inspection of the coefficients in Table 13.6. (See the downloadable file "ch13_2.do" for details on how the moving averages were constructed.)

Inspecting Figures 13.4 and 13.5, we see that marriage rates for Blacks differ substantially from those for non-Blacks, with Blacks much less likely than non-Blacks to marry at all. Moreover, non-Black females (especially those whose mothers have only a high school education) marry at disproportionately high rates at ages nineteen through twenty-five; non-Black males marry a bit later and with less concentration in a short period. Black marriage rates, by contrast, are spread out over a much longer period, but with an upsurge in marriage rates for males in their thirties, especially those whose mothers are high school educated. For both Blacks and non-Blacks, males tend to marry later than females, with male rates exceeding those of females beginning around age thirty. Finally, among all race-by-sex groups, those whose mothers are high school graduates are more likely to marry than are those whose mothers are college graduates.

If I were preparing these results for publication, I would present only a subset of the rather large set of tables and graphs we have just marched through. The intent here, of

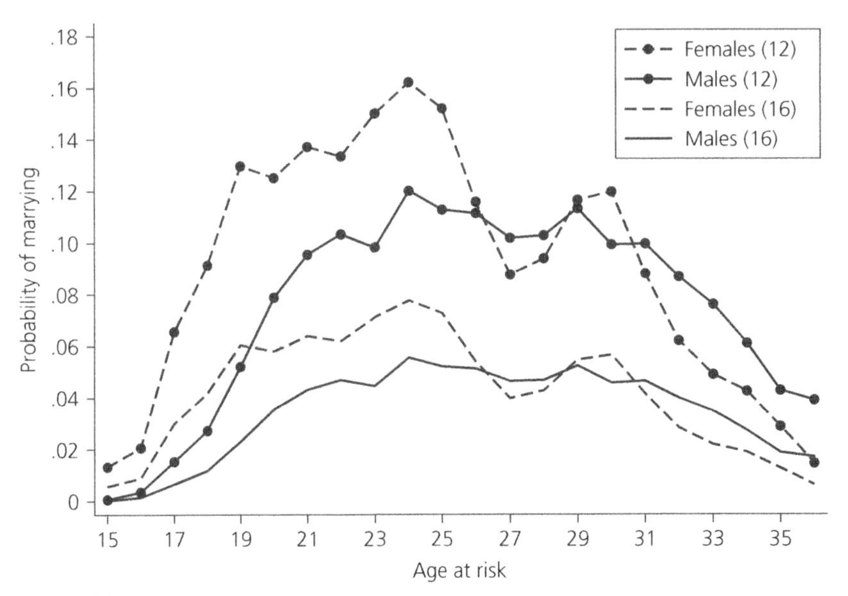

FIGURE 13.4. *Expected Probability of Marrying for the First Time by Age at Risk, Sex, and Mother's Education (Twelve and Sixteen Years of Schooling), Non-Black U.S. Adults, 1994.*

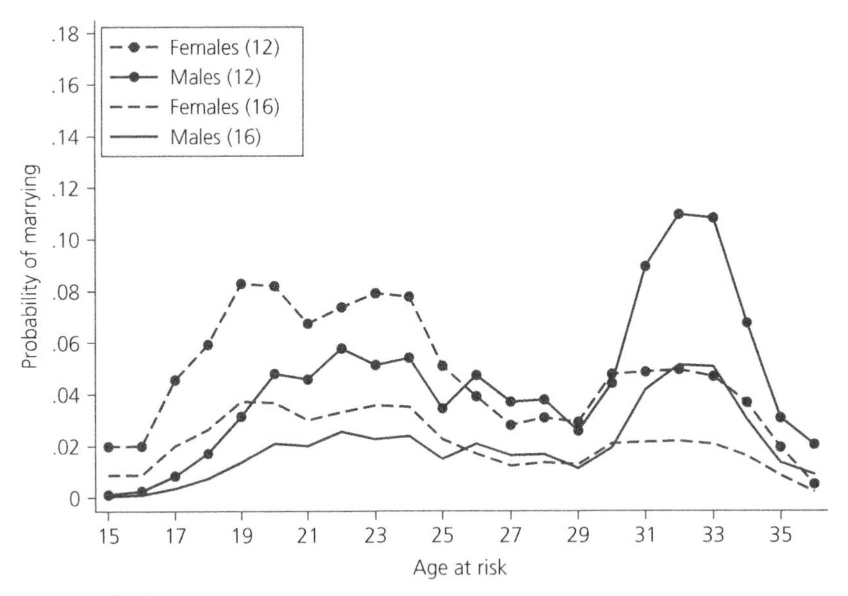

FIGURE 13.5. *Expected Probability of Marrying for the First Time by Age at Risk, Sex, and Mother's Education (Twelve and Sixteen Years of Schooling), Black U.S. Adults, 1994.*

course, is to provide alternatives for you to consider when presenting your own analyses. Examples of the application of discrete-time hazard-rate models include Astone and others (2000), Dawson (2000), Lewis and Oppenheimer (2000), and Sweeney (2002).

A FOURTH WORKED EXAMPLE (CASE-CONTROL MODELS): WHO WAS APPOINTED TO A *NOMENKLATURA* POSITION IN RUSSIA?

When a dependent variable is a rare event, it is inefficient to draw a representative sample of the population at risk for the event, because the sample size would have to be extremely large to obtain enough "positive" cases to analyze. This is a frequent occurrence in epidemiological research, where the events of interest are diseases, but it also occurs in the social sciences. For example, if we are interested in studying what determines who gets elected to Congress, we could hardly do this by drawing a representative sample of the population and looking for the congressmen in it. We have similar problems in studying crime, crime victimization, homosexuality, and various other relatively uncommon phenomena. One solution to this problem is to sample on the dependent variable (that is, to draw a sample of congressmen, criminals, or homosexuals), collect information on that sample, collect corresponding information on a representative sample of the population that has not experienced the rare event (becoming congressmen, criminals, or homosexuals), combine the two samples, and model the odds of experiencing the rare event. This is known as *case-control sampling* in the epidemiological literature (for an excellent review of the statistical procedures involved, see Breslow [1996]).

Case-control sampling exploits the fact that odds ratios are invariant under shifts in the distribution of the data. This extremely important feature of odds ratios makes it possible to combine samples with very different distributions on the independent and dependent variable in order to model rare events. This capability is not possible with OLS regression because OLS coefficients *are* affected by the distributions of the variables in the model.

To see how case control procedures work in practice, let us consider what factors affected the odds of becoming a member of the Russian political elite at the end of the Communist era. From *Social Stratification in Eastern Europe after 1989* (Treiman and Szelényi 1993), we have two representative samples from Russia: a probability sample of the adult population (N = 5,002) and a random sample of persons who were in *nomenklatura* positions as of January 1988 (N = 850). (See Appendix A for a description of the data and information on how to obtain them.) *Nomenklatura* positions were those that required the approval of the Central Committee of the Communist party. They ranged from very high government officials (for example, members of the politburo) down to heads of sensitive organizations—for example, rectors of universities, editors in chief of important newspapers, and heads of large industrial enterprises.

The general population sample departs in two ways from compliance with the assumptions underlying case-control sampling, but neither deviation is important from a practical standpoint. First, it is a probability sample of the 1993 population rather than the 1988 population. However, the sampling frame is based on the 1989 census, and the sample therefore probably represents the 1988 population nearly as well as it does the 1993 population, the main (but probably not very large) discrepancy resulting from differential mortality by socioeconomic status between 1988 and 1993. Second, the general population sample does not strictly meet the requirement that it be drawn from the population of non-*nomenklatura* members. However, membership in the *nomenklatura* is so rare in Russia (about 10,000 persons, from an adult population of nearly 100 million persons, which implies an expected value of about 0.5 persons in the general population sample) that the violation is of trivial importance; and in fact there was no one in a *nomenklatura* position in the general population sample.

Data for both samples were collected in 1993, using substantially identical questionnaires. I restricted the analysis to those age twenty to sixty-four in 1988, for whom complete data were available on all variables (777 members of the 1988 *nomenklatura* and 2,369 persons in the general population sample). I then merged the two samples into a single sample and carried out a logistic regression predicting the log odds of *nomenklatura* membership from a standard set of independent variables: education, gender, and age, and father's education, occupational status, and Communist Party membership. I weighted the general population data to correct for differential household size and certain biases in the sample design (see Treiman 1994, Section I.G). Because the *nomenklatura* sample is a random sample of 1988 *nomenklatura* members, no weighting was necessary—that is, all *nomenklatura* cases were weighted 1. No Stata log is shown, because once the two samples are combined we have a conventional binomial logistic regression. The results of the computations are shown in Table 13.7.

TABLE 13.7. Coefficients for a Model of Determinants of *Nomenklatura* Membership, Russia, 1988.

	b	Robust Standard Error	e^b
E: Years of school completed	0.546	0.030	1.726
M: Gender (1 = male, 0 = female)	2.84	0.28	17.12
A: Age	0.132	0.009	1.141
E_F: Father's years of school	0.108	0.029	1.114
S_F: Father's occupational status (ISEI)	0.00581	0.00596[a]	1.006
P_F: Father ever CP member	0.402	0.188[b]	1.495

[a] $p = .330$.
[b] $p = .033$. All other coefficients are significant at beyond the .0005 level.

Before turning to interpretation of the results, we should note the one difference between case-control analysis and ordinary binomial logistic regression: in case-control analysis the intercept is not meaningful. This should be obvious from the fact that the intercept in logistic regression indicates the proportion of the sample that is "positive" with respect to the dependent variable. However, in case-control designs this proportion is fixed by the sample design, and thus the coefficient adds no information.

Inspecting the coefficients in Table 13.7, we see very large effects and few surprises. Each year of schooling increases the odds of becoming a member of the *nomenklatura* by more than 70 percent. Thus, all else equal, university graduates (who typically have 15 years of schooling in Russia) are more than 15 times as likely as high school graduates (with 10 years of schooling) to be appointed to *nomenklatura* positions (precisely, $15.32 = 1.726^{(15-10)}$). The effect of gender is astronomical: males are more than 17 times as likely as females to be appointed to *nomenklatura* posts. The effect of age is also extremely strong: all else equal, the odds of being appointed to a *nomenklatura* position increase about 14 percent per year. Thus, for example, a 50-year-old is more than 7 times as likely to secure a *nomenklatura* position as is a 35-year-old (precisely, $7.23 = 1.141^{(50-35)}$). Perhaps more interesting, the effect of social origins, even among those equally well educated, is far from trivial. Coming from a family in which one's father was a member of the Communist Party improves one's chances of a *nomenklatura* appointment by about half, all else equal. Also, each year of father's schooling increases the odds of *nomenklatura* appointment by about 11 percent—this in the worker's paradise!—so that the offspring of the university-educated intelligentsia (15 years of school) are about three times as likely as the offspring of those with only a primary education to secure *nomenklatura* appointments, irrespective of their own educational achievement (precisely, $2.94 = 1.114^{(15-5)}$). Alone among the variables we have considered, father's occupational status has no impact on the odds of appointment to a *nomenklatura* post.

WHAT THIS CHAPTER HAS SHOWN

In this chapter we have seen how to estimate and interpret binary logistic regression models, which are widely used to model dichotomous outcomes such as whether people vote, are employed, or are members of a particular organization. We have seen that although the estimation procedures are quite different, the interpretation of the coefficients of such models is similar to that of OLS regression, except that the coefficients represent net effects of each independent variable on the log odds of an outcome.

Because log odds are not intuitive quantities, we have considered two nonlinear transformations to more readily interpretable coefficients—odds and expected probabilities—and have also seen how to graph net relationships, a form of regression standardization for logistic regression. Finally, we considered three extensions of the basic logistic regression model: education progression ratios, discrete-time hazard-rate models, and case-control models. A notable feature of logistic regression models is that they are invariant with respect to the distributions of variables in the sample, which is what makes case-control procedures legitimate in the logistic regression context but not in the OLS regression context.

APPENDIX 13.A SOME ALGEBRA FOR LOGS AND EXPONENTS

For those who have forgotten their school algebra, here are some useful equivalencies involving natural logarithms and antilogs (exponents):

$$e^{\ln(X)} = X$$

$$\ln(X * Y) = \ln(X) + \ln(Y)$$

$$\ln(X / Y) = \ln(X) - \ln(Y)$$

$$X * Y = e^{\ln(X)} e^{\ln(Y)} = e^{(\ln(X) + \ln(Y))}$$

$$e^{(X+Y)} = e^X * e^Y$$

$$e^{(X-Y)} = e^X / e^Y$$

$$\ln(X^P) = P * \ln(X)$$

$$X^P = (e^{\ln(X)})^P = e^{P * \ln(X)}$$

Note that $Y = \ln(X)$ and $X = e^Y$ are equivalent.

APPENDIX 13.B INTRODUCTION TO PROBIT ANALYSIS

As noted at the beginning of this chapter, an alternative to logistic regression as a model for predicting model binary responses is the *probit* model, which is defined as

$$\Pr(Y = 1 \mid \mathbf{x}) = \Phi(\beta'\mathbf{x}) = \Phi(\beta_0 + \sum_{i=1}^{k} \beta_i x_i) \qquad (13.B.1)$$

where Φ is the standard cumulative normal distribution and there are k predictor variables. From this definition it is evident that the βs are z-scores, and that the associated probability can be determined by finding the area under the normal curve corresponding to a particular z-score. This can be done by invoking Stata's `-normal-` function.

Consider the example used in the chapter to illustrate the interpretation of logistic regression models—the determinants of the likelihood of being threatened by a gun or being shot at. Table 13.B.1 shows the probit coefficients—the bs—corresponding to the logistic coefficients shown for Models 2 and 4 in Table 13.3. Note that the probit and logit models yield similar conclusions except that in Model 4 the interaction term is marginally significant when estimated using a logit model, and not even that when estimated by using a probit model.

Because probits are z-scores, they indicate the expected change, in standard deviation units, in the latent dependent variable (what StataCorp 2007 [*Reference I-P,* 620] calls the "probit index"), resulting from a one-unit change in the associated predictor variable. However, it is a property of probits, in common with logits, that the variance of the latent

TABLE 13.B.1. Effect Parameters for a Probit Analysis of Gun Threat (Corresponding to Models 2 and 4 of Table 13.3).

Independent Variable	b	Standard Error	p	Marginal Effect
Model 2				
Male	0.8022	.0255	.000	.2135
Education	−0.0111	.0038	.004	−.0029
Year	0.0062	.0022	.004	.0016
Black	0.2586	.0397	.000	.0729
Intercept	−1.7095	.1820	.000	-
Predicted probability				.1754
Model 4				
Male	0.8126	.0274	.000	-
Education	−0.0114	.0038	.003	-
Year	0.0062	.0022	.004	-
Black	0.2994	.0545	.000	-
Black*Male	−0.0806	.0721	.264	-
Intercept	−1.7117	.1810	.000	-

variable changes as additional variables are introduced into a model. This means that it is not appropriate to compare corresponding probits across equations to assess the effect of mediating variables, as we do with metric OLS coefficients. Rather, we must first standardize the latent dependent variable by dividing by the variance of the latent distribution. This produces Y^*-standardized coefficients, which can then be directly compared across equations with differing numbers of predictor variables. To see how to accomplish this, consult the discussion of Y^* standardization of ordinal logit coefficients in the next chapter. (Note that in Equation 14.9, var(μ) = 1 for the probit model.)

Because neither probits nor Y^*-standardized probits have intrinsic metrics, they are difficult to interpret. Thus probits typically are transformed in one of two ways: by finding the expected probability of a positive outcome for a given configuration of values of the predictor variables or by interpreting the *marginal effect* of a change in each predictor variable on the probability of a positive outcome.

Again, consider the worked example. In the chapter we saw how to evaluate the expected probabilities by race and sex implied by the logit coefficients in Model 4. To calculate the corresponding probabilities from a probit model, we get the predicted z-scores and then transform them into probabilities using the cumulative normal transformation. Recall that we evaluated the logit equation for people with twenty years of schooling in 1994. To evaluate the probit equation for the same values of education surveyed in the same year requires that we compute a new intercept: $a' = a + b_E{*}20 + b_Y{*}94 = -1.7117 - 0.0114{*}20 + 0.0062{*}94 = -1.3569$ (where b_E is the probit coefficient for education and b_Y is the probit coefficient for year of survey). Then we write out the expected z scores by race and sex (where b_M is the coefficient for *MALE*, b_B is the coefficient for *BLACK*, and b_{BM} is the coefficient for the interaction term) and transform them using Stata's `-normal-` function:

	Non-Blacks	*Blacks*
Females	$\Phi(a')$	$\Phi(a' + b_B)$
Males	$\Phi(a' + b_M)$	$\Phi(a' + b_B + b_M + b_{BM})$

Substituting the numerical values of these coefficients, we have

	Non-Blacks	*Blacks*
Females	$\Phi(-1.3569) = 0.0874$	$\Phi(-1.3569 + 0.2994)$ $= \Phi(-1.0575) = 0.1451$
Males	$\Phi(-1.3569 + 0.8126)$ $= \Phi(-0.5443) = 0.2931$	$\Phi(-1.3569 + 0.8126 + 0.2994$ $-0.0806) = \Phi(-0.3255) = 0.3724$

Note that, multiplied by 100, these are extremely close to the percentages predicted by the logit model, which are, respectively, for non-Black women, 8.8; for Black women, 14.6; for non-Black men, 29.3; and for Black men, 37.2. (See the paragraph following Equation 13.16.)

Now let us consider the marginal effect. We might ask how big a change in the probability we could expect for a small change in a particular independent variable. However, because the relationship between the probit index and the probability is nonlinear, the answer depends on the values of the independent variables at which we evaluate the change. Unless we have a reason for doing otherwise, evaluating the marginal effect of each variable relative to the expected value when all independent variables are set at their means would seem most reasonable, and this is the approach Stata takes for continuous variables. However, there is an exception—it makes little sense to evaluate marginal changes in dummy variables relative to their means. A better approach for dummy variables is to compute the *discrete change*—the difference in the expected probability for those scored 1 and 0 on the dummy variable, with all other variables (including any other dummy variables in the equation) set at their means. Thus, for example, we would want to know the expected difference in the probability of males and females having been threatened, among people who are at the mean with respect to the other variables. For continuous variables, however, we want to know the effect of a small change relative to the mean for all variables. Thus for continuous variables the marginal effect is defined as the slope of the probability function at the mean, extrapolated to a unit increase.

The marginal effects for Model 2 are shown in the rightmost column of Table 13.B.1. Note that I do not show marginal effects for Model 4. This is because when we have interaction terms, the effects of the variables included in the interaction cannot be separated. Thus when we have a model involving interactions, it is best to evaluate the probabilities for various combinations of variables, as in the logit example.

The first thing to note is the predicted probability, 0.1753, which tells us the expected probability that the average person in the data set has ever been threatened by a gun or shot at. It is reassuring that the predicted value is close to the observed value—19.5 percent of our sample has been threatened. This gives us confidence in the correctness of the model.

Now note the marginal effect for males. Because sex is a dichotomous variable, this coefficient gives the difference in the expected probability of having ever been threatened for males and females who are at the mean with respect to the other characteristics included in the model; among such people, males are predicted to be 21 percent more likely than females to have experienced a gun threat. We also see that, at the mean, a one-year increase in schooling would be expected to reduce the probability of having been threatened by 0.0029. What would, say a ten-year increase in schooling bring? Note that here we cannot simply extrapolate the marginal effect. For example, it is not correct to say that a ten-year increase in schooling would result in a 0.029 decrease in the expected probability of having been threatened. Rather, we need to compare the cumulative normal transformations at the mean and at the mean plus ten years:

$$\begin{aligned}
&\Phi(\beta_0 + \beta_1\bar{M} + \beta_2(\bar{E}+10) + \beta_3\bar{Y} + \beta_4\bar{B}) - \Phi(\beta_0 + \beta_1\bar{M} + \beta_2\bar{E} + \beta_3\bar{Y} + \beta_4\bar{B}) \\
&= \Phi(-1.710 + 0.802 * 0.451 - 0.0111 * (12.39+10) + 0.0062 * 84.47 + 0.259 * 0.111) \\
&\quad - \Phi(-1.710 + 0.802 * 0.451 - 0.0111 * 12.39 + 0.0062 * 84.47 + 0.259 * 0.111) \\
&= .1482 - .1753 \\
&= -.0272
\end{aligned} \tag{13.B.2}$$

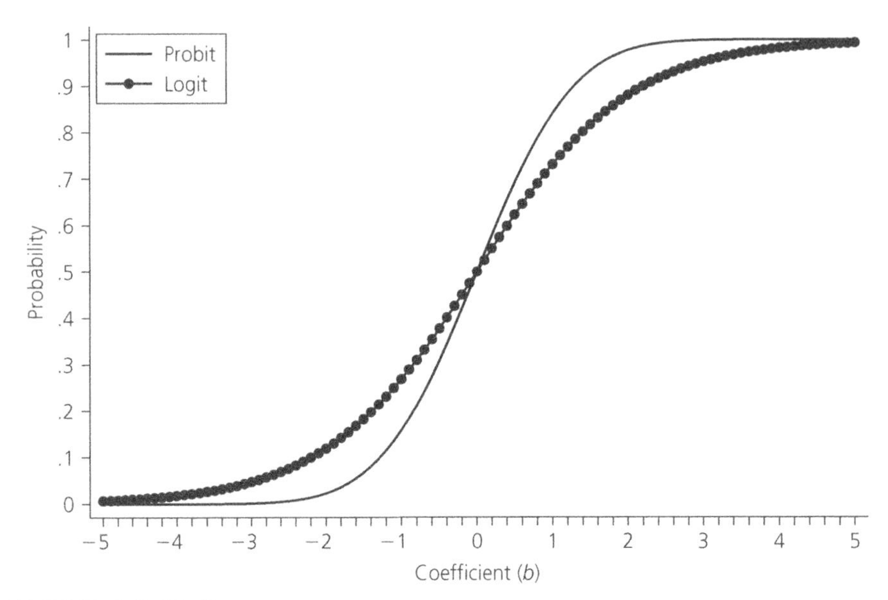

FIGURE 13.B.1. *Probabilities Associated with Values of Probit and Logit Coefficients.*

A final point to note is that the logit and probit models have similar shapes, except that probit coefficients more quickly reach probabilities asymptotically close to zero or one than do logit coefficients, as is evident from Figure 13.B.1. For this reason, logit models are more sensitive when dealing with rare events or with predicted probabilities close to zero or one. But with this exception, the two models almost always yield similar substantive conclusions.

For further discussion of the binomial probit model, see Petersen (1985), Long (1997, 40–84), Powers and Xie (2000, Chapter 3), Long and Freese (2006), Wooldridge (2006, 583–595), and the `-probit-`, `-probit postestimation-`, `-svy:probit-`, and `-svy:probit postestimation-` entries in StataCorp (2007). For an interesting application see Manski and Wise (1983).

The Stata commands used to create the worked example for the probit model and the output are shown as the last part of downloadable files "ch13_1.do" and "ch13_1.log."

CHAPTER

MULTINOMIAL AND ORDINAL LOGISTIC REGRESSION AND TOBIT REGRESSION

WHAT THIS CHAPTER IS ABOUT

In this chapter we consider models for three additional types of limited dependent variables:

- categorical variables with more than two categories, for which multinomial logistic regression is appropriate
- ordinal variables, for which ordinal logistic regression is appropriate
- truncated, or censored, dependent variables, where observations are not observed below or above some level, for which tobit regression is appropriate

In each case we see how the model is specified and then work through an illustrative substantive analysis.

MULTINOMIAL LOGIT ANALYSIS

Sometimes we wish to analyze categorical dependent variables with more than two categories. In this case, we have available a natural extension of binomial logistic regression: multinomial logistic regression. The procedure involves simultaneously estimating a set of logistic regression equations, of the form

$$\text{In}\left(\frac{P(Y=1|X)}{P(Y=0|X)}\right)=a_1+\sum_{k=1}^{K}b_{k1}X_k$$

$$\text{In}\left(\frac{P(Y=2|X)}{P(Y=0|X)}\right)=a_2+\sum_{k=1}^{K}b_{k2}X_k \qquad (14.1)$$

$$\cdots$$

$$\text{In}\left(\frac{P(Y=m|X)}{P(Y=0|X)}\right)=a_m+\sum_{k=1}^{K}b_{km}X_k$$

Here, one category of the dependent variable is omitted and becomes the reference category. The estimation procedure yields, for a set of $m + 1$ categories of some dependent variable, m logistic regression equations, each of which predicts the log odds of a case falling into a specific category rather than into the reference category (here designated by $Y = 0$). Note, however, that although the interpretation is similar to the binomial case, the estimation procedure is *not* equivalent to estimating a set of binomial logistic regression equations in which the odds of being in a particular category versus not being in that category are predicted. In general, the estimates will differ and the binomial estimates will be incorrect.

This can easily be appreciated by imagining that we are interested in what factors determine whether, in 1988 Poland, a person was a Communist Party official, a Communist Party member but not an official, or neither a member nor an official. If we estimated a binomial logistic regression predicting ordinary party membership (without office holding) and another binomial logistic regression equation predicting party office holding, we would be in trouble with respect to the first equation because the negative category (not an ordinary party member) would include those who were neither party members nor officials *and also those who were party officials.* In consequence, the resulting coefficients would be misleading. For example, it is likely that a coefficient relating education to party membership would be very weak because party officials are likely to be better educated than mere members, whereas party members are likely to be better educated than nonmembers.

The appropriate way to handle this problem would be to estimate a multinomial logistic regression model with three categories: nonmember, ordinary member, and official. Doing so would result in two equations, one contrasting ordinary members versus nonmembers and the other contrasting officials versus nonmembers, which are then interpreted in the ordinary way. An alternative would be to do a sequential logit analysis in which first membership versus nonmembership is modeled, and then office holding versus ordinary membership is modeled for party members only. The choice between these alternatives would depend on how the process of becoming a party member or a party official occurs. (See the brief discussion at the end of the chapter in the section on "Other Models.")

A Worked Example: Foreign-Language Competence in the Czech Republic

To see how this procedure works in practice, let us analyze the factors that account for competence in English and Russian in the Czech Republic. The data used here were collected in 1993 from a representative national probability sample of 5,496 Czechs age twenty to sixty-nine, as part of the survey *Social Stratification in Eastern Europe After 1989* (Treiman and Szelényi 1993; see Appendix A for details on this survey and how to obtain the data set and documentation). Here we consider four groups:

- those who speak neither English nor Russian
- those who speak English but not Russian
- those who speak Russian but not English
- those who speak both languages

To be classed as a speaker of a language, a respondent had to report that he speaks the language "fairly well" or "very well"; those who reported that they speak the language "only a little" or "not at all" or who failed to answer the question were classified as nonspeakers of the language. Because the survey was conducted in Czech, everyone interviewed spoke Czech. A few may also have spoken a second language other than Russian or English, but this possibility is not analyzed here.

My expectation is that professionals and technicians would be more likely than other occupation groups to speak English because English is now the international language of science, technology, and scholarship, and hence the ability to speak English is important for professional advancement. Those who were ever Communist Party members, and especially those who were government or party officials, would be more likely than other occupation groups to speak Russian because Russian was necessary for political advancement in the Eastern Bloc. It is less clear whether or to what extent being a manager would increase the odds of speaking English (perhaps necessary for international business dealings) or Russian (perhaps necessary for Eastern Bloc dealings).

To identify those who potentially needed Russian for their careers, I classify respondents by their 1988 occupation and create four dummy variables for 1988 occupation, each scored 1 for those in the category and scored 0 otherwise: officials, other managers, professionals and technicians, and others. (This variable was constructed by recoding the expanded version of ISCO 88 shown in Treiman [1994, Appendix C]. "Officials" include codes 1000 to 1166, "other managers" include codes 1200 to 1320, "professionals and technicians" include codes 2000 to 3480, and "others" include codes 4000 to 9333. Those not reporting an occupation in 1988 were excluded from the analysis.) In addition to these variables, I include education as a control variable because it is clear that those who are educated are more likely to speak foreign languages in general.

The data were weighted to adjust for differential household size and to bring the sample characteristics into conformity with population distributions (see Treiman 1994, Section I.G, for details). However, standard errors were not adjusted for clustering. In the sample design, census tracts were divided into eight strata on the basis of size, and households were randomly sampled within strata. Because the stratum identification is not given in the documentation, there is no alternative but to treat the sample as a simple (weighted) random sample. Given the probable lack of systematic association between the size of census tracts and other characteristics, the lack of adjustment for stratification is likely to be of little consequence. The results are reported in Table 14.1 for the 3,945 people with a job in 1988 for whom complete information was available. (Downloadable file "ch14_1.log" shows the Stata log for the analysis, and "ch14_1.do" shows the -do- file used to obtain the results.)

Inspecting the coefficients in Table 14.1, we see that, as expected, the odds of speaking either Russian or English, or both, improved substantially with education. The odds multipliers in the second panel tell us that each additional year of schooling increased the odds of speaking Russian by 25 percent, the odds of speaking English by 36 percent, and the odds of speaking both languages by 51 percent—all in contrast to speaking neither Russian nor English. Thus, for example, net of other factors, the odds that a Czech university graduate could speak Russian but not English (in contrast to speaking neither Russian nor English) are nearly two and one half times as high as the odds that a high school graduate could do so (because $1.248^{(16-12)} = 2.43$). The odds that a university graduate could speak English but not Russian are more than three times the odds for a high school graduate (because $1.363^{(16-12)} = 3.45$). The odds that a university graduate could speak both Russian and English are more than five times the odds for a high school graduate (because $1.508^{(16-12)} = 5.17$).

Note that we are not restricted to comparisons with the omitted reference category. By subtracting the coefficients for the log odds (or, alternatively, taking the ratio of the odds multipliers), we can compare the categories for which we have explicit coefficients. Thus, for example, each year of school increases the odds that a Czech could speak English instead of Russian by about 9 percent (because $e^{(.3096-.2213)} = 1.363/1.248 = 1.092$). Hence, the odds that a university graduate could speak English and not Russian (rather than Russian and not English) are more than 40 percent greater than the corresponding odds for high school graduates (because $e^{4(.3096-.2213)} = (1.363/1.248)^4 = 1.423$. (Note that in contrast to our usual rule of thumb that three significant digits are sufficient, it probably is best to report four digits for the coefficients because they often are used in subsequent calculations. Too much rounding error is introduced when only three digits are reported, so that the mathematical relationships implied by the coefficients shown in downloadable file "ch14_1.log" appear no longer to hold.)

Continuing with our substantive comparison, we note that, as expected, membership in the Communist Party increased the odds of speaking Russian and decreased the odds of speaking English but had no impact on the odds of speaking both Russian and English. All else equal, the odds that Communist Party members spoke Russian but not English are

TABLE 14.1. Effect Parameters for a Model of the Determinants of English and Russian Language Competence in the Czech Republic, 1993 (N = 3,945). (Standard Errors in Parentheses; *p*-values in Italic.)

Variable	Russian	English	Both
Logits (*b*)			
Years of school completed	0.2213 (.0247) *.000*	0.3096 (.0404) *.000*	0.4107 (.0429) *.000*
Ever a Communist Party member	0.3020 (.1488) *.042*	−0.8965 (.3332) *.007*	0.0484 (.2778) *.862*
Government or CP official in 1988	1.5591 (.7169) *.030*	−29.2975 (.6097) *.000*	−29.3602 (.7039) *.000*
Other manager in 1988	0.9941 (.2725) *.000*	0.8010 (.4844) *.098*	0.8534 (.5330) *.109*
Professional or technician in 1988	0.9943 (.1548) *.000*	1.120 (.2990) *.000*	1.3856 (.3577) *.000*
Intercept	−5.5378 (.3021) *.000*	−8.1541 (.5036) *.000*	−10.1965 (.5866) *.000*
Odds multipliers (e^b)			
Years of school completed	1.248	1.363	1.508
Ever a Communist Party member	1.353	0.408	1.050

(Continued)

TABLE 14.1. Effect Parameters for a Model of the Determinants of English and Russian Language Competence in the Czech Republic, 1993 (N = 3,945). (Standard Errors in Parentheses; *p*-values in Italic.) *(Continued)*

Variable	Russian	English	Both
Government or CP official in 1988	4.754	0.000	0.000
Other manager in 1988	2.702	2.228	2.348
Professional or technician in 1988	2.703	3.064	3.997

about a third higher than the odds that they spoke neither language, whereas the odds that Communist Party members spoke English but not Russian are only about 40 percent as great as the odds that they spoke neither language. Thus the odds that Communist Party members spoke Russian but not English are more than three times as great as the odds that they spoke English but not Russian (because $e^{(.3020-(-.8965))} = 1.353/0.408 = 3.316$). The same is true of service as a government or Communist Party official. Here, as expected, officials were nearly five times as likely to speak Russian (in contrast to speaking neither Russian nor English) than were those who were neither managers nor professionals or technicians (recall that the reference category is all other occupations). The odds that government officials spoke English or both Russian and English are effectively zero—which they should be because not one of the sixteen officials in the sample spoke English. Finally, we see that being a professional or technician in 1988 roughly triples the odds of speaking Russian only or English only, and quadruples the odds of speaking both English and Russian, relative to speaking neither English nor Russian. By contrast, being a manager in 1988 nearly triples the odds of speaking Russian only, relative to speaking neither. But the effect of being a manager on the odds of speaking English or of speaking both English and Russian were both somewhat smaller than the effect of being a manager on the odds of speaking Russian. Also, the coefficients are only marginally significant—at about the 0.1 level.

Although for this example I settled on a single model in advance, model selection for multinomial logit models is carried out in exactly the same way as for binomial logit models—by taking the ratio of the difference in L^2 *s* (Model X^2 s) to the difference in the degrees of freedom for any two models, to determine whether one model fits the data significantly better than the other model (but recall that this procedure is not possible when robust estimation is used—that is, when the data are weighted or clustered; rather, a Wald test should be used to compare models).

Independence of Irrelevant Alternatives

In the multinomial logit model, the relative odds of being in two categories are assumed to be independent of the other alternatives included in the model. This follows from Equation 14.1, from which we can derive the difference in log odds for two categories, *d* and *c*, as

$$\ln\left(\frac{P(Y=d|X)}{P(Y=c|X)}\right)=\left(a_d+\sum_{k=1}^{K}b_{kd}X_k\right)-\left(a_c+\sum_{k=1}^{K}b_{kc}X_k\right) \tag{14.2}$$

Note that only the two categories being compared enter the equation. If, however, the relative odds do depend on what the alternatives are, the model produces misleading estimates. To see this clearly, consider McFadden's (1974) well-known example of transportation choice. Suppose people can travel to work by bus or by car and that half choose to go by car and half by bus. Now suppose a competing bus company establishes buses with the same routes and schedule, so we no longer have, say, only blue buses but also red buses. Presumably, the half that traveled by car would continue to do so, but the half that traveled by bus would divide equally between the red and blue buses, taking whichever bus showed up first at the bus stop. Thus the odds ratio for car versus blue-bus ridership would change from 1:1 to 2:1, violating the assumption of the model.

Now consider another example. Suppose there are two restaurants in a neighborhood, a Mexican and an Italian restaurant, and that the Mexican restaurant gets 60 percent of the total business. Then a new Chinese restaurant opens in the neighborhood and draws off 20 percent of the business of the Mexican restaurant and 20 percent of the business of the Italian restaurant. The Mexican restaurant's share of the total is now 48 percent, and the Italian restaurant's share of the total is 32 percent. Here the independence-of-irrelevant-alternatives (IIA) assumption holds because 60/40 = 48/32 = 3/2.

Because the multinomial model is misleading when the IIA assumption is violated, McFadden suggests that multinomial (and conditional) logistic regression models should be estimated only when the outcome categories "can plausibly be assumed to be distinct and weighed independently in the eyes of each decision maker" (1974, 113).

A formal test of the IIA property is available, implemented in Stata 10.0 as `-suest-` ("seemingly unrelated estimation," a generalization of an earlier command, `-hausman-`). The `-suest-` test compares models that do and do not include presumably irrelevant outcomes. If the resulting parameters for the restricted and unrestricted models are similar, the additional outcomes can be assumed to be irrelevant. Applying these ideas to our current example, we might ask whether the odds that people speak English are affected by including "Russian" as an alternative in the model. In this case the test strongly suggests that the IIA condition is not satisfied. Thus we might consider estimating a sequential logit model in which we successively consider two questions: whether a respondent speaks either Russian or English versus speaking neither language, and for each of the two subsets of respondents—those speaking Russian and those speaking English—whether they speak the other language as well.

For further discussion of the IIA assumption and its consequences, see McFadden (1974), Hausman and McFadden (1984), Hoffman and Duncan (1988), Zhang and Hoffman (1993), Long (1997, 182–184), Powers and Xie (2000, 245–247), Long and Freese (2006), and the `-hausman-` and `-suest-` entries in StataCorp (2007). Additional examples of the application of multinomial logit models include Aly and Shields (1991), Haynes and Jacobs (1994), Tomaskovic-Devey and Skaggs (1999), and Breen and Jonsson (2000).

ORDINAL LOGISTIC REGRESSION

Often in the social sciences we have ordinal dependent variables, where the response categories can be ordered on some dimension but where the distance between categories is unknown. Most attitude variables are of this sort. For example, if people are asked to say how happy they are, and the response categories include "very happy," "pretty happy," and "not too happy," there is no ambiguity in assuming that those who say they are "pretty happy" are less happy than those who say they are "very happy" and are more happy than those who say they are "not too happy." However, there is no basis for assuming that the distance between "not too happy" and "pretty happy" is the same as the distance between "pretty happy" and "very happy." Many other attitude scales have similar properties. In such cases we could predict the scale score using ordinary least-squares regression. However, to do so would be tantamount to assuming that the distance between response categories is uniform. (For a useful discussion of this and other points, see Winship and Mare [1984].)

An alternative is to estimate an *ordinal logit* equation, which makes use of the ordered property of the response categories on the dependent variable but makes no assumptions at all about the relative distances between categories. The basic assumption of the ordered logit model is that there is an unobserved continuous dependent variable, Y^*, which is a linear function of a set of independent variables:

$$Y^* = a + \sum_{j=1}^{J} b_j X_j + \mu \tag{14.3}$$

However, what is observed is a set of ordered categories, $Y = 1 \ldots I$, such that

$$\begin{aligned} Y &= 1 \textit{ if } -\infty \leq Y^* < k_1 \\ &= 2 \textit{ if } k_1 \leq Y^* < k_2 \\ &\ldots \\ &= I \textit{ if } k_{I-1} \leq Y^* < \infty \end{aligned} \tag{14.4}$$

where the k_i are "cutting points" on the unobserved, or latent, underlying variable. Now, because we observe $Y = 1$ when $Y^* < k_1$, observe $Y = 2$ when $k_1 \leq Y^* < k_2$, and so on, it follows that

$$\Pr\left(Y = i \middle| X\right) = \Pr\left(k_{i-1} \leq Y^* < k_i \middle| X\right) \tag{14.5}$$

Substituting from Equation 14.3 and imposing the constraint that $a = 0$, which is necessary to identify the equation, we have

$$\Pr\left(Y = i \middle| X\right) = \Pr(k_{i-1} \leq \sum_{j=1}^{J} b_j X_j + \mu_i < k_i | X) \tag{14.6}$$

Then, subtracting within the inequality and noting that the probability that a random variable falls between two values is the difference between the cumulative density functions evaluated at these values, we have

$$\Pr(Y = i|X) = \frac{1}{1+e^{(-k_i+\sum b_j X_j)}} - \frac{1}{1+e^{(-k_{i-1}+\sum b_j X_j)}} \tag{14.7}$$

That is, the expected probability that an observation will have a particular value is the difference between the probability associated with reaching the upper-bound cutting point and the probability associated with reaching the lower-bound cutting point, where these probabilities are estimated from logistic functions known as *cumulative logits* because they give the log odds of reaching each cutting point. (Note that for the extreme categories one of the terms of Equation 14.7 drops out because the probabilities associated with $-\infty$ and ∞ are zero.)

A Worked Example: Political Party Identification in the United States, 1998

Consider the following substantive problem. Suppose we wish to assess what factors lead people to place themselves toward the Republican end rather than toward the Democratic end of a scale of political party identification. Here is the item and the ordinal response categories in the 1998 GSS:

Generally speaking, do you usually think of yourself as a Republican, Democrat, Independent, or what?

IF REPUBLICAN OR DEMOCRAT: Would you call yourself a strong (Republican/Democrat) or not a strong (Republican/Democrat)?

IF INDEPENDENT, NO PREFERENCE, OR OTHER: Do you think of yourself as closer to the Republican or Democratic Party?

This set of questions and responses yielded seven response categories:

- Strong Democrat
- Not strong Democrat
- Independent, near Democrats
- Independent
- Independent, near Republicans
- Not strong Republican
- Strong Republican

On the ground that the Republican Party is increasingly the party of nonurban affluent non-Black males, especially those from the South, I predict the score on a continuum underlying the listed response categories from the following variables:

- size of place (people living in large [population more than 250,000] central cities of Standard Metropolitan Statistical Areas [SMSAs], other people living in SMSAs, and people living outside SMSAs)
- income (with categories recoded to their midpoints and the open-ended upper category, $110,000 and over, recoded to $150,000)
- gender (male versus female)

- region of residence (South versus other)
- race (Black versus non-Black)

Survey estimation procedures were used to take account of clustering and differential household size. The Stata commands that carry out this analysis are shown in downloadable file "ch14_2.do" and the results are shown in "ch14_2.log."

A property of ordinal logistic regression (which applies also to binomial logistic regression models of the sort discussed in the previous chapter) is that the variance of the latent variable presumed to underlie the observed outcome variable changes as variables are added to the prediction equation. Thus, it is not appropriate to directly compare corresponding coefficients across models, as is commonly done with OLS models (see Chapter Six). Rather, the latent dependent variable must first be standardized. To illustrate how to carry out such a standardization and how to interpret the resulting coefficients, I estimate two models—Model 1, which omits race, and Model 2, which includes race.

First consider Model 1, shown in the left panel of Table 14.2. Here we see that all of the variables have the expected signs—a positive sign means a shift toward Republican identification. However, Southern residence is not at all significant. (To assess the joint significance of the two "urban" coefficients, I use Stata's `-test-` command in the usual way. Doing so, I conclude that the urban distinctions are significant at well beyond conventional levels.) Now consider Model 2. Once Blacks are included in the model, the effect of Southern residence becomes marginally significant (at the .048 level). This is as we would expect, considering that Blacks are more likely than non-Blacks to reside in the South (53 percent of Blacks versus 33 percent of non-Blacks) and are also much more likely to identify as Democrats (63 percent of Blacks versus 30 percent of non-Blacks). When race is not included in the model, the large fraction of Southern Black Democrats suppresses the positive effect of Southern residence on Republican leaning. Once we control for race, this effect emerges clearly.

Converting the Logits to Y*-Standardized Form Inspecting the coefficients, it appears that the inclusion of race in the model dramatically increases the effect of Southern residence, from .050 to .187. However, this comparison is inappropriate because the variance of the latent "Republicanism" variable changes when additional variables are included in the model. Thus, before comparing coefficients, it is necessary to standardize the coefficients. Although there are several ways to do this, a particularly appealing approach is to standardize only the latent dependent variable, so that the resulting (Y^*-standardized) coefficients indicate the expected change in the standard deviation of the latent variable for a one-unit change in the independent variable. An important advantage of Y^*-standardization over full standardization is that, as we saw in Chapter Six, fully standardized coefficients are not appropriate for categorical variables because for such variables they are affected by the relative size of the category as well as by the size of the metric effect.

An additional reason for standardizing the coefficients, even when we do not want to compare corresponding coefficients across models, is that the latent dependent variable

TABLE 14.2. Effect Parameters for an Ordered Logit Model of Political Party Identification, U.S. Adults, 1998 (N = 2,443).

	Model 1				Model 2			
	b	Standard Error	p	Y*-Std. Coeff.	b	Standard Error	p	Y*-Std. Coeff.
Substantive variables								
Annual income (0000s)	0.0914	0.0105	.000	.028	0.0764	0.0100	.000	.023
SMSA, not in large center	0.517	0.105	.000	.156	0.400	0.092	.000	.120
Residence outside SMSA	0.769	0.155	.000	.232	0.594	0.138	.000	.178
Male	0.360	0.081	.000	.108	0.334	0.081	.000	.100
Southern residence	0.050	0.100	.620	.015	0.187	0.093	.048	.056
Black					−1.414	0.164	.000	−.423

(*Continued*)

TABLE 14.2. **Effect Parameters for an Ordered Logit Model of Political Party Identification, U.S. Adults, 1998 (N = 2,443).** *(Contiued)*

	Model 1	Model 2
Cutting point 1	–0.90	–1.33
Cutting point 2	0.38	0.02
Cutting point 3	0.95	0.62
Cutting point 4	1.66	1.36
Cutting point 5	2.12	1.84
Cutting point 6	3.55	3.29

has no intrinsic metric, which makes the size of the unstandardized coefficients meaningless. (Recall from Equation 14.3 that the coefficients shown in Table 14.2 represent the effect of a unit change in each of the independent variables on the unobserved, or latent, dependent variable, Y^*, holding constant all other independent variables.) However, because it is possible to estimate the variance of Y^*, we can divide the coefficients by the standard deviation of Y^* to get semistandardized, that is, Y^*-standardized, coefficients, which are then interpreted as the number of standard deviations of difference in Y^* expected for two individuals who differ by one unit on the given independent variable. That is,

$$\beta_i = \frac{b_i}{\sqrt{\text{var}(Y^*)}} \tag{14.8}$$

where b_i is the coefficient associated with the ith variable and β_i is the Y^*-standardized coefficient. To get the variance of Y^*, I follow Long (1997, 129):

$$\text{var}(Y^*) = \mathbf{B'VB} + \text{var}(\mu) \tag{14.9}$$

where **B** is a vector of coefficients, **V** is the variance-covariance matrix of the independent variables, and var(μ) is $\pi^2/3$. (See the downloadable file "ch14_2.do" for how to estimate these coefficients, which are reported in the rightmost column of each panel of Table 14.2.)

Consider Model 2. As we see, net of all other factors, Blacks are nearly a half standard deviation lower than non-Blacks with respect to Republican orientation. No other variable has nearly so strong an impact. In particular, although positive, the effect of Southern residence is weak, only about half as strong as the effect of gender and about a third as strong as the effect of nonmetropolitan residence. Family income also has only a modest effect. For example, two individuals would have to differ in income by about $184,000 per year, net of all other factors, to be about as far apart in Republican tendencies as are Blacks and non-Blacks, who are identical in other respects (precisely, 0.423 = 0.023*18.39).

Getting Predicted Percentages Another way to assess the magnitude of the effects is to evaluate the prediction equation for particular values of the independent variables. To do this we need to take account both of the coefficients associated with each of the independent variables and of the *ancillary parameters*, the *cut points*, which correspond to the *k*s in Equation 14.4. Together, these two sets of coefficients model the categorization. For example, from Equation 14.7 we can estimate (from the Model 2 coefficients) the probability that a non-Black man earning $40,000 to $50,000 per year and living in a central city of an SMSA outside the South is categorized as a "strong Democrat":

$$\frac{1}{1+e^{(-(-1.33)+.0764*4.5+.334)}} = .118 \tag{14.10}$$

Similarly, the probability that such a person is a "not strong Democrat" is

$$\frac{1}{1+e^{(-(.02)+.0764*4.5+.334)}}-\frac{1}{1+e^{(-(-1.33)+.0764*4.5+.334)}}=.223 \tag{14.11}$$

Although these probabilities may be computed by hand, it is easy to have Stata do the work. The trick is to use the `-predict-` command to get the predicted distribution for persons with the desired profile of characteristics (see the Stata `-log-`). Table 14.3 shows the predicted distribution of party identification for Black and non-Black males earning $40,000 to $50,000 per year and living in central cities of SMSAs. (Of course, I could equally well estimate predicted probability distributions for any combination of characteristics. Indeed, it is possible to get estimates for combinations of variables not found in the sample by creating a new data set containing these combinations. (See the discussion of `-predict-` in StataCorp 2007.) As we see, non-Blacks are substantially more Republican than otherwise similar Blacks.

Constructing Odds Ratios Still another way to assess the net effect of an independent variable is to compute its contribution to the ratio of the odds of being below any given value in the ordinal scale to the odds of being at or above that value. Because of the way the logits are derived, their contributions to odds ratios are constant regardless of the cutting point, and it can be shown (Long 1997, 139) that they are equal to e^{-b_k}, for the kth independent variable. Thus, for example, the ratio of the odds that males and females will be strong Democrats versus less-than-strong Democrats is just $e^{-0.334}=0.72$; or, putting it more naturally, net of other factors women are about 40 percent more likely than men (precisely, 1.39 = 1/0.72) to be strong Democrats (rather than anything closer to the Republicans). Similarly, women are about 40 percent more likely than men to be any kind of Democrat (compared to Independents plus Republicans).

***Comparisons to Other Estimating Procedures:* `-gologit2-`** As we have just seen, an important constraint embedded in the `-ologit-` estimation procedure is what is known as the *proportional odds assumption*—that the explanatory variables have the same effect on the odds that the dependent variable is below any dividing point. On the face of it, there is often little reason to assume that the odds are proportional. Why should we assume, for example, that gender has the same effect in distinguishing strong Democrats from all others and in distinguishing anyone who is Democratic-leaning from Independents and Republicans; and the same for each other independent variable? A user-written `-ado-` file, `-gologit2-` (for *Generalized Ordered Logit Model*) relaxes this assumption, allowing the odds to vary across cutting points. Reestimating Model 2 of Table 14.2 using `-gologit2-` rather than `-ologit-` yields the coefficients shown in Table 14.4.

As we can see, the effects of each variable differ substantially from category to category. For example, Southern residence distinguishes neither strong Democrats nor both kinds of Democrats from those who lean more toward Republicans, nor does it distinguish strong Republicans from others; but it does significantly affect the remaining distinctions. Similarly, nonmetropolitan residence matters rather more in the middle of the distribution than at either extreme. Still, the pattern of distinctions does not appear to be very systematic.

TABLE 14.3. Predicted Probability Distributions of Party Identification for Black and Non-Black Males Living in Large Central Cities of Non-Southern SMSAs and Earning $40,000 to $50,000 per Year.

	Black	Non-Black
Strong Democrat	.356	.119
Not strong Democrat	.325	.224
Independent, near Democrat	.114	.144
Independent	.095	.179
Independent, near Republican	.038	.097
Not strong Republican	.053	.169
Strong Republican	.018	.069
Total	**0.999**	**1.001**

In my judgement, because the gologit model is rather more complex than the ologit model, two criteria need to be satisfied to justify substituting `-gologit2-` for `-ologit-` estimates: first, that the proportional odds assumption be shown to be inadequate; and second, that the coefficients for the gologit model be interpretable and informative. To determine whether the proportional odds assumption is inadequate, we estimate the gologit model and test the equality of corresponding coefficients for each of the cutting points. In the present case we reject the null hypothesis that the coefficients are equal (χ^2, with 30 *d.f.* = 147; $p < .0000$). However, I am hard pressed to arrive at a coherent interpretation of the variations in the coefficients across cutting points. I would

ESTIMATING GENERALIZED ORDERED LOGIT MODELS WITH STATA A routine to estimate generalized ordered logit models was first written by Vincent Fu when he was a graduate student at UCLA (Fu 1998). Fu's routine has been enhanced by Williams (2006). Williams's `-ado-` file, `-gologit2-`, can be downloaded from within Stata. Type "net search gologit2," click the first entry, and then select "Click here to install."

TABLE 14.4. Effect Parameters for a Generalized Ordered Logit Model of Political Party Identification, U.S. Adults, 1998.

	b	Standard Error	p
Strong Democrat versus other			
Annual income (000s)	0.0732	0.0195	.000
Residence in SMSA, not in large center	0.391	0.107	.000
Residence outside SMSA	0.494	0.198	.012
Male	0.206	0.143	.149
Southern residence	−0.094	0.149	.528
Black	−1.361	0.189	.000
Intercept	1.504	0.172	.000
Strong or not strong Democrat versus other			
Annual income (000s)	0.0461	0.0122	.000
Residence in SMSA, not in large center	0.345	0.095	.000
Residence outside SMSA	0.445	0.150	.003
Male	0.395	0.094	.000
Southern residence	0.074	0.108	.493
Black	−1.287	0.179	.000
Intercept	0.151	0.116	.193

Democrat or Democratic-leaning Independent versus other

Annual income (000s)	0.0582	0.0111	.000
Residence in SMSA, not in large center	0.469	0.135	.001
Residence outside SMSA	0.700	0.182	.000
Male	0.239	0.096	.013
Southern residence	0.238	0.117	.043
Black	−1.579	0.201	.000
Intercept	−0.591	0.158	.000

Other versus Republican-leaning Independent or Republican

Annual income (000s)	0.0941	0.0127	.000
Residence in SMSA, not in large center	0.574	0.142	.000
Residence outside SMSA	0.820	0.182	.000
Male	0.383	0.093	.000
Southern residence	0.345	0.114	.002
Black	−1.724	0.225	.000
Intercept	−1.696	0.171	.000

(*Continued*)

TABLE 14.4. Effect Parameters for a Generalized Ordered Logit Model of Political Party Identification, U.S. Adults, 1998. *(Continued)*

	b	Standard Error	p
Other versus not-strong or strong Republican			
Annual income (000s)	0.0919	0.0118	.000
Residence in SMSA, not in large center	0.534	0.184	.004
Residence outside SMSA	0.845	0.210	.000
Male	0.340	0.099	.001
Southern residence	0.281	0.114	.014
Black	−1.416	0.228	.000
Intercept	−2.119	0.212	.000
Other versus strong Republican			
Annual income (000s)	0.0847	0.0192	.000
Residence in SMSA, not in large center	0.249	0.268	.353
Residence outside SMSA	0.451	0.289	.119
Male	0.408	0.159	.010
Southern residence	0.253	0.161	.116
Black	−1.517	0.455	.001
Intercept	−3.264	0.319	.000

thus be inclined to settle for the conventional ordinal logit model on the grounds of parsimony.

Ordinary Least Squares as an Alternative Finally, we could treat the dependent variable as an interval variable and estimate an ordinary least-squares equation. This amounts to assuming that the distance between any pair of adjacent categories is identical. As it turns out, in the present case the coefficients yielded by the OLS model, shown in Table 14.5, are quite similar to those yielded by the ologit model. Thus, we might be as well served by simply estimating an OLS model, which is much simpler to estimate and to interpret than is the ologit model. The difficulty is that unless we carry out the analysis both ways we really do not know whether the results will be similar in any particular instance. Thus, a reasonable strategy is, indeed, to carry out the analysis both ways and, if the results prove to be similar, to present the OLS results but to add a note indicating that you did the analysis both ways and got similar results. Of course, if the results differ enough to affect the conclusions, the ordinal logit model is to be preferred over OLS because it is less restrictive; that is, because it does not assume that the categories are equidistant.

TOBIT REGRESSION (AND ALLIED PROCEDURES) FOR CENSORED DEPENDENT VARIABLES

Often we have dependent variables that are *censored* in the sense that the recorded values do not represent the entire range of the true underlying variable. The classic case is that studied by the economist James Tobin (1958)—hence the name *tobit regression* (coined by econometrician Arthur Goldberger when he described "Tobin's probit")—where a consumer good was purchased if the desire was high enough, with "desire" measured by the dollar amount spent on the good. From this definition of "desire," it is evident that the measure is "censored" at zero, because all those not making the purchase are recorded as having "zero" desire, whereas in reality some might have been close to making it and might have done so had the price been a little lower. Others might have had no desire at all and would never have made the purchase regardless of the price, and still others might have wavered in between. That is, there actually is variability in the relative desire of those recorded as having zero desire.

An underlying variable is censored in many other situations as well. The classic case is where many values are below a threshold that would lead to action; for example, the number of extramarital affairs (Fair 1978), the number of infant deaths experienced by mothers (Wood and Lovell 1992), the number of killings committed by police in different jurisdictions (Jacobs and O'Brien 1998), the number of arrests after release from prison (Witte 1980), the number of scientific publications (Stephan and Levin 1992), the number of protests in a nation (Walton and Ragin 1990), and the number of hours worked per year (Rosen 1976, Keeley and others 1978, Quester and Greene 1982). But we also can imagine other kinds of cases: attitude variables that fail to offer enough options, income coded in categories with a top code that is too low, censoring that occurs because the length of time to an event is analyzed only for those to whom the event has occurred

TABLE 14.5. Effect Parameters for an Ordinary Least-Squares Regression Model of Political Party Identification, U.S. Adults, 1998.

	b	Standard Error	p
Annual income (000s)	0.0803	.0100	.000
Residence in SMSA, not in large center	0.412	.100	.000
Residence outside SMSA	0.620	.142	.000
Male	0.337	.082	.000
Southern residence	0.212	.096	.029
Black	−1.386	.146	.000
Intercept	1.981	.123	.000

JAMES TOBIN (1918–2002), a Nobel Laureate in Economics (1981), is best known among social scientists other than economists as the creator of the "tobit model," a procedure for estimating models with censored dependent variables. But his major work, for which he won the Nobel Prize, was his analysis of financial markets and their relations to consumption and investment decisions, employment, production, and prices. He made major contributions to the analysis of how households and firms actually determine the composition of their assets, developing what is known as "portfolio selection theory." The result was a description and analysis of financial markets and flows in the economy.

Tobin grew up in a liberal household in Champaign, Illinois, where his father was a journalist who worked as publicity director for the University of Illinois athletic program and his mother was a social worker. He attended the university's laboratory high school where, as he notes in his Nobel lecture, he cast the only vote for Roosevelt in a 1932 presidential straw vote. He did his undergraduate and graduate work in economics at Harvard, where he earned his PhD in 1947, his graduate studies having been interrupted, first by a job in Washington, D.C., and then by service in the Navy as an officer on a destroyer. After three years as a Harvard Junior Fellow (a very prestigious fellowship), which he used in part to study econometric developments he had missed during the war, he then spent his entire academic career at Yale.

(Daula, Smith, and Nord 1990). Other substantive applications include Mare and Chen (1986); Saltzman (1987); Roncek (1992); and Treno, Alaniz, and Gruenewald (2000).

The Tobit Model

The obvious question is what to do in the case where we have censored observations—for example, observations scored zero (or some other constant) when we think that there is variability in the true underlying value of the censored observations. One solution is to simply carry out an OLS regression of the entire data set. But this produces inconsistent estimates (Long 1997, 188–190). Another solution is to discard the censored cases and carry out an OLS estimation of the relationship in the noncensored cases—for example, determinants of how many hours people work among those who work at least some hours. But this approach, which amounts to truncating the distribution, also produces inconsistent estimates (Long 1997, 188–190). Tobin's solution was to divide observations into two sets: uncensored and censored observations. Formally, for observed values of dependent variable *Y*, censored at some value τ, we have

$$Y_i = \begin{cases} Y_i^* = a + \sum_{k=1}^{K} b_k X_{ik} + \varepsilon_i \;\; if\; Y_i^* > \tau \\ \tau_Y \qquad\qquad\qquad\qquad\quad if\; Y_i^* \leq \tau \end{cases} \tag{14.12}$$

That is, the observed value of *Y* is equal to the "true" value of *Y*, Y^*, if Y^* is above the value at which observations are censored and is equal to some constant value (usually, but not necessarily the value at which observations are censored) if the true value is at or below the value at which observations are censored. For the first set, estimates are derived in the same way as in ordinary least-squares estimation. For the second set, it is possible to estimate the probability that an observation is censored, conditional on the values of the independent variables, and to use this probability to estimate the likelihood. These estimates are then combined to produce expected values for all observations, conditional on the values of the independent variables:

$$\begin{aligned} E(Y_i|\mathbf{X_i}) = &[\Pr(uncensored|\mathbf{X_i}) * (E(Y_i|Y_i > \tau, \mathbf{X_i})] \\ &+ [\Pr(censored|\mathbf{X_i}) * \tau_Y] \end{aligned} \tag{14.13}$$

where

$$\mathbf{X_i} = a + \sum_{k=1}^{K} b_k X_{ik}$$

For an accessible exposition of the mathematics involved, see Long (1997, Chapter 7).

The tobit model has been extended and generalized in a number of ways:

- to allow for right censoring and both left and right censoring (that is, censoring at low values and high values of a distribution)
- to allow for the possibility that different observations are censored at different values (for example, income when several years of the GSS are pooled)
- to allow for situations in which an underlying continuous variable is coded as a set of categories (in many surveys income is coded this way)
- to correctly estimate effects where observations are truncated
- to deal with sample-selection problems

In the following section I provide a worked example that illustrates many of these extensions. (For estimation details, see the Stata downloadable files "ch14_3.do" and "ch14_3.log.")

A Worked Example: Frequency of Sex

The 2000 GSS included the question "About how often did you have sex during the last twelve months?" The response categories (shown with codes to be used later) are detailed in Table 14.6.

Clearly, these data are censored both below and above. Those who have not had sex at all in the last year include those who have never ever had sex and those who have simply been unlucky in the past year, with others in between. At the other extreme, coding "more than three times a week" as four times a week, or five times a week, may understate

TABLE 14.6. **Codes for Frequency of Sex in the Past Year, U.S. Adults, 2000.**

	Midpoint	Lower Bound	Upper Bound
Not at all	0	—	0
Once or twice	1.5	1	2
About once a month	12	12	12
2 or 3 times a month	30	24	36
About once a week	52	52	52
2 or 3 times a week	130	104	156
More than 3 times a week	208	208	—

the prowess of newlyweds and other sexual athletes. Finally, some categories include a range, which might or might not be optimally represented by the midpoint.

To illustrate the effect of censoring, let us consider a simple model in which frequency of sex is predicted from age, gender, and marital status (currently married versus not). In fact, in this and most analyses involving age, it would be better to include a squared term. However, I do not do so just yet because including only linear terms makes the exposition easier.

Table 14.7 shows the results for four estimates:

- ordinary least-squares estimates with the categories coded at their midpoints but with an arbitrary top code of 208 for "more than 3 times a week" (= 52*4)
- tobit estimates with censoring from below
- tobit estimates with censoring both below and above
- interval regression estimates with censoring both below and above

Comparing the coefficients in the two left columns, we see that the effect of censoring from below is severe. Failure to take proper account of such censoring results in an

TABLE 14.7. Alternative Estimates of a Model of Frequency of Sex, U.S. Adults, 2000 (N = 2,258). (Standard Errors in Parentheses; All Coefficients Are Significant at .001 or Beyond.)

	Model 1: OLS	Model 2: Tobit, Left Censored	Model 3: Tobit, Left and Right Censored	Model 4: Interval, Left and Right Censored
Married	22.0 (2.7)	41.6 (3.4)	43.2 (3.7)	43.1 (3.7)
Age	−1.41 (0.09)	−2.16 (0.12)	−2.31 (0.13)	−2.30 (0.13)
Male	10.5 (3.0)	15.7 (3.5)	16.9 (3.8)	16.8 (3.8)
Intercept	103.5 (5.1)	113.4 (6.2)	119.2 (6.8)	118.4 (6.8)
R^2	0.154			
s.e.e.		65.9	71.7	71.4

underestimate of the effect of marital status on frequency of sex by about half and also very substantial underestimation of the effects of age and of being male. Interestingly, taking account of censoring from above as well as below hardly changes the coefficients, suggesting that marital status, age, and gender have little impact on the probability of being extremely sexually active. Inspection of the probability of censorship from above confirms this supposition: even among the most sexually active group, young married men, no more than about 15 percent have sex more than three times per week. By contrast, there is great variability by marital status, sex, and especially age in the probability of never having had sex in the last year, ranging from about 3 percent of young married men to about 90 percent of elderly unmarried women.

Apart from the probabilities, three predictions are of interest: the linear prediction from the model, the censored prediction, and the truncated prediction. Graphs of these predicted values for Model 4 are shown in Figure 14.1 by age, for married women. The *linear prediction* is the latent prediction from the model, which tells us that, net of other factors, the frequency of sex per year declines by about 2.3 occasions per year of age. The graph tells us that for married women the frequency of sex declines to less than once a year by about age seventy. Although negative observed values make no sense, the linear prediction gives the values of a *latent,* or underlying, variable. We can think of this variable as the propensity for sex, which declines steadily with age (because, of course, we have modeled the frequency of sex as a linear function of age).

The *censored prediction* equals the latent prediction when the dependent variable is observed and equals the censoring value when the dependent variable is censored. (Somewhat confusingly, Stata calls censored predictions the "ystar" option, although Y^* is

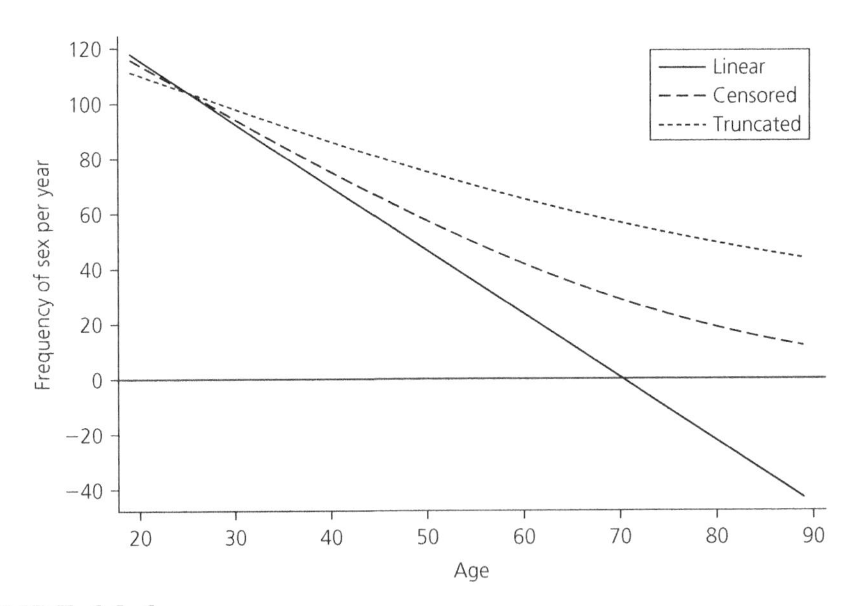

FIGURE 14.1. *Three Estimates of the Expected Frequency of Sex per Year, U.S. Married Women, 2000 (N = 552).*

usually taken to indicate the latent variable, as it is in Equation 14.12.) Thus in this case, we assume that 0 and 208 are true values for those in the lowest and highest categories. By construction, censored predictions must fall within the range of the uncensored observations.

The *truncated prediction* is defined only for those observations that are not censored. In this case the truncated prediction gives the predicted frequency of sex among those who had any sex at all in the last year. Note that neither the censored prediction nor the truncated prediction is linear. Thus, these predictions must be evaluated at specific levels of the independent variables. Most commonly we will be interested in the linear prediction.

Now that we see how to interpret tobit coefficients, let us extend the analysis slightly to make it more substantively plausible. I do this by adding a squared term for age and allowing interactions between age, gender, and marital status. As it happens, it is not necessary to posit three-way interactions among marital status, gender, and, respectively, age and age squared; a model positing the three-way interactions does not fit significantly better than a model with the two sets of two-way interactions, between gender and, respectively, age and age squared, and between marital status and, respectively, age and age squared. The coefficients for this model are shown in downloadable file "ch14_3.log." Because they are difficult to interpret directly, I have graphed (in Figure 14.2) the relationship between age and the frequency of sex for each gender–marital status combination.

Inspecting the graph, we see that—no surprise—married people have more active sex lives than do currently unmarried people of the same age and gender, and that sexual activity declines at an increasing rate with age.

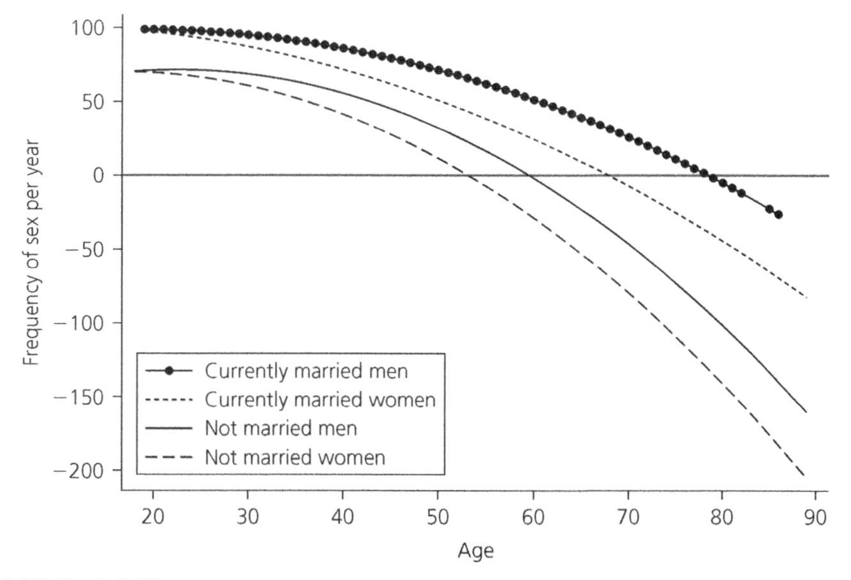

FIGURE 14.2. *Expected Frequency of Sex Per Year by Gender and Marital Status, U.S. Adults, 2000 (N = 2,258).*

Interestingly, in both marital status categories, men report more active sex lives than do women of the same age and marital status. The reason for the gender discrepancy within marital status categories is not completely clear but probably reflects a tendency for men to overreport and (or) for women to underreport their sexual activity. Note that, considering only heterosexual activity, both the average number of sexual encounters and the average number of partners must be identical for males and females. Thus there clearly is biased reporting; differential nonresponse (for example, the likelihood that, say, women with many sexual partners—for example, prostitutes—are underrepresented in the GSS); or more reported homosexual activity among men than among women.

Married men and women both average about one partner (precisely, 1.03 and .99), which suggests that for both married men and married women their spouse is usually their only partner, which in turn would imply that the average number of sexual encounters should be the same for currently married men and women, adjusting for the three-year average difference in age. However, inspection of Figure 14.2 shows a difference larger than can be explained by the age gap (if the age gap were the full explanation, the lines would be parallel for married men and women, and a line segment of three-years' length drawn to the left of the male line and parallel to the x-axis should just touch the female line). This suggests the possibility that either or both married men and married women distort their reports of the frequency of sexual activity in a socially desirable direction—men claiming sexual prowess and women claiming sexual modesty. The likelihood of distortion is substantially greater among the unmarried: unmarried men on average report about twice as many partners in the last year as do unmarried women (1.85 compared to .90), which—given that this discrepancy is far too large to be accounted for by differential homosexual activity—suggests the possibility that unmarried men and women distort both the number of partners and the frequency of sexual activity in the socially desirable direction. Another possibility is that the propensity for women to be younger than their male partners partly accounts for the gender difference in reported sexual activity among the unmarried. Adjudicating among these possibilities would require more analysis than is warranted here.

OTHER MODELS FOR THE ANALYSIS OF LIMITED DEPENDENT VARIABLES

This introduction by no means exhausts the variety of procedures available for the analysis of limited dependent variables. Stata 10.0 includes commands to carry out a number of procedures, including

- Conditional logistic regression and mixed models, where outcomes depend on features of the outcomes as well as on characteristics of the individuals. For examples, see Boskin (1974), Hoffman and Duncan (1988), White and Liang (1998), and Yanovitzky and Cappella (2001).
- Nested logistic regression, which extends conditional logit analysis by dividing outcomes into a hierarchy of levels. For examples, see Cameron (2000), Soopramanien and Johnes (2001), and South and Baumer (2001).

- Probit regression, an alternative to logistic regression. For a brief introduction, see Appendix 13.B.
- Poisson regression, used to model counts, the number of occurrences of an event. A classic example is von Bortkiewicz's 1898 study of the number of soldiers kicked to death by horses in the Prussian army. Applications in the social sciences include Long (1990), Greenberg (1991), Rasler (1996), Chattopadhyay and others (2006), and Weitoff and others (2008). The definitive statistical treatment of poisson regression is Cameron and Trivedi (1998).

Long (1997), Hosmer and Lemeshow (2000), and Powers and Xie (2000) provide excellent introductions to many of these procedures that, with a bit of diligence, are accessible to social scientists who have a modest statistical background. Long and Freese (2006) provide a guide to using the procedures in Stata. For a useful overview, see Gould (2000).

WHAT THIS CHAPTER HAS SHOWN

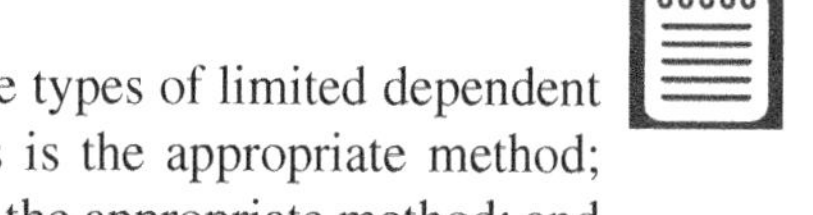

In this chapter we have seen how to estimate models for three types of limited dependent variables: ordinal variables, for which ordinal logit analysis is the appropriate method; polytomous variables, for which multinomial logit analysis is the appropriate method; and censored variables (where values above or below some cutting point are not observed), for which tobit modeling is the appropriate method.

CHAPTER

IMPROVING CAUSAL INFERENCE: FIXED EFFECTS AND RANDOM EFFECTS MODELING

WHAT THIS CHAPTER IS ABOUT

In this chapter we consider two closely related techniques for coping with omitted variable bias. Recall from Chapter Six that omitted variable bias occurs when we have failed to include in our model variables that affect the outcome and that are correlated with one or more of the predictor variables. The techniques discussed in this chapter for estimating unbiased coefficients are known as *fixed effects* and *random effects* models. These models use information on the same individuals from two or more time points or information on two or more individuals within groups (families, schools, firms, communities, or similar groups) to purge the estimating equation of all characteristics, measured or unmeasured, that are constant over time or constant within groups. The result is that the characteristics we are able to measure are unbiased by unobserved time-invariant factors. For useful introductions to these techniques, see Allison (2005) and Wooldridge (2006, Chapters 13 and 14), both of which I draw on in this chapter.

INTRODUCTION

As we have seen at many points in this book, the nonexperimental methods we have been studying are vulnerable to *omitted variable bias*: the possibility that unmeasured factors affect both the predictor and outcome variables. In this case the coefficients we estimate through OLS or logistic regression will be incorrect. To appreciate this most fully, it is helpful to contrast the linear model approaches we have been studying with randomized experiments.

In the classic randomized experiment, individuals are randomly assigned to two groups; members of the treatment group are exposed to some sort of intervention while members of the control group are not, and differences in one or more outcomes are measured. (This design can be generalized to include several different treatment groups, but the logic remains unchanged.) Because the treatment and control groups are, within the limits of sampling error, identical on average in their pretreatment attributes—or, to state the same point differently, receipt of treatment is uncorrelated with pretreatment attributes of individuals—any difference in average outcomes may be assumed to be caused by the treatment.

With linear model approaches, we attempt to approximate randomized experiments by statistically controlling for as many confounding factors—that is, factors correlated with both the predictor and outcome variables—as possible. For example, if we observe that men earn more than women, we might wonder whether this is due to discrimination. However, before we accepted such a conclusion we would want to consider whether, at least in part, the pay gap is due to the fact that men are more likely to have technical training, to enter high-paying fields, to have more work experience, and to work longer hours. We would then statistically control for these variables and assess the effect of gender on earnings among people who are identical with respect to the control variables. If we still found a gender difference in pay, we might then be willing to attribute the remaining difference to discrimination. However, we would be vulnerable to the claim that we had not included other crucial factors that could result in pay differences. For example, women may not bargain as effectively as men and may therefore accept jobs at lower salary levels than men. If we omit a measure of bargaining prowess (or if we measure bargaining prowess imperfectly so that true bargaining prowess remains partly unmeasured), then any effect of prowess would be captured by the error term. However, if bargaining prowess is correlated with gender, the assumption of OLS (and other linear models) that error is uncorrelated with the predictor variables would be violated, producing biased coefficients.

So what can we do? It turns out that if we have measurements on the same individuals for at least two points in time, we can get unbiased estimates of the effects of variables that, at least for some individuals being studied, change over time. We do this by predicting change in our outcome variable from changes in our predictor variables, which has the effect of purging from our prediction equation those factors, measured and unmeasured, that do not change over time. But there is no such thing as a free lunch. The cost of this method, known as *fixed effects* (FE) modeling, is twofold: (1) We are unable to estimate the "main effects" of predictors that do not vary over time for individuals, for example, sex and race (although we are able to estimate interaction effects involving these variables and variables that do change over time—we will return to this point in the context of our gender pay gap example—and we also are able to estimate *effects* that change over time for time-constant variables). (However, recent work by Bollen and Brand [2008] has shown how, with suitable assumptions about unobserved latent factors, it is possible to obtain effects of time-constant predictors within a structural equation modeling (SEM) framework—a set of techniques briefly discussed in the next chapter). (2) When we are analyzing limited dependent variables, we usually will have a substantial reduction in sample size because in FE logistic regression, individuals who do not change over time on the outcome variable are dropped from the analysis. However, under some circumstances, and with some additional assumptions, we can recover our sample size by resorting to what is known as *random effects* (RE) modeling. We will consider this approach later in the chapter.

FIXED EFFECTS MODELS FOR CONTINUOUS VARIABLES

To see how FE works for continuous outcome variables, let us write a prediction equation:

$$y_{it} = \mu_t + \beta x_{it} + \gamma z_i + \alpha_i + \varepsilon_{it} \quad i=1, \quad ,n; \quad t=1, \quad ,T \tag{15.1}$$

where y_{it} is the value of the outcome variable for the ith individual at time t; μ_t is an intercept that is allowed to vary with time; x_{it} is a vector of variables that vary both over individuals and, for each individual, over time; z_i is a vector of variables that vary over individuals but, for each individual, not over time; α_i represents unmeasured differences between individuals, that is, differences not accounted for by the γz_i, that are fixed over time; and ε_{it} represents idiosyncratic factors that vary both over time and across individuals.

To simplify the discussion, assume that $T = 2$, although the same conclusions hold when $T > 2$. Now suppose we simply pooled observations from the two time points and estimated our outcome through OLS. Clearly, insofar as omitted variables are correlated with the variables in the model (as in our example involving bargaining prowess), doing this will produce biased estimates because the fundamental assumption of OLS, that the error term (which in this case is the sum of $\alpha_i + \varepsilon_{it}$ because α_i is unobserved) is uncorrelated with the predictor variable, will be violated.

Fundamental FE Equation

However, suppose we write separate equations for each time period and subtract one from the other. Subtracting

$$y_{i1} = \mu_1 + \beta x_{i1} + \gamma z_i + \alpha_i + \varepsilon_{i1}$$

from (15.2)

$$y_{i2} = \mu_2 + \beta x_{i2} + \gamma z_i + \alpha_i + \varepsilon_{i2}$$

yields

$$y_{i2} - y_{i1} = (\mu_2 - \mu_1) + \beta(x_{i2} - x_{i1}) + (\varepsilon_{i2} - \varepsilon_{i1}) \tag{15.3}$$

Notice that both γz_i, the time-constant predictor variables, and α_i, the effect of unobserved variables, have been "differenced out" of Equation 15.3, which is why equations of this sort are known as *first-differenced equations*. That is, Equation 15.3 has been purged of *all* unmeasured factors that are constant over time, as well as any measured factors that are constant over time. Thus Equation 15.3 solves the omitted-variable-bias problem—assuming that there are no unmeasured factors whose effects change over time; this is a nontrivial assumption that is often overlooked. Equation 15.3 can be estimated through OLS provided (1) that there are at least some x_i for which $x_{i2} \neq x_{i1}$ and x_{i2} and x_{i1} are not perfectly correlated (thus, for example, ruling out age as a candidate variable) and (2) that the observed predictor variables are uncorrelated with the idiosyncratic error terms, ε_{i1} and ε_{i2}, at both time points; that is, that the observed predictor variables are *strictly exogenous*—crucially, that they do not depend on the outcome observed for an earlier time point.

Allowing the Slopes of the Xs to Vary

Notice that in Equation 15.3 it is assumed that the effects of the predictor variables, the xs, are constant over time. This assumption can be tested by estimating a first-differenced equation in which the slopes of the xs are allowed to vary. To see this, consider the following pair of equations:

$$y_{i1} = \mu_1 + \beta_1 x_{i1} + \gamma z_i + \alpha_i + \varepsilon_{i1}$$

and (15.4)

$$y_{i2} = \mu_2 + \beta_2 x_{i2} + \gamma z_i + \alpha_i + \varepsilon_{i2}$$

Here, subtracting the first equation of 15.4 from the second yields

$$\begin{aligned} y_{i2} - y_{i1} &= (\mu_2 - \mu_1) + \beta_2 x_{i2} - \beta_1 x_{i1} + (\varepsilon_{i2} - \varepsilon_{i1}) \\ &= (\mu_2 - \mu_1) + \beta_2 (x_{i2} - x_{i1}) + (\beta_2 - \beta_1) x_{i1} + (\varepsilon_{i2} - \varepsilon_{i1}) \end{aligned} \quad (15.5)$$

That is, to test the hypothesis that the slope of any of the xs differs over time, we include both the time 1 variable and the difference score. Then, if the coefficient for the time 1 variable differs significantly from zero, we can conclude that the slopes are not equal and can get the value of the time 1 slope by subtracting the coefficient for the time 1 variable in Eq. 15.5 $(\beta_2 - \beta_1)$ from the coefficient for the difference score (β_2).

Testing Whether the Effects of the Time-Invariant Variables Vary over Time

We also can allow the coefficients for the time-invariant variables to change over time. To see this, consider the following pair of equations:

$$y_{i1} = \mu_1 + \beta x_{i1} + \gamma_1 z_i + \alpha_i + \varepsilon_{i1}$$

and (15.6)

$$y_{i2} = \mu_2 + \beta x_{i2} + \gamma_2 z_i + \alpha_i + \varepsilon_{i2}$$

Subtracting the first equation of 15.6 from the second yields

$$y_{i2} - y_{i1} = (\mu_2 - \mu_1) + \beta(x_{i2} - x_{i1}) + (\gamma_2 - \gamma_1)z_i + (\varepsilon_{i2} - \varepsilon_{i1}) \tag{15.7}$$

From Equation 15.7 we see that it is possible to assess the claim that the effects of the z_i do not vary over time, by testing the significance of the coefficients associated with the z variables. Note that these coefficients do not show the effects of the zs but rather the *differences* in the effects of the zs between time 2 and time 1.

Interactions Between Time-Constant and Time-Varying Variables

As noted previously, we generally cannot get the effect of time-constant variables from the FE model (but see Bollen and Brand [2008]). However, we can get the effect of the interaction of the time-constant variables with the time-varying variables, the xs. To see this, consider the following pair of equations:

$$y_{i1} = \mu_1 + \beta x_{i1} + \gamma z_i + \delta x_{i1} z_i + \alpha_i + \varepsilon_{i1}$$

and (15.8)

$$y_{i2} = \mu_2 + \beta x_{i2} + \gamma z_i + \delta x_{i2} z_i + \alpha_i + \varepsilon_{i2}$$

Subtracting the first equation of 15.8 from the second yields

$$y_{i2} - y_{i1} = (\mu_2 - \mu_1) + \beta(x_{i2} - x_{i1}) + \delta z_i(x_{i2} - x_{i1}) + (\varepsilon_{i2} - \varepsilon_{i1}) \tag{15.9}$$

For example, returning to the effect of gender on income, the FE model does not allow a direct assessment of the role of gender in creating income differences. However, it does allow us to determine, say, gender differences in the effect of changes in performance evaluation scores on changes in income. Suppose gender is coded 1 for males and 0 for females, and x (now designating a single variable rather than a vector of variables) is a performance evaluation measure. Then we would have:

for females: $y_{i2} - y_{i1} = \beta(x_{i2} - x_{i1}) +$

(15.10)

for males: $y_{i2} - y_{i1} = (\beta + \delta)(x_{i2} - x_{i1}) +$

Analyzing More than Two Time Points

When we have three or more measurements per individual—which we increasingly do because a number of multiple-wave data sets are now in the public domain—there are several possibilities, of which two are simple extensions of the methods we have just discussed. Consider each of these.

First, we may analyze two waves at a time, computing first differences between successive waves. This approach has the limitation that, unless we turn to advanced methods such as generalized least squares, we cannot get a single set of coefficients for all waves combined, but rather $T - 1$ sets of coefficients for T waves of data. Thus the successive-waves approach tends to be of greatest interest when the number of waves is small, say three or four.

An alternative is to pool the data over waves; then, for each variable in the model, compute the average value for each individual across waves; compute the difference between the wave-specific observation for the individual and that individual's over-time average; and use the resulting difference scores in a conventional OLS regression equation. That is, from an equation of the form of Equation 15.1, compute

$$\bar{y}_i = \frac{1}{n_{it}} \sum_t y_{it} \quad \text{and} \quad \bar{x}_i = \frac{1}{n_{it}} \sum_t x_{it} \tag{15.11}$$

where n_{it} is the number of observations for person i. Then, for each variable, subtract the person-specific mean from the observed value:

$$y'_{it} = y_{it} - \bar{y}_i \quad \text{and} \quad x'_{it} = x_{it} - \bar{x} \tag{15.12}$$

This yields an equation of the form

$$y'_{it} = \sum_{t=2}^{T} \mu_t D_t + \beta x'_{it} + \varepsilon'_{it} \tag{15.13}$$

where the D_t are dummy variables that allow the intercepts to differ by time. Notice that the zs and α in Equation 15.1 drop out of Equation 15.13 because, since they are constant within individuals, the deviation scores are all zero. Equation 15.13 can be estimated through OLS except that the standard errors will be incorrect. The reason for this is that Equation 15.13 is the equivalent of pooling the data but, instead of taking deviation scores, including a dummy variable for each individual in the sample—that is, representing the unobserved individual differences, α_i, by dummy variables. Such equations can be estimated, with correct standard errors, by using Stata's `-areg-` command. But the optimal way to estimate FE models is to use one of the `-xt-` commands.

The various elaborations of FE models described in this chapter also can be readily adapted to the analysis of more than two time points and need not be further discussed here.

Fixing Effects Across Individuals Rather Than over Time

So far we have discussed methods for purging data of the effects of unobserved variables that are constant over time. Exactly the same logic can be applied when we have observations on multiple individuals within groups—families, schools, firms, communities, or similar groups. For example, if in a sample of individuals we observe a positive relationship between education and income, we might suspect that the relationship is due, at least in part, to characteristics of families that lead offspring both to go far in school and to be successful in the job market. One way to control for such unobserved characteristics of families would be to compare siblings, estimating differences in income as a function of differences in the level of education. Ashenfelter and Krueger (1994) carried out such an analysis on a sample of 149 identical twins and showed that the effects of education were in fact slightly stronger than in a corresponding OLS analysis that controlled for gender, age, and race.

PANEL SURVEYS IN THE PUBLIC DOMAIN Major U.S. panel studies of interest to social scientists include

- Panel Study of Income Dynamics (PSID): http://psidonline.isr.umich.edu
- National Longitudinal Surveys (NLS): http://www.bls.gov/nls
- Wisconsin Longitudinal Study (WLS): http://www.ssc.wisc.edu/wlsresearch
- Health and Retirement Study (HRS): http://hrsonline.isr.umich.edu
- National Longitudinal Study of Adolescent Health (Add Health): http://www.cpc.unc.edu/addhealth

Important foreign panel studies include

- China Health and Nutrition Survey (CHNS): http://www.cpc.unc.edu/china
- German Socio-Economic Panel Study (SOEP): http://www.diw.de/english/sop/index.html
- Indonesia Family Life Survey (IFLS): http://www.rand.org/labor/FLS/IFLS
- Mexican Family Life Survey (MxFLS): http://www.radix.uia.mx/ennvih/main.php?lang=en
- Mexican Health and Aging Study (MHAS): http://www.mhas.pop.upenn.edu/english/home.htm

Many additional panel surveys more or less comparable to the PSID are listed at http://psidonline.isr.umich.edu/Guide/PanelStudies.aspx.

Now consider a second example. In an analysis of Indonesian data, Frankenberg and Mason (1995) studied the effect of maternal education on behaviors conducive to children's health, including sanitation and hygiene practices such as the source and treatment of drinking water, waste disposal practices, and so on. However, in developing nations such as Indonesia, both a mother's level of education and the possibility of easily obtaining safe water or protecting against contamination from human waste tend to vary across communities, depending on their level of development. In this situation one would want to purge the association between maternal education and child health–related practices of the confounding influences of community characteristics. This is what Frankenberg and Mason did by fixing community characteristics and relating differences in health practices to differences in maternal education among women in the same communities. In this way they were able to show a causal effect of maternal education on behavior conducive to child health.

Limitations of Fixed Effects Approaches and Cautions to Keep in Mind

Like all other statistical procedures, FE approaches carry a set of assumptions and requirements. When these are violated, FE coefficients may be worse (more biased) than simply pooling data and obtaining OLS estimates. Unfortunately, often these assumptions are untestable. Here are some cautions:

- If unmeasured effects do change over time (or, in the cross-sectional application just discussed, do vary across individuals), FE estimation does not solve the bias problem. It is thus necessary to think carefully about whether the assumption of time-constant unmeasured effects is tenable. The same point holds even more strongly for family or community fixed effects—one has to assume that none of the unmeasured factors affecting the outcome varies across individuals within families or communities. This is often dubious, especially within families. To convince yourself of this, think of recent U.S. presidents and their ne'er-do-well siblings; or simply consider variations among siblings in families you know. Could such differences account for differences in the kinds of outcomes studied with family FE models? This is a crucial question, often ignored by researchers. (Of course, unmeasured effects that change over time also bias OLS coefficients. So resorting to OLS regression in such cases is no solution.)
- The predictor variables must be strictly exogenous, conditional on the unobserved variables. That is, we must assume that once we control for the unobserved variables, there is no remaining correlation between the predictor variables and the idiosyncratic errors, the X_{it}s and the ε_{it}s. One common way strict exogeneity is violated is when one or more of the predictor variables depends on the outcome variable measured at a previous point in time. For example, if we were studying how the crime rate responds to changes in the size of the police force, and the size of the police force were determined by the crime rate in the previous year, the strict exogeneity assumption would be violated.
- Relative to variability in the outcome variable, there must be sufficient variability over time in the predictor variables (or across individuals in the cross-sectional FE approach). What is sufficient? This is difficult to quantify. Still, it is obvious that predictor variables that hardly vary can have little impact on the outcome, just as in OLS analysis one cannot predict a variable from a constant and will do a poor job of trying to predict a variable from a near constant.
- A corollary of the previous point is that variables that differ only by a linear transformation over time are regarded as unchanged over time. Thus, for example, age cannot be included in an over-time FE analysis because age at time 2 is identical to age at time 1 plus a constant. It then follows that variables that differ over time by a near linear transformation create problems.
- The predictor variables must be reliably measured. As Wooldridge notes, "Differencing a poorly measured regressor reduces its variation relative to its correlation with the differenced error caused by classical measurement error, resulting in a potentially sizable bias" (2006, 475).

RANDOM EFFECTS MODELS FOR CONTINUOUS VARIABLES

Because FE models do not allow us to assess the size of time-invariant variables (or, in family, organizational, or community applications, variables that are invariant across individuals within units), there has been a strong incentive to find models that do yield such estimates. Among these, a frequently used approach is the *random effects* (RE) model. Like the FE model, the RE model can be written by starting with Equation 15.1. However, the assumptions are different. Whereas the FE model assumes that the α_i represent a set of fixed parameters, which are purged from the model by differencing, the RE model assumes that each α_i is a normally distributed random variable with a mean of zero and constant variance and that it is independent of z_i, x_{it}, and ε_{it}. This is a strong assumption. Fortunately, it can be tested, using a test proposed by Hausman (1978). The strategy is to estimate corresponding FE and RE models and to compare the similarity of the coefficients using the Hausman test. If the null hypothesis of no difference is not rejected, we can conclude that the independence of the α_i is supported, which means that the RE model yields unbiased coefficients. Because the RE model yields estimates of the effects of the z_i, the RE model is to be preferred if the independence assumption is satisfied. If it is not satisfied, we must settle for the FE model and forgo estimates of the effects of the z_i. The Hausman test is quite restrictive and often does not support the RE model. Bollen and Brand (2008) offer a range of alternative statistics for comparing FE and RE models and also procedures for forming hybrid models. Bollen and Brand's procedures are based on structural equation modeling, which is beyond what is covered in this book but is briefly discussed in the next chapter.

How can we estimate the RE model? The details are beyond what can be considered here, but it is possible to sketch the general approach. Because, by assumption, α_i is uncorrelated with the explanatory variables, the coefficients of these variables could be consistently estimated from a single cross-section. However, doing so would ignore at least half of the data (or more, for more than two time periods). Pooling the data and estimating the coefficients through OLS also would yield consistent estimates. However, neither procedure yields the correct standard errors. The reason for this is that the errors will be serially correlated over time. We can easily see this by replacing the two error terms in Equation 15.1 with a single term for the composite error:

$$\nu_{it} = \alpha_i + \varepsilon_{it} \tag{15.14}$$

Because α_i is included in the composite error for each time period, the ν_{it} are serially correlated over time, with the correlation given by

$$corr(\nu_{it}, \nu_{is}) = \sigma_\alpha^2 / (\sigma_\alpha^2 + \sigma_\varepsilon^2),\ t \neq s \tag{15.15}$$

where σ_α^2 = Var(α_i) and σ_ε^2 = Var(ε_{it}). However, it is possible to derive a generalized least-squares transformation that eliminates the serial correlation in the errors. Defining

$$\lambda = 1 - (\sigma_{\varepsilon}^2 / (\sigma_{\varepsilon}^2 + T\sigma_{\alpha}^2))^{1/2} \qquad (15.16)$$

we can write

$$y_{it} - \lambda \bar{y}_i = \beta_0(1-\lambda) + \beta_1(x_{it1} - \lambda \bar{x}_{i1}) + \quad + \beta_k(x_{itk} - \lambda \bar{x}_{ik}) + (\nu_{it} - \lambda \bar{\nu}_i) \qquad (15.17)$$

Note the similarity of Equation 15.17 to Equation 15.13. Whereas the FE estimator (Equation 15.13) subtracts the within-individual over-time average from each observation, the RE transformation subtracts a fraction of the over-time average, where the size of the fraction depends on σ_{ε}^2, σ_{α}^2, and T. Once λ is estimated (which can be done in several ways that need not concern us), Equation 15.17 can be estimated through OLS from the pooled over-time data to yield consistent estimates of the coefficients and the correct standard errors.

Finally, we can see the relationship between FE and RE by rewriting the error term in Equation 15.17 as

$$\nu_{it} - \lambda \bar{\nu}_i = (1-\lambda)\alpha_i + \varepsilon_{it} - \lambda \bar{\varepsilon}_i \qquad (15.18)$$

which follows directly from Equation 15.14. Equation 15.18 makes it clear that the errors in Equation 15.17 weight the unobserved effect by $1 - \lambda$. Thus as λ approaches 1, the RE estimates approach the FE estimates and the bias approaches zero. By contrast, as λ approaches 0, a larger fraction of the unobserved effect is left in the error term and hence, by definition, the bias increases.

A WORKED EXAMPLE: THE DETERMINANTS OF INCOME IN CHINA

To see how to estimate and interpret FE and RE models for continuous dependent variables, I consider the determinants of family income in China. In China, as elsewhere, income differs substantially across communities. Communities higher in the Chinese urban hierarchy (a seven-category scheme that ranges from rural villages to province-level cities: Beijing, Chongqing, Shanghai, and Tianjin) tend to have higher average family income. But they also have populations that are better endowed with both the human capital characteristics and the jobs associated with high income. We see this in Table 15.1, which shows the mean years of schooling, occupational status (ISEI), and per capita family income for each of the seven levels of the urban hierarchy. Thus the question arises as to what extent the association between human capital and other factors, on the one hand, and family income, on the other, simply reflects community differences in labor market and other conditions that affect income—for example, the tendency of those with university education to disproportionately move to the capital.

To study the determinants of family income, I estimate community-level FE (and RE) models as a way of purging unmeasured community characteristics from the analysis. This analysis is based on the 1996 Chinese national sample survey used in several previous chapters. The sample design for this survey included one hundred rural villages and one hundred urban neighborhoods, in each of which information was obtained from about thirty households. (See Appendix A for details on the study design and for information on how to obtain the data.)

TABLE 15.1. Socioeconomic Characteristics of Chinese Adults by Size of Place of Residence, 1996.

	Mean Years of School	Mean Occ. Status (ISEI)	Median Family Income[a]	Median Fam. Inc. per Worker[a]	N
Village	6.7	23.8	5,500	2,100	2,973
Township or town	8.2	37.9	9,000	4,000	410
County seat	10.2	50.5	10,600	5,000	542
County-level city	10.4	49.8	10,680	5,000	606
Prefecture-level city	9.2	47.0	10,000	5,000	866
Provincial capital	10.5	48.1	13,000	6,000	495
Province-level city	10.5	49.4	15,600	8,700	198
Total	**7.6**	**29.9**	**7,000**	**2,775**	**6,090[b]**

[a]The survey was conducted during the summer of 1996. The income question was "Now, from all sources, what was your family income in the past year?" During the relevant period (mid-1995 to mid-1996), the RMB was worth $0.12, with hardly any fluctuation.

[b]No data are missing for size of place or years of schooling. For the remaining columns, missing data on that variable are excluded.

In this analysis I predict family income (in RMB) from the education, occupational status, and age of the respondent or spouse (whichever is higher), the number of people in the household who are employed, and whether the household is engaged in various types of family enterprises. (Because in the survey used here no variable identifies the head of the household, I used the higher of the respondent's and spouse's characteristics as a proxy for the characteristics of the household head. This variable will be incorrect insofar as our respondents are other relatives of the household head—for example, adult children or siblings. In a serious analysis, I would develop a more sophisticated decision rule for deciding who is the head or how to characterize the socioeconomic status of the household. But for our present purposes, this proxy is adequate.) We would, of course, expect the education and occupational status of household members to affect household income. In addition, household income is likely to increase with age, which can be taken as a proxy for experience. I have no clear predictions regarding the effect of engaging in crop production, agricultural sidelines, or nonagricultural sidelines, but I suspect that these aspects of entrepreneurship affect income.

TABLE 15.2. Comparison of OLS and FE Estimates for a Model of the Determinants of Family Income, Chinese RMB, 1996 (N = 5,342).

	β	s.e.	t	p	95% Confidence Interval
Community FE estimates					
Education (years)	440	181	2.43	.015	85 to 795
Occupational status (ISEI)	97	38	2.53	.011	22 to 172
Age	83	45	1.86	.063	−5 to 172
Number of family workers	4,529	480	9.44	.000	3,588 to 5,470
Grow crops?	8,851	2,117	4.18	.000	4,701 to 13,001
Agricultural sideline?	134	764	0.18	.860	−1,363 to 1,632
Nonagricultural sideline?	−5,613	1,224	−4.58	.000	−8,014 to −3,533
Intercept	−12,879	4,767	−2.70	.007	−22,225 to −3,533
OLS estimates					
Education (years)	302	119	2.54	.012	68 to 537
Occupational status (ISEI)	105	34	3.09	.002	38 to 171
Age	37	42	0.88	.379	−46 to 120
Number of family workers	4,346	1,682	2.58	.010	1,029 to 7,664
Grow crops?	10,570	4,440	2.38	.018	1,815 to 19,325
Agricultural sideline?	456	461	0.99	.324	−453 to 1,365
Nonagricultural sideline?	−8,475	3,206	−2.64	.009	−14,796 to −2,153
Intercept	−7,027	5,162	−1.36	.175	−17,206 to 3,153

Table 15.2 shows the results of estimating an FE model and also, for comparison purposes, the results of estimating an OLS model of the kind we have encountered previously. The Stata files that created these results are in downloadable files "ch15_1.do" and "ch15_1.log."

Notice that the FE and OLS results differ substantially. Interestingly, the effect of education is substantially stronger in the FE model than in the OLS model, somewhat contrary to my expectations. I would have expected community differences in income to be correlated with community differences in education in such a way as to upwardly bias the education effect in the OLS analysis. In addition, age has a marginally significant effect in the FE analysis but no effect at all in the OLS analysis, perhaps as a result of the fact that young people tend to migrate to high-income areas, thus suppressing the positive effect of age on income that emerges in the FE analysis. Both analyses show that, net of sociodemographic characteristics, families that grow crops tend to earn substantially more than families that do not grow crops, whereas families engaged in nonagricultural enterprises earn substantially less. Although the signs and significance levels are not qualitatively different in the OLS and FE analyses, the effects are substantially reduced in the FE model relative to the OLS model. In sum, it is evident here that FE analysis produces results that are substantially different from those produced by OLS analysis. Because the FE approach purges the results of bias caused by the correlation between community factors that affect both the predictor variables and the outcome variable, this approach clearly is to be preferred to OLS regression.

But can we do even better? What about an RE approach? Recall that a major limitation of FE models is that characteristics that are invariant across individuals within units (or within individuals over time) cannot be studied, because they are swept out by the differencing procedure. RE models do not suffer this limitation. Thus, for example, using an RE model we can study characteristics of communities such as their position in the urban hierarchy. However, as noted previously, RE coefficients are valid only if the unobserved effects, the α_i, are independent of the remaining variables—the observed variables that change over time or across individuals within units, the x_{it}; the observed variables that are constant over time or across individuals within units, the z_i; and the idiosyncratic errors, the ε_{it}. This is a testable claim, by means of a Hausman test that compares the similarity in corresponding coefficients from FE and RE models. (See downloadable file "ch15_1.do" for the implementation of this test using Stata.) It turns out that the hypothesis of similarity in coefficients is rejected, which means that the RE model estimates are biased, and hence we must settle for the FE estimates. You should keep in mind that the failure to reject the null hypothesis of no difference between FE and RE coefficients may mean either that the two sets of coefficients are quite similar or that the standard errors are too large to permit much of an inference. That is, you should be cautious about too readily accepting the legitimacy of RE estimates even when they appear to pass the Hausman test. As noted earlier, for advanced procedures that permit greater flexibility in the specification of the problem, see Bollen and Brand (2008).

FIXED EFFECTS MODELS FOR BINARY OUTCOMES

The logic of FE modeling with binary outcome variables is similar to that for continuous variables. However, the procedures are somewhat different. Let us start by considering FE modeling for binomial logistic regression with two time points, writing an equation analogous to Equation 15.1:

$$\ln\left(\frac{p_{it}}{1-p_{it}}\right) = \mu_i + \beta x_{it} + \gamma z_i + \alpha_i \quad i = 1, \quad ,n; \quad t = 1,2 \tag{15.19}$$

where p_{it} is the probability that $y_{it} = 1$ rather than 0, and the remaining terms are defined as in Equation 15.1. In addition, we need to assume that within individuals, y_{i1} and y_{i2} are independent. Then it follows that

$$\begin{aligned} \Pr(y_{i1} = 0, y_{i2} = 0) &= (1 - p_{i1})(1 - p_{i2}) \\ \Pr(y_{i1} = 1, y_{i2} = 0) &= p_{i1}(1 - p_{i2}) \\ \Pr(y_{i1} = 0, y_{i2} = 1) &= (1 - p_{i1})p_{i2} \\ \Pr(y_{i1} = 1, y_{i2} = 1) &= p_{i1}p_{i2} \end{aligned} \tag{15.20}$$

Because our goal is to estimate μ_t and β while controlling for the time-invariant covariates, we use only variation within individuals to estimate these parameters. Thus, because individuals for whom the outcome variable y_{it} does not change between time 1 and time 2 contribute no information, we drop them from the sample. We are left with the two middle rows of Equation 15.20. We take the log of the ratio of these probabilities to get an equation that "differences out" the z_i and a_i:

$$\begin{aligned} \ln\left(\frac{\Pr(y_{i1} = 0, y_{i2} = 1)}{\Pr(y_{i1} = 1, y_{i2} = 0)}\right) &= \ln(1 - p_{i1}) + \ln p_{i2} - \ln p_{i1} - \ln(1 - p_{i2}) \\ &= \ln\left(\frac{p_{i2}}{1 - p_{i2}}\right) - \ln\left(\frac{p_{i1}}{1 - p_{i1}}\right) \end{aligned} \tag{15.21}$$

Substituting the right-hand terms of Equation 15.19, we have

$$\ln\left(\frac{\Pr(y_{i1} = 0, y_{i2} = 1)}{\Pr(y_{i1} = 1, y_{i2} = 0)}\right) = (\mu_2 - \mu_1) + \beta(x_{i2} - x_{i1}) \tag{15.22}$$

Notice that the outcome variable in Equation 15.22 is the equivalent of the log odds of a "positive" outcome at time 2 for those whose outcomes differ between time 1 and time 2. Thus Equation 15.22 reduces to a conventional binomial logistic regression equation in which the predictor variables are the difference scores for the xs. However, because Equation 15.22 is estimated only for individuals whose outcome has changed, there is usually a large reduction in the sample size relative to the full sample. Keep this limitation in mind when interpreting FE logistic regression results.

FE estimation can be generalized to permit observations on more than two time points per person (or more than two people within a unit) by resorting to conditional maximum likelihood estimation. That is, when there are more than two observations per unit, the problem becomes a conditional logistic regression analysis. (The algebra involved is sketched in Allison [2005, 57–59]; see also the entry for the `-clogit-` command in StataCorp [2007] and the references cited there.)

RANDOM EFFECTS MODELS FOR BINARY OUTCOMES

As in the continuous case, we can estimate RE binary logistic regression models as an alternative to FE models. RE models for binary outcomes not only have the advantage of allowing estimates of the effects of variables that are constant across observations within units but also are not restricted to observations for which the outcome varies over time. However, logistic regression RE models require the same assumptions about unobserved effects as in the continuous case: that they have an expected value of zero, are normally distributed with constant variance, and are independent of both the time-varying and time-constant observed variables. As in the continuous case, the assumption of independence can be tested with a Hausman test.

A WORKED EXAMPLE WITH A BINARY OUTCOME: THE EFFECT OF MIGRATION ON SCHOOL ENROLLMENT AMONG SOUTH AFRICAN BLACKS

To illustrate how to derive and interpret FE and RE models for binary outcomes, I present a portion of the analysis Lu and Treiman (2007) carried out in their study of the effect of labor migration and remittances on children's schooling in South Africa. As a consequence of apartheid-era restrictions on residential rights, many South African Blacks were forced to live in rural "homelands" carved out of the least productive land in the nation. As a result, many people, mainly men but sometimes women, left their families behind and sought employment in "White" South Africa. In a majority of cases labor migrants sent remittances home to their families.

The question Lu and Treiman addressed was whether remittances benefited the children left behind, by improving the odds that they would enroll in school. It might be argued that the extra income provided by remittances increases the likelihood of school enrollment. But suppose parents are committed to keeping their children in school and hence decide to go out for work to make this possible. That is, suppose that the same unmeasured characteristics of families determine both the migration decision and the school enrollment decision. If this is the case, the coefficient relating remittances to school enrollment will be biased. However, an over-time FE analysis can control for this (and all other unmeasured characteristics that can be assumed to be constant over time).

Using data for the South African Black population (which constitutes 78 percent of the total population) from the September 2002 and September 2003 South African Labour Force Survey, Lu and Treiman studied changes in school enrollment between 2002 and 2003 as a function of changes in the migration-and-remittance status of the household and other time-varying household attributes (household income, the highest level of education attained by household members, the number of children in the household, whether the household was female-headed, and the year of the survey). In addition, they included the age of the child as a predictor variable. Although this variable is regarded as time constant because the time-2 value is an exact linear transformation of the time-1 value ($age_2 = age_1 + 1$), recall from Equation 15.7 that such variables can be included in an FE equation to test the possibility that the *effects* differ over time; the coefficients associated with such variables give the *difference* in the size of the effect for time 2 relative to time 1. In the present case, we might suspect that for South African Blacks the effect of children's age on school enrollment has gone down year by year in the postapartheid period as schooling has become more readily available.

Table 15.3 shows three sets of estimates: for an FE model, for an RE model, and for an RE model estimated on the full sample rather than the reduced sample of those who changed enrollment status as required by the FE model. (See downloadable files "ch15_2.do" and "ch15_2.log" for details on how the analysis was carried out.)

Consider the FE results first. They provide substantial support for the central hypothesis of the study—when labor migrants send remittances home, the odds that children enroll in school increase by 50 percent, net of other factors, relative to the odds for children living in nonmigrant households (precisely, $1.49 = e^{.399}$). However, it is important to keep in mind exactly what has been demonstrated. Recall that what we are predicting is the odds of school enrollment for those whose school enrollment status changed between 2002 and 2003, which is only 20 percent of the sample of all South African Black children. Thus the finding applies to this selected subgroup only. Moreover, for this subgroup nothing else has much effect on the odds of enrollment. Interestingly, the difference between 2002 and 2003 in the effect of a child's age on school enrollment is negative, which indicates that, as hypothesized, age mattered slightly less in 2003 than in 2002.

The next step is to estimate an RE model corresponding to the FE model shown in the first two columns. Recall that random effects are legitimately interpreted only if the unmeasured effects are independent of the measured effects and the idiosyncratic error. To determine whether this is the case, we perform a Hausman test of the similarity of corresponding coefficients in the FE and RE models. Because we are interested only in the similarity of the estimates of migration-remittance effects, we restrict the Hausman test to the two migration-remittances dummy variables. We cannot reject the null hypothesis ($p = 0.47$) and thus conclude that the FE and RE analyses yield similar results with respect to migration-remittance status. This allows us to interpret the RE model.

It turns out that the results are similar to those for the FE model. Of course, this is necessarily the case for the two migration-remittance variables because otherwise the null hypothesis of similarity would have been rejected by the Hausman test. For the restricted sample used for the FE analysis, the RE estimates show that the odds that children living in households receiving remittances will attend school are more than 40 percent as large as the corresponding odds for those living in households without migrants. Moreover, household income and the child's age have similar effects in the two models. However, in the RE analysis we are able to include two time-constant variables: gender and urban residence. Interestingly, males appear to be more likely than females to be enrolled in 2003. But place of residence (urban versus rural) has no effect.

However, the question remains as to whether an RE model estimated for the full sample rather than being restricted to the 20 percent of the sample whose enrollment status changed would yield similar results. To determine this, we estimate the RE model for all South African Black children. Again, to determine whether the full-sample RE model is acceptable, we compare the consistency of coefficients from this model and the FE model. And again we fail to reject the null hypothesis ($p = .72$).

TABLE 15.3. Comparison of OLS and FE Estimates for a Model of the Effect of Migration and Remittances on South African Black Children's School Enrollment, 2002 to 2003. (N(FE) = 2,408 Children; N(full RE) = 12,043 Children.)

	Fixed Effects		Random Effects (Restricted Sample)		Random Effects (Full Sample)	
	β	p	β	p	β	p
Household migration-and-remittance status (reference category: no out-migrants)						
Out-migrants, no remittances	0.190	.244	0.177	.222	0.022	.798
Out-migrants, remittances	0.399	.006	0.325	.003	0.373	.000
ln (HH income, excluding remittances)	0.030	.017	0.026	.023	0.045	.000
Highest education level of household adults (reference category: no schooling)						
Primary	−0.570	.442	−0.347	.601	2.42	.000
Some secondary	−0.464	.537	−0.376	.569	2.93	.000
Completed secondary or more	0.045	.953	−0.152	.818	3.53	.000
Number of school-age children	0.079	.073	0.001	.962	0.074	.000
Female-headed household	−0.195	.167	−0.121	.183	0.014	.788
Survey year (2003)	−1.74	.000	−3.67	.000	−1.54	.000
Age of child	−0.029	.000	−0.019	.000	−0.063	.000
Male			0.169	.050	0.210	.000
Urban			0.003	.974	0.330	.000

Still again, we are led to the same qualitative conclusion regarding the effect of remittances—they increase the odds of enrollment by nearly 45 percent. However, for the full sample all other effects are significant as well, with two exceptions: households in which there are migrants but not remittances are indistinguishable from those in which there are no migrants; and, net of other factors, children in female-headed households are no more and no less likely than others to be enrolled in school. Moreover, most of the effects are larger than for the other two models. The odds of school enrollment increase monotonically with the education level of the household; the effect of household income is twice as large as in the restricted models; the number of school-age children has a *positive* effect on school enrollment; and both males and urban residents are more likely to be enrolled than are female or rural residents.

A BIBLIOGRAPHIC NOTE

This chapter draws heavily on Allison (2005) and Wooldridge (2006), both of which are excellent introductions to FE and RE models. Other useful discussions of FE and allied models can be found in Chamberlain (1980), Hamerle and Ronning (1995), and Halaby (2004). Instructive examples of the applications of such models to substantive problems of interest to social scientists include Geronimus and Korenman (1992), Ashenfelter and Krueger (1994), Frankenberg and Mason (1995), Budig and England (2001), Campbell and Lee (2005), Hotz and Xiao (2005), Hotz, Mullin, and Scholz (2005), Buttenheim (2006a, 2006b), Nobles and Frankenberg (2006), and Lu and Treiman (2007).

WHAT THIS CHAPTER HAS SHOWN

In this chapter we have considered two techniques—fixed effects (FE) and random effects (RE) models—for coping with omitted variables that may bias estimates. The essence of these procedures is to predict differences in the outcome as a function of differences in the predictor variables over time or, equivalently, differences across individuals within groups (families, communities, and so on). The assumptions underlying each method were discussed. Two worked examples were presented, one involving a continuous dependent variable and the other involving a dichotomous dependent variable, because the two cases are somewhat different. FE models for dichotomous outcomes typically are estimated on only a subset of the original sample because cases for which the outcome variable does not change over time or differ across members of a group must be dropped from the analysis. Moreover, FE models do not permit estimation of the effects of either measured or unmeasured time-constant variables, whereas RE models do. For these two reasons, when the assumptions of the RE model are met, the RE model is to be preferred.

CHAPTER 16

FINAL THOUGHTS AND FUTURE DIRECTIONS: RESEARCH DESIGN AND INTERPRETATION ISSUES

WHAT THIS CHAPTER IS ABOUT

In this chapter I review various aspects of research design, some of which we have encountered in previous chapters and some of which are new. In the course of doing this, I also briefly discuss a number of advanced statistical techniques and procedures, which you will need to complete the "tool kit" required for state-of-the-art data analysis in the social sciences, and which you are now in a position to tackle, given the foundation provided by what has been covered in this book. I then comment on the importance of probability sampling and ways to think about populations. I conclude with some advice about good research practice.

RESEARCH DESIGN ISSUES

In this section I consider some issues regarding appropriate analytic designs to answer research questions using nonexperimental data.

Comparisons Are the Essence

Not infrequently term paper proposals take the following form: I want to study caregivers, and I have found a sample of caregivers to use for my analysis; or I want to evaluate a new educational program instituted in a school, and I have a sample of students from that school. The problem with these proposals is that you cannot study a constant. If you want to know, for example, whether caregivers are particularly prone to depression, you need a sample of caregivers and noncaregivers. Similarly, if you want to know what factors lead people to migrate, you need a sample of both those who do and those who do not migrate. And if you want to evaluate the efficacy of a program, you need a sample of places where the program has and has not been implemented (or data before and after implementation—although there are special problems in making comparisons over time, which will be dealt with a bit later in the chapter). This is an extremely simple point but one that is often ignored in data collection efforts. If, for example, you have a sample of migrants or delinquents or caregivers, all you can do is study internal variations among different types of migrants or delinquents or caregivers, which presumably is not what you are really interested in.

If you have sampled only the population of interest, you are forced to rely on data external to your study to make comparisons, which often entails trying to compare not-quite-comparable data. Sometimes in such situations, researchers compare their data for a special population to patterns assumed to hold for some standard population. For example, a recent study of schooling available to migrant children in Beijing (Chen and Liang 2007) is based on a survey of migrant households with school-age children. From these data the researchers calculated and reported the proportion of such children not enrolled in school. The implicit contrast—and it was only implicit—is that all nonmigrant Beijing children attend school. But there is no particular reason to presume this. Indeed, often the assumptions social scientists make about their own societies prove to be false. Thus, explicitly comparative data are strongly to be preferred.

If comparisons are the essence of analysis, the obvious next question is, what sorts of comparisons are appropriate for what purposes?

Population Subgroups, Populations, and Historical Periods A common research question in the social sciences is whether population subgroups (males versus females, ethnic groups, and so on) differ with respect to some outcome and the factors determining that outcome. We saw in Chapter Six, in the section "A Strategy for Comparisons Across Groups," how to approach this kind of analytic question. Here I briefly review the strategy.

To determine whether a relationship between a set of X predictors and some outcome variable Y holds for all subgroups of a population or whether it differs across subgroups, we estimate three prediction equations (which may be OLS equations or those appropriate for some other linear model—for example, some form of logistic regression):

$$\hat{Y} = a + \sum_{i=1}^{I} b_i X_i \tag{16.1}$$

$$\hat{Y} = a' + \sum_{i=1}^{I} b'_i X_i + \sum_{j=2}^{J} c_j G_j \tag{16.2}$$

$$\hat{Y} = a'' + \sum_{i=1}^{I} b''_i X_i + \sum_{j=2}^{J} c'_j G_j + \sum_{j=2}^{J} \sum_{i=1}^{I} d_{ij} X_i G_j \tag{16.3}$$

In Equations 16.1 through 16.3, the X_i are predictor variables and the G_j are population subgroups, with each subgroup (except the first) represented by a dummy variable coded 1 for those in the subgroup and coded 0 otherwise. We then contrast Model 3 (Equation 16.3) against Model 1 (Equation 16.1) to determine whether we need to posit different relationships between the X_i and Y for the different groups. (We do this by assessing the significance of the increment in R^2 or, equivalently, assessing whether the c'_j and the d_{ij} in Model 3 are collectively not significantly different from zero.) If Model 3 fits significantly better than Model 1, we conclude that the social process being studied differs among the groups and ask the subsidiary question: is the difference only in the intercepts, or is it in the slopes as well? (We do this by assessing the significance of the increment in R^2 between Model 3 and Model 2 or, equivalently, assessing whether the d_{ij} in Model 3 are collectively not significantly different from zero.) Note that this strategy is only appropriate when the groups can be taken to be exogenous to the outcome under study, which holds for gender, ethnicity, and so on. When selection into the group is correlated with the outcome, net of the other predictor variables in the model, the assumption of OLS regression that the predictor variables are uncorrelated with the error is violated, and endogenous switching regression procedures, discussed later in the chapter, must be employed to yield unbiased estimates of effects.

If Model 3, or Model 2, proves to be the preferred model, it is then possible to decompose the differences between groups in their average outcomes, using the procedures for decomposing differences in means discussed in Chapter Seven. Note that the decomposition procedure was discussed in Chapter Seven in the context of OLS regression. The same procedure can be used to decompose differences in logged dependent variables (see Treiman and Roos [1983, 636–640] for an example) or in log odds, albeit without quite the same intuitive appeal.

A variant on the strategy for assessing group differences is to start with an equation of the form

$$\hat{Y} = a + \sum_{j=2}^{J} c_j G_j \tag{16.4}$$

Because for Equation 16.4 the predicted values for Y are simply the means of Y for each subgroup, the question being addressed by contrasting Equation 16.3 with Equation 16.4 (or Equation 16.2 with Equation 16.4) is to what extent group differences with respect to the outcome can be explained by group differences in the other predictor variables.

Exactly the same procedures can be used to make comparisons over time. For example, we might want to know whether the relationship between political attitudes (liberalism versus conservatism) and acceptance of abortion was the same in the 1970s, when *Row v. Wade* was first decided, and in the 2000s, when opposition to abortion apparently became obligatory for Republican presidential candidates. In this case, time becomes the G variable and political attitudes the X variable in Equations 16.1 through 16.3. And, of course, the same logic holds for comparisons of changes over time in group differences, albeit with an increase in complexity because of the need to consider three-way interactions. For example, the analysis of the interaction between education and religious denomination in determining acceptance of abortion in 1974, which I used in Chapter Six to introduce the strategy for group comparisons, could be replicated in 2006 to assess how the "abortion wars" over the last thirty-two years have affected attitudes.

Some cross-temporal comparisons are vulnerable to estimation problems stemming from the fact that data from different time periods may not be independent. This is true of aggregate measures such as the average level of schooling. The value of such a variable computed for, say, the United States in 2005 will hardly differ from the value for 2000 because both computations are based on more or less the same population. Thus, the two observations are not independent. Procedures for coping with the nonindependence of observations, known as *autocorrelation,* and with other special features of *time-series data* are well developed; see the Stata manual *Time Series* [*TS*] (StataCorp 2007). Time-series procedures are widely used in economics. Another kind of data, widely used in other social science fields as well as economics, derives from panel studies, in which the same individuals are surveyed two or more times, typically several months or years apart. Data with this structure provide one means of carrying out FE and RE analysis of the kind discussed in the previous chapter. These and other techniques for dealing with the nonindependence of observations are known as XT (cross-sectional time series) models. Such models go beyond what we have been able to consider in this book. For standard introductions, consult the Stata 10.0 manual *Longitudinal/Panel Data* [*XT*] (StataCorp 2007) and texts by Sayrs (1989), Wooldridge (2002), Hsiao (2003), Baltagi (2005), and Greene (2008). The Sayrs text is quite accessible, the Greene text reasonably so, and the other three fairly formidable.

Both cross-sectional comparisons and cross-temporal comparisons may, of course, be extended to more than two comparisons (more than two groups or more than two time points), and groups may be subpopulations of a single nation or populations of different nations. For examples of the latter, see Erikson and Goldthorpe (1987a, 1987b).

The reason for carrying out cross-population or cross-temporal comparisons is to test some hypothesis, or hypotheses, about how populations or population subgroups differ or time periods vary. If you have a priori hypotheses, this is a reasonable strategy. However, you are vulnerable to the counterclaim that the differences you posit and observe are spurious, because they reflect differences between groups or over time that affect both the independent and dependent variables. Binary comparisons are particularly vulnerable to such claims, because any number of other factors could account for the differences.

Indeed, any graduate student in the social sciences can invent a post hoc explanation for *any* observed difference! If you do not believe me, try this simple test on your friends: invent a finding about some society or population they do not know about or, better yet, report a finding but reverse the sign or otherwise change the outcome. Then wait to see what plausible explanation they give you. I used to do this at cocktail parties when I discovered that everybody thought that my finding of essential invariance in occupational prestige hierarchies around the world (Treiman 1977) was a case of documenting the obvious. I started telling people that prestige hierarchies were quite different in, say, Russia, and got all sorts of interesting explanations about why it was obvious that this had to be so (even though it was not!). Comparisons of three groups (time points) are far more constraining, and comparisons of still more groups (time points) even more so.

As a case in point, consider historical comparisons. Nee (1989, 1996) has argued that the shift toward a market economy in China reduced the power of cadres and enhanced the power of "direct producers." The difficulty, as Walder points out in a critique (1996, 1,064), is that many things change over time:

> *The problem with time [as a measure] is that many other changes conceptually distinct from the spread of markets, and which may also affect the distribution of power and income, also occur through time and at different rates in different regions. Some emerging market economies grow rapidly while others do not; state policy may provide windfall gains to grain producers in one period only; private enterprise may thrive in some regions but remain marginal in others; capital may be highly concentrated in some regions, more dispersed or absent in others. All of these processes affect the distribution of power and income; any time-dependent measure of market allocation must carefully control for them.*

This difficulty, sometimes referred to as the "too many degrees of freedom" problem, because there are too many plausible explanations for whatever is observed, is generic to two-case comparisons, both cross-temporal and cross-sectional. For this reason, comparisons of a small number of cases are more helpful in demonstrating similarity than in explaining differences. Sometimes it is helpful to show that a finding in one society or at one point in time also holds in a different time and place. If so, we can have more confidence that we have identified a general phenomenon and not just an idiosyncratic result.

By contrast, consider a test of the "fetal origins" hypothesis (Barker 1998) by Almond (2006). The hypothesis states that adverse events experienced by pregnant women can have long-term consequences for their offspring. Almond studied this claim by analyzing the consequences of the 1918 flu pandemic for educational attainment, occupational status, income, disability, and other outcomes measured in the 1960, 1970, and 1980 censuses. He finds strong effects, one of which is shown in Figure 16.1, male disability rates in 1980 by quarter of birth in 1918 through 1920. Because disability rates were elevated only for those who were in utero during the pandemic and returned to the trend line for

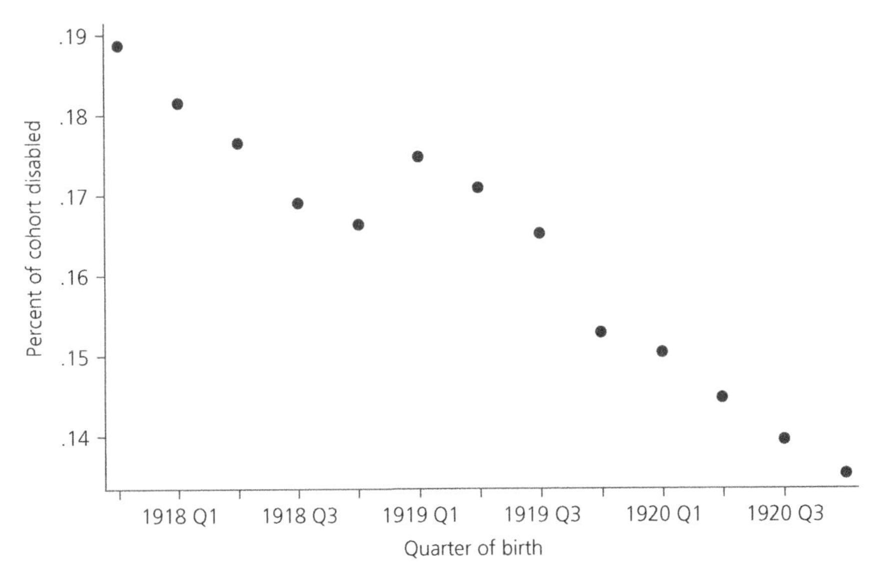

FIGURE 16.1. *1980 Male Disability by Quarter of Birth (Prevented from Work by a Physical Disability).*

Source: Almond 2006, Figure 2.

those born later, we can rule out the possibility that some other, unmeasured, change coincided with the onset of the pandemic. More precisely, any alternative explanation would have to show a pattern of timing that exactly coincided with the pandemic, which in this case is not remotely plausible.

Natural Experiments Analyses of this kind are particularly compelling because they constitute *natural experiments*. As we have seen, the difficulty with almost all nonexperimental work is the possibility that we have omitted variables that affect both the outcomes and the predictor variables, thus biasing our estimates. Natural experiments reduce or eliminate omitted variable bias by focusing on natural events that plausibly can be argued to be distributed randomly in the population. Because the 1918 pandemic struck without warning in October 1918 and had largely dissipated by early 1919, it is reasonable to regard those in utero during the months of the pandemic as a treatment group and those in utero just before and just after these months as a control group. Because there are no plausible differences between these groups, except that one had the misfortune to be in utero at the time of the pandemic, it is reasonable to infer that differences in outcomes are due to exposure to the pandemic. Of course, not all pregnant mothers became infected. But we know that about one-third of childbearing-age women did become infected, which is a large enough group to reveal differences in outcomes if they exist. Almond's paper, which also exploits state-to-state variations in the severity of the pandemic, is a model of how to do analysis of this kind.

For other examples of natural experiments, albeit some more persuasive than others in how thoroughly they overcome potential omitted-variable bias, see Deng and Treiman (1997), Ansolabehere, Snyder, and Stewart (2000), Abadie and Gardeazabal (2003), Lassen (2005), Oster (2005), Treiman (2007a; see also the discussion of this example in Chapter Seven), and Lu and Treiman (2008).

Multilevel Analysis When you have many comparison groups (many time points, many nations, and so on), it makes sense to shift from treating each group as a discrete point (by including a set of dummy variables representing groups) to scoring each group with respect to various dimensions (for example, characterizing nations by their level of economic development, the degree of urbanization, and so on). The optimal way to do this is to carry out the analysis at two or more levels. In the latter case, macrosocial "contexts" are defined (for example, classrooms or schools, or both, in educational studies; societies in cross-national studies; birth cohorts or historical periods in cross-temporal comparisons; and so on). Then a microequation (one representing some social process) is estimated separately for each context, and variations in the coefficients representing the microprocess are predicted from characteristics of the contexts.

For example, suppose you wish to test the hypothesis that the negative effect of the number of siblings on educational attainment is stronger where school fees as a fraction of total family incomes are higher. Here is the typical setup for such an analysis:

$$Y_{ij} = a_j + b_j X_{ij} + \varepsilon_{ij} \tag{16.5}$$

with

$$\begin{aligned} a_j &= \eta_{00} + \eta_{01} G_j + \alpha_{0j} \\ b_j &= \eta_{10} + \eta_{11} G_j + \alpha_{1j} \end{aligned} \tag{16.6}$$

where $j = 1, 2, \ldots, J$ denotes contexts, or level-2 units (school systems in the example), and $I = 1, 2, \ldots, n_j$ denotes individuals within contexts (level-1 units). The level-2 equations assert that the intercepts and slopes of the level-1 equations vary over contexts as linear functions of G (or, in the example, that the [negative] slope associated with the number of siblings is greater [in absolute value] when school fees are higher; in this example I have provided no hypothesis regarding the intercepts of the level-1 equations, although it might be plausible to expect that the level of educational attainment is lower when school fees are high). Of course, the level-1 equations may include more than one X, and the level-2 equations may include more than one G.

To appreciate the substantive payoff of multilevel analysis, consider a 1989 paper by Lee and Bryk that analyzed the effect of differences in the academic organization of high schools on the distribution of achievement in standardized mathematics tests. This early paper by one of the authors of the standard text on multilevel analysis (Raudenbush and Bryk [2002]) not only provides a compelling substantive example but lays out the technical issues in a very clear way. Using data from the *High School and Beyond* study (10,187 students in 160 high schools), Lee and Bryk showed that achievement scores tend to be highest and differentiation among students by race and socioeconomic status lowest in schools with a standardized academic curriculum required of all students, in contrast to "shopping mall" schools with a wide variety of courses and many electives. They cite this difference in the organization of the curriculum as the main reason Catholic schools tend to be more successful, a widely noted but not previously well-understood phenomenon.

There are many ways to carry out a multilevel analysis, including the meta-analysis approach used by Treiman and Yip (1989) in the example presented in Chapter Ten to illustrate regression diagnostic procedures. The technical details go beyond what we have been able to consider in this book. Good introductions include a paper by DiPrete and Forristal (1994), which emphasizes substantive applications and provides a good sense of what this body of techniques can be used to do. A paper by Mason (2001) focuses on various multilevel analysis procedures as a way of coping with observations that are not independent because they are clustered within higher-level units (for example, children in families, pupils in classrooms). Several book-length treatments also address multilevel analysis, of which Raudenbush and Bryk (2002) is the standard but is technically demanding, as is Goldstein (2003); Snijders and Bosker (1999) may be more accessible. For additional substantive examples, see Entwisle and Mason (1985) on the relationship between level of socioeconomic development and fertility; DiPrete and Grusky (1990a, b) and Grusky and DiPrete (1990) on temporal variations in socioeconomic attainment in the United States; and Sampson, Raudenbush, and Earls (1997) on the role of neighborhood efficacy in reducing crime.

Endogeneity, Sample-Selection Bias, and Other Threats to Correct Causal Inference

In this section I discuss several closely related circumstances that require special treatment to avoid biased estimates. I also provide brief introductions to some of the standard solutions not addressed in previous chapters.

Treatment Effects *Endogeneity* refers to situations in which one or more of the predictor variables is correlated with unobserved variables that affect the outcome. Because the effects of unobserved variables are relegated to the error term, the coefficients of predictor variables correlated with such unobserved variables are biased. A common situation in which endogeneity problems may arise is when we wish to assess the effect of a "treatment" that itself depends on unmeasured factors that affect the outcome. For example, if in a developing nation midwives are placed in villages with the worst health outcomes, an assessment of the effect of midwives on health that fails to control for this nonrandom assignment will be downwardly biased (see Frankenberg and Thomas [2001] for an analysis of Indonesian data that uses a fixed effects [difference-in-difference] analysis, of the kind discussed in the previous chapter, to derive correct estimates). Similarly, if workers less able to command high wages (for unmeasured reasons) are more likely to join a union, OLS estimates of the effect of union membership—here regarded as the "treatment"—on wages will be downwardly biased. For an interesting example of how to carry out an analysis using a treatment-effects approach, see Brand's 2006 study of the effect of job displacement on the quality of subsequent jobs.

The most obvious way to correct for endogeneity problems is to measure all the factors thought to affect the outcome. We encountered this idea in our consideration of ordinary least-squares regression in Chapter Six, when we discussed the presentation of several models, with successively more variables, to assess how newly introduced variables mediate the effects of variables already in the model. From this we see that endogeneity bias is a form of omitted-variable bias.

However, it is not always possible to measure all the potential influences on an outcome, either because we are reanalyzing already collected data or because the analyst may not be able to identify a priori all the potential influences on an outcome that are correlated with variables explicitly measured—for example, all the factors that might both lead individuals to join a union and be correlated with their capacity to command high wages as individual employees. Thus we need ways to correct for endogeneity bias (and its close cousin, sample-selection bias). We have already covered one approach, fixed effects or random effects modeling, which is possible when we have measurements for individuals at more than one point in time or measurements for different individuals within groups (for example, families or classrooms or communities). When such data are not available, several other analytic strategies may be considered, all of which go beyond what it has been possible to cover in this book. For useful discussions of what is entailed in establishing causality, see Holland (1986), together with comments by Rubin, Cox, Glymour, and Granger, and Holland's rejoinder; and Winship and Morgan (1999).

Instrumental Variables Regression An approach to coping with endogeneity that is popular among economists is *instrumental variable* (IV) estimation. If a variable (Z) can be found that is uncorrelated with unobserved variables (u) that affect the outcome (Y), is correlated with the variable (X) in the model thought to be correlated with the unobserved variables, and is conditionally unrelated to the outcome variables net of the effect of both the observed and unobserved variables, Z can be used as an instrument for X to yield relatively unbiased estimates of the effect of X on Y.

For example, consider a 1990 paper by Angrist studying the effect of service in the military during the Vietnam War on lifetime earnings. The difficulty with estimating an OLS equation is that the decision to join the military might well have been correlated with unmeasured factors that affect earnings. Angrist exploited the fact that for much of the war period, a lottery system was used to determine who would be drafted into service. Although there were many exceptions, the increased probability of being drafted for those with low numbers makes the assignment of lottery numbers a kind of natural experiment—one's lottery number was correlated with the likelihood of serving but not with other factors related both to service and to subsequent income. Thus the lottery number is a good instrument for adjusting the effect of Vietnam veteran status on income.

Another situation where IV estimation may be helpful is where the causal order is ambiguous. Suppose we observe that women who work are less likely to be depressed than women who do not work. Can we conclude that employment protects against depression? Perhaps. But the causal order might go the other way: depressed women may be less likely to seek or retain employment. One way to address this problem would be to find an instrument for employment. A reasonable choice might be whether the mother worked. It is known that the daughters of mothers who worked are more likely to work themselves. But there is no particular reason to believe that, net of her own employment, a woman's mother's employment affects the likelihood that the woman herself experiences depression (the example is from Ettner [2004]). Thus, mother's employment would satisfy the conditions for an instrumental variable.

A final circumstance in which IV approaches can be helpful is to estimate simultaneous equation models or, as they are sometimes called, reciprocal causation models. Wooldridge (2002, 555) provides a useful example: in a sample of cities, we might expect the murder rate to depend on the size of the police force—the more police per capita, the lower the expected per capita murder rate. But we might also expect the size of the police force to depend on the murder rate—the higher the (anticipated) murder rate, the greater the incentive to increase the size of the police force. Because we observe only the equilibrium condition—a particular murder rate and a police force of a particular size—specifying a simultaneous equation model amounts to asking the counterfactual questions: what would be the murder rate if the size of the police force were different? What would be the size of the police force if the murder rate were different? IV methods provide a way of estimating such models.

Another case that usually involves reciprocal causation and hence might be handled by IV models (or by structural equation models of the kind discussed later) is when one attitude is thought to affect another but they are measured at the same time. Rather than assuming that somehow one causally precedes the other, it usually is more sensible to treat them as either both dependent on a third variable (seemingly unrelated regression, encountered in Chapter Eleven, can be helpful in such cases) or as having reciprocal effects.

The difficulty with IV estimation is that it often is difficult to find good instrumental variables, and poor instruments often produce results worse (more biased) than using no instruments at all. For a good introduction to IV estimation and its dangers, see Wooldridge (2006, Chapters 15 and 16). Other useful references include Baum (2006), the entry for the `-ivregress-` command in StataCorp (2007), and Green (2008).

Sample-Selection Bias Sample-selection bias arises when unmeasured factors correlated with an outcome determine whether an individual is included in the sample. For example, a woman may enter the labor force only if she can command a reasonably high wage. Thus selection into the sample (people in the labor force) is nonrandom but depends on unmeasured characteristics correlated with the outcome variable (wages). Analyzing only those with wages then results in biased estimates.

Consider another example. Many surveys in China are restricted to the de jure urban population. As Wu and Treiman (2007) have shown, in such studies estimates of intergenerational mobility are overstated because those of rural origins who obtain urban registration are not a random sample of the rural population but rather are disproportionately the "best and the brightest," who have experienced long-range upward social mobility. If the entire population is included in the analysis, estimates of the extent of mobility are much more modest.

Heckman Selection Model A standard approach to correcting for sample-selection bias (in cases where it is not possible to redefine the population as Wu and Treiman did) is to use a *Heckman correction* (see Heckman 1979). The procedure involves predicting (using a binary probit equation) the probability of being in the sample (or, equivalently, of having an observed outcome), calculating the expected error for each observation, and using these errors as regressors in an equation predicting the outcome of interest. See Winship and Mare (1992) for a very clear exposition of this and other models for sample-selection bias, and see Dubin and Rivers (1989) for an extension of these procedures to models with binary outcomes.

The Stata entry for the `-heckman-` command (StataCorp 2007) offers another very clear example and exposition of the method, using the canonical example, women's earnings. In the example, earnings (for women who have earnings) are predicted from education and age, and the probability of having earnings is predicted from marital status, the number of children at home, education, and age (and implicitly—through the inclusion of education and age, which predict the outcome—of the expected wage itself). Note that the assumption here is that marital status and the number of children at home do not affect earnings but only the probability of having earnings. We might well question this assumption because married women, and particularly women with children at home, may choose to take lower-paying jobs that more readily accommodate their dual careers as workers and mothers.

This example thus reveals a major limitation of the procedure. To yield robust results, the predictors in the selection equation should strongly affect the probability of being selected but should have no net effect on the outcome. (Heckman corrections can be made even when there are no such variables, by relying on the functional form of the equation to identify the model. However, the results are often neither robust nor substantively compelling.) Suitable variables are often difficult to find. Note the similarity to IV estimation discussed previously.

For instructive applications of corrections for sample-selection bias, see Mare and Winship's 1984 study of employment trends for young Black and White men; Hagan's studies of factors influencing the severity of punishment for convicted criminals (Peterson and Hagan 1984; Hagan and Parker 1985; Zatz and Hagan 1985); Manski and Wise's (1983) study of the determinants of graduation from college; and Hardy's (1989) study of occupational mobility in the nineteenth century based on matching data across censuses, which takes account of selection due to deaths, emigration, and name changes.

Endogenous Switching Regression Note that the Heckman procedure also can be used to analyze endogenous treatment effects, as an alternative to IV estimation. However, a separate command, `-treatreg-`, is also available in Stata, in addition to `-heckman-`. The problem of an endogenous treatment effect—that is, where there is a nonzero correlation between assignment to a "treatment" group and unmeasured factors affecting the outcome—can in turn be generalized to the case in which the parameters of a model linking treatments to outcomes differ across treatment groups and assignment to treatment groups is endogenous. For example, Gerber (2000) asks whether the fact that former Communist Party members do better in post-Soviet Russia than do others is due to residual social capital (the fact that connections continue to favor former party members) or rather to unmeasured factors that affect both the likelihood that people became party members during the Soviet era and size of their earnings in the post-Soviet period.

This kind of problem can be addressed using methods that are similar to those for treatment effects and sample-selection problems—specifically, *endogenous switching regression* models. Endogenous switching regression models are used in situations where one outcome, Y_1, is observed if a selection variable, $Z_1 = 0$, but a different outcome, Y_2, is observed if $Z = 1$. Using this method, Gerber concludes that the advantage enjoyed by former communists is due entirely to unmeasured characteristics associated with becoming a member of the Communist Party and that there is no lingering effect of Soviet-era social or political capital. (See also the critique by Rona-Tas and Guseva [2001] and the rejoinder by Gerber [2001].)

Good descriptions of the technique and of how to implement it can be found in Mare and Winship (1988) and Powers (1993). For additional applications see Willis and Rosen (1979); Gamoran and Mare (1989); Long (1990); Sakamoto and Chen (1991); Manski and others (1992); Tienda and Wilson (1992); Powers and Ellison (1995); Smock, Manning, and Gupta (1999); Hofmeyr and Lucas (2001); Lichter, McLaughlin, and Ribar (2002); Sousa-Poza (2004); and Prouteau and Wolff (2006).

Propensity Score Matching Another threat to correct causal inference occurs when the predictor variable of interest occurs only rarely in the sample and is highly correlated with other independent variables. For example, what is the effect of attending an elite university on subsequent occupational status? The usual way of approaching such a question is to carry out a multiple regression of occupational status on attendance at elite versus other universities plus a set of variables controlling for family background, high school performance, and so on. The difficulty is that attending an elite university tends to be so highly correlated with the control variables that controlling for confounding factors fails to hold them constant, because there are few people with low values on the control variables who attend elite universities. Apart from the conceptual problem this creates about the meaning of "holding constant," there is a serious statistical problem—"unbalanced treatments" tend to in ate standard errors (Rosenbaum and Rubin 1983, 48), making problematic the rejection of the null hypothesis of no effect. To cope with this problem, analysts sometimes resort to matching pairs of cases that differ with respect to the variable of interest (the "treatment" variable) but that are identical on a set of covariates. However, as Smith notes (1997, 326–327), until recently matching studies often have been resisted on the ground that they involve "throwing away" a lot of data. Moreover, it often is difficult to find good matches for more than a small number of variables because for a linear increase in the number of covariates there is a geometric increase in the number of matches required.

However, advances in the statistical theory of matching—the seminal article is by Rosenbaum and Rubin (1983)—have led to the development of a procedure that replaces the large set of discrete matches required by classical matching procedures with a *propensity score*, a scalar summary of the degree of similarity between cases with respect to a large number of covariates. The procedure involves predicting the treatment variable from covariates and then matching each "treatment" case with the control case that has the nearest propensity score (or sometimes with several control cases; see Morgan and Winship [2007] for a useful discussion of the technical issues involved). The resulting sample is then analyzed in one of several ways: focusing on outcome differences between matched treatment and control cases, ignoring the unmatched cases; stratifying the sample into strata with similar propensity scores and comparing outcomes within strata (for an interesting application, see Brand and Xie [2007]); or using the propensity score directly in a regression equation to get an estimate of the effect of the treatment net of the propensity to be in the "treatment" group. The essential insight is that by comparing cases that have a similar propensity to be in the treatment group, we create a quasi experiment. That is, we can think of matched cases as being, in effect, randomly assigned to either the treatment or the control group because they have the same probability of being in either group, given their covariates.

Consider the example presented by Smith (1997) in his illuminating exegesis of matching methods. He was interested in comparing the mortality rate in two types of hospitals, ordinary hospitals (N = 5,053) and "magnet" hospitals (N = 39)—hospitals with organizational practices that enhanced their reputations as good places to practice nursing. Contrasting an OLS analysis with a propensity matching procedure, he showed that the two methods yielded similar estimates of the difference in mortality rates in the two types of hospitals, but the latter method had far smaller standard errors, yielding a statistically significant reduction in mortality in the magnet hospitals compared to ordinary hospitals, a conclusion not yielded by the OLS analysis because of the large standard errors resulting from the unbalanced design.

There is by now a substantial literature on both the statistical theory underlying propensity score matching and practical procedures for implementing the method. The 1997 Smith paper is a good place to start and also has a useful bibliography. Becker and Ichino (2002), Abadie and others (2004), and Becker and Caliendo (2007) discuss the implementation of propensity score matching in Stata. Dehejia and Wahba (2002) and Brand and Halaby (2006) provide useful evaluations and worked examples. Harding (2003) is a particularly instructive application. For other applications, see Berk and Newton (1985), Stone and others (1995), Keating and others (2001), Lu and others (2001), Morgan (2001), Black and Smith (2003), Lundquist (2004), and Cohen (2005). One limitation of propensity score matching is that it may not balance unobserved covariates. Thus if you suspect endogeneity, you will need to resort to one of the methods discussed here or in the previous chapter that are specifically designed to handle such problems.

Structural Equation Models

Structural equation modeling (SEM) is a technique (or, more precisely, a set of techniques) that permits the estimation of systems of equations, often involving unmeasured or latent constructs. Consider a simple example, Blau and Duncan's (1967, 170) classic model of status attainment, shown in Figure 16.2. When we think about how occupational status is transmitted from one generation to the next, it becomes evident that this is a multistep process: men whose fathers are well educated and have high-status jobs tend to achieve more schooling; those who achieve high levels of schooling tend to obtain

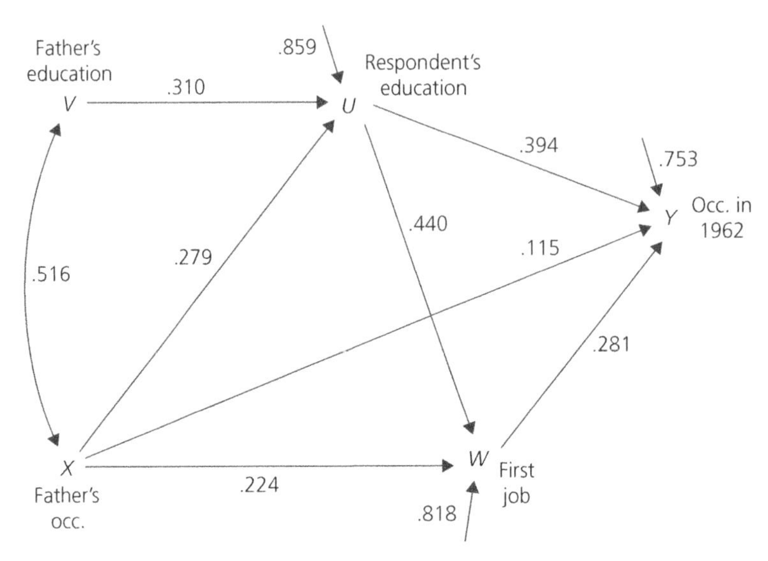

FIGURE 16.2. *Blau and Duncan's Basic Model of the Process of Stratification.*

Source: Blau and Duncan 1967, 170.

high-status first jobs (but their social origins may also help); and those who have high-status first jobs are likely to be able to parlay them into high-status current jobs (but their education and even their social origins may continue to matter). The various "paths" by which fathers' occupational status is transmitted to their sons are shown in the figure, known as a "path diagram." The paths can be represented by a set of equations predicting each of the outcomes in turn. The relationships among the equations can be explored to yield insights regarding the relative importance of different paths linking the two variables. Moreover, under some circumstances (typically, if the size of particular coefficients is fixed, usually but not necessarily at zero, or two or more coefficients are constrained to be equal—that is, if the model is *overidentified*), the goodness of fit of the model can be assessed.

In the model just discussed, there is only one indicator for each variable. However, often the analyst has available repeated measures or a set of measures thought to represent a single underlying or latent construct. In such cases, it is possible to use SEMs to assess and correct for measurement error. See Bielby, Hauser, and Featherman (1977) and Hauser, Tsai, and Sewell (1983) for two early but instructive examples. Still another use of SEMs is to estimate processes involving reciprocal causation, even involving latent variables. See Duncan, Haller, and Portes (1968) for such an example. (Note that the applications just cited are all very old. This is not due to a lack of recent work—the current literature is vast—but rather to the fact that the early work was much more explicit

OTIS DUDLEY DUNCAN (1921–2004), known as Dudley by his friends, fundamentally transformed the study of social stratification and social mobility. Called by statistician-sociologist Leo Goodman "the most important quantitative sociologist in the world in the latter half of the twentieth century," Duncan was responsible for introducing path analysis (a version of structural equation models) into sociology. He used *path analysis* as the technical apparatus to reconceptualize intergenerational social mobility as a multistep process in which status attributes (such as education, occupational status, and income) are modeled as depending not only on parental status but also on the prior statuses of individuals. Duncan also contributed importantly to our understanding of racial differences in socioeconomic attainment, spatial and racial inequalities within cities, and, late in his career, attitude measurement.

Although lacking advanced mathematical training, Duncan probably made better use of the statistical tools at his disposal than any other social scientist, through the combination of an unusual ability to think through a problem in advance and great clarity about how to represent sociological ideas in statistical models. It is striking, and telling, that because of then-extant rules governing access to Current Population Survey data, all of the tabulations and estimates in Duncan's landmark book *The American Occupational Structure* (Blau and Duncan 1967) were specified in advance, without the analysts having seen a single coefficient. Interestingly, Duncan himself regarded his late book, *Notes on Social Measurement* (1984), as his most important contribution, a judgment not widely shared by the many researchers strongly influenced by his substantive contributions.

Born in Nocona, Texas, Duncan spent most of his precollege years in Stillwater, Oklahoma, where his father, Otis Durant Duncan, also a sociologist, was a professor at Oklahoma State University. Duncan did his undergraduate work at Louisiana State University, obtained an MA at the University of Minnesota, served three years in the U.S. Army during World War II, and then completed his PhD at the University of Chicago in 1949. He taught at Pennsylvania State University, the University of Wisconsin, the University of Chicago, the University of Michigan, the University of Arizona, and the University of California at Santa Barbara. Duncan enjoyed a second career as a composer of electronic music and was famous among people who had no idea that he was a distinguished social scientist.

about the models being estimated than much of the literature that followed, after structural equation modeling became widely used. Thus for didactic purposes the early papers are more useful.)

The strategy for estimating SEMs is to exploit the fact that the posited relationships among the variables (observed and latent) implies a particular covariance structure (that is, a set of relationships among the variances and covariances of the observed variables), which is why the technique is sometimes called *covariance structure modeling*. Goodness of fit is assessed by comparing the covariance structure implied by the model with the covariance structure observed in the data set being analyzed.

Structural equation modeling has a long and diverse history, beginning with the creation of *path analysis* by the population geneticist Sewall Wright in 1918. It was introduced into sociology (and other social sciences) by Otis Dudley Duncan (see Duncan [1975] for an early exegesis); credit for discovering the work of Wright properly belongs to Robert W. Hodge, when he was a student of Duncan's at the University of Chicago. At the same time, economists were beginning increasingly to use simultaneous equation models. As will be evident from perusal of the applications cited here, the specification and estimation of structural equation models, even with the aid of path diagrams as a graphic representation of the assumed relationships in the system, was a tedious and error-prone process. The situation changed with the introduction of computer software specifically designed to estimate covariance structural models. The first such software package was introduced by Jöreskog (1970) and was known as LISREL (for *Linear Structural Relations*). Subsequent software included EQS by Bentler and Wu, AMOS by Arbuckle, and MPlus by Muthén and Muthén. Many of the newer packages allow the analyst to easily draw a path diagram, which is then converted by the software into a system of equations, and coefficients are estimated. Check the internet for the most recent versions of these and other software packages, including modules for general-purpose statistical software packages.

SEWALL WRIGHT (1889–1988) is known to social scientists as the creator of *path analysis,* which he introduced in 1918. But this was actually a tangential accomplishment, deriving from his work as a population geneticist. He was born in Melrose, Massachusetts; did his undergraduate work in mathematics at Lombard College, where his father then taught, and completed his PhD in biology from Harvard. After an early position at the U.S. Department of Agriculture, he joined the Department of Zoology at the University of Chicago in 1925 and remained there until his retirement in 1955, after which he moved to the University of Wisconsin–Madison. At Wisconsin he continued to do research and publish. His final paper, his 211th publication, appeared in 1988, a few days before his death from a fall at age ninety-nine.

Wright's primary contribution to science was as a cofounder, together with R. A. Fisher and J.B.S. Haldane, of theoretical population genetics. The work of Fisher, Wright, and Haldane was a major step in the development of the "modern evolutionary synthesis" of genetics with evolution. Wright also made major contributions to mammalian genetics and biochemical genetics.

James Steiger, a psychologist and contributor to structural equation modeling, takes a somewhat skeptical view of these developments in a 2001 paper, suggesting that the new ease in estimating SEMs has resulted in applications (particularly in psychology) that are inattentive to the assumptions underlying the method and its limitations—most importantly the fact that structural equation modeling does not somehow magically overcome the simple point that correlation does not imply causation. Rather, it is a procedure that enables the analyst to explore the implications of whatever model the analyst posits on a priori grounds. Thus structural equation modeling is best seen as an interpretative procedure, with the added feature that in some cases it is possible to determine whether a particular model is consistent with observed data. Used properly in this way, SEM can be a valuable tool. (The best introduction remains the 1989 text by Bollen, which, although somewhat demanding, is intended for and accessible to social scientists. See also a collection of papers on technical issues, edited by Bollen and Long [1993]; Bollen and Curran's 2006 book using SEMs to estimate latent curve models; and Bollen and Brand's 2008 paper using SEMs to estimate random and fixed effects models.)

THE IMPORTANCE OF PROBABILITY SAMPLING

To generalize from a sample to a population—which is what social scientists are almost always interested in doing, whether we admit it or not—it is necessary to sample cases from the population of interest in such a way that each individual in the population has a known probability of being included in the sample. Only under this circumstance do the principles of statistical inference apply.

Nonetheless, many studies violate this principle, drawing "convenience" or "casual" samples. Chinese social surveys are particularly egregious in this respect, often sampling a set of provinces or cities that are said to be typical of particular types of places; this is true even of high-quality surveys such as the Chinese Health and Nutrition Survey (Henderson and others 1994). The difficulty is that there is no way of knowing to what extent and in what ways the chosen places are indeed similar to the places that are not chosen but are purported to be represented by the chosen places. In sum, samples of "typical" places are no substitute for probability samples. It is well worth the extra cost—in the sampling effort and, often, in the fieldwork—to design a sample in such a way that it can be generalized to the population of interest.

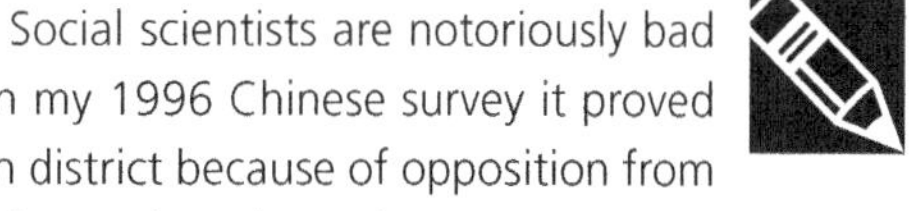

ASK A FOREIGNER TO DO IT Social scientists are notoriously bad at characterizing their own societies. A case in point: in my 1996 Chinese survey it proved impossible to do the fieldwork in one county-level urban district because of opposition from local officials. Instead of asking me to provide a substitute place from the same stratum (recall from Chapter Nine that there were twenty-five urban strata, based on the level of education in the population), my Chinese colleagues simply substituted another district from the same city that they said was very similar to the omitted district. However, it turned out that whereas the omitted district was in the eighteenth stratum, the substitute was in the twenty-third stratum, clearly a violation of the stratified sampling design.

The truth is that if you want a clear-headed characterization of a society, you should ask a foreigner to render it. This essential point was understood by the Carnegie Corporation, which commissioned the Swedish economist and sociologist Gunnar Myrdal to head a study of race relations in the United States in the 1930s. The result was the classic monograph An American Dilemma (Myrdal 1944, vi–vii).

Superpopulations

Sometimes, however, we wish to generalize beyond the population that was sampled. Single cross-sectional surveys exemplify this situation. Although the survey is conducted in a single year, the analyst usually wants to generalize beyond that year. To invoke my 1996 Chinese survey again, my interest was in generalizing to "the Chinese population toward the end of the twentieth century" and not to "the Chinese population in the summer of 1996." Typically, this sort of generalization is left implicit and becomes an issue only when a critic suggests that what was then true is no longer true. The response to the critic usually takes the form, sometimes explicitly but often implicitly, of invoking the "ethnographic present," suggesting that what is being described is Chinese society in 1996 "and all societies in which the same conditions prevail." That is, the concept of a *superpopulation* is invoked—that the population sampled is representative of a conceptual superpopulation of similar populations.

Another example is the set of societies studied by anthropologists for which data were compiled by Murdock and his colleagues (see, for example, Murdock and Provost [1973]). Because not all small and isolated societies have been studied by ethnographers, the Murdock compilation is hardly a probability sample of such societies. Nonetheless, users of this data set treat it as if it were a probability sample of societies, on the assumption that the set of societies studied by nineteenth- and twentieth-century ethnographers is representative of the superpopulation of all such societies.

GEORGE PETER MURDOCK (1897–1985) was an anthropologist who, rather against the grain of American anthropology, promoted quantitative cross-societal comparisons as a way of testing hypotheses and establishing general laws of social organization and human behavior. Born into a fifth-generation farming family in Connecticut, he earned a BA in American history at Yale and began Harvard Law School. He quit law school in his second year and took a long trip around the world. His journey sparked his interest in anthropology, which—after he was refused admission to Columbia, the preeminent department at the time—led him to Yale, still under the influence of Sumner's evolutionary ideas (in contrast to the societal particularism of Boas, at Columbia, that came to dominate anthropology). He earned his PhD at Yale in 1925 and remained on the faculty there through most of his career. Murdock published his first comparative paper in 1938 at about the same time he began a project to compile data on all known human societies. Ten years later (after naval service in World War II, where he and his students wrote ethnographic handbooks on Micronesian cultures, working out of an office at Columbia University), he decided that his compilation would be more valuable if it were widely available. He persuaded the Social Sciences Research Council to fund an interuniversity organization, what came to be known as the Human Relations Area Files (HRAF), to promote access to his collection. In 1950 Murdock moved to the University of Pittsburgh, where he founded the journal Ethnology and continued his comparative data compilation and research.

Still another example can be found in institutionally based studies—for example, studies of hospitals, clinics, and their catchment areas, which are often used in public health research. The justification for what amount to convenience samples is that the particular hospitals or clinics being studied are representative of all similar places.

When is it legitimate to invoke the concept of a superpopulation? I suggest that when data for a population exist from which a probability sample can be drawn, convenience samples are a poor substitute and do not meet current scientific standards—claims of graduate student poverty, lack of time, and so on, notwithstanding. However, when the population is unknown and unknowable, as in the case of Murdock and Provost's ethnographic sample, use of the available data and generalization to a superpopulation are legitimate. In the case of single cross-sectional surveys based on probability samples of the population at the time of the survey, we are on firm ground in characterizing the society as it was at the time of the survey but are increasingly on shaky ground as we try to generalize over time. It can be done, but it must be justified.

Pooling Data from Multiple Surveys Invoking the concept of a superpopulation has considerable practical use when it can be justified. A particularly compelling application is when comparable data are available over time, as in the U.S. GSS and other repeated cross-sections. If it can be shown that relationships among the variables of interest do not vary over time, data from several years may be pooled to increase the size of the sample available for analysis. This can be a particularly useful strategy when any one year yields insufficient data to sustain reliable comparisons, for example of race differences in the United States. The basic test is a variant on the strategy for group comparisons discussed earlier in this chapter and also in Chapter Six (see also the discussion of trend analysis in Chapter Seven). There are two steps. First, estimate an equation of the form

$$\hat{Y} = a + \sum_{i=1}^{I} b_i X_i + \sum_{j=2}^{J} c_j T_j + \sum_{i=1}^{I}\sum_{j=2}^{J} d_{ij} X_i T_j \qquad (16.7)$$

where the X_i are predictor variables and the T_j are cross-sectional replicates of the survey (with the first omitted to avoid linear dependency). Second, test whether the c_j and the d_{ij} are collectively zero. If so, you can conclude that all the samples are drawn from a single population and happily proceed to pool your data. But even if there are year-to-year variations in the level of Y (significant differences among the c_j) or in the relationship of one or more of the Xs to Y (significant differences among the d_{ij}), you may still wish to pool your data but your model should then include the dummy variables and interaction terms necessary to capture the way the social process you are studying changes over time. This has the advantage of simultaneously permitting an analysis of change and increasing statistical power for assessing the relationships that do not vary over time. (For some recent examples of the use of this strategy, see Barkan and Greenwood [2003], Chen and Guilkey [2003], Powers [2003], Fitzgerald and Ribar [2004], Kelly and Kelly [2005], and Tavits [2005].)

An alternative use of repeated cross-sections, which has much to recommend it, is to use data from one survey to develop a preferred model, modifying the model in light of relationships unanticipated by your theory but observed in the data. Then estimate your preferred model using data from a replicated cross-section, for example, in the case of the GSS, the survey conducted in the preceding or following year (recall the discussion of this strategy in Chapter Seven).

A final possibility, in cases where information is collected for more than one individual within a household (either by interviewing more than one household member or by asking a respondent about the characteristics of other household members), is to expand the sample by treating each individual for whom information is available as a separate case. However, in such cases it is necessary to take account of the fact that observations are not independent, by adjusting for clustering within households using survey estimation procedures or by adopting the kind of multilevel modeling strategy proposed by Mason (2001) cited earlier. Moreover, when information is available for a restricted subset of others, for example, spouses, it is important to be attentive to the consequences—for example, by carrying out sensitivity analysis of the differences in conclusions yielded by a sample of married people and a sample of all adults (see the discussion of sensitivity analysis in the next section).

A FINAL NOTE: GOOD PROFESSIONAL PRACTICE

Now that we have considered various issues facing quantitative data analysts and have had a brief introduction to advanced techniques worthy of study, I close by offering several points regarding good professional practice—the things that make a difference between mediocre and superior quantitative data analysis. These are simple principles, available to any analyst; their application does not require particular brilliance or insight or mathematical facility. But attention to them is sure to improve the quality of your work.

Understand the Properties of Your Data

Whether working with data you acquired from another analyst or data from an archive or data you yourself collected, you should thoroughly understand how the data were created and also should explore their properties. Pay particular attention to the sample design to determine whether survey estimation is possible and how to implement it. For the same reason, you need to understand how any weight variables were constructed and how to use them. Often the weights provided by the original investigators are poorly documented. It is entirely appropriate to write to the investigators to ask them how they constructed their weights. You should not regard this as an imposition, even if they do, because one of the responsibilities of those who make their data available for public use is to provide adequate documentation.

You also should calculate and inspect univariate frequency distributions for every variable in the data set, or at least every variable pertinent to your analysis. This is easy to do by using Stata's `-codebook-` command. With respect to each variable, ask yourself whether the observed distribution is plausible, given what you know about the population being studied. It is surprising how informative the inspection of univariate frequency distributions can be. The next step is to create cross-tabulations or tables of means that show the association between each of your central dependent variables and all

IN THE UNITED STATES, PUBLICLY FUNDED STUDIES MUST BE MADE AVAILABLE TO THE RESEARCH COMMUNITY It is now a requirement of both the National Science Foundation (NSF) and the National Institutes of Health (NIH) that sample surveys funded by these agencies be made available for public use in a timely way. The current NIH policy reads, "NIH endorses the sharing of final research data . . . and expects and supports the timely release and sharing of final research data from NIH-supported studies for use by other researchers. 'Timely release and sharing' is defined as no later than the acceptance for publication of the main findings from the final data set" (http://grants.nih.gov/grants/policy/nihgps_2003/NIHGPS_Part7.htm#_Toc54600131, accessed December 9, 2007). The NSF policy statement is less precise but conveys the same principle: "NSF expects . . . investigators to share with other researchers, at no more than incremental cost and within a reasonable time, the data, samples, physical collections and other supporting materials created or gathered in the course of the work. It also encourages awardees to share software and inventions or otherwise act to make the innovations they embody widely useful and usable" (http://www.nsf.gov/pubs/2001/gc101/gc101rev1.pdf, accessed December 9, 2007). Providing adequate documentation is part of the requirement.

the candidate predictor variables. This too can be extremely informative, revealing both deficiencies in the data and deficiencies in your a priori assumptions.

I still recall, with some embarrassment, an incident forty-five years ago when I was a beginning graduate student at the University of Chicago. I worked as a research assistant at the National Opinion Research Center (NORC), and Peter Rossi was the director of NORC. I ran into him one evening as he was leaving the building and carrying a great stack of computer printout—cross-tabs from the study we were working on. I made some snide remark about why should we bother with cross-tabs now that we could do regressions by computer, and he gave me a withering look and said something like, "Live and learn, kid." Of course, he was completely correct. There is a lot to be learned by getting a feel for the data before rushing to estimate fancy, or even not-so-fancy, models.

Explore Alternatives to Your a Priori Hypotheses

One of the features of truly strong research papers is that the author anticipates and explores all of the alternative explanations for the observed phenomenon or relationship that a critic might propose. In nonexperimental work the search for alternative explanations often amounts to assessing the possibility of spurious association due to the failure to include variables that affect both the independent and dependent variables in the model. Thus you need to ask yourself, is there an alternative explanation for the association I observe? In particular, might some other variable be causing both the outcome I observe and the values of my predictor variables? Then, if possible, include the candidate variables in your model, or do a side analysis (even using a different data set) to investigate the association of these variables with variables already in your model.

A nice example of the use of this strategy is a paper by Miller (2007) that explored whether granting women the vote early in the twentieth century resulted in an increase in public health spending and hence a reduction in child mortality. He finds strong evidence in support of his claim. But he recognizes that before accepting the causal argument as valid, he needs to rule out the possibility that suffrage legislation was endogenous to factors that also resulted in an increase in public health spending. He thus devotes a section of his paper (24–28) to various "validity tests" designed to rule out the possibility of alternative explanations for his results.

Where the strategy just discussed is not possible because the potential confounding variables have not been observed, it may be possible to rule them out by noting that if they were operative, their predicted effects would differ from what is observed. For example, in a paper analyzing the determinants of literacy in China (Treiman 2007a), I argued that a divergence in the level of literacy with age for manual and nonmanual workers confirmed "the hypothesis that nonmanual work reinforces literacy over the life course while manual work suppresses it" (146). However, before accepting this conclusion, I had to rule out the possibility that age differences, measured in 1996, simply reflected historical changes in China that produced differences in literacy by cohort. I did this by pointing out that if the quality of education increased (decreased) over time, we should expect an increase (decrease) in literacy for both manual and nonmanual workers rather than the observed pattern of divergence. In a similar way I ruled out the possibility that as the nonmanual sector grew, the average "quality" of both nonmanual and manual workers declined.

Still another option, available under some circumstances, is to sweep unmeasured potential confounders out of the analysis by estimating fixed effects or random effects models as we did in the previous chapter. Still another possibility is to adjust for the effect of potential confounders by using one of the methods for coping with endogeneity and sample-selection bias discussed earlier in this chapter.

Conduct Sensitivity Analysis

Another way to gain confidence—and inspire confidence on the part of your reader—that your results are robust is to conduct *sensitivity analysis,* exploring different functional forms by which you represent relationships in a general linear model framework, different cutting points when carrying out tabular analysis, and—more generally—different ways of representing your concepts. Like consideration of potential omitted-variable bias, this sort of exploration also may require going beyond the data set being analyzed. See, for example, Treiman and Roos (1983, 620–621), who assessed the adequacy of a standard proxy for labor force experience (= age minus years of schooling minus six) by comparing estimates from the *Quality of Employment Survey* based on two alternative measures: the proxy measure and actual labor force experience.

Your hope, of course, is that different specifications yield similar results. But even if not, you need to report all that you have discovered. Remember: the goal is not to "prove" a hypothesis but to discover how the social world actually works. Sometimes this means that we must conclude that our data are not informative because our results are not robust to alternative specifications.

Hout and Hauser (1992) critiqued Erikson and Goldthorpe's important comparative study of social mobility, the *Constant Flux* (1992b), showing that Erikson and Goldthorpe's results are not robust to changes in the model specification, in the statistical procedure used, or in the level of aggregation of their occupational classification. See also Erikson and Goldthorpe's (1992a) response. The exchange provides an illuminating example of why it is important to carry out sensitivity analysis yourself before a critic does it for you. For a striking example of a tendentious and sloppily argued analysis that was thoroughly demolished by the long knives of critics, see Herrnstein and Murray (1994) and important critiques by Heckman (1995), Fischer and others (1996), and Hauser and Huang (1997).

One useful approach is to "bracket" your results, reporting not a point estimate but a range of estimates derived under different assumptions. For example, if it is not clear to you whether, in an attitude scale, a "don't know" response should be coded "missing" or given the middle value, intermediate between a positive and a negative attitude, try it both ways and assess—and, of course, report—the results.

Document Your Work

You should carry out all your analysis using command files (`-do-` files in Stata) and producing a log of your commands and results each time you execute your command file (a `-log-` file in Stata). Moreover, you should use extensive comments in your command files, saying for each bit of analysis what you are doing and why you are doing it. In my own work I go further, adding comments on the results.

This practice has several advantages. First, it provides a record of what you have done. The process of research production in the social sciences from initial idea to published paper often covers a period of several years. Even if you are an efficient person who does one thing at a time and thus are able to execute your analysis from start to finish in a matter of a few weeks, you then have to submit your paper to a journal, which typically will take several months to get back to you, often with a request for revision and resubmission that entails doing additional analysis. At this point you do not want to be in the embarrassing position of not remembering exactly how you carried out the computations to produce the statistics that appear in your tables and graphs and, worse still, not being able to replicate them. If you have a well-documented command file, you will be able to figure out what you have done, and why.

Moreover, you will be able to modify your analysis and create a new set of computations efficiently. Suppose, for example, that the referees suggest that you control for an additional variable. This is a trivial task if you have an existing command file. You simply add the variable to your model and execute the command file. This is far preferable to redoing an entire section of your analysis.

Finally, you will make it possible for others to replicate—or challenge—your work, by archiving your log file so that it is available on demand. You may be tempted to obscure the details of your work so that no one else can discover errors in it. But this is not how science progresses—far better to be clear (even if wrong) than vague. If you are clear about your procedures, you make it possible for others to exactly replicate what you have done and perhaps to figure out how to do it better. Remember, the aim of the game is to advance our collective understanding of social structure and process.

Of course, the gold standard for the production of research papers is that they contain all of the information necessary to exactly replicate the research. Your goal should be to document your work thoroughly enough so that if you handed your paper and your data set to a competent analyst, he or she could reproduce every number in your paper. As laudable as this goal is, however, it tends to be frustrated by journal editors who insist on shortening papers by omitting technical detail. So in addition to describing your technical procedures as clearly as possible in your paper, archiving your log file is very good professional practice.

Do a Last Check for Errors

The last thing you should do before you submit a paper for publication (or as a term paper or a dissertation chapter or post it in a working paper series) is to execute your command

AN "AVAILABLE FROM AUTHOR" ARCHIVE Because claims in published papers that additional materials are "available from author" usually prove to be false, at least after a few months, the California Center for Population Research (CCPR) at UCLA recently implemented a mechanism by which additional materials, for example, `-do-` and `-log-` files, can be attached to papers posted in its Population Working Paper archive. Other research centers are to be encouraged to do the same.

file and then to check *every single number in your paper* against the corresponding numbers in your log file. You will be amazed at the number of discrepancies you find. Because producing a professional paper is typically a lengthy process, it is extremely easy for inconsistencies to creep in. Your goal should be to produce a single command file that contains *all* the computations required for an analysis. Even in cases where you are analyzing more than one data set, you would be well advised to incorporate all your commands into a single file. In this way, you create a single document that produces and explains all of your work. You also minimize the chance that portions of the analysis will fail to be documented or that the documentation will be lost. For the same reason, you should incorporate side computations, even hand computations, into your command file. (In Stata the `-display-` command accommodates this by functioning like a hand calculator.)

The standard to be emulated—at least partly—is the lab notebook conventionally kept in a chemistry lab. Lab notebooks record the conditions under which an experiment was conducted, including the temperature and humidity of the room, whether a reagent was spilled on the floor that day (together with the exact time and description of what was spilled and where), and the outcome of each procedure, whether successful or not. We need not go that far. Nothing much is gained by recording the errors we made in the process of getting our file to execute. But we *should* record analytic dead ends, hypotheses that did not pan out, assumptions that proved to be incorrect, and so on. You will find such commentary enormously helpful when you return to analysis after an absence of months or years, which, as I have noted, is not an uncommon gap. Moreover, by documenting and archiving your analytic dead ends, you may help other analysts.

WHAT THIS CHAPTER HAS SHOWN

In this chapter I have reviewed some general points regarding good research design; have briefly introduced a number of advanced statistical techniques and procedures, which you should pursue in further course work or independent study; have emphasized the value of probability sampling; and have concluded with some advice about good research practice. On the basis of the material we have covered in this book, you are well prepared to do high-quality and rigorous analysis of sample survey and other data. But you should not stop here, because statistical methodology in the social sciences is advancing rapidly, and a first course in data analysis is no longer sufficient to master state-of-the-art techniques, many of which I reviewed in this chapter. I thus urge you to think of this book as the beginning of a career-long commitment to continually expand your tool kit, just as I have done in the more than forty years since completing my PhD. If an old dog like me can learn new tricks, so can you! Have fun!

APPENDIX

DATA DESCRIPTIONS AND DOWNLOAD LOCATIONS FOR THE DATA USED IN THIS BOOK

This appendix describes all of the data sets used to create the worked examples in the book. A common feature of all these surveys is that they are household samples, which means that the data need to be weighted by the reciprocal of the number of adults in the household to convert them into person samples—see the discussion of this issue in Chapter Nine. They are all based on probability samples of households, and the details of the design are given in other sources, indicated in the references included in this appendix.

CHINA

The 1996 survey, *Life Histories and Social Change in Contemporary China* (Treiman, Walder, and Li 2006), was conducted by faculty and students of People's University, Beijing, with funding from the U.S. National Science Foundation (SBR-9423453), the Ford Foundation—Beijing, and the Luce Foundation. The principal investigators were Donald J. Treiman (UCLA), Andrew G. Walder (Stanford), and Qiang Li (then at People's University and now at Qinghua University, Beijing). The survey collected extensive information on respondents' socioeconomic characteristics and educational, occupational, and family histories, and also information on their spouses, parents, children, and other family members.

The survey was based on a stratified national probability sample of the population of China age twenty to sixty-nine, yielding 6,090 cases plus a special sample of 383 village leaders (*village cadres*). Details of the sample design are given in Treiman (1998).

The data and accompanying documentation can be downloaded from the UCLA Social Science Data Archive at http://www.sscnet.ucla.edu/issr/da. Click *Catalog, Index, Asia-China,* and then *Life Histories and Social Change in Contemporary China, 1996.*

EASTERN EUROPE

The survey of *Social Stratification in Eastern Europe after 1989* (Szélenyi and Treiman 1994) consists of six general population surveys, based on probability samples of the adult populations of Bulgaria, the Czech Republic, Hungary, Poland, Russia, and Slovakia, with all the surveys conducted in 1993 except the Polish survey, which was carried out in 1994, and all surveys using an essentially identical questionnaire. Each survey sampled approximately 5,000 adults using a multistage stratified national probability sample design (except that the Polish sample was smaller, approximately 3,500 adults). Details on the survey design can be found in Treiman (1994). These surveys were funded by the U.S. National Science Foundation (SES-9111722 and SBR-9310395), the U.S. National Council for Soviet and Eastern European Research (806-29), the Dutch National Science Foundation (NWO), and various Eastern European governmental agencies. The principal investigators were Ivan Szélenyi and Donald J. Treiman, at that time both at UCLA.

The focus of the data collection was on the effect of the collapse of communism on life chances. Extensive information was collected on respondents' socioeconomic characteristics and educational, occupational, residential, and family histories, and also information on their spouses, parents, children, and other family members. In addition, a good deal of political information was collected, as well as information that permitted a contrast between 1988 and 1993.

The data and accompanying documentation can be downloaded from the UCLA Social Science Data Archive at http://www.sscnet.ucla.edu/issr/da. Click *Catalog, Index, Europe-Bulgaria,* and then *Social Stratification in Eastern Europe After 1989: General Population Survey.*

The survey of elites that was carried out in each of the six nations at the same time as the general population survey and analyzed in Chapter Thirteen is not currently available for public distribution due to the difficulty of protecting the confidentiality of responses. The difficulty with an elite survey, of course, is that individuals are fairly readily identifiable from details of their biographies.

SOUTH AFRICA

The Survey of Socioeconomic Opportunities and Achievement in South Africa (Treiman, Moeno, and Schlemmer 1994) is a multistage national probability sample survey of all races in "greater South Africa" carried out in the early 1990s in several stages between 1991 and 1994. Greater South Africa refers to what was historically, and is currently, the South African nation; that is, it includes the "TVBC States" that at the time of data collection were nominally independent puppet states hived off by the *apartheid* regime (see Treiman [2007b] for a brief history). The sample consists of a general population sample of 8,714 adults and a Black elite sample of 372 adults. The principal investigators were Donald J. Treiman and two South African sociologists, Sylvia N. Moeno and Lawrence Schlemmer. See Treiman, Lewin, and Lu (2006) for details on the survey design.

Extensive information was collected on respondents' socioeconomic characteristics and educational, occupational, residential, and family histories, and also information on their spouses, parents, children, and other family members.

The data and accompanying documentation can be downloaded from the UCLA Social Science Data Archive at http://www.sscnet.ucla.edu/issr/da. Click *Catalog, Index, Africa-South Africa*, and then *Survey of Socioeconomic Opportunities and Achievement*.

U.S. GENERAL SOCIAL SURVEY

Funded by the U.S. National Science Foundation, the *General Social Survey* (GSS; Davis, Smith, and Marsden 2007) is a repeated cross-sectional survey, with data collected from a national multistage probability of U.S. adults—about 1,500 people approximately each year from 1972 through 1991 and then, beginning in 1994, about 3,000 people every other year. The principal investigators are James A. Davis and Tom W. Smith and, in recent years, Peter V. Marsden. Appendix B provides details on the sample design as it has changed over the years.

The GSS is intended to be an all-purpose survey, to permit analysis of the attitudes, behavior, and characteristics of the U.S. population by those who cannot afford the massive resources required to mount a national probability sample survey. As it has matured, it has become an increasingly valuable vehicle for the study of social change, especially changes in attitudes. The strategy of the GSS is to repeat a substantial portion of the questionnaire year after year to permit the analysis of changes over time but also to incorporate new questions that are responsive to changing conditions and concerns.

The data may be downloaded from the National Opinion Research Center at the University of Chicago, the producer of the GSS: http://www.norc.org/GSS+Website/Download. It is also possible to do data analysis using this site, without actually downloading the data. An alternative site, which also permits both data analysis and downloading, is the SDA Archive at the University of California-Berkeley: http://sda.berkeley.edu/archive.htm. For information regarding how to download or purchase the documentation, see http://www.gss.norc.org.

APPENDIX

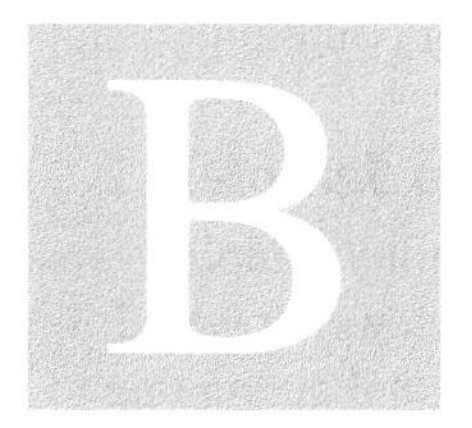

SURVEY ESTIMATION WITH THE GENERAL SOCIAL SURVEY

INTRODUCTION

The *General Social Survey* (GSS) uses a stratified multistage probability sample (described in Davis, Smith, and Marsden [2007, Appendix A], and the sources cited there), which means that correct estimates of standard errors require survey estimation procedures. Unfortunately, the GSS documentation is not complete, presumably to maintain confidentiality; only the primary sampling units (PSUs) are identified by the *SAMPCODE* variable; neither the strata nor the secondary sampling units are identified. This is unfortunate because it precludes fully exploiting Stata's procedures for adjusting for multistage stratified sampling. Also, information is not provided that would permit a finite population correction, although this limitation is not important, given the large number of units in the population sampled at each stage. Moreover, the sample design has changed over the years, with a new sampling frame created each decade based on the decennial census results and additional major changes in 1976, from a block quota design to a full probability sample design; in 2004, with the introduction of a partial list-based sample using the U.S. Postal Service address list, an aggressive effort to convert a subset of initial nonresponses, and post-enumeration adjustments for differential nonresponse; and in 2006, with the introduction of a Spanish language sample. Finally, in 1982 and 1987 Blacks were oversampled. These changes complicate pooling data across years.

ANALYZING DATA FROM A SINGLE YEAR

Here are my recommendations regarding the use of survey estimation under various scenarios. I start with suggestions for analyzing data from a single year and then, in the following section, offer suggestions for combining data from more than one year.

The 1972 to 1976 Block Quota Samples

In the early years of the GSS, probability sampling was carried out down to the block level. Then, within each block, the interviewer traveled in a specified way around the block, approaching each household in turn and attempting to achieve interviews with a specified number of people with particular combinations of age, sex, and employment status. If a person meeting the criteria was found, that person was interviewed, and the interviewer then moved on to the next household (that is, only one interview was permitted per household), continuing until the required number of interviews in each combination of categories was completed. In 1975 and 1976, the sampling procedure was split: approximately half the interviews were conducted using the block quota method, and half were conducted using a full probability sample. Subsequent to 1976, a full probability sample design has been employed. Using the 1975 and 1976 data, Stephenson (1979) compared the two sampling procedures and concluded that the block quota sample substantially under-represented fully employed men and somewhat over-represented individuals living in single-adult households (which, in turn, resulted in under-representation of married people and Catholics). He found little evidence of bias with respect to the many other variables in the GSS.

Although procedures for doing statistical inference with quota samples exist, either using information from repeated surveys or doing within-survey comparisons (Stephan and McCarthy 1963, Chapter 10), these are somewhat cumbersome to implement. Because the GSS block quota samples have average design effects of about 1.5, approximately as large as GSS multistage probability samples (Davis and Smith 1992; Davis, Smith, and Marsden 2007, 2,097), a reasonable approach is to treat the block quota samples as if they are probability samples and apply the same survey estimation procedures as for the true probability samples subsequent to 1976.

With respect to the bias in coverage in the GSS block quota samples, you can choose to ignore it or to do post-enumeration weighting based on the distributions observed in the 1970 census. My suggestion is that you not bother to do post-enumeration weighting. Rather, simply include gender and employment status in your analysis, which will yield unbiased estimates of the effects of other variables. However, you do need to be cautious in your interpretation of descriptive statistics, investigating whether the attributes you are describing differ between employed males and others and, if so, perhaps bracketing the summary statistics by weighting the data to inflate the number of employed males. Then report the original and inflated estimates as an estimate of the range within which the true values of the statistics fall.

The 1977 to 2002 Surveys, Except 1982 and 1987

With the exception of the 1982 and 1987 surveys, which included over-samples of Blacks, the full probability samples used in the 1977–2002 surveys can all be treated in a standard way. You need to adjust for the fact that the GSS, like most household surveys, is a probability sample of households rather than people. But the eligible population consists of all adults (people age eighteen and over) who are capable of responding to an interview. Because households are randomly sampled within small areas but only one randomly chosen adult per household is interviewed, adults living in households with many adults have a smaller chance of being included in the sample than do adults living in households with few adults. A reasonable way to convert the sample of households into a sample of people is to weight each respondent by the ratio of the number of adults in the household to the mean number of adults in all households in the sample. This can be accomplished in Stata by constructing a household weight variable, *HHWT:*

```
egen adultm = mean(adults)

gen hhwt = adults/adultm
```

and weighting your data by this variable. (In fact, because Stata renorms probability weights to the original sample size, you can simply use the *ADULTS* variable as your weight variable unless you are using this variable as a component in a more complex weight variable—shown in the next section—in which case you should use *HHWT* as the component.)

Although the GSS is a multistage sample with two, and for some PSUs three, stages, the documentation only provides information on the primary sampling units (metropolitan areas and nonmetropolitan counties) and no information on strata (based on region, size of place, and race/ethnicity). This means that we can go only part way to adjusting for clustering in the GSS sample design. Here are the Stata commands that will accomplish this, using the GSS PSU variable, *SAMPCODE:*

```
svyset sampcode [pweight=adults]
```

or

```
svyset sampcode [pweight=hhwt]
```

Note that this command also adjusts for differential household size.

The 1982 and 1987 Surveys with Oversamples of Blacks

If you are analyzing the 1982 or 1987 surveys and want to compute descriptive statistics, you need to adjust for the fact that Blacks were oversampled. To adjust for both the oversampling of blacks and differential household size, create a new weight variable that is the product of the *OVERSAMP* weight variable provided by the GSS and the weight variable you constructed to correct for differential household size—that is

```
gen newwt = hhwt*oversamp
```

(Note that the mean of this new variable is 1.0.) Then set your data for survey analysis:

```
svyset sampcode [pweight=newwt]
```

The 2004 and 2006 Surveys

In the 2004 GSS, NORC introduced a radically new sampling procedure that exploits the availability of a list of addresses maintained by the U.S. Postal Service, which in 2004 covered 72 percent of households (O'Muircheartaigh 2003). For areas covered by the Postal Service list, it was possible to go directly from PSUs to small areas—in essence, from the PSU to the tertiary sampling unit. A second innovation was an aggressive effort to convert a random half of initial nonrespondents to respondents. The second innovation necessitated a change in the way the GSS data were weighted to make them representative of the population—the converted cases had to be weighted by twice the weight of the remaining cases because only half were followed up. The variable *WTSS* adjusts both for this and for differential household size.

Note that in the original version of the 2004 data, this variable is named *WTSS2004.* In the 2006 data and in the 1972–2006 cumulative file, this variable appears twice, as *WTSS* for years 2004 and 2006 and as *WTSSALL* for all years; for 2004 and 2006 the two variables are identical. Thus, depending on which version of the earlier files you are using, you may have to rename *WTSS2004* or *WTSS* to *NEWWT* (or whatever other name you give to your constructed weight variable) to have a comparable weight variable for all years.

If you are carrying out analysis using only the 2004 or the 2006 data, or pooled data for the two years, you may want to use *WTSS2004NR* or *WTSSNR*, which adjusts the 2004 and 2006 data for differential nonresponse across geographic areas. Note that by using this weight variable, you are implicitly assuming that nonrespondents are identical to respondents from the same geographical area, which may or may not be reasonable. If you think that differential nonresponse is due mainly to differences across areas in the quality of the NORC field operation, such an adjustment would be reasonable. But if you think it is due to characteristics of respondents (for example, the difference between an area with many restricted access buildings such as Manhattan, or many people who work at night or on the weekend, or many people who live alone, and the remaining places), you might conclude that nonrespondents cannot be assumed to be similar to respondents. In this case, you would not want to use the ". . . NR" weights. The latter is the more reasonable assumption.

As noted previously, in 2006, interviews were conducted in Spanish as well as in English, whereas in previous years those unable to manage an interview in English were excluded. About six percent of respondents were interviewed in Spanish, and eighty-five percent of these were judged by interviewers to be unable to communicate effectively in English. The variable SPANENG identifies the language of the interview.

The FORMWT Variable

In some years (1978, 1980, and 1982–1985), some questions were asked only of a subsample of respondents. Although the intent was to administer the questions to a random subset, this was not always realized (Smith and Peterson 1986). Thus, the GSS offers a correction weight, *FORMWT*. My recommendation is that you not use this variable but rather do multiple imputation (see Chapter Eight) to create a complete data set that includes all respondents.

POOLING SURVEYS FROM MORE THAN ONE YEAR

When we pool surveys from more than one year, it is reasonable to treat *YEAR* as the stratum variable because the surveys from each year are independent, and *YEAR* is a fixed variable. The Stata command to accomplish this is

```
svyset sampcode [pweight=newwt],strata(year)
```

REFERENCES

Abadie, Alberto, and Javier Gardeazabal. 2003. The economic costs of conflict: A case study of the Basque country. American Economic Review 93(1):113–132.

———, David Drukker, Jane Leber Herr, and Guido W. Imbens. 2004. Implementing matching estimators for average treatment effects in Stata. Stata Journal 4(3):290–311.

Agresti, Alan. 2002. Categorical data analysis. 2nd ed. New York: Wiley-Interscience.

Allison, Paul D. 2001. Missing data. Sage university papers series on quantitative applications in the social sciences, 07-136. Thousand Oaks, CA: Sage.

———. 2005. Fixed effects regression methods for longitudinal data using SAS. Cary, NC: SAS Institute Inc.

Almond, Douglas. 2006. Is the 1918 influenza pandemic over? Long-term effects of in utero influenza exposure in the post-1940 U.S. population. Journal of Political Economy 114(4):672–712.

Aly, Hassan Y., and Michael P. Shields. 1991. Son preference and contraception in Egypt. Economic Development and Cultural Change 39(2):353–370.

Anderson, Andy B., Alexander Basilevsky, and Derek P. J. Hum. 1983. Missing data: A review of the literature. In Handbook of survey research, ed. Peter H. Rossi, James D. Wright, and Andy B. Anderson, 415–494. New York: Academic Press.

Andrews, Frank M., James N. Morgan, John A. Sonquist, and Laura Klem. 1973. Multiple classification analysis: A report on a computer program for multiple regression using categorical predictors. 2nd ed. Ann Arbor: University of Michigan, Institute for Social Research.

Angrist, Joshua D. 1990. Lifetime earnings and the Vietnam era draft lottery: Evidence from Social Security Administration records. American Economic Review 80(3):313–336. (See also the errata, 80[5]:1284–1286.)

Anscombe, F. J. 1973. Graphs in statistical analysis. American Statistician 27(1):17–22.

Ansolabehere, Stephen, James M. Snyder Jr., and Charles Stewart III. 2000. Old voters, new voters, and the personal vote: Using redistricting to measure the incumbency advantage. American Journal of Political Science 44(1):17–34.

Ashenfelter, Orley, and Alan Krueger. 1994. Estimates of the economic return to schooling from a new sample of twins. American Economic Review 84(5):1157–1173.

Astone, Nan Marie, Robert Schoen, Margaret Ensminger, and Kendra Rothert. 2000. School reentry in early adulthood: The case of inner-city African Americans. Sociology of Education 73(3):133–154.

Baltagi, Badi H. 2005. Econometric analysis of panel data. 3rd ed. New York: Wiley.

Barkan, Steven D., and Susan F. Greenwood. 2003. Religious attendance and subjective well-being among older Americans: Evidence from the General Social Survey. Review of Religious Research 45(2):116–129.

Barker, D.J.P. 1998. Mothers, babies, and health in later life. Edinburgh: Churchill Livingstone.

Barnes, Samuel H., and Max Kaase. 1979. Political action: An eight nation study, 1973–1976. Machine-readable data file. Samuel H. Barnes and Max Kaase [principal investigators]. Ann Arbor, MI: Inter-University Consortium for Political and Social Research [distributor].

Baum, Christopher F. 2006. An introduction to modern econometrics using Stata. College Station, TX: Stata Press.

Becker, Howard S. 1986. Writing for social scientists: How to start and finish your thesis, book, or article. Chicago: University of Chicago Press.

Becker, Mark P., and Clifford C. Clogg. 1989. Analysis of sets of two-way contingency tables using association models. Journal of the American Statistical Association 84(405):142–151.

Becker, Sascha O., and Marco Caliendo. 2007. Sensitivity analysis for average treatment effects. Stata Journal 7(1):71–83.

———, and Andrea Ichino. 2002. Estimation of average treatment effects based on propensity scores. Stata Journal 2(4):358–377.

Berelson, Bernard. 1979. Romania's 1966 anti-abortion decree: The demographic experience of the first decade. Population Studies 33(2):209–222.

Berk, Richard A. 1983. An introduction to sample selection bias in sociological data. American Sociological Review 48(3):386–398.

———. 1990. A primer on robust regression. In Modern methods of data analysis, ed. John Fox and J. Scott Long, 292–324. Newbury Park, CA: Sage.

———, and Phyllis J. Newton. 1985. Does arrest really deter wife battery? An effort to replicate the findings of the Minneapolis Spouse Abuse Experiment. American Sociological Review 50(2):253–262.

———, and Subhash C. Ray. 1982. Selection biases in sociological data. Social Science Research 11(4):352–398.

Bielby, William T., Robert M. Hauser, and David L. Featherman. 1977. Response errors of Black and Nonblack males in models of the intergenerational transmission of socioeconomic status. American Journal of Sociology 82(6):1242–1288.

Bishop, Yvonne M. M., Stephen E. Fienberg, and Paul W. Holland. 1975. Discrete multivariate analysis: Theory and practice. Cambridge, MA: MIT Press.

Black, Dan A., and Jeffrey A. Smith. 2003. How robust is the evidence on the effects of college quality? Evidence from matching. Journal of Econometrics 121(1–2):99–124.

Blau, Peter M., and Otis Dudley Duncan. 1967. The American occupational structure. New York: Wiley.

Blinder, Alan S. 1973. Wage discrimination: Reduced form and structural estimates. Journal of Human Resources 8(4):436–455.

Bollen, Kenneth A. 1989. Structural equations with latent variables. New York: Wiley.

———, and Jennie E. Brand. 2008. Fixed and random effects in panel data using structural equation models. Population Working Paper PWP-CCPR-2008-003, California Center for Population Research, University of California, Los Angeles.

———, and Patrick J. Curran. 2006. Latent Curve models: A structural equation perspective. New York: Wiley.

———, and Robert W. Jackman. 1990. Regression diagnostics: An expository treatment of outliers and influential cases. In Modern methods of data analysis, ed. John Fox and J. Scott Long, 257–291. Newbury Park, CA: Sage.

———, and J. Scott Long, eds. 1993. Testing structural equation models. Newbury Park, CA: Sage.

Boskin, Michael J. 1974. A conditional logit model of occupational choice. Journal of Political Economy 82(2, Part 1):389–398.

Brand, Jennie E. 2006. The effects of job displacement on job quality: Findings from the Wisconsin Longitudinal Study. Research in Social Stratification and Mobility 24(3):275–298.

———, and Charles Halaby. 2006. Regression and matching estimates of the effects of elite college attendance on educational and career achievement. Social Science Research 35(3):749–770.

———, and Yu Xie. 2007. Who benefits most from college? Evidence for negative selection in heterogeneous economic returns to higher education. Population Working Paper PWP-CCPR-2007-035, California Center for Population Research, University of California, Los Angeles.

Breen, Richard. 1996. Regression models: Censored, sample selected, or truncated data. Sage university papers series on quantitative applications in the social sciences, 07-111. Thousand Oaks, CA: Sage.

———, and Jan O. Jonsson. 2000. Analyzing educational careers: A multinomial transition model. American Sociological Review 65(5):754–772.

Breslow, Norman E. 1996. Statistics in epidemiology: The case-control study. Journal of the American Statistical Association 91(433):14–28.

Brick, J. Michael, and Graham Kalton. 1996. Handling missing data in survey research. Statistical Methods in Medical Research 5(3):215–238.

Budig, Michelle J., and Paula England. 2001. The wage penalty for motherhood. American Sociological Review 66(2):204–225.

Burgess, Eric. 1978. To the red planet. New York: Columbia University Press.

Buttenheim, Alison M. 2006a. Microfinance programs and contraceptive use: Evidence from Indonesia. Population Working Paper PWP-CCPR-2006-020, California Center for Population Research, University of California, Los Angeles.

———. 2006b. Flood exposure and child health in Bangladesh. Population Working Paper PWP-CCPR-2006-022, California Center for Population Research, University of California, Los Angeles.

Cameron, A. Colin, and Pravin K. Trivedi. 1998. Regression analysis of count data. Econometric Society Monographs, 30. New York: Cambridge University Press.

Cameron, Lisa. 2000. The residency decision of elderly Indonesians: A nested logit analysis. Demography 37(1):17–27.

Campbell, Cameron, and James Z. Lee. 2005. Deliberate fertility control in late imperial China: Spacing and stopping in the Qing imperial lineage. Population Working Paper PWP-CCPR-2005-041, California Center for Population Research, University of California, Los Angeles.

Campbell, Donald T. 1957. Factors relevant to the validity of experiments in social settings. Psychological Bulletin 54(4):297–312.

———, and David A. Kenny. 1999. A primer on regression artifacts. New York: Guilford Press.

———, and H. Laurence Ross. 1968. The Connecticut crackdown on speeding: Time-series data in quasi-experimental analysis. Law and Society Review 3(1):33–53.

———, and Julian C. Stanley. 1966. Experimental and quasi-experimental designs for research. Chicago: Rand McNally.

Carmines, Edward G., and Richard A. Zeller. 1979. Reliability and validity assessment. Sage university papers series on quantitative applications in the social sciences, 07-017. Beverly Hills, CA: Sage.

Chamberlain, G. 1980. Analysis of covariance with qualitative data. Review of Economic Studies 47(1):225–238.

Chattopadhyay, Arpita, Michael J. White, and Cornelius Debpuur. 2006. Migrant fertility in Ghana: Selection versus adaptation and disruption as causal mechanisms. Population Studies 60(2):189–203.

Chen, Susan, and David K. Guilkey. 2003. Determinants of contraceptive method choice in rural Tanzania between 1991 and 1999. Studies in Family Planning 34(4):263–276.

Chen, Yiu Por, and Zai Liang. 2007. Educational attainment of migrant children: The forgotten story of China's urbanization. In Education and reform in China, ed. Emily Hannum and Albert Park, 117–132. Oxford: Routledge.

Clark, T. G., and D. G. Altman. 2003. Developing a prognostic model in the presence of missing data: An ovarian cancer study. Journal of Clinical Epidemiology 56(1):28–37.

Clogg, Clifford C. 1982. Using association models in sociological research: Some examples. American Journal of Sociology 88(1):114–134.

Cohen, Gidon. 2005. Propensity score methods and the Lenin School. Journal of Interdisciplinary History 36(2):209–232.

Cohen, Jacob, and Patricia Cohen. 1975. Applied multiple regression/correlation analysis for the behavioral sciences. Hillsdale, NJ: Lawrence Erlbaum Associates.

Daula, Thomas, D. Alton Smith, and Ray Nord. 1990. Inequality in the military: Fact or fiction? American Sociological Review 55(5):714–718.

Davis, James A. 1971. Elementary survey analysis. Englewood Cliffs, NJ: Prentice Hall.

———. 1974. Hierarchical models for significance tests in multivariate contingency tables: An exegesis of Goodman's recent papers. Sociological Methodology 5:189–231.

———, and Tom W. Smith. 1992. The NORC General Social Survey: A user's guide. Guides to major social science data bases 1. Thousand Oaks, CA: Sage.

———, Tom W. Smith, and Peter V. Marsden. 2007. General Social Surveys, 1972–2006 cumulative file [computer file]. Principal investigator, James A. Davis; director and coprincipal investigator, Tom W. Smith; coprincipal investigator, Peter V. Marsden. Chicago: National Opinion Research Center [producer]; Storrs, CT: The Roper Center for Public Opinion Research, University of Connecticut; Ann Arbor, MI: Inter-University Consortium for Political and Social Research [distributors].

Dawson, Deborah A. 2000. The link between family history and early onset alcoholism: Earlier initiation of drinking or more rapid development of dependence? Journal of Studies on Alcohol 61(5):637–646.

Dehejia, Rajeev H., and Sadek Wahba. 2002. Propensity score-matching methods for nonexperimental causal studies. Review of Economics and Statistics 84(1):151–161.

Deng, Zhong, and Donald J. Treiman. 1997. The impact of the Cultural Revolution on trends in educational attainment in the People's Republic of China. American Journal of Sociology 103(2):391–428.

DiPrete, Thomas A., and Jerry D. Forristal. 1994. Multilevel models: Methods and substance. Annual Review of Sociology 20:331–357.

———, and David B. Grusky. 1990a. Structure and trend in the process of stratification for American men and women. American Journal of Sociology 96(1):107–143.

———, and David B. Grusky. 1990b. The multilevel analysis of trends with repeated cross-sectional data. Sociological Methodology 20:337–368.

Domanski, Henryk. 2008. A new dimension of social stratification in Poland? Class membership and electoral voting in 1991–2001. European Sociological Review 24(2):169–182.

Downey, Douglas B. 1995. When bigger is not better: Family size, parental resources, and children's educational performance. American Sociological Review 60(5):746–761.

Dubin, Jeffrey, and Douglas Rivers. 1989. Selection bias in linear regression, logit and probit models. Sociological Methods and Research 18(2–3):360–390.

Duncan, Beverly. 1965. Family factors and school dropout: 1920–1960. Cooperative Research Project 2258 (with the U.S. Office of Education). Ann Arbor, MI: Population Studies Center.

Duncan, Otis Dudley. 1968. Inheritance of poverty or inheritance of race? In On understanding poverty: Perspectives from the social sciences, ed. Daniel Patrick Moynihan, 85–110. New York: Basic Books.

———. 1975. Introduction to structural equation models. New York: Academic Press.

———. 1979. How destination depends on origin in the occupational mobility table. American Journal of Sociology 84(4):793–803.

———. 1984. Notes on social measurement: Historical and critical. New York: Russell Sage Foundation.

———, Archibald O. Haller, and Alejandro Portes. 1968. Peer influences on aspirations: A reinterpretation. American Journal of Sociology 74(2):119–137.

Eliason, Scott R. 1993. Maximum likelihood estimation: Logic and practice. Sage university papers series on quantitative applications in the social sciences, 07-096. Newbury Park, CA: Sage.

Eltinge, John L., and William M. Sribney. 1996. sv1: Some basic concepts for design-based analysis of complex survey data. In Stata technical bulletin reprints, vol. 6, ed. H. Joseph Newton, 208–213. College Station, TX: Stata Corporation.

Entwisle, Barbara, and William M. Mason. 1985. Multilevel effects of socioeconomic development and family planning programs on children ever born. American Journal of Sociology 91(3):616–649.

Erikson, Robert, and John H. Goldthorpe. 1987a. Commonality and variation in social fluidity in industrial nations. Part I: A model for evaluating the "FJH hypothesis." European Sociological Review 3(1):54–77.

———. 1987b. Commonality and variation in social fluidity in industrial nations. Part II: The model of core social fluidity applied. European Sociological Review 3(2):145–166.

———. 1992a. The CASMIN project and the American dream. European Sociological Review 8(3):283–305.

———. 1992b. The constant flux: A study of class mobility in industrial societies. Oxford: Clarendon.

Ettner, Susan. 2004. Methods for addressing selection bias in observational studies. Text version of a slide presentation at a National Research Service Award Trainees Research Conference. Agency for Healthcare Research and Quality, Rockville, MD. http://www.ahrq.gov/fund/training/ettnertxt.htm (accessed 25 December 2007).

Evans, M.D.R., and Jonathan Kelley. 2004. Australian economy and society 2002: Religion, morality, and public policy in international perspective, 1984–2002. Sydney: Federation Press.

Evans, M.D.R., Jonathan Kelley, Joanna Sikora, and Donald J. Treiman. 2005. Scholarly culture and educational success in 27 nations. Revised version of a paper presented at the World Congress of Sociology, Brisbane, Australia, July 2002.

Fair, Ray C. 1978. A theory of extramarital affairs. Journal of Political Economy 86(1):45–61.

Featherman, David L., and Robert M. Hauser. 1978. Opportunity and change. New York: Academic Press.

Fienberg, Stephen E. 1980. The analysis of cross-classified categorical data. 2nd ed. Cambridge, MA: MIT Press.

Fischer, Claude S., Michael Hout, Martín Sánchez-Jankowski, Samuel R. Lucas, Anne Swidler, and Kim Voss. 1996. Inequality by design: Cracking "The Bell Curve" myth. Princeton, NJ: Princeton University Press.

Fisher, Ronald A. [1925] 1970. Statistical methods for research workers. 14th ed. Edinburgh: Oliver and Boyd.

Fitzgerald, John M., and David C. Ribar. 2004. Welfare reform and female headship. Demography 41(2):189–212.

Fox, John. 1991. Regression diagnostics. Sage university papers series on quantitative applications in the social sciences, 07-079. Newbury Park, CA: Sage.

———. 1997. Applied regression analysis, linear models, and related methods. Thousand Oaks, CA: Sage.

———, and Georges Monette. 1992. Generalized collinearity diagnostics. Journal of the American Statistical Association 87(417):178–183.

Frankenberg, Elizabeth, and William M. Mason. 1995. Maternal education and health-related behaviors: A preliminary analysis of the 1993 Indonesian family life survey. Journal of Population 1(1):21–44.

———, and Duncan Thomas. 2001. Women's health and pregnancy outcomes: Do services make a difference? Demography 38(2):253–265.

Fu, Vincent Kang. 1998. sg88: Estimating generalized ordered logit models. In Stata technical bulletin reprints, vol. 8, ed. H. Joseph Newton, 160–164. College Station, TX: Stata Corporation.

———. 2001. Racial intermarriage pairings. Demography 38(2):147–159.

Gamoran, Adam, and Robert D. Mare. 1989. Secondary school tracking and stratification: Compensation, reinforcement, or neutrality? American Journal of Sociology 94(5):1146–1183.

Ganzeboom, Harry B. G., Paul de Graaf, and Donald J. Treiman. 1992. An international scale of occupational status. Social Science Research 21(1):1–56.

———, Ruud Luijkx, and Donald J. Treiman. 1989. Intergenerational class mobility in comparative perspective. Research in Social Stratification and Mobility 8:3–84.

———, and Donald J. Treiman. 1996. Internationally comparable measures of occupational status for the 1988 international standard classification of occupations. Social Science Research 25(3):201–239.

Gaziano, Cecilie. 2005. Comparative analysis of within-household respondent selection techniques. Public Opinion Quarterly 69(1):124–157.

Gelman, Andrew, and Donald B. Rubin. 1995. Avoiding model selection in Bayesian social research. Sociological Methodology 25:165–173.

Gerber, Theodore P. 2000. Membership benefits or selection effects? Why former Communist Party members do better in post-Soviet Russia. Social Science Research 29(1):25–50.

———. 2001. The selection theory of persisting party advantages in Russia: More evidence and implications. Social Science Research 30(4):653–671.

Geronimus, Arline T., and Sanders Korenman. 1992. The socioeconomic consequences of teen childbearing reconsidered. Quarterly Journal of Economics 107(4):1187–1214.

Gilbert, G. Nigel. 1981. Modeling society: An introduction to loglinear analysis for social researchers. London: George Allen and Unwin.

Goldberger, Arthur S. 1968. Topics in regression analysis. London: Macmillan.

Goldstein, Harvey. 2003. Multilevel statistical models. 3rd ed. London: Arnold.

Goodman, Leo A. 1972. A general model for the analysis of surveys. American Journal of Sociology 77(6):1035–1086.

———. 1978. Analyzing qualitative/categorical data: Log-linear models and latent-structure analysis. Cambridge, MA: Abt Books.

———. 1979. Simple models for the analysis of association in cross-classifications having ordered categories. Journal of the American Statistical Association 74(367):537–552.

———. 1984. The analysis of cross-classified data having ordered categories. Cambridge, MA: Harvard University Press.

———, and Michael Hout. 1998. Statistical methods and graphical displays for analyzing how the association between two qualitative variables differs among countries, among groups, or over time: A modified regression-type approach. In Sociological Methodology 1998, ed. Adrian E. Raftery, 175–230. Washington, DC: American Sociological Association. (See also the comments by Xie and Yamaguchi, and the reply.)

Gould, William W. 1993. sg19: Linear splines and piecewise linear functions. In Stata technical bulletin reprints, vol. 3, ed. Sean Becketti, 98–104. College Station, TX: StataCorp.

———. 2000. sg124: Interpreting logistic regression in all its forms. In Stata technical bulletin reprints, vol. 9, ed. H. Joseph Newton, 257–270. College Station, TX: StataCorp.

———, and William Scribney. 1999. Maximum likelihood estimation with Stata. College Station, TX: Stata Press.

Greenberg, David F. 1991. Modeling criminal careers. Criminology 29(1):17–46.

Greene, William H. 2008. Econometric analysis. 6th ed. Upper Saddle River, NJ: Prentice Hall.

Grusky, David B., and Robert M. Hauser. 1984. Comparative social mobility revisited: Models of convergence and divergence in 16 countries. American Sociological Review 49(1):19–38.

Grusky, David B., and Thomas A. DiPrete. 1990. Recent trends in the process of stratification. Demography 27(4):617–637.

Hagan, John, and Patricia Parker. 1985. White-collar crime and punishment. American Sociological Review 50(3):302–316.

Halaby, Charles N. 2004. Panel models in sociological research: theory into practice. Annual Review of Sociology 30:507–544.

Hamerle, A., and G. Ronning. 1995. Panel analysis for qualitative variables. In Handbook of statistical modeling for the social and behavioral sciences, ed. Gerhard Arminger, Clifford C. Clogg, and Michael E. Sobel, 401–451. New York: Plenum.

Hamilton, Lawrence C. 1992a. srd1: How robust is robust regression? In Stata technical bulletin reprints, vol. 1, ed. Joseph Hilbe, 169–175. Santa Monica, CA: Computing Resource Center.

———. 1992b. Regression with graphics: A second course in applied statistics. Belmont, CA: Duxbury Press.

———. 2006. Statistics with Stata (updated for version 9). Belmont, CA: Brooks/Cole.

Hanushek, Eric A., and John E. Jackson. 1977. Statistical methods for social scientists. New York: Academic Press.

Harding, David. 2003. Counterfactual models of neighborhood effects: The effect of neighborhood poverty on dropping out and teenage pregnancy. American Journal of Sociology 109(3):676–719.

Hardy, Melissa A. 1989. Estimating selection effects in occupational mobility in a 19th-century city. American Sociological Review 54(5):834–843.

———. 1993. Regression with dummy variables. Sage university papers series on quantitative applications in the social sciences, 07-093. Thousand Oaks, CA: Sage.

Hargens, Lowell L. 1976. A note on standardized coefficients as structural parameters. Sociological Methods and Research 5(2):247–256.

Hauser, Robert M. 1978. A structural model of the mobility table. Social Forces 56(3):919–953.

———. 1980. Some exploratory methods for modeling mobility tables and other cross-classified data. Sociological Methodology 11:413–458.

———. 1995. Better rules for better decisions. Sociological Methodology 25:175–83.

———, and Min-Hsiung Huang. 1997. Verbal ability and socioeconomic success: A trend analysis. Social Science Research 26(3):331–376.

———, Shu-Ling Tsai, and William H. Sewell. 1983. A model of stratification with response error in social and psychological variables. Sociology of Education 56(1):20–46.

Hausman, Jerry. 1978. Specification tests in econometrics. Econometrica 46(6):1251–1271.

———, and Daniel McFadden. 1984. Specification tests for the multinomial logit model. Econometrica 52(5):1219–1240.

Haynes, Stephen E., and David Jacobs. 1994. Macroeconomics, economic stratification, and partisanship: A longitudinal analysis of contingent shifts in political identification. American Journal of Sociology 100(1):70–103.

Heckman, James J. 1979. Sample selection bias as a specification error. Econometrica 47(1):153–161.

———. 1995. Review: Lessons from The Bell Curve. Journal of Political Economy 103(5):1091–1120.

Henderson, Gail, John Akin, Li Zhiming, Jin Shuigao, Ma Haijiang, and Ge Keyou. 1994. Equity and the utilization of health services: Report of an eight-province survey in China. Social Science and Medicine 39(5):687–699.

Herrnstein, Robert J., and Charles Murray. 1994. The bell curve: Intelligence and class structure in American life. New York: Free Press.

Hendrickx, John. 1999. dm73: Using categorical variables in Stata. In Stata technical bulletin reprints, vol. 9, ed. H. Joseph Newton, 51–59. College Station, TX: Stata Corporation.

———. 2000. dm73.1: Contrasts for categorical variables: Update. In Stata technical bulletin reprints, vol. 9, ed. H. Joseph Newton, 60–61. College Station, TX: Stata Corporation.

———. 2001a. dm73.2: Contrasts for categorical variables: Update. In Stata technical bulletin reprints, vol.10, ed. H. Joseph Newton, 9–14. College Station, TX: Stata Corporation.

———. 2001b. dm73.3: Contrasts for categorical variables: Update. In Stata technical bulletin reprints, vol. 10, ed. H. Joseph Newton, 14–15. College Station, TX: Stata Corporation.

Hildebrand, David K., James D. Laing, and Howard Rosenthal. 1977. Analysis of ordinal data. Sage university papers series on quantitative applications in the social sciences, 07-008. Beverly Hills, CA: Sage.

Hoffman, Saul D., and Greg J. Duncan. 1988. Multinomial and conditional logit discrete-choice models in demography. Demography 25(3):415–427.

Hofmeyr, Julian F., and Robert E. Lucas. 2001. The rise in union wage premiums in South Africa. Labour 15(4):685–719.

Holland, Paul W. 1986. Statistics and causal inference (with comments by Rubin, Cox, Glymour, and Granger, and a rejoinder by Holland). Journal of the American Statistical Association 81(396):945–970.

Hosmer, David W., and Stanley Lemeshow. 2000. Applied logistic regression. 2nd ed. New York: Wiley.

Hotz, V. Joseph, Charles H. Mullin, and John Karl Scholz. 2005. Examining the effect of the earned income tax credit on the labor market participation of families on welfare. Population Working Paper PWP-CCPR-2005-065, California Center for Population Research, University of California, Los Angeles.

———, and Mo Xiao. 2005. The impact of minimum quality standards on firm entry, exit and product quality: The case of the child care market. Population Working Paper PWP-CCPR-2005-063, California Center for Population Research, University of California, Los Angeles.

Hout, Michael. 1983. Mobility tables. Sage university papers series on quantitative applications in the social sciences, 07-031. Thousand Oaks, CA: Sage.

———. 1984. Status, autonomy, and training in occupational mobility. American Journal of Sociology 89(6): 1379–1409.

———, and Robert M. Hauser. 1992. Symmetry and hierarchy in social mobility: A methodological analysis of the CASMIN model of class mobility. European Sociological Review 8(3):239–266.

Hsiao, Cheng. 2003. Analysis of panel data. 2nd ed. New York: Cambridge University Press.

International Labour Office. 1969. International standard classification of occupations: Revised edition 1968. Geneva: International Labour Office.

———. 1990. International standard classification of occupations (ISCO88). Geneva: International Labour Office.

Jacobs, David, and Robert M. O'Brien. 1998. The determinants of deadly force: A structural analysis of police violence. American Journal of Sociology 103(4):837–862.

Jahoda, Marie, Paul F. Lazarsfeld, and Hans Zeisel. [In German, 1933] 1971. Marienthal: The sociography of an unemployed community. Trans. by the authors, with John Reginall and Thomas Elsaesser. Chicago: Aldine, Atherton.

Jasso, Guillermina. 1985. Marital coital frequency and the passage of time: Estimating the separate effects of spouses' ages and marital duration, birth and marriage cohorts, and period influences. American Sociological Review 50(2):133–149.

———. 1986. Is it outlier deletion or is it sample truncation? Notes on science and sexuality. American Sociological Review 51(5):738–742.

Jencks, Christopher, Susan Bartlett, Mary Corcoran, James Crouse, David Eaglesfield, Gregory Jackson, Kent McClelland, Peter Mueser, Michael Olneck, Joseph Schwartz, Sherry Ward, and Jill Williams. 1979. Who gets ahead? The determinants of economic success in America. New York: Basic Books.

———, Marshall Smith, Henry Acland, Mary Jo Bane, David Cohen, Herbert Gintis, Barbara Heyns, and Stephan Michelson. 1972. Inequality: A reassessment of the effect of family and schooling in America. New York: Basic Books.

Johnson, Robert, and Reynolds Farley. 1985. On the statistical significance of the index of dissimilarity. Proceedings of the Social Statistics Section. Washington, DC: American Statistical Association.

Jones, F. Lancaster, and Jonathan Kelley. 1984. Decomposing differences between groups: A cautionary note on measuring discrimination. Sociological Methods and Research 12(3):323–343.

Jones, Michael P. 1996. Indicator and stratification methods for missing explanatory variables in multiple linear regression. Journal of the American Statistical Association 91(433):222–230.

Jöreskog, Karl G. 1970. A general method for analysis of covariance structures. Biometrika 57(2):239–251.

Judson, D. H. 1992. smv5: Performing loglinear analysis of cross-classifications. In Stata technical bulletin reprints, vol. 1, ed. Joseph Hilbe, 139–152. Santa Monica, CA: Computing Resource Center.

———. 1993. smv5.1: Loglinear analysis of cross classifications, update. In Stata technical bulletin reprints, vol. 2, ed. Joseph Hilbe, 162–163. Santa Monica, CA: Computing Resource Center.

Kahn, Joan R., and J. Richard Udry. 1986. Marital coital frequency: Unnoticed outliers and unspecified interactions lead to erroneous conclusions. American Sociological Review 51(5):734–737.

Kaufman, Robert L. 1983. A structural decomposition of Black-White earnings differentials. American Journal of Sociology 89(3):585–611.

———, and Paul G. Schervish. 1986. Using adjusted crosstabulations to interpret log-linear relationships. American Sociological Review 51(5):717–733.

Keating, Nancy L., Jane C. Weeks, Mary Beth Landrum, Catherine Borbas, Edward Guadagnoli. 2001. Discussion of treatment options for early-stage breast cancer: Effect of provider specialty on type of surgery and satisfaction. Medical Care 39(7):681–691.

Keeley, Michael, Philip Robins, Robert Spiegelman, and Richard West. 1978. The labor supply effects and costs of alternative negative income tax programs. Journal of Human Resources 13(1):3–36.

Kelly, Nathan J., and Jana Morgan Kelly. 2005. Religion and Latino partisanship in the United States. Political Research Quarterly 58(1):87–95.

Kim, Jae-On, and Charles W. Mueller. 1976. Standardized and unstandardized coefficients in causal analysis: An expository note. Sociological Methods and Research 4(4):423–438.

King, Gary. 1989. Unifying political methodology: The likelihood theory of statistical inference. New York: Cambridge.

Kish, Leslie. 1965. Survey sampling. New York: Wiley.

Kitagawa, Evelyn M. 1955. Components of a difference between two rates. Journal of the American Statistical Association 50(272):1168–1194.

———, and Philip M. Hauser. 1973. Differential mortality in the United States: A study in socioeconomic epidemiology. Cambridge, MA: Harvard University Press.

Knoke, David, and Peter J. Burke. 1980. Log-linear models. Sage university papers series on quantitative applications in the social sciences, 07-020. Beverly Hills, CA: Sage.

Kraus, Vered. 1986. Group differences: The issue of decomposition. Quality and Quantity 20(2–3):181–190.

Lassen, David Dreyer. 2005. The effect of information on voter turnout: Evidence from a natural experiment. American Journal of Political Science 49(1):103–118.

Lazarsfeld, Paul F. 1955. Interpretation of statistical relations as a research operation. In The language of social research: A reader in the methodology of social research, ed. Paul E. Lazarsfeld and Morris Rosenberg, 115–124. Glencoe: Free Press.

Lee, Valerie E., and Anthony S. Bryk. 1989. A multilevel model of the social distribution of high school achievement. Sociology of Education 62(3):172–192.

Lewis, Susan K., and Valerie K. Oppenheimer. 2000. Educational assortative mating across marriage markets: Non-Hispanic Whites in the United States. Demography 37(1):29–40.

Lichter, Daniel, Diane McLaughlin, and David Ribar. 2002. Economic restructuring and the retreat from marriage. Social Science Research 31(2):230–256.

Little, Roderick J. A. 1992. Regression with missing x's: A review. Journal of the American Statistical Association 87(420):1227–1238.

———, and Donald B. Rubin. 2002. Statistical analysis with missing data, 2nd ed. New York: John Wiley and Sons.

Long, J. Scott. 1990. The origins of sex differences in science. Social Forces 68(4):1297–1316.

———. 1997. Regression models for categorical and limited dependent variables. Thousand Oaks, CA: Sage.

———, and Jeremy Freese. 2006. Regression models for categorical dependent variables using Stata. 2nd ed. College Station, TX: Stata Press.

Lu, Bo, Elaine Zanutto, Robert Hornik, and Paul R. Rosenbaum. 2001. Matching with doses in an observational study of a media campaign against drug abuse. Journal of the American Statistical Association 96(456):1245–1253.

Lu, Yao. 2005. Sibship size, family organization, and children's education in South Africa: Black-White variations. Population Working Paper PWP-CCPR-2005-045, California Center for Population Research, University of California, Los Angeles.

———, and Donald J. Treiman. 2007. The effect of labor migration and remittances on children's education among Blacks in South Africa. Population Working Paper PWP-CCPR-2007-001, California Center for Population Research, University of California, Los Angeles.

———, and Donald J. Treiman. 2008. Forthcoming. The effect of sibship size on educational attainment in China: Cohort variations. American Sociological Review 73(5).

Lundquist, Jennifer Hickes. 2004. When race makes no difference: Marriage and the military. Social Forces 83(2):731–757.

Manski, Charles F., Sarah S. McLanahan, Daniel Powers, and Gary D. Sandefur. 1992. Alternative estimates of the effects of family structure during adolescence on high school graduation. Journal of the American Statistical Association 87(417):25–37.

———, and David A. Wise. 1983. College choice in America. Cambridge, MA: Harvard University Press.

Maralani, Vida. 2004. Family size and educational attainment in Indonesia: A cohort perspective. Population Working Paper PWP-CCPR-2004-017, California Center for Population Research, University of California, Los Angeles.

Mare, Robert D. 1980. Social background and school continuation decisions. Journal of the American Statistical Association 75(370):295–305.

———. 1981. Change and stability in educational stratification. American Sociological Review 46(1):72–87.

———. 1991. Five decades of educational assortative mating. American Sociological Review 56(1):15–32.

———. 1995. Changes in educational attainment and school enrollment. In State of the union: America in the 1990s, vol. 1, Economic trends, ed. Reynolds Farley, 155–213. New York: Russell Sage.

———, and Meichu D. Chen. 1986. Further evidence on sibship size and educational stratification. American Sociological Review 51(3):403–412.

———, and Christopher Winship. 1984. The paradox of lessening racial inequality and joblessness among black youth: Enrollment, enlistment, and employment, 1964–1981. American Sociological Review 49(1):39–55.

———, and Christopher Winship. 1988. Endogenous switching regression models for the causes and effects of discrete variables. In Common problems/proper solutions: Avoiding error in quantitative research, ed. J. Scott Long, 132–160. Newbury Park, CA: Sage.

Marx, Gary T. 1967a. Religion: Opiate or inspiration of civil rights militancy among Negroes. American Sociological Review 32(1):64–72.

———. 1967b. Protest and prejudice: A study of belief in the Black community. New York: Harper and Row.

Mason, William M. 2001. Multilevel methods of statistical analysis. In International encyclopedia of the social and behavioral sciences, ed. Neil J. Smelser and Paul B. Baltes, 14,988–14,994. Amsterdam: Elsevier Science.

McFadden, Daniel. 1974. Conditional logit analysis of qualitative choice behavior. In Frontiers of econometrics, ed. Paul Zarembka, 105–142. New York: Academic Press.

McIvery, John, and Edward G. Carmines. 1981. Unidimensional scaling. Sage university papers series on quantitative applications in the social sciences, 07-024. Beverly Hills, CA: Sage.

Miller, Grant. 2007. Women's suffrage, political responsiveness, and child survival in American history. Paper presented in the California Population Research Workshop series, California Center for Population Research, University of California, Los Angeles, 5 December. http://www.ccpr.ucla.edu/seminars/Seminar%20Papers/Miller-suffrage.pdf.

Miller, Herman. 1966. Income distribution in the United States. Washington, DC: U.S. Government Printing Office.

Miller, Jane E. 2004. The Chicago guide to writing about numbers. Chicago: University of Chicago Press.

______. The Chicago guide to writing about multivariate analysis. Chicago: University of Chicago Press.

Mincer, Jacob, and Solomon Polachek. 1974. Family investments in human capital: Earnings of women. Journal of Political Economy 82(suppl.):S76–S108.

Morgan, Stephen L. 2001. Counterfactuals, causal effect heterogeneity, and the Catholic school effect on learning. Sociology of Education 74(4):341–374.

———, and Christopher Winship. 2007. Counterfactuals and causal inference: Methods and principles for social research. Cambridge, MA: Cambridge University Press.

Müller, Walter, and Yossi Shavit. 1998. The institutional embeddedness of the stratification process: A comparative study of qualifications and occupations in thirteen countries. In From school to work: A comparative study of educational qualifications and occupational destinations, ed. Yossi Shavit and Walter Müller, 1–48. Oxford: Clarendon Press.

Murdock, George P., and Caterina Provost. 1973. Measurement of cultural complexity. Ethnology 12(4):379–392.

Myrdal, Gunnar. 1944. An American dilemma: The Negro problem and modern democracy. New York: Harper and Brothers.

Nee, Victor. 1989. A theory of market transition: From redistribution to markets in state socialism. American Sociological Review 54(5):663–681.

———. 1996. The emergence of a market society: Changing mechanisms of stratification in China. American Journal of Sociology 101(4):908–949.

Netemeyer, Richard G., William O. Beardon, and Subhash Sharma. 2003. Scaling procedures: Issues and applications. Thousand Oaks, CA: Sage.

Nieuwbeerta, Paul, and Harry B. G. Ganzeboom. 1996. International social mobility and politics file: Documentation of an integrated dataset of 113 national surveys held in 16 countries, 1956–1991. Amsterdam: Steinmetz Archive.

Nobles, Jenna, and Elizabeth Frankenberg. 2006. Mother's community participation and child health. Population Working Paper PWP-CCPR-2006-016, California Center for Population Research, University of California, Los Angeles.

Nordholt, Eric Schulte. 1998. Imputation: Methods, simulation experiments and practical examples. International Statistical Review 66(2):157–180.

Nunnally, Jum, and Ira H. Bernstein. 1984. Psychometric theory. 3rd ed. New York: McGraw-Hill.

Oaxaca, Ronald. 1973. Male-female wage differentials in urban labor markets. International Economic Review 14(3):693–709.

O'Muircheartaigh, Colm. 2003. There and back again: Demographic survey sampling in the 21st century. Keynote address, Federal Committee on Statistical Methodology, 2003 conference. http://www.fcsm.gov/events/papers2003.html.

Oster, Emily. 2005. Hepatitis B and the case of the missing women. Journal of Political Economy 113(6):1163–1216.

Panis, C. 1994. sg24: The piecewise linear spline transformation. In Stata technical bulletin reprints, vol.3, ed. Sean Becketti, 146–149. College Station, TX: StataCorp.

Park, Hyunjoon, and Jeroen Smits. 2005. Educational assortative mating in South Korea: Trends 1930–1998. Research in Social Stratification and Mobility 23:103–127.

Paul, Christopher, William M. Mason, Daniel McCaffrey, and Sarah A. Fox. 2008. Forthcoming. What should we do about missing data? (A case study using logistic regression with missing data on a single covariate). Statistical Methods and Applications 17.

Petersen, Trond. 1985. A comment on presenting results from logit and probit models. American Sociological Review 50(1):130–131.

Peterson, Ruth, and John Hagan. 1984. Changing conceptions of race: Towards an account of anomalous findings of sentencing research. American Sociological Review 49(1):56–70.

Pisati, Maurizio. 2001. sg142: Uniform layer effect models for the analysis of differences in two-way associations. In Stata technical bulletin reprints, vol. 10, ed. H. Joseph Newton, 169–187. College Station, TX: StataCorp.
Powers, Daniel A. 1993. Endogenous switching regression models with limited dependent variables. Sociological Methods and Research 22(2):248–273.
———, and Christopher G. Ellison. 1995. Interracial contact and Black racial attitudes: The contact hypothesis and selectivity bias. Social Forces 74(1):205–226.
———, and Yu Xie. 2000. Statistical methods for categorical data analysis. San Diego, CA: Academic Press.
Powers, Elizabeth T. 2003. Children's health and maternal work activity: Estimates under alternative disability definitions. Journal of Human Resources 38(3):522–556.
Prouteau, Lionel, and Francois-Charles Wolff. 2006. Does volunteer work pay off in the labor market? Journal of Socio-Economics 35(6):992–1013.
Quester, A., and W. Greene. 1982. Divorce risk and wives' labor supply behavior. Social Science Quarterly 63:16–27.
Radelet, Michael L., and Glenn L. Pierce. 1985. Race and prosecutorial discretion in homicide cases. Law and Society Review 19(4):587–622.
Raftery, Adrian E. 1986. Choosing models for cross-classifications. American Sociological Review 51(1):145–146.
———. 1995a. Bayesian model selection in social research. Sociological Methodology 25:111–163.
———. 1995b. Rejoinder: Model selection is unavoidable in social research. Sociological Methodology 25:185–195.
Ransom, Michael R. 2000. Sampling distributions of segregation indexes. Sociological Methods and Research 28(4):454–475.
Rasler, Karen. 1996. Concessions, repression, and political protest in the Iranian revolution. American Sociological Review 61(1):132–152.
Raudenbush, Stephen W., and Anthony S. Bryk. 2002. Hierarchical linear models: Applications and data analysis methods. 2nd ed. Thousand Oaks, CA: Sage.
Rona-Tas, Akos, and Alya Guseva. 2001. The privileges of past Communist Party membership in Russia and endogenous switching regression. Social Science Research 30(4):641–652.
Roncek, Dennis W. 1992. Learning more from tobit coefficients: Extending a comparative analysis of political protest. American Sociological Review 57(4):503–507.
Roberts, John M., Jr., and Garry Chick. 2007. Culture and behavior: Applying log-linear models for transitions between offices in a Mexican festival system. Social Science Research 36(1):313–328.
Rosen, Harvey. 1976. Taxes in a labor supply model with joint wage-hours determination. Econometrica 44(3):485–507.
Rosenbaum, Paul R., and Donald B. Rubin. 1983. The central role of the propensity score in observational studies for causal effects. Biometrika 70(1):41–55.
Rossi, Alice. 1966. Public views on abortion. Unpublished paper, Committee on Human Development, University of Chicago.
Royston, Patrick. 2004. Multiple imputation of missing values. Stata Journal 4(3):227–241.
———. 2005a. Multiple imputation of missing values: Update. Stata Journal 5(2):188–201.
———. 2005b. Multiple imputation of missing values: Update of ice. Stata Journal 5(4):527–536.
———. 2007. Multiple imputation of missing values: Further update of ice, with an emphasis on interval censoring. Stata Journal 7(4):445–464.
Rubin, Donald B. 1987. Multiple imputation for nonresponse in surveys. New York: Wiley.
———, and Nathaniel Schenker. 1986. Multiple imputation for interval estimation from simple random samples with ignorable nonresponse. Journal of the American Statistical Association 81(394):366–374.
Sakamoto, Arthur, and Meichu D. Chen. 1991. Inequality and attainment in the dual labor market. American Sociological Review 56(3):295–308.
Saltzman, Gregory M. 1987. Congressional voting on labor issues: The role of PACs. Industrial and Labor Relations Review 40(2):163–179.
Sampson, Robert J., Stephen W. Raudenbush, and Felton Earls. 1997. Neighborhoods and violent crime: A multilevel study of collective efficacy. Science 277(5328):918–924.
Savage, Leonard J. 1976. On rereading R. A. Fisher. Annals of Statistics 4(3):441–500.
Sayrs, Lois W. 1989. Pooled time series analysis. Sage university papers series on quantitative applications in the social sciences, 07-070. Newbury Park, CA: Sage.
Schafer, Joseph L. 1997. Analysis of incomplete multivariate data. London: Chapman and Hall.

———. 1999. Multiple imputation: A primer. Statistical Methods in Medical Research 8(1):3–15.
Schwartz, Christine R., and Robert D. Mare. 2005. Trends in educational assortative marriage from 1940 to 2003. Demography 42(4):621–646.
Schenker, Nathaniel, Donald J. Treiman, and Lynn Weidman. 1993. Analysis of public use decennial census data with multiply imputed industry and occupation codes. Applied Statistics 42(3):545–556
Sloan, John H., A. L. Kellermann, D. T. Reay, J. A. Ferris, T. Koepsell, F. P. Rivara, C. Rice, L. Gray, and J. LoGerfo. 1988. Handgun regulations, crime, assaults, and homicide: A tale of two cities. New England Journal of Medicine 319(19):1256–1261.
Smith, Herbert L. 1997. Matching with multiple controls to estimate treatment effects in observational studies. Sociological Methodology 27:325–352.
Smith, Patricia L. 1979. Splines as a useful and convenient statistical tool. American Statistician 33(2):57–62.
Smith, Tom W. 1979. Sex and the GSS: Nonresponse differences. GSS Methodological Report No. 9. Chicago: National Opinion Research Center.
———, and Bruce L. Peterson. 1986. Problems in form randomization on the General Social Surveys. GSS Methodological Report No. 36, July. Chicago: National Opinion Research Center.
Smock, Pamela J., Wendy D. Manning, and Sanjiv Gupta. 1999. The effect of marriage and divorce on women's economic well-being. American Sociological Review 64(6):794–812.
Snijders, T., and R. Bosker. 1999. Multilevel analysis. London: Sage Publications.
Sobel, Michael E., Michael Hout, and Otis Dudley Duncan. 1985. Exchange, structure, and symmetry in occupational mobility. American Journal of Sociology 91(2):359–372.
Soopramanien, Didier, and Geraint Johnes. 2001. A new look at gender effects in participation and occupation choice. Labour 15(3):415–443.
Sorokin, Pitirim A. [1927] 1959. Social and cultural mobility. Glencoe: Free Press.
Sousa-Poza, Alfonso. 2004. Is the Swiss labor market segmented? An analysis using alternative approaches. Labour 18(1):131–161.
South, Scott J., and Eric P. Baumer. 2001. Community effects on the resolution of adolescent premarital pregnancy. Journal of Family Issues 22(11):1025–1043.
StataCorp. 2007. Stata statistical software: Release 10. College Station, TX: Stata Press.
Steele, Claude. 1997. A threat in the air: How stereotypes shape intellectual identity and performance. American Psychologist 52(6):613–629.
Steiger, James H. 2001. Driving fast in reverse: The relationship between software development, theory, and education in structural equation modeling. Journal of the American Statistical Association 96(453):331–338.
Stephan, Frederick J., and Philip J. McCarthy. 1963. Sampling opinions: An analysis of survey procedure. New York: Wiley Science Editions.
Stephan, Paula E., and Sharon G. Levin. 1992. Striking the mother lode in science: The importance of age, place, and time. New York: Oxford University Press.
Stephenson, C. Bruce. 1979. Probability sampling with quotas: An experiment. Public Opinion Quarterly 43(4):477–496.
Stine, Robert. 1990. An introduction to bootstrap methods: Examples and ideas. In modern methods of data analysis, ed. John Fox and J.Scott Long, 325–373. Newbury Park, CA: Sage.
Stoltzenberg, Ross. 1974. Estimating an equation with multiplicative and additive terms. Sociological Methods and Research 2(3):313–331.
———. 1975. Education, occupation, and wage differences between White and Black men. American Journal of Sociology 81(2):299–323.
———, and Daniel A. Relles. 1997. Tools for intuition about sample selection bias and its correction. American Sociological Review 62(3):494–507.
Stone, Roslyn A., D. Scott Obrosky, Daniel E. Singer, Wishwa N. Kapoor, Michael J. Fine, Pneumonia Patient Outcomes Research Team Investigators. 1995. Propensity score adjustment for pretreatment differences between hospitalized and ambulatory patients with community-acquired pneumonia. Proceedings of the Conference on Measuring the Effects of Medical Treatment, April. Medical Care 33(4):AS56–AS66.
Stouffer, Samuel A. 1949. The American soldier. 2 vols. Vol. 1, Adjustment during army life. Vol. 2, Combat and its aftermath. Princeton, NJ: Princeton University Press.
———. 1955. Communism, conformity, and civil liberties: A cross-section of the nation speaks its mind. Garden City, NY: Doubleday.
———. 1962. Social research to test ideas: Selected writings. New York: Free Press of Glencoe.

Sudman, Seymour. 1976. Applied sampling. New York: Academic Press.

Sweeney, Megan M. 2002. Two decades of family change: The shifting economic foundations of marriage. American Sociological Review 67(1):132–147.

Szelényi, Iván, and Donald J. Treiman. 1994. Social stratification in Eastern Europe after 1989 (computer file). Principal investigators, Iván Szelényi and Donald J. Treiman. Produced by a consortium of research groups in the nations involved. Social Science Data Archive, University of California, Los Angeles [distributor].

Tavits, Margit. 2005. The development of stable party support: Electoral dynamics in post-communist Europe. American Journal of Political Science 49(2):283–298.

Thomas, Duncan, Elizabeth Frankenberg, Jed Friedman, Jean-Pierre Habicht, Nathan Jones, Christopher McKelvey, Gretel Pelto, Bondan Sikoki, James P. Smith, Cecep Sumantri, and Wayan Suriastini. 2004. Causal effect of health on labor market outcomes: Evidence from a random assignment iron supplementation intervention. Population Working Paper PWP-CCPR-2004-022, California Center for Population Research, University of California, Los Angeles.

Tienda, Marta, and Franklin F. Wilson. 1992. Migration and the earnings of Hispanic men. American Sociological Review 57(5):661–678.

Tobin, James. 1958. Estimation of relationships for limited dependent variables. Econometrica 26(1):24–36.

Tomaskovic-Devey, and Sheryl Skaggs. 1999. Degendered jobs? Organizational processes and gender segregated employment. Research in Social Stratification and Mobility 17:139–172.

Treiman, Donald J. 1970. Reply to Geschwender on "Status discrepancy and prejudice." American Journal of Sociology 76(1):162–168.

———. 1977. Occupational prestige in comparative perspective. New York: Academic Press.

———. 1994. Social stratification in Eastern Europe after 1989: General population survey. Provisional codebook (revised), 8 December 1994. Social Science Data Archive, University of California, Los Angeles.

———, ed. 1998. Life histories and social change in contemporary China: Provisional codebook. Social Science Data Archive, University of California, Los Angeles.

———. 2007a. Growth and determinants of literacy in China. In Education and reform in China, ed. Emily Hannum and Albert Park, 135–153. Oxford: Routledge.

———. 2007b. The legacy of apartheid: racial inequalities in the new South Africa. In Unequal chances: Ethnic minorities in Western labour markets, ed. Anthony Heath and Sin Yi Cheung, 401–447. Oxford: Oxford University Press.

———, William Bielby, and Man-Tsun Cheng. 1988. Evaluating a multiple-imputation method for recalibrating 1970 U.S. Census detailed industry codes to the 1980 standard. Sociological Methodology 18:309–345.

———, William Cumberland, Xilai Shi, Zhongdong Ma, and Shaoling Zhu. 1998. A sample design for the Chinese life history survey. In Life histories and social change in contemporary China: Provisional codebook, ed. Donald J. Treiman, Appendix D.1.a. Social Science Data Archive, University of California, Los Angeles.

———, and Heidi I. Hartmann, eds. 1981. Women, work, and wages: Equal pay for jobs of equal value. Washington, DC: National Academy Press.

———, and Hye-Kyung Lee. 1996. Income differences among 31 ethnic groups in Los Angeles. In Social differentiation and social inequality: Essays in honor of John Pock, ed. James Baron, David Grusky, and Donald J. Treiman, 37–82. Boulder, CO: Westview Press.

———, Alisa Lewin, and Yao Lu. 2006. Survey of socioeconomic opportunity and achievement in South Africa: Codebook. Social Science Data Archive, University of California, Los Angeles.

———, William M. Mason, Yao Lu, Yi Pan, Yaqiang Qi, and Shige Song. 2006. Observations on the design and implementation of sample surveys in China. Social Transformations in Chinese Societies 1:81–101.

———, Matthew McKeever, and Eva Fodor. 1996. Racial differences in occupational status and income in South Africa, 1980–1991. Demography 33(1):111–132.

———, Sylvia N. Moeno, and Lawrence Schlemmer. 1994. Survey of socioeconomic opportunity and achievement in South Africa [computer file]. Principal investigator, Donald J. Treiman; coprincipal investigators, Sylvia N. Moeno and Lawrence Schlemmer. Human Sciences Research Council, Pretoria, South Africa [producer]; Social Science Data Archive, University of California, Los Angeles [distributor].

———, and Patricia A. Roos. 1983. Sex and earnings in industrial society: A nine-nation comparison. American Journal of Sociology 89(3):613–650.

———, and Iván Szelényi. 1993. Social stratification in Eastern Europe after 1989. In Transformation processes in Eastern Europe (Proceedings of a workshop held at the Dutch National Science Foundation [NWO], 3–4 December 1992), 163–78. The Hague: NWO.

———, and Kermit Terrell. 1975. The process of status attainment in the United States and Great Britain. American Journal of Sociology 81(3):563–583.
———, Andrew G. Walder, and Qiang Li. 2006. Life histories and social change in contemporary China [computer file]. Principal investigator, Donald J. Treiman; coprincipal investigators, Andrew G. Walder and Qiang Li. Department of Sociology, People's University, Beijing [producer]; Social Science Data Archive, University of California, Los Angeles [distributor].
———, and Kazuo Yamaguchi. 1993. Trends in educational attainment in Japan. In Persistent inequality: Changing educational attainment in thirteen countries, ed. Yossi Shavit and Hans-Peter Blossfeld, 229–250. Boulder: Westview Press.
———, and Kam-Bor Yip. 1989. Educational and occupational attainment in 21 countries. In Cross-national research in sociology, ed. Melvin L. Kohn (ASA Presidential Series), 373–394. Beverly Hills, CA: Sage.
Treno, Andrew J., Maria L. Alaniz, and Paul J. Gruenewald. 2000. The use of drinking places by gender, age and ethnic groups: An analysis of routine drinking activities. Addiction 95(4):537–551.
Tukey, John W. 1977. Exploratory data analysis. Reading, MA: Addison-Wesley.
Upton, Graham J. G. 1978. The analysis of cross-tabulated data. New York: Wiley.
Van Buuren, S., H. C. Boshuizen, and D. L. Knook. 1999. Multiple imputation of missing blood pressure covariates in survival analysis. Statistics in Medicine 18(6):681–694.
Vermunt, Jeroen K. 1997. *l*em: A general program for the analysis of categorical data. Tilburg, Netherlands: Tilburg University.
Walder, Andrew G. 1996. Markets and inequality in transitional economies: Toward testable theories. American Journal of Sociology 101(4):1060–1073.
Walton, John, and Charles Ragin. 1990. Global and national sources of political protest: Third world responses to the debt crisis. American Sociological Review 55(6):876–890.
Weakliem, David L. 1999. A critique of the Bayesian Information Criterion for model selection. Sociological Methods and Research 27(2):359–397.
Weitoft, Gunilla Ringback, Anders Hjern, Ilija Batljan, and Bo Vinnerljung. 2008. Health and social outcomes among children in low-income families and families receiving social assistance: A Swedish national cohort study. Social Science and Medicine 66(1):14–30.
White, Michael J., and Zai Liang. 1998. The effect of immigration on the internal migration of the native-born population, 1981–1990. Population Research and Policy Review 17(2):141–166.
Williams, Richard. 2006. Generalized ordered logit/partial proportional odds models for ordinal dependent variables. Stata Journal 6(1):58–82.
Willis, Robert J., and Sherwin Rosen. 1979. Education and self-selection. Journal of Political Economy 87(suppl.):S507–S536.
Winsborough, H. H., and Peter Dickinson. 1971. Components of Negro-White income differences. In Proceedings of the American Statistical Association. Social statistics section, ed. Edwin G. Goldfield, 6–8. Washington, DC: American Statistical Association.
Winship, Christopher. 1999a. Editor's introduction to the special issue on the Bayesian Information Criterion. Sociological Methods and Research 27(3):355–358.
———, ed. 1999b. Special issue on the Bayesian information criterion. Sociological Methods and Research 27(3):355–443.
———, and Robert D. Mare. 1984. Regression models with ordinal variables. American Sociological Review 49(4):512–525.
———, and Robert D. Mare. 1992. Models for sample selection bias. Annual Review of Sociology 18:327–350.
———, and Stephen L. Morgan. 1999. The estimation of causal effects from observational data. Annual Review of Sociology 25:659–706.
Witte, A. 1980. Estimating an economic model of crime with individual data. Quarterly Journal of Economics 94(1):57–84.
Wood, Charles H., and Peggy A. Lovell. 1992. Racial inequality and child mortality in Brazil. Social Forces 70(3):703–724.
Wooldridge, Jeffrey M. 2002. Econometric analysis of cross section and panel data. Cambridge, MA: MIT Press.
———. 2006. Introductory econometrics: A modern approach. 3rd ed. Mason, OH: Thomson South-Western.
Wright, Erik Olin. 1985. Classes. London: New Left.
———, Cynthia Costello, David Hachen, and Joe Sprague. 1982. The American class structure. American Sociological Review 47(6):709–726.

———, and Bill Martin. 1987. The transformation of the American class structure, 1960–1980. American Journal of Sociology 93(1):1–29.
Wright, Sewall. 1918. On the nature of size factors. Genetics 3(4):367–374.
Wu, Xiaogang, and Donald J. Treiman. 2004. The household registration system and social stratification in China: 1955– 1996. Demography 41(2):363–384.
———. 2007. Inequality and equality under Chinese socialism: The hukou system and intergenerational occupational mobility. American Journal of Sociology 113(2):415–445.
Xie, Yu. 1992. The log-multiplicative layer effect model for comparing mobility tables. American Sociological Review 57:380–395.
Yamaguchi, Kazuo. 1987. Models for comparing mobility tables: Toward parsimony and substance. American Sociological Review 57(3):380–395.
Yanovitzky, Itzhak, and Joseph N. Cappella. 2001. Effect of call-in political talk radio shows on their audiences: Evidence from a multi-wave panel analysis. International Journal of Public Opinion Research 13(4):377–397.
Zatz, Marjorie S., and John Hagan. 1985. Crime, time, and punishment: An exploration of selection bias in sentencing research. Journal of Quantitative Criminology 1(1):103–126.
Zeisel, Hans. [1947] 1985. Say it with figures. Rev. 6th ed. New York: Harper and Row.
Zhang, Junsen, and Saul D. Hoffman. 1993. Discrete-choice logit models: Testing the IIA property. Sociological Methods and Research 22(2):193–213.
Zuckerman, Harriet. 1977. Scientific elite: Nobel laureates in the United States. New York: Free Press.

INDEX

Q

R

S

X

Y

Z

homogenizing forces of globalization (Hettinger 2011). As someone who has followed both the biological and the cultural critics, I am unpersuaded by both. This is a false binary and for the most part a moot point. Globalization has been ongoing for millennia, plants and animals have been moved around for just as long—hybridity and multiplicity is all around us. Also, globalization does not automatically dilute distinctiveness in either the cultural or biological world. New configurations, new identities, new possibilities emerge (W. O'Brien 2006). This is not to suggest that cosmopolitanism—of the floral, faunal, or human variety—is easy or unproblematic (Jamieson 1995, Pollan 1994, Soulé 1990, Paretti 1998). Blind cosmopolitanism, like blind multiculturalism, is likely to bring along its own sets of problems.

But this is not to suggest that globalization has been always good or productive. Movements of humans, plants, and animals have always been in reaction to particular political, economic, and social forces, and one can trace these circulations. Alien plants and animals share a common set of natural, cultural, and political contexts, and all of us are sometimes caught up in geopolitical circuits of power. As Karen Cardozo and I have argued elsewhere, some plants, animals, and humans share historical, economic, and political histories. Using the term *Asian American* as a geopolitical frame, we demonstrate how the term should be understood as one that is not just human but rather one that is multispecies (Cardozo and Subramaniam 2013). Like Asian American humans, Asian American plants and animals also share in its complex geopolitics, its colonial legacies, the eras of trade, and its cultural and culinary circulations. For example, the much maligned and invasive Asian longhorned beetle (ALB) owes its origins to China's policies to combat soil erosion and deforestation, which resulted in the country planting massive rows of monoculture poplar plants as wind-breakers. The ALB favors poplar trees, and these monocultures allowed an explosion in growth of ALB populations. Soon after, when trade exploded in the 1970s and 1980s and there was need for wooden crates, China was well equipped to provide poplar wooden crates, which of course now carried the larvae of ALBs. These wooden crates were shipped worldwide, enabling the extensive circulation of these beetles. We need to remember the current problems with ALB are part of these complex circuits of ecological trade, environmental, and commerce policies (Alsop 2009, MacAusland and Costello 2004). And this is the context we should recognize as we are inundated with "wanted" posters in newspapers, magazines, movie theaters, billboards, and i-phone apps that call for its eradication and destruction.

The much famous Georgia peach also has its origins in China, being brought to the United States through the travels of USDA agents (Kaplan 1991). Of course, we want to domesticate and appropriate the peach as our own, even calling it the Georgia peach, while the ALB is remembered for its destructive foreignness. We can tell similar stories of many other plants and animals, as we can about humans. Thus all of us, creatures of this earth, are caught up in complex and unexpected networks of entanglements.[1] As chapter 5 shows, it is astonishing sometimes how the rhetoric, ideology, and politics from one sphere quickly engulf another. The xenophobic rants of anti-immigrant activists can quickly be heard in environmentalist circles. It is a keen awareness of these kindred subjectivities, of broad interdisciplinary approaches, that allows us to trace these dense and vibrant circulations.

Nationalism, Race, and Xenophobia: Disciplinarity and the Circuits of Knowledge

Just as the geopolitics of Asia and America have generated a richly textured world of multiple species of Asian Americans, so have other forms of geopolitics. Conversely, this work allows us to recognize the limits of disciplinarity. The narrow lenses of the biological sciences see invasive species purely as a "biological" problem, and the narrow lenses of the humanities see xenophobia or racism as a human problem. Each fails to see how their disciplinary frames need be so much broader and more complex. There are deep links between the nationalization of nature and the naturalization of nation (Sivaramakrishnan 2011). In discussing the case of Europe, Zimmer notes that "as politicized nature, particular landscapes evolved into integral parts of historicism's search for national pedigrees, something that happened across Europe in the late 18th and early 19th century" (Zimmer 1998: 641). Tracing the rhetoric of invasive species brings the natural world squarely within race and immigration politics. As we have seen before, it is not accidental that historical moments of strong xenophobia in human cultures have been associated with panics about foreign plants, animals, or germs. Indeed, Olwig argues that such ideas of nationalist landscapes/cultures are drawn from a common epistemological template that can be traced back to the Renaissance, when methods of surveying and cartography were rediscovered (Olwig 2003). The links between plant/animal control and human control are well documented; perfection in gardens and peoples is rooted in an ongoing struggle against "difference" (Mottier 2008). And the vast resources of the humanities have amply demonstrated how Hitler's vision of pure human populations was accompanied by visions of pure gardens,[2] demonstrated in this quote by the German landscape architect Willey Lange:

> Our feelings for our homeland should be rooted in the character of domestic landscapes; therefore it is German nature that must provide all ideas for design of gardens. They can be heightened by artistic means, but we must not give up the German physiognomy. Thus, our gardens become German if the ideas for the design are German, especially if they are borrowed for the landscape in which the garden is situated. (trans. and qtd. in Groning and Wolschke-Bulmahn 2003: 79)

As Rodman notes, applying the native/alien binary runs the risk of "the precarious utopia of a racially pure Reich" (Rodman 1993: 152). In a context of racism, xenophobia, and nationalism, the ideology of "blood and soil" made deep links between pure humans and pure gardens. The ideas of national cultures and gardens emerge again and again and have appeal at various times across countries (Groning and Wolschke-Bulmahn 2003).[3] And indeed, these kindred analogies between humans and plants were also embraced by American landscape architects such as Jens Jensen, Wilhelm Miller, and Frank Waugh.[4] For example, Jensen is quoted in a 1937 article as saying:

> The gardens that I created myself shall . . . be in harmony with their landscape environment and the racial characteristics of its inhabitants. They shall express the spirit of America and therefore shall be free of foreign character as far as possible . . . The Latin and the Oriental crept and creeps more and more over our land, coming from the South, which is settled by Latin people, and also from centers of mixed masses of immigrants. The Germanic character of our race, of our cities and settlements was overgrown by foreign character. Latin spirit has spoiled our lot and still spoils things every day. (qtd. in Groning and Wolschke-Bulmahn 2003: 85)

The fact that American eugenicists pointed explicitly to the forest and gardens of Germany as the birthplace of the instinct for democracy is a chilling reminder of the origins of U.S. nationalism and nativism. The idea of "native" nations and plants and animals has a long history and is deeply embedded in a politics of purity. It should come as no surprise that with the rampant fear of immigration in the United States, there was a 53 percent growth in housing units in gated communities between 2001 and 2009. "Stand your ground" laws show one response to the calls for assimilation (Benjamin 2012).

Thinking natureculturally on the politics of purity, I see the connections of xenophobic rhetoric of invasive species to those that sometimes permeate discourses against new reproductive technologies, genetically modified organisms, or the movements for local foods. One of the reasons cited against foreign plants is preservation of the "genetic integrity" of native and nations. The fear that foreign species might interbreed with natives yields discourses that have a familiar ring of anti-miscegenation policies (Smout 2003). Yet, as Forrest and Fletcher argue, when pressed for a definition of genetic integrity, "there is generally some reluctance on the part of those employing such terminology to come to the point, although emotive issues connected with the archival 'preservation of our priceless heritage' and perhaps a variety of 'ethnic cleansing' are seldom far from the surface" (Forest and Fletcher 1995: 99).

ROOTS OF COINCIDENCE: THE POLITICS OF PURITY

Fears of impurity usually grow alongside fears of pernicious sexuality, global miscegenation, and unbounded migrations. And indeed, living in the United States, I began to note the politics of fears in multiple sites, all feeding on familiar tropes of race, and nation (B. Hartmann et al. 2005). What is fascinating is that despite deep ideological differences, in three current issues— invasive species, GMOs (genetically modified organisms), and NRTs (new reproductive technologies)—positions of the political left and the right converge. Whether some individuals are looking for a return to an imagined nostalgic past or for a future without foreigners, individuals across the political spectrum make similar arguments. Some environmentalists want a pure nature. Some feminists are critical of reproductive technologies and their impact on women. Some religious conservatives are afraid that we are taking the place of "God." And some conservative environmentalists are afraid that we are destroying God's creations. Conversely, some environmentalists are critical about the purity discourse while conservatives embrace technology and the free flow of flora and fauna in the name of free markets and the free flow across borders. What do we make of such a convergence? Are the roots of these fears and anxieties the same? Does the common rhetoric belie an anachronistic political similarity between the right and the left? Or is it entirely coincidental? Briefly, I want to suggest four ways in which I believe the arguments of the right and the left converge as they express their opposition to alien biota, GMOs, and new reproductive technologies.

First, the creation of the "other." In each of the three cases, the resulting product—the proliferating invasive species, the genetically modified organism, or the technologicalized mother/baby—is viewed with deep suspicion and as "foreign." In the case of invasive species, the term *invasive* literally becomes synonymous with exotic/alien/foreign species—ruling out the possibility of native invasive species, which seldom get any publicity. Signaling the ultimate monster, GMOs are often even dubbed "Frankenfood" (Egan 2011). The many "accidents" of new reproductive technologies—where white women carry black babies, black women carry white babies, grandmothers are pregnant with grandchildren, and women routinely carry multiple pregnancies—all warn of the "bizarre" and the creation of possibilities that are unlikely to occur in our peculiarly gendered and raced world. It valorizes women's bodies as the sacred site of motherhood. In each of these cases, the language "reinscribes" particular notions of the "other," simultaneously reinforcing the normative as the native, the natural, the pure.

Second, the "other," most often the female, is often attributed with "hypersexual" fertility. Invasive species are routinely ascribed with superfertility. Consider, for example, the title of an article on invasive species: "They Came, They Bred, They Conquered." Within the GMO literature, one sees the fear of transgenes quickly moving to other plants and even crossing species boundaries to create "superweeds" with superfertility. The case of NRT is more complex and interesting because the object of superfertility and primary beneficiary of NRT is the white woman—whose fertility our culture deeply desires.

Third, linked to this is a valorization of nature and the "natural." Fundamentally in this framework, "nature" is a realm that is seen as removed from human interference but also human-friendly, safe, and trustworthy, that is, products of nature are safe for humans. By tinkering with nature, humans are argued to assume an unparalleled arrogance and are accused of playing "God." In this vision, if respected and left undisturbed, nature nurtures native species, nature produces "wholesome natural products" with pristine seeds that are good for you. If, however, we disturb this co-evolved nature with its own checks and balances, we are at risk of unleashing monsters. Invasive species transcend the "natural" order by moving where they do not belong. GMOs enable unnatural gene mixing and NRTs threaten the "natural" process of women and reproduction. Interestingly, companies that produce transgenic plants and GM food have moved to use the same rhetoric to celebrate GM food because it will allow the "natural" to be more "natural." They suggest that producing varieties with higher yields will save biodiversity and ultimately conserve native forestland. Similarly, transgenes that bring pesticide and weed resistance to plants will reduce the use of pesticides and herbicides and ultimately create more "pure" nature and sustainable agricultural practices.

Fourth, the rhetoric of purity is striking in each of these discourses—"pure" nature, "pure species," "pure women," species fidelity. Anxieties abound about native and exotic species cross-breeding, thus "contaminating" the native gene pool and gene purity. The rhetoric emphasizes purity by highlighting "leaking genes," "genetic pollution," and "contamination." Activists and policy makers have created a purity index and have developed standards to measure "seed purity."

Similarly, the vast resources of biology also remind us of the complex interactions that make ecosystems. We have observed the profound impact of ecological managements when new species have been introduced to control pests or weeds and have caused more harm than good. Or at other times when foreign species have evolved to form new communities where native species have become dependent on the foreign species (Woods and Moriarty 2001). Disciplinary thinking fails to illuminate these interconnections and entanglements.

The same sentiments of purity inhabit disciplines as they systematically dismiss methods, methodologies, and theories from outside their disciplines as trivial or sloppy. Thinking naturecultrally and interdisciplinarily unleashes the vast resources of the humanities, arts, social sciences, and sciences. It gives us access to the vast repertoire of tools, theories, methods, and histories. It opens up our imagination to the vast possibilities of the universe—of poets and naturalists, of fiction writers and science writers, of rhetoricians and physicists. It reminds us yet again that words and language are powerful—not transparent and apolitical but powerful tools that can be used toward divisive and violent ends or toward egalitarian, peaceful ones. For those working toward a better world, thinking naturecultrally opens up the insights of conservationists and social activists to each other. It allows us to see that the multiple strategies of various social activists—pro-immigrationists, ecologists, humanists, internationalists, and globalists—can be learned and shared.

The Politics of Assimilation

Alongside a politics of purity, one also begins to see a call for a politics of assimilation. Biologists have long recognized the "naturalization" process of plants and animals and that many foreign species become "culturally native," deeply implicated in local cultural geographies. Some of the calls warning about the dangers of a rigid native/alien distinction come from those who recognize the positive contributions of many foreign species. In addition to many crops, vegetables, and economically important plants and animals, alien species have often naturalized into the biotic world in productive and intricate ways. The case of the eucalyptus tree that is now important to native monarch butterflies as well as several bird species reminds us that a vast number of foreign species have been in the United States for centuries, even millennia. These organisms have evolved to create new communities and new biotic interactions, at times creating new equilibria. The wholesale eradication of alien species is particularly unproductive in the face of climate change that is fundamentally transforming our Planet and local environmental contexts. What does it mean to harp on a return to a past when the environmental contexts the plants evolved into no longer exist? It would appear that we need to move away from a clear separation of culture and environment and toward thinking in terms of relational geographies of plant/human interactions as global environmental change refigures circuits of plant, animal, and human migration, adaptation, adaptability, ecology, and evolution (Head and Atchison 2009).

All of these factors have pushed many to contemplate a less rigid and more flexible approach to the environment, driven less by arbitrary categories of native/alien and more by the empirical realities on the ground. In recent years, a productive site of dealing with invasive species has to do with modifying and expanding our culinary habits. Environmentalists have been getting people to combat invasive species by eating them. For example, in recent years they have tried to harvest crayfish in Lake Tahoe in order to improve the water quality through commercial harvesting. In Nevada, mostly made up of desert, availability of local seafood is exciting. "This is where science stops and you need people to step in and make a decision to improve the lake," a local scientist argued (Onishi 2012). While Asian carp, a delicacy in China, has been overfished there, it is reviled in the Midwest. Companies in the Midwest are now exporting carp to China (Frazier 2010). Locavore movements are popularizing recipes of invasive species (often considered delicacies in their homelands) within the United States. After all, Chilean sea bass, now a delicacy, was less palatable as the Patagonia toothfish! In similar moves, companies in Illinois are trying to convert carp into organic fertilizer as well as introduce products where fish meat is ground into products such as salami, bologna, and even jerky (*PBS Newshour* 2012). A recent *New York Times* story featured Ms. Wong, who having given up fighting her weeds, has instead turned to eating most of them. Moving beyond the narrow offerings of commercial vegetables and herbs, she has taken to growing weeds to wide acclaim and now is popularizing them through a new cookbook (Raver 2012).[5]

A more spiritual take on invasive species comes from those who see invasive species as healing a destroyed Planet. Arguing that many of the species considered invasive are in fact medicinal plants in their home country, they argue that invasive plants are in fact a great resource and boon. Rather than respond with toxic pesticides to control them as some restoration ecologists do, invasive plants can grow on damaged land and perform an essential ecological function to heal both the land and the humans who live on it. In such a view invasive plants transcend the good/evil binary to become the healers of a damaged and sick Planet and world (T. Scott and Buhner 2010).

The Politics of Knowledge and Knowledge Production

Is paying attention to this kinship between my foreign status and that of my objects of study dangerous? After all, isn't objectivity one of the key cornerstones of science? Despite all the claims of objectivity, we have seen again and again in history that science is far from objective. As we have seen in the introduction to this book, a vast literature in feminist science and technology studies shows us that objectivity is an illusion, a mirage that obscures the ways in which science is deeply embedded in its historical and political contexts. That said, a kindred subjectivity is not about a relativistic world where everything goes or where subjectivity comes to stand in for some idea of absolute truth. Rather, thinking natureculturally, thinking with and through kindred subjectivities, forces us to think reflexively. It enables interdisciplinary thinking, compelling us to consider the complex circulations of knowledge. It moves us away from dualities such as labeling all of one category of plants and animals as "evil." Rather, it allows us to understand how we have all come to this country because of complex histories. It allows us to understand that weekend campaigns to go pull out the latest undesirable species from the local pond will not solve the underlying problem. It forces us to acknowledge that we have to think more broadly about national and international environmental policy and how our local problems are connected to larger national policies of energy, development, and globalization. We need to work against an unproblematic scientism in environmental policy and take seriously the institutions of governance, systems of values, and ways of knowing (Jamieson 1995).

The world itself is not disciplinary—plants, animals, and humans are connected through complex histories and geographies. The interdisciplinary experiment I was involved in allowed me to show how moving past disciplinary thinking is helpful in our interdisciplinary interconnected world. I was able to transfer ideas across disciplines to explore the power of language and rhetoric; to bring insights from colonial and postcolonial studies into botany and zoology; to trace the historical, geographic histories of plants, animals, and humans simultaneously; to stretch narrative theory into telling new stories about our nonhuman co-inhabitants. Understanding this means opening ourselves up to a naturecultural world and the vast resources of interdisciplinarity. We cannot understand it any other way. Challenging the nationalist nativist landscapers we read earlier, the Jewish writer Rudolf Borchardt, who was persecuted by National Socialists and who wrote this in 1938, puts it particularly well:

> If this kind of garden-owning barbarian became the rule, then neither a gillyflower nor a rosemary, neither a peach-tree nor a myrtle sampling nor a tea-rose would ever have crossed the Alps. Gardens connect people, time and latitudes. If these barbarians rule, the great historic process of acclimatization would never have begun and today we would horticulturally still subsist on acorns . . . The garden of humanity is a huge democracy. It is not the only democracy which such clumsy advocates threated to dehumanize. (trans. and qtd. in Groning and Wolschke-Bulmahn 2003: 86)

All this is not to suggest that a nonimmigrant could not come to these insights. Indeed, many have. Nor is it to suggest that all immigrants would see these connections. Many do not. Rather, it is to suggest that our life histories, our experiences, our identities, that the vagaries of life can at times open up the world in particular and surprising ways. We should regard these as wonderful and rich opportunities to understand the world in new ways. I would never have seen these connections, never have been able to explore these connections, if not for the rich opportunities of a broad and interdisciplinary training in biology, environmental studies, feminist studies, ethnic studies, critical race studies, postcolonial studies, and queer studies. If our training in the humanities and the sciences could train our eyes, ears, and minds to be open to the world outside the discipline, our theories and knowledge about the world would be that much richer. Scientists and humanists are not removed from the context of their lives. Denying these connections is not objectivity, but rather a lost opportunity.

Toward a Multispecies, Multidisciplinary View of Life

If we open ourselves to a naturecultural world, we have new stories to tell, new narratives, new histories, new cartographies of knowledge. Listening to poets and scientists alike, the naturecultural world is teeming with insight and possibility. The atoms and subatomic particles connect all life and nonlife in this universe. Life on earth is connected through complex evolutionary histories, literally sharing a material connection through the helices of our DNA. And indeed, we are not individuals but multispecies entities ourselves, each of our cells a collection of multiple species that have over the centuries symbiotically evolved to create individual cells and subsequently whole organisms (Margulis 1998, Marguis and Sagan 2002). Yet, our biologies and cultural ideas resolutely center the individual (Gilbert, Sapp, and Tauber 2012). And what about science? We tell a unitary story of "western science" as a linear, progressive story of knowledge produced exclusively in the west (Teresi 2001). Yet, science itself is the product of multiple miscegenations. Appropriating, embracing, and accumulating knowledge as it traveled through complex histories of colonialism, trade, and empire, science is the ultimate mutt. These pedigrees erased and forgotten, the mythologies of a pure "western" science continue to be told in colonial and colonized worlds. We can bemoan the colonial legacies, but if we continue to tell these stories of some pure entity called "western science" in the twenty-first century, we should also see it as an unfortunate and lost opportunity to tell new stories.

These are the mythologies we tell about science, about ourselves. Divided always, specialized always, each neatly packaged in our disciplinary boxes. Yet, we are connected through molecules, through the helices of our DNA, through histories, through geographies, through colonial travel, through vibrant trade, connected through an ever global, ever connected world. Oliver, a character in Ruth Ozeki's *A Tale for the Time Being,* creates an art project, a botanical intervention he calls the Neo-Eocene, a collaboration with time and place. He argues that the rapid onset of climate change will radically expand the term *native* to include formerly and even prehistorically native species. After all, native depends a great deal on how you define it. We might well have a return to older times!

Like the peach, the kudzu, the snakehead, and the carp, I find my own journeys and travels caught within these circuits of global capital. My own value and worth is tied to vast naturecultural assemblages. I began with noting how in moving from India to the United States, I was transformed from a native to an alien. Yet, my exotic status as the only South Asian graduate student in my department twenty years ago is now a rarity. Today, South Asians and Chinese are overrepresented in science and engineering departments. What was exotic once is now being transformed through the new geopolitical realities into the fear of India and China, the emerging powers of the future. My identity, my fortunes, and indeed all of ours are tied to these global networks and circuits of power. We cannot escape them. Aliens of the world unite! We are all aliens, we are all natives, in this vast entangled naturecultural multiverse.

PART III

Biographies of Variation

The Case of Women in the Sciences

The limits of variation are really much wider than anyone would imagine from the sameness of women's coiffure and the favorite love stories in prose and verse. Here and there a cygnet is reared uneasily among the ducklings in the brown pond, and never finds the living stream in fellowship with its own vary-footed kind. Here and there is born a Saint Theresa, foundress of nothing, whose loving heartbeats and sobs after an unattained goodness tremble off and are dispersed among hindrances instead of centering in some long recognizable deed.

—George Eliot, prelude to *Middlemarch*

CHAPTER SEVEN

Through the Prism of Objectivity

Dispersions of Identity, Culture, Science

> Although the universe is under no obligation to make sense, students in pursuit of the Ph.D. are.
>
> —Robert Kirshner, qtd. in Ferris, *The Whole Shebang*

> There was an arid precision to life that bothered me. In fact, science was like table manners, a ritual correctness in life where morality yielded to routine.
>
> —Shiv Vishwanathan, "The Laboratory and the World"

> I guess transformations—the really important ones—require more than time and distance, and even desire.
>
> —Chitra Banerjee Divakaruni, *Arranged Marriages*

My memories of second grade largely come from a wastepaper basket. In my second-grade class, whenever a student talked too much, usually when the teacher's back was turned and there were more interesting things to talk about than the lesson at hand, the teacher would empty the wastepaper basket. Then she would make the student stand in the wastepaper basket in the corner of the room to observe the rest of the class. My year in second grade is filled with images from that corner, my feet in the basket. My teacher added a comment to my report card, "too talkative." Despite the fact that I was being disciplined, my self-confidence and self-worth flourished from the wastebasket corner as my friends would make faces and signs at me when the teacher was not looking.

This sense of creativity, self-confidence, assertiveness, and rebellion from boredom that punctuated my educational experience in India is one I pondered often during my years in graduate school. I ascribed it to a sense of privilege I felt as a middle-class Indian girl born into an upper-caste family belonging to the religious majority, schooled mostly in urban all-girls' schools. As a member of the majority, all social markers worked in my favor. Marginality—minority status in ethnicity, nationality, religion, caste, educational background, the "otherness" that breeds insecurity and marks one's body and identity—were alien to me. Entering a graduate program in biology in the United States, I literally became that alien, officially, a nonresident alien. The science classroom in my graduate years contributed to a growing sense of marginality, insecurity, and invisibility. I desperately wanted to conform and belong. I struggled to find a voice to articulate a word, a sentence. I longed for my wastepaper basket years!

When I began taking courses in women's studies as a graduate student in the sciences, the experience was completely different. The very social categories that marked my "otherness" in the sciences were now tools of knowledge making, providing rich standpoints from which to interrogate U.S. culture, academia, feminist theories, and the educational settings I found myself in. These new perspectives were enabled by theories and contexts of feminist scholarship and the pedagogical tools in these particular women's studies classrooms. Feminist scholarship gave me the tools by which to interrogate my experiences on the other side of campus. What is it about the U.S. graduate education system and the sciences in particular that can turn a relatively self-confident third world woman into an insecure, marginalized one? How are race, gender, nation, and sexuality coded into the enculturation of scientists and the educational process?

During my graduate training in the sciences, I was an avid journal writer. In this chapter I use these memories to explore and theorize what these experiences can tell us about the culture of science. To be sure, experiences are at once rich in their depth and detail yet particular and idiosyncratic as a source for theory. To be sure "experience" is by now a fraught and much critiqued site for theory making in feminist scholarship (Joan Scott 1991). In using my experiences in graduate school, my goals are less about codifying an authentic or definitive narrative of being a woman in the sciences. Experience is not the "origin of our explanation, but that which we want to explain. This kind of approach does not undercut politics by denying the existence of subjects; it instead interrogates the processes of their creation" (Joan Scott 1991: 797). In this book, I am interrogating the processes that shaped my experiences. I explore questions of genealogy, geography, and biography and their interrelations in the sciences through my experiences in the hallways of science. These experiences, while individual, point to how the histories, cultures, and epistemologies of science are operationalized within scientific life reproduced through the generations, sometimes with surprising fidelity. My decision to use this form comes from three main gaps I see in feminist science and technology studies. First, personal narratives and memoirs are a highly underutilized mode in feminist science studies. Personal narratives can reveal the personal, professional, and institutional connections that are underdeveloped in the field.[1] Women are a diverse group, and I explore differences among women of color with the complexities of gender, race, class, and nation. Second, feminist science studies has largely focused on the construction of scientific knowledge rather than the production of scientists (Traweek 1992), but I see value in doing these simultaneously. To me, feminist frameworks and critiques of objectivity were immensely useful as I negotiated my life in science. I explore how presumptions of objectivity and rationality constrain and curtail originality and innovation in the culture of science, shaping normative expectations and scientific norms. Feminist scholarship and its critical reflexivity can enrich the lives and lived experiences of scientists as well as science.

Finally, and most important, I want to argue that graduate education is a critical juncture in the educational ladder where students are "enculturated" into their professional identity as scientists (Subramaiam and Wyer 1998). Graduate education is not only about learning scientific methods and methodologies, but also about developing a professional identity. After all, graduate school is structured around enabling a remarkable transformation in one's relationship to the faculty—from "student" when one enters to a "colleague" when one leaves. Yet the power dynamics that inform this transformation are hardly benign though seldom discussed. Graduate education is an undertheorized and relatively unexplored area, a rich site through which to understand how scientists are trained and produced.

It is striking that in higher education the term *selection* is used in reference to sorting students. In undergraduate education, large introductory science courses are often called "weeder" courses—where weak students are weeded out so as to "select" for the scientifically gifted students (Barton 2001). Similarly, in graduate school, ideas about who is "gifted" or can "cut it" or has the "spark" are rampant. Still, being academically gifted does not ensure the makings of a great scientist—something else is necessary—and so my graduate student colleagues and I often pondered the mythologies about this *je ne sais quoi* factor. It was striking to me how in my graduate program the frame of "natural selection" of plants and animals permeated similar frames of the "selection" of students. There is thus a deep structural resonance between our genealogies of variation (the research practices that produce certain science and scientists), the geographies of variation (the processes that brought new species into the west and helped the global circulation of floral, fauna, people, and knowledges), and the biographies of knowledge (the practices that have "selected" a strikingly homogeneous scientific institution and ivory tower). As I hope is clear, genealogy, geography, and biography are part of a "naturecultural selection" that shapes the sciences—its practitioners, cultures, and knowledges. George Eliot in the opening epigraph to this section reminds us of the many brilliant women, people of color, and third world individuals who were excluded from science and scientific inquiry and of those who persisted, how their achievements were rendered invisible or marginalized in "lost history." Anyone who has worked with graduate students today will recognize the persistence of these patterns. The many ghosts of talented individuals roam the hallways of science. We need to listen to them carefully.

What follows represents my efforts to hear the ghosts, prompted by flashes of memories captured in my graduate school journals.

Beginnings

I have arrived! The excitement of leaving one's country is tremendous. The dream of enrolling in a graduate program in the sciences is unfolding before me. I am all eyes and ears. Large, sparkling buildings, well-equipped labs, and exciting research. It is amazing to be in the heart of the scientific enterprise. On the social front, everyone smiles and says "Hi!" Coming from a city, the friendliness is infectious. I feel loved and accepted. It is wonderful. Of course, they never get my name right. Oh why, oh why do I have such a cumbersome name! They all tell me in astonished tones how wonderfully I speak English. I am puzzled by what their impressions of India must be.

I happened on the United States with what all scientists dream of—a null hypothesis! I believed that there was no racism, sexism, and classism here. I came into a biology department where the number of men and women graduate students had been equal for more than a decade (although not at the faculty level). Thinking back, I had a naive, exuberant openness that I still find incredible yet heartening. My training in the sciences in postcolonial India was strongly grounded in assumptions of objectivity, rationality, and value neutrality. Thanks to a traditional education (British style), I fervently believed that the sciences provided a haven where identity ceased to matter. I reveled in the possibility of belonging to a profession where social and cultural prejudices were irrelevant and where hard work and merit would win the day. I was an individual like every other. Such was the legacy of colonialism in British India, the enlightenment postcolonial subject!

Growing up in post-independent, secular, urban India meant growing up with the promise of science. Science was central to my image of modernity. Science, as it was taught to me in school, as it was represented in the books I read and the popular culture I watched, was "western" science. Indigenous forms of science and medicine had not been (and are still not) integrated into the rubric and authority of science. Even though Indian scholars have made rich contributions to western science and mathematics, I was never taught this. Religious orthodoxy was in my eyes associated with discrimination, "backward" thinking, superstition, and blind faith. When my family would consult the astrological charts to look for auspicious times for a move or tell me I should not sleep with my head facing north, I scoffed at them. When I saw families separating girls and women during their menstrual days, I was outraged. I ridiculed silly superstition, laughed at irrational tradition, and became enraged when I saw discriminatory or hateful practices against any man or woman. Growing up in postcolonial India meant having access to a vibrant and visible feminist movement. To my young, modern, and urban feminist self, my feminism and politics were linked to ideas of modernity, and modernity was linked with claims of reason, and reason was linked with objectivity and rationality of science. Science could not have found a more committed or ardent citizen!

Very early in life, I was passionate about the sciences. I was drawn to their call for logic, reason, rationality, and objectivity. Science was a meritocratic world where my identity as a woman, Indian, third-worlder was irrelevant. The white men (dead and alive) who inhabited my textbooks were my role models, and I was quite oblivious to my brown skin or my sex. A large poster of Charles Darwin hung above my desk. It did not occur to me that, with the exception of C. V. Raman and J. C. Bose, there were no Indian men in my science textbooks and certainly no Indian women. The hope for the world rested squarely with science. It came as no surprise that after an undergraduate education in biology in India, I should cross the ocean and come to the United States for a graduate degree in evolutionary biology full of visions and dreams of being a model scientist.

Throughout postcolonial India, "western science" was the science that the Indian state supported; alternate forms of science and medicine have remained in the periphery. As Susantha Goonatilake suggests, modern science in the third world has always been defined by the center, that is, the west, and any creativity that has emerged has come from indigenous and peripheral practices (1984). Western science had been transplanted into India and embraced as the central force for modernity and development. For a young woman who grew up in such a climate, being part of a graduate program in the sciences in the west was a dream fulfilled. I had indeed arrived!

Enculturation

AN ODE TO THE FURNITURE WORLD

It has been a month. I cannot put my finger on what I am doing wrong, but I have a deep sense of loneliness. People tell me about culture shock. When people can talk to each other, what can go wrong? It must be me. No one around seems to understand the alienation I feel. It is this funny combination of feeling invisible and sometimes too visible! How should I act? Who should I be to fit in? I solve it by a tremendous affinity for the furniture world. So I have begun to play this game every day where I become the table, the chair, the wall, or any inanimate object that catches my fancy, and watch and absorb everything—gestures, words, phrases. It is all new. There are assumptions, expectations, which I cannot fathom or figure out. Being entirely stripped of all context is a very frightening experience. On the social front, all conversations are new information. So, I listen, absorb . . . We go out to eat. I have no clue what is on the menu and no one around me seems to have a sense of my predicament—the ability to eat spaghetti is not an inborn human trait! And so yet again, I watch, emulate, absorb . . . Then, there is living in a technological society. Coke machines, copy machines, bank machines—I stare stupefied, paralyzed. Then comes the absolutely awesome experience of walking into an American grocery store. Shiny fruit and vegetables—all sterile, waxed, and polished (which I am discovering is what they taste like). The choices are mind-boggling . . . I have become quite good at it. The immobility, the invisibility, are like second nature. I have a strong solidarity with all that is wood or steel!

At one level, these sentiments are likely shared by many foreign students. Encountering a new culture, a new educational system is decidedly and necessarily disorienting. But having talked to students across the disciplines, I'm convinced that the particularities of scientific culture differ from those of other disciplines. In hindsight the framework of science that I brought from middle-class urban postcolonial India was unlike that of my U.S. counterparts. Many of them had worked in labs, read scientific papers, and done research projects before they came to graduate school. The world of science was not entirely a figment of their imagination fed by Jacques Cousteau, Richard Attenborough, and science fiction movies! Furthermore, most had taken courses in the humanities and the social sciences as undergraduates. One of the legacies of British colonialism was an educational system that specializes very early. After tenth grade, I chose to follow the "science" route (as opposed to arts or commerce). During my undergraduate years, I chose a "zoology" degree. With little background in the humanities or social sciences, I had no theory of culture. Watching and emulating others would do the trick. These were superficial things—nothing to do with science. I assumed that my sense of loneliness and cluelessness in class had more to do with my educational background and capacity than with the culture that shaped the social relations I was part of—who talked to whom, who helped whom. There was little recognition that such social networks would be crucial to learning. As I discovered, graduate and higher education, at least in the sciences, has less to do with book learning than with intellectual community. Community networks can make or break you. In this, I was ill prepared for graduate school because I presumed my knowledge of nature was all that mattered. The graduate program I entered was ill prepared for me, as well. There was a kind of cultural heterogeneity—there certainly were a lot of different people getting along with one another. After all, the conservative Texan felt he had little in common with the Northeast liberal, but nevertheless they worked together. All in all, I came from a little farther away. While the graduate students were almost exclusively white and from the United States, the wide U.S. geographic differences reinforced the robust rhetoric that difference was irrelevant to the pursuits of science. Initially, I did not realize the degree to which assumptions about the third world, and visions of "the Orient" in particular, shaped their comments and their impressions of me. But at one point I had to recognize that I felt exotic/"other"/marginal in constantly being asked to be the spokesperson for another part of the world. In hindsight, the level of ignorance (theirs and mine) is rather striking. There was no vocabulary for discussing the stereotypes they were using and so for me, as the only "other," it felt futile, even hazardous, to confront them. In the absence of understanding the social dynamics as relations of exclusions and privilege, I criticized myself, internalizing the messages as reflections of my lack of abilities/capacities. In hindsight, I wish I had some understanding of social relations and how cultures worked. It would have made this translation so much more bearable.

So, instead of being accepted as another distinctive member of a heterogeneous community, I got the clear message that I should strip myself of cultural markers. Be like everyone else. In a culture where my differences grew by the day, I wanted to do everything to minimize these differences. I did not want to make waves, did not want to be labeled. It was either that or be the cantankerous bitch, the primitive third-worlder, the ignorant Indian, the hopelessly middle-classer, and so on. Not surprisingly, assimilation was the path I chose.

The Culture of the Classroom

With the "absorbing" attitude I have adopted the "I will work hard" resolution. I am ready, willing, all eager, and inspired to be the evolutionary biologist I have always dreamed of.

. . . I did not realize how different this system is and how much education about education was in store for me! First, I needed to learn the currency of coursework: credits, units, hours. Then it was off to classes. It is difficult to figure out notions of "authority" and "power" in the classroom in the United States. Superficially, the dynamics and the professor/student relationship seem quite egalitarian. You call the professors by their first names. Students assume an air of familiarity with the professor and the light banter is quite refreshing. At first, calling professors by their first names seemed almost blasphemous. Once I got used to that, came my cultural notions of what first names meant—friends. A couple of times I know I crossed the boundary of professor/student and I was told in no uncertain terms that this was inappropriate. Despite appearances there was not much difference in authority structures. I've begun to hear about assumptions made about South Asians and Indians. Friends have told me about professors who talk about Indians lacking the skill to write or think critically; other friends have been told that they do not "act" like scientists. An Indian woman is assumed to be passive, quiet, and obedient.

I keep wondering how I am doing. But it seems impossible without parading my insecurity, which no one else seems to do. Course performance is a secret. In India, it is common to ask your friends for their grades. Here it is secret. I feel I am left floating and therefore am never sure about how I am doing relative to others.

Classroom dynamics have been a nightmare. I feel invisible. I would desperately try to catch the professor's eye to be included. This exclusion and alienation was paralyzing, and slowly, with time, I find that I have withdrawn, first in mind and then in spirit. This social "disconnectedness" manifests itself in the classroom. People talk before and after class and I often have nothing to say. I find myself well and truly the "other"—a nonhuman amorphous being. The problem is clearly me. I am the problem.

Many of my classes involve discussions. The motto seems to be—be aggressive, opinionated, and above all: TALK! And the object of talk is to dismember, tear to shreds if possible, the papers we read. There is nothing healthy about this. It has a vicious, cannibalistic air about it. In addition, an obvious display of self-glorification and egotism surrounds each discussion. Not only are you tearing to shreds someone else's work, you score a point with each tear and anyone else in the room you refute or humiliate. I just heard that a student recently flung a paper across the room because he just did not like it! There is little room for doubt or ignorance—graduate education appears to be training in camouflage or better known as "fake it till you make it!"

The connection between the social relations within science and that of academic work was a difficult one to make. If it was difficult to work it out in my social life, it was even more difficult to make that connection in academic performance. Despite a strong ethos of egalitarianism, the academic world I was experiencing systematically reinforced traditional norms of authority and power. It was the professor who made up the exam and decided on the final grade. It was the professor who decided how I performed in class discussions. At first I did not realize how crucial it was to make a good impression. In India, unknown professors graded the final exams and each student paper had a number and not a name, but here the professor had a face and an opinion of each student while grading. Notions of gender, race, ethnicity, and nationality were necessarily an active part of this process. The assumptions by many that I did not speak English were just the tip of the iceberg.[2]

My new academic world, like the larger culture, also subscribed strongly to notions of privacy. Information about grades, performance, stipends, and salaries was considered private.[3] There was a strong secrecy surrounding performance, and it was virtually impossible for me to figure out how I was doing relative to others. This bred deep levels of insecurity that I later discovered everyone around me shared although it certainly was not apparent in daily interactions. This was deeply connected to issues of dedication. There was a strong sense that students and faculty were there because we loved science and learning. Yet, the culture was very competitive. While no one spoke about their grades, daily interactions were peppered with more subtle indicators. We were keenly aware of who got to go for lunch/coffee/beer with faculty, who was invited to do research, work, and hang out with faculty, as well as who received compliments from faculty. Reports of compliments they had received slipped into students' conversations. It was a culture that bred competitiveness and a deep insecurity. The individualizing and privatization of experience and emotions was the mechanism behind the "divide and conquer" strategy that prevented us from understanding what we had in common and made possibilities of alliances and coalitions very difficult.

The competitive, macho culture was most apparent in discussions. It was a culture of vicious criticism. The majority of the time was spent on what was wrong with papers and little time was spent on what might be interesting or useful about them. This was perhaps the single most important "skill" we were taught and expected to learn. The better we perfected this, the better we were regarded as students. What was ironic was that science was considered "unemotional." And yet, I have rarely seen more emotional or passionate people. It was not that people around me did not tolerate emotions at all, but rather they did not tolerate a particular kind of emotions—those associated with the feminine: crying, displays of insecurity, confession, self-effacement, timidity, giggling. I noticed, however, that "strong" emotions (read as masculine) were coded as revealing one's passion for the work. Faculty could yell, scream, or throw things in the labs in fury yet still be considered excellent scientists.

Class discussions did not have moderators or facilitators, only leaders. There was absolutely no focus on creating a classroom climate in which everyone felt comfortable speaking. There was no attempt to work with students who said little or nothing. Saying nothing was assumed to be an indication of having nothing to say and by extension being a poor student. There was a profound assumption of meritocracy—if students have something to say, they will talk, and what they say will be evaluated as demonstrating their quality of mind. In short, science faculty had no apparent regard for the actual skills of facilitation, teaching, and mentoring that would create a classroom dynamic in which all could participate and thrive.

Scientific Temperament

It is curious how people who initially assume that you cannot speak the language assume so much once they discover you can speak it! But I feel that no one appreciates or understands where I'm coming from. Without seeming ignorant and ill prepared, how do I explain that I had never read a scientific article or even seen the major journals in my field? How do I explain that only one of the faculty at my undergraduate university had a Ph.D. and that none had a research program? Handling expensive equipment is terrifying! Most labs in India are poorly equipped and any expensive equipment is protected. Brought up with this reverence for technology, I was taken aback when someone told me last week that I am seen as "not curious." I did not realize that my hesitation and fear of damaging expensive equipment is perceived as a lack of curiosity and interest. I'm growing to see how different my background is. I realize that growing up in Indian cities all my life, images of nature largely lie in the alien realm of David Attenborough and Jacques Cousteau. I have never had close access to forests. I see that most of my peers have grown up exploring the woods, hiking, visiting national parks or the stream behind their house. Many are naturalists or aspire to be. I do not share this socialization and it feels ominous not to know a single species of plant or animal around. Not a great start for a budding ecologist.

I met with several universalist assumptions about science and scientific ability that interfered with my learning. In my undergraduate education, I had never read a scientific journal or seen many research labs. Even large Indian universities do not have many resources compared to institutions in the United States. None of the professors in my college had a research agenda (most did not have Ph.D.s). There was very little technology, and what was available was fervently protected. The sole copy machine at the college was housed in its own room in the library with a paid operator who controlled access. In the United States, in contrast, the ethos of science was one of play. It was routine to see undergraduates playing with equipment (and often breaking it) and they were provided with expensive chemicals and restriction enzymes that many research labs in India could not afford for the most experienced scientists. It has now become apparent to me that ideas about what constitutes a scientific temperament and "good" scientific practice are very culture-/nation-specific, deeply influenced by the particularities of historical, economic, and political contexts. I wish I had known.

Initiations

POPGEN (short for population genetics), an informal discussion group, meets every Thursday night. It involves a gathering of professors and graduate students. It was started in the "good old days," we are told, in the true masculinist tradition of talking shop over beer into the early morning hours (while, of course, the "girls" cooked and cared for the babies). I know all the people and I like most of them, but put them all together in a room!

The session traditionally starts with announcements and then the speaker is introduced; the object here is to be witty and male. The degree to which this is accomplished depends on the status of the introducee (just how much of an "old boy" s/he is) and the introducer (just how steeped in tradition s/he is). On a really good night you might hear something like, "What I really like about X is that he always gets his priorities right. When not working, it's coffee, beer, and sex and when he's working it's much of the same." Loud chuckles, laughs. Then the speaker starts talking and can be interrupted anytime, and then it is a free for all. The culture is violent, insular, familiar, steeped in inside jokes. An informal history is passed down the generations of all the ones POPGEN "got" and the "you won't believe what he said . . ." quotes. The ones POPGEN "got" involve the historic sessions, the speaker usually being an outsider.

The sessions also serve as an "initiation" rite for students. Anyone in the field "has" to present their work as an unwritten rule. So you prepare in dread and everyone looks at you with empathy as you pass the halls the week before, quite akin to the sacrificial lamb being led to slaughter. With time you are the seasoned veteran with a clique to sit with, nudge, crack jokes in the back, and incite laughter. Most sessions are, however, fairly cordial with an undertone now and then that clearly reminds you that the "old boys" are alive and kicking.

POPGEN was one of the more fun and also defining experiences of my graduate school days. It was a space where you built community, where everyone in the field gathered together one evening a week to informally discuss science. Everyone who worked in the area of population genetics was there. Most of the faculty also made it a point to attend. Very often we got people from neighboring institutions as well as our own. These sessions were intellectually engaged and exciting, and I certainly learned a great deal from them. It was a ritual I attended every Thursday night, and I was initiated into it as well. POPGEN was the in-group, the defining collection of individuals, where multiple lab groups came together. The idea of informal but engaged intellectual exchange was wonderful with tremendous potential. The spirit of engagement, collegiality, and informality was philosophically wonderful, but the ethos betrayed its cultural assumptions—sometimes in clear and terrifying terms. Underlying notions of machismo, power, and masculinity pervaded the culture as a consistent theme in subtle and unsubtle comments, actions, and conventions.

Sharon Traweek in her work on high energy physics suggests that the culture of science is the "culture of no culture" because scientists claim not to have a culture. POPGEN was a moment where I was most struck by this. The culture was obvious and well defined. POPGEN had a clear ethos. It was about exhibiting one's intellectual/scientific rigor through a display of aggression, competition, machismo, and criticism. It was clear what was deemed funny and witty. Cliques were full of innuendo and history. Being "nice" was frowned upon. The initiation ritual was clearly about learning to face the "firing squad" (a word often used). This rigor was proportional to the degree to which one could tear apart others' arguments, humiliate them to the point that they were speechless or in tears. The necessary response was also clear—fight back. Breaking down into tears was a sure way to display your scientific unworthiness, of just not being "cut out" for the profession. It was ultimately about being one of the "boys." In retrospect, what POPGEN represented for me was the universalization of very particular cultural norms: American-ness, whiteness, maleness, heterosexuality, argumentation versus dialogue, along with the assumptions that these norms grounded good and rigorous science.

Dispersions of Identity

Despite all my emulations of furniture and attempts to blend in, I did not succeed. Rather than blend in, my identities seemed to take on an increasingly significant role. Like rays of light dispersing into colors of vibrant spectra on passing through a prism, my travels through the prism of objectivity in halls of science have dispersed into a dizzying and fragmented array of multiple identities. Never in my life had my identities seemed so pregnant with meaning and symbolism, defining and determining as it were issues of professional life and death.

The "prism" has been very useful in thinking through my experiences. Coming to this country, I did not affiliate with any "identity." The culture of the United States, academe, graduate education, and the sciences nonetheless all proved to be prism-like—dispersing and fragmenting the "I." The rest of this chapter tracks the process of my growing political consciousness. To be sure, now I know that we all have multiple identities, belonging to particular or multiple genders, races, ethnicities, classes, and nations at the same time. Yet when no one around me shared these multiple identities, my growing realization of the importance of identity came in fragmented ways—coming to identify the individual colors of a prism. I explore my growing realization of my multiple identities. What I hope these narratives show is that identities "no matter how strategically deployed are not always chosen, but in fact are constituted by relations of power always historically determined" (Viswesaran 1994: 6).

Gender

The silences of insecurity are growing deafening and I've begun talking to friends about the deep sense of insecurity I feel. And to my astonishment, I am not alone! Some of us have decided to start a discussion group on "Women in Science." It has grown into a format of women bringing their experiential accounts and the group attempting an analysis. To our surprise some men have joined. Is this experience gender specific? Don't we all feel insecure and as imposters? So, in the true scientific tradition three of us decided to test the hypothesis. Becky Dunn, Lynn Broaddus, and I devised and administered a short questionnaire on the effects of graduate school—on self-confidence, ranking of self, perceived perception of peers, considering quitting, strengths, weaknesses, future goals, and so on. We were quite astonished by the striking results. Gender differences seemed to be consistent and at times remarkable. And so, we presented our results in an open seminar to the two departments (Subramaniam, Dunn, and Broaddus 1992). The room was packed. The response was hardly surprising. Our study was interesting! The audience was deeply reluctant to accept the underlying suggestion that institutional structures discriminated along lines of gender. There was far too little diversity in the departments to include race, class, or sexual orientation. "Had we considered age? That might explain it"; "The problem was not so much women's lack of self-confidence but overconfidence of men"; "Why were we worried about this? At least one-fourth of the students had expressed aspirations of going to a large research-oriented university like ours. Surely we did not want more!" The comments ranged from questioning, to accepting, to legitimizing our data. The results and process were, however, very empowering for many of us. First of all, our perceived insufficiencies and doubts were no longer individual but formed a collective. A pattern had been documented and comments like "Oh! that happens elsewhere, but not here!" were not valid anymore. With the survey, and, above all with the individual work of women in the two departments, things began to change a little. If nothing else we felt legitimized in treating it as an issue.

Our analysis grew. For example, we observed faculty searches. Graduate students have a representative on the search committee and also meet with the applicant as a group. Graduate student opinions are presented to the faculty, who then vote. Various individuals' impressions of the candidates of course float through the grapevine. Women are traditionally either the "mouse" ("she's much too quiet," "she has no opinions") to the proverbial bitch ("she is always dying to pick a fight," "she's too argumentative," and on one occasion "she wore pants to the interview!"). We've had "nice country girls" to "she's really nice and I really like her, but . . ."

What are the criteria of the process, we wondered? Superficially it would seem that objective criteria exist—of papers published, seminars presented, ability to relate, broadness of interests in biology. Once in a while a candidate appears who fits all the above criteria and who also seems to be a radical and you are thrilled. Clearly the candidate is best qualified. But as we listened more carefully we heard that his or her thinking was "diffuse," "fuzzy." We asked what that meant and usually got a shrug and an "I don't know quite how to put it, well . . ." With time, we realized that the criteria were changing with the searches. Suddenly we were marginalized for having an "objective, non-discriminating" voice. Ultimately, the candidate that was "best" suited was someone who would be a good colleague and the "system" would replicate itself, and thus, the written and unwritten rules and norms of academia were legitimized and perpetuated.

I still remember how powerful this moment was. On the one hand the reaction was depressing. There were some whom we could never convince, never mind what data we produced. There were many who refused to believe that the data captured something about scientific culture. It was a powerful moment for many of us, however, as it led to changes in departmental practices. There was increased representation of students on faculty searches. And suddenly faculty candidates were asked about their perspectives on women in science. For a little while at least the issue was alive.

In retrospect it was entirely a project about culture and climate. What was striking about the results was that all students were having a rough time. All experienced a severe reduction in self-confidence and self-worth, although the decrease was much more dramatic and significant among the women. Even our own understandings were very much within the framework of the sciences. There were some "bad apples" who were making the life of students miserable. Women needed to be better represented among the faculty and in search committees for future faculty. Students needed better advice and faculty needed to be held accountable. None of us knew about the feminist studies of science. We did not recognize that our work had a history with which we could connect our present struggles.

This is where my future would lie. As the feminist studies of science has repeatedly argued, addressing the perspectives of those underrepresented is not a special interest issue but renders visible privileging structures that had been previously normalized—and which impact *everyone.* I am reminded of well-publicized anecdotes such as one from the neurobiologist Ben Barres who after transitioning from a female to a male, from Barbara Barres to Ben Barres, recalls hearing a male colleague praising his work as "Ben Barres" while simultaneously disparaging the work of his sister, "Barbara"; these are powerful reminders of the climate for women in science (Barres 2006, Baty 2010, Begley 2006). Attending to how scientific cultures function to normalize some bodies and behaviors while rendering others aberrant impacts all scientists, men and women, especially future scientists who desire something other than the prototypical disembodied, a-cultural fiction of scientific rationality.

Race/Ethnicity

Resolving my Indian and third world identity was more difficult and it took longer. It was clear how white masculinity was the normative mode of engagement. Most department members were from the United States and white. Whiteness thus became an unmarked, neutral category for science. Since there was little diversity in the department, I tried my level best to ignore that it was a problem until it was so blatant that it stared me in the face. As with gender, the most obvious moments came from the occasional faculty of color who came for talks or presentations. The stereotypes that often surfaced in the department shocked me. In one case, for example, a candidate was introduced for his job by detailing all the violence he had encountered because of his brown skin and presumed nationalities! Candidates are usually introduced with a history of their scientific accomplishments. I could see how innuendos emerged and how they slowly grew to become facts of sorts. At times with male scientists of color, the old trope of the oversexualized male of color surfaced. Watching how social innuendo translated into claims of scientific unsuitability was fascinating. In many cases this felt personal, and I felt hypervisible as a nonwhite, non-U.S. student. But it was these rare moments that gave some of us a window into the racialized world of science. After all, I came to realize that science was inhabited by scientists and that the same prejudice and biases that pervade mainstream culture are likely to suffuse the hallways of science.

In most of these cases, I did not confront my colleagues, and they seemed oblivious enough. And I was not alone since some white colleagues also saw the process for what it was. It was through this process that other memories came flooding through. I had thought them curious and filed them away. I had felt uneasy about the comments but had never confronted the racism, and they all rushed into my consciousness. The professor who said, "She came with a stone on her forehead and look what we've done to her"; a fellow student who read an article somewhere about India where Indians during some religious festival stood in line and jumped down into a river below and drowned. Was it true? Volunteering for pro-choice events, the rationale was often "this sort of thing (illegal abortions) happens in third world countries, but we are part of the industrialized west, it is unacceptable here . . ." It was coming back, all fitting together.

There were no other Indians in the department or indeed in women's studies and not many other third-worlders or students of color in either location. My Indian and international friends were mostly in other fields, often in other universities. North Carolina is particularly stark when it comes to race. All janitors in my department were African American and all faculty but one were white. Similar patterns confronted me in stores and fast-food restaurants. This lack of diversity did give me pause, although it was never personalized. Still, I had become a racialized subject.

Observing the reproduction of racialized subjects of science, I came to understand what was ultimately an act of translation: I was learning to encode cultural difference within the language of meritocracy. A racialized subject is never a simple category. I saw that the African American candidate for a job is read very differently than the Middle Eastern candidate or the South Asian one. Men and women are racialized very differently. Someone like me who grew up middle class is read very differently than someone who might have grown up poor in the United States. Thus different histories of race, I observed, were salient within my scientific culture even though they were seemingly unnamed. Unacceptable or divergent cultural manifestations were reworked as deficiencies of scientific ability or practice. Ironically, it was these same assumptions of race and national origins that helped me pursue my interests in women's studies. Taking courses outside the biological sciences was frowned upon (especially if it was women's studies). When I decided to start taking courses in women's studies, my interest was supported because some faculty felt that they ought to foster and support my interest in "feminism." After all I was from the third world and in need of such influence and education. Indeed, some of my white U.S. counterparts were not so lucky and were asked to put more time in the laboratory instead of "wasting" time in irrelevant courses outside their area of research.

Sexuality

I was entirely oblivious to the heteronormative culture of my graduate training. It was only when I began working more substantively in women's studies that I recognized that assumptions about sexuality pervaded the culture. I still remember the moment when this hit me. I was completing my doctoral work in evolutionary biology while beginning a position in a neighboring institution where I had been hired to establish a "women in science" program. Every morning, just before sunrise, I would stop by the greenhouses to collect data on my plants and subsequently head to my women's studies job. In the evening, I returned to the greenhouse and my office in biology before driving home. One day while en route from the greenhouse to my job I caught myself in the midst of a transformation. I realized that I had begun responding to my bifurcated life by quite unconsciously transforming my persona in the sciences into my persona in women's studies. Earrings went on, plus a waistcoat for color and to cover up the dirt stains, to sanitize my mediations with the natural world. When I drove back from women's studies to the greenhouses, I realized that I enacted the reverse transformation. In the sciences, I had learned to strip myself of all markers of my sex or sexuality. Not only could I not be marked "female," but I also needed to appear to be oblivious about my appearance—a practiced casual grunge. In field biology, clean clothes are the surest sign of not working! The two worlds required phenotypic transformations that required deliberate and practiced performances. Like gender, disciplinarity should also be understood as a performance.

One of the attractions of the sciences was precisely the absence of pressure in having to look "nice." But I was surprised by the unofficial rule and unarticulated pressure to not look "nice." For field biologists, sex and sexuality were very important. What you wore, how you carried yourself, whether and how you dealt with your sexuality (with peers and faculty) were filled with meaning and had profound consequences from comments in the hallways to being "advised" on propriety. Any visible articles of femininity usually elicited numerous comments about one's appearance. I suppose I had enough battles to fight that I never took this one on. In hindsight I remember that I stripped myself of all markers of my feminine self. I stopped wearing earrings, gradually threw away dresses or any clothes that could be construed as feminine, started cutting my hair short. It is rather extraordinary to me to realize the power that climate had over me; in fact, I had little awareness of what I was doing. It is only the power of hindsight and analysis that allows me to recognize the profound impact of the culture on my day-to-day existence. This fit between a scientific identity and a masculine one has profound implications for the training of graduate women since it suggests that being at once a scientist and a woman is not possible (E. Keller 1987). With its roots in a Christian clerical tradition, science emerges from an "exclusively male—and celibate, homosocial, and misogynous—culture, all the more so because a great many of its early practitioners belonged to the ascetic mendicant orders" (Noble 1992: 163). Like the clerical coat, the lab coat signified devotion and dedication. In my case, I made a deliberate choice, informed by my feminist consciousness, to perform my appropriate and expected gender role, in order to avoid the consequences of "gender trouble" (Butler 2006) and any ensuing disciplinary trouble! The rules of dress, however, were entirely different at social gatherings, parties, and receptions. Here, assumptions of compulsory heterosexuality and American femininity were desirable. Thus there were expectations of masculinity within the walls of science and of femininity outside those bounds (E. Keller 1987). To remain a whole and coherent individual required phenotypic transformations or a deliberate feminist consciousness where one deliberately performed one's appropriate and expected gendered roles or battled the consequences of one's "gender trouble" (Butler 2006).

Dissembling

GAMES ACADEMICS PLAY

And so I have begun to explore exactly what about the environment around me nurtured my baseline paranoia—that because I was different, everyone was out to get me. What about the culture made me insecure? I began to observe behaviors, attitudes, underlying presumptions of people's comments, actions, and beliefs.

The "You don't work as hard as me type": This is an extremely insidious type. They constantly have to let you know how late they stayed up the previous night. "God! you're taking time off? Lucky you." It almost seems like an unwritten rule that if you are a graduate student you must have a tired, harried, longsuffering look about you! Being visible around the department is crucial and students develop strategies of advertising their presence.

The I am buddy/buddy with faculty type: These types are constantly reinforcing their faculty connections. "Oh, I went for a beer with . . ." or "While I was chatting with . . ." The implication is that they are fraternizing with "the powers that be" and are on their way up.

Things are going fantastically well type: These are the perpetual optimists constantly stating how well they are doing irrespective of their actual performance. Often the stories through the grapevine tell you otherwise.

The "It is your fault" type: They espouse a singular dedication to the system. If you present them with a critique or criticism of the system, the fault is yours—"You asked for it," "You are misinterpreting." They make up more tasks for you. These are lethal.

The "You're so emotional" type: This is a perpetual problem. "Oh, he did not mean that . . . ," "You're overreacting," "Chill out." The basic premise is that you are feminine, and therefore unqualified.

Other notions of gender and the tightly knit history of masculinity and objectivity also invaded. In a class for which I was a teaching assistant, the professor misspelled a word and knew it was wrong and stood staring. I corrected him. He laughed off his ignorance as "I cannot spell, therefore I became a scientist!" thereby implying that because I could, I should not be one?

You may or may not find this typology useful or true to your experience, but I am sounding a warning. I believe that the socializing and the game playing are an active part of maintaining the "old boy network." Most significantly, what

I am suggesting is that the acceptance of these value systems will not get you into the network. Instead, the value system is designed to exclude the ones who cannot stay as well as to condition the ones who can. For example, spending all your time in the department is valued not because you are getting more work done, but because you are more easily accepted as "serious" or "committed." This structure is never innocent—making it easier to exclude those who cannot stay around the clock and who may even work more productively because of their lives and commitments outside the labs.

Culture and Politics of Science

Coming to political consciousness profoundly transformed my perception of the world around me. A world that once seemed about producing "good science" suddenly also became a world about producing "good scientists." I came to learn that graduate education was about "making" scientists and reproducing a set of cultural practices and behaviors about good scientists. As I observed the world around me, the expectations and norms of the culture of science emerged. It reminds me of Toni Morrison's wonderful observation on learning about racism in American life: "it's as if I had been looking at the fish and suddenly I saw the fishbowl" (Morrison 1992:17). Indeed, the tools of women's studies enabled me to understand science as a historical and social set of structures, practices, and epistemologies of knowledge. Graduate school for me was a place of policing these boundaries—the weeding out of those who did not participate in these cultural practices. And the clashing of cultures of a middle-class, Indian, postcolonial feminist and those of western science was precisely what I encountered—the disjunctures and dissonances of these cultural practices. My love of science was grounded in a fascination of understanding the natural world around me. But what I failed to recognize was that all knowledge is produced by a set of scientific practices generated, agreed upon, and regulated by a community of scientists. The crucial step for me was the recognition that "good" scientific practices were less about good science than about reifying historically generated and reproduced practices.

While learning and doing science, graduate school was also teaching me the set of cultural practices as an important (but unannounced) part of science. And these cultural practices were hardly universal. They were (and still are) a set of practices bound by the historical roots of science in western, white, male, heterosexual culture. And this was precisely what I realized the good students were so good at—aligning themselves with the markers of "good" scientists until it was soon almost naturalized, subsumed, and ingrained. These norms and rules are hardly totalizing, however. Students and faculty resisted them in all kinds of ways—not all of the rules all of the time but rather some of the rules some of the time. Figuring out the outer bounds of acceptance was crucial. And these bounds varied depending on the race, class, gender, nationality, and sexuality of the individual. It was about negotiating a complicated set of social relations—playing it well enough to remain credible and little enough to stay sane!

And so I came to recognize these scientific rituals and postures—when they had to be practiced and when not. I came to recognize when people were trying to put me down, pull rank, appear smarter, point to my insufficiencies, or blame me as the problem. This new perspective created for me a space where I began to enjoy "doing science." Ironically, it took a political consciousness to begin to enjoy doing science again. I had at this point in some ways traveled a full circle—returning to the fascination of science I once felt, although it was more complicated and nuanced. Identities, cultures, and expectations began to fall into place. And it showed. Students began to come to me for advice. How was it, they asked, that I was doing so well? It felt so good!

As I think back on my graduate education, I wonder what it might have been like if women's studies had been an interdisciplinary option of study instead of a crisis management strategy. What if we could break this cycle of silence? Science and the prism of objectivity split my individual self into an awareness of the dispersions of my identities. The tools of feminist scholarship allowed me to refocus these multiple identities to feel whole again. Interdisciplinarity has enriched my intellectual development more than I could have ever imagined. This is the stuff dreams are made of!

CHAPTER EIGHT

Resistance Is Futile! You Will Be Assimilated

Gender and the Making of Scientists

> What was so new about these projects of docility . . . ? [An] uninterrupted, constant coercion, supervising the process of the activity rather than its result and it is exercised according to a codification that partitions as closely as possible time, space, movement.
>
> —Michel Foucault, *Discipline and Punish*

> Making visible the experiences of a different group exposes the existence of repressive mechanisms, but not their workings or logic; we know that difference exists, but we don't understand it as relationally constructed. For that we need to attend to the historical processes that, through discourse, position subjects and produce their experience. It is not individuals who have experience, but subjects who are constituted through experience.
>
> —Joan Scott, *The Evidence of Experience*

During the hey day of the television series *Star Trek: The Next Generation,* I remember being transfixed by the episodes on the Borg.[1] Their repeated and relentless "Resistance is futile; you will be assimilated" was a catchy and unforgettable slogan. Steeped in a project on the culture of science, I was struck by the resonance of what I was hearing from the Borg and from graduate students in the sciences about the culture of science. The singular focus, the complete dedication, the all-absorbing culture, the strict adherence to rules and order, the intolerance of deviance of any kind—these were some of the broad descriptions that emerged about the sciences. The Borg shared the same passions. I use the title not as a way to embrace the antitechnology or anticollective action implied in the Borg episodes, but instead to refer to assimilation in its most literal sense—about the requirement for compliance to gender norms and rules in the culture of science, to a set of norms and rules that refuse to go away despite the entry of women in significant numbers into the hallways of science.

Over the past two decades, the social and especially the feminist studies of science have developed a provocative theoretical framework in which to locate the processes by which cultural constructions of gender are related to the creation of scientific knowledge. Sharon Traweek's path-breaking work reveals that like all institutions, science also has a "culture," albeit a "culture of no culture"! Scientific practice, she notes, requires an objective, rational, asocial, decontextualized researcher, and a person immune from context, from culture (Traweek 1992). David Noble provides a sustained analysis of how the history of science has developed and shaped a culture that has systematically excluded and marginalized women and people of color. He points out that scientific culture originated within the western Christian clerical tradition. The history of science is a history where women and people of color have been systematically excluded and/or marginalized. These histories have consequences and have shaped the contemporary culture of science. In tracing the roots of western science, he observes that

> several habits and characteristics of modern science have often been noted: the strict separation of subject and object, the priority of the objective over the subjective, the depersonalized and seemingly disembodied discourse, the elevation of the abstract over the concrete, the asocial self-identity of the scientists, the total commitment to the calling, the fundamental incompatibility between scientific career and family life, and of course, the alienation from and dread of women. (1992: 281–82)

The belief that science does not have a culture, the resoluteness with which the subjective and experiential are relegated to the personal, are ultimately intellectual and epistemological moves. Science need not attend to the experiences because it claims a "depersonalized and disembodied discourse." Scientists' individual identity does not matter because scientists are interchangeable, all independent nodes in the production of knowledge.

Reflecting on the consequences of science's construction of "nature as female" and "mind as male," Evelyn Fox Keller points out that "this poses a critical problem of identity: any scientist who is not a man walks a path bounded on one side by inauthenticity and on the other by subversion" (Keller 1985: 174). These works suggest the importance of a more sustained analysis of how contemporary scientific culture functions and how variables such as gender, race, class, and sexuality shape this culture.

This fit between a scientific identity and a masculine one has profound implications for the training of graduate women since it suggests that being at once a scientist and a woman is not possible (Keller 1987). A woman would have to be trained in such a way as to render her gender identity invisible or always conflicted and difficult. Appropriate and adequate mentoring in this context would have to make this a necessary component of developing a professional identity. Therefore, mentoring, rather than helping students grow into a new sense of self, is directed at requiring women students to give up a core and familiar sense of self—what Mary Wyer and I have called a (de)mentoring because it removes rather than contributes (Subramaniam and Wyer 1998). This is to suggest that women need to be "untrained" as women and "retrained" as scientists. How do faculty in the sciences achieve this untraining of women and retraining as scientists? How can feminists intervene in this process? What new visions of mentoring can we imagine that do not render "woman" and "science" as incompatible identities?

Armed with these insights, we ventured into science land on an NSF-funded project to open conversations between faculty and students in the sciences about the culture of science and the mentoring of women graduate students. In the absence of a well-developed research base on women in graduate education, we hoped to begin formulating a theoretically informed exploration of graduate women's experiences, one that could explain their underrepresentation across most science and engineering disciplines. Again, as with the previous chapter, some of these observations are likely generalizable to graduate education across fields.

The project I describe was based on an innovative project and methodology developed by Mary Wyer for a previous NSF project (Wyer 1993). One of our first discoveries was that graduate education is structured less around the classroom and more around a model of apprenticeship, that is, a protégé-master model. In such a one-on-one model, interpersonal communication and relationships are critical, and social markers of gender, class, ethnicity, and sexuality ubiquitous. Yet, as Traweek suggests, talking about interpersonal communication, relationships, and social markers is forbidden. At the same time, what the literature on mentoring reveals is the implicit kinship basis of the model—the word *mentor* comes from the *Odyssey,* where Mentor stepped in to guide Telemachus when his father (Odysseus) was away. So the problem of protégé-master is that it is a reproductive model, reproducing the master in the protégé who cannot handle difference; resistance is futile; the mentee's only hope of success is to adopt the master's identity.

We also came to realize that graduate education is a unique training ground in the educational ladder. One enters graduate school as a "student" clearly subordinate to the faculty and in search of training from them. A few years later, however, one is expected to leave as a "colleague" to the very same faculty. While during undergraduate years we learn about science and might perhaps even learn how to do experiments and interpret data, it is during the graduate school years that students learn how to "be" "scientists." For this, they must learn to present themselves as credible professionals—how to design and carry out research projects, choose interesting and productive research topics, plan and carry out the logistics of experiments, give talks, network, discuss science with colleagues, procure grants, publish results, recruit and motivate good students. Graduate school is thus a critical phase in the "enculturation" of scientists and a place for the reproduction of a particular professional "identity." The enculturation of scientists in our project ended up being the production of knowledge makers, who value *not talking about* and *not recognizing* the social world they create, maintain, and reproduce (Subramaniam and Wyer 1998). How does this culture function? How does it reinscribe particular notions of gender, race, and class with the next generation of aspiring scientists?

In order to establish a dialogue between faculty and students, we began with facilitating a conversation between faculty (men and women) and women graduate students about the strengths and limitations of graduate education for women. We worked with four departments (Chemistry, Molecular and Cellular Biology, Ecology and Evolutionary Biology, and Mathematics) at a large southwestern public university. We chose these departments because they had supportive chairs and because they represented different organizational environments for doing research (Fox 2000). We were particularly interested in identifying distinct faculty and student concerns around issues of gender in graduate education.

Following Wyer's (1993) method, we developed four groups: two groups of students and faculty each, in order to replicate our findings. Two groups were composed of women graduate students, with ten students in each group. The other two groups were composed of ten men and women faculty. Groups, like individuals, can often develop idiosyncratic behaviors; two groups therefore made the project more robust, within the limitations of the money and time available to the personnel of the project. There were several criteria for choosing students for this project among the many in the department. These included creating a diverse group (across ethnicity), year at graduate school, experiences in graduate school thus far, kind of research, and small and large lab groups. We chose students who were open-minded, articulate, interested, and committed to being part of the project. Faculty were chosen to create a diverse group along the same lines as students, for their reputation of being supportive of women graduate students, their commitment to seeing an increase in the recruitment and retention of women in science, and their commitment to improving graduate education in general. Two facilitators facilitated five sets of meetings with the four groups, for a total of twenty two-hour sessions.[2] All sessions were taped and subsequently transcribed.

A central concern of the research was to attend to the power inequities between faculty and students. The faculty and students came from the same department, and in a few cases faculty members in the project served as dissertation advisors to some students in the student group. It was clear that this would not produce an open and honest dialogue if students or faculty knew their advisor or student was participating. An innovative aspect of Wyer's research design was to create a dialogue between faculty and students through facilitators without them knowing each other's identity (Wyer 1993). The student and faculty groups met independently and heard about the responses through the work of two facilitators without meeting each other. Therefore, the identity of participants in any group remained anonymous to the other groups.

The Encounter, the Silences: Some General Observations

We began the discussions by asking graduate women to describe their experiences, their interactions with faculty, and their departmental cultures. We encouraged participants to share anecdotes about their experiences with graduate education. The experiences women graduate students shared revealed a great deal about the place of graduate students in the larger fabric of scientific culture and the frustrations and conflicting messages they received from faculty.

What was striking when talking to the two groups was their stark ignorance about each other. It seemed to us that there must be profound silences that accompany their daily lives where although they inhabit the same world, they rarely talk to each other about their daily experiences. It seemed as though faculty and students came from different worlds even though many shared the same laboratory. The issues that were persistent in student experiences were the lack of and the need for greater communication between faculty and students. While there was departmental variation, on the whole, students felt there were not enough occasions for faculty-student interactions and that overall they did not believe faculty cared.

Given that graduate education is largely structured around a one-on-one mentoring system, individual interactions with graduate advisors take on critical significance. In our project, gender and ethnic identities became very important in individual interactions and in departmental cultures. While there were a few examples of blatant objectification of women, in seminars and lectures, these were not prevalent in these departments.

The bulk of examples from students were subtler. Most spoke about small, daily interactions with faculty. For example, a student recounted a committee member who returned a proposal to her telling her basically, "It sucks. You'd better start from the beginning." The student described herself as being deeply depressed for weeks, unable to look at the proposal she had put a lot of work into. She then found the professor to discuss the proposal, and during the conversation it emerged that he thought that apart from the introduction, the rest of the proposal looked fine. She remarked, "He got so pissed off reading the introduction that he put a horrible comment at the beginning and returned it." She pointed out that had his comments been more precise, it would have saved her a lot of anxiety and dejection.

What seemed significant to us as facilitators of these meetings was the disjunction between the two groups—students' conviction that their anecdotes were real and very common, and faculty's insistence that students were overinterpreting or overreacting or just plain wrong. Students insisted that gender and ethnicity played a salient role in student-faculty interactions. While there was variation among and between men and women faculty, it was striking that faculty as a group held more similar views with each other than women faculty and graduate students. It was remarkable that faculty, who were once graduate students, could remember so little of their graduate student experiences. Yet, it is important to note that women faculty did recognize their gendered experiences as women. Some attempted to educate their women students, to soften the harshness of the expectations, while others worked hard to "toughen" women students for what lay ahead. It was apparent that the culture of science enabled a deep conformity, so it was not obvious that women graduate students always found more supportive climates in laboratories run by women.

LABORATORY CULTURE

For disciplines such as chemistry and biology, the laboratory proved to be a crucial location where the negotiation of social relations reigned supreme. Here, graduate students were confronted with interacting with the same men and women day after day. Each lab group seemed to have its own set of issues, often dependent on the lab advisor. Some lab advisors were very involved with dynamics in the lab, often setting rules for research and conduct—such as lab meetings and cleanup protocols. Other lab directors were absent, letting students and post-docs negotiate their ways around the lab. Often this meant that students had to take on the role of developing sign-up sheets and organizing the labs and cleanups. These advisors often discouraged students from coming to them to ask for mediation of a dispute. In such contexts women students described dynamics where they took on the role of mediators of the lab. For example, one woman in an all-male lab discussed how her advisor expected her to be the mediator and confidante in the lab, thus fulfilling the traditional "female" role. The advisor would insist on telling her all about his personal life and problems. The man would not discuss anything personal with the male graduate students. Instead, he would pump the woman for information on her peers. Where had this male graduate student gone? How long would he be gone? The woman had to play the role of mediator and interpreter.

SELF-CONFIDENCE

Another important theme that emerged was the issue of self-confidence and self-esteem. Students described their perception that graduate school was very hard on student self-confidence. Students acknowledged that graduate school was really different than undergraduate life. They felt that it took a lot of time and effort to figure out exactly what was expected of them in graduate school and the fact that these expectations varied within the faculty. They insisted that it ought to be the graduate mentor's role to teach students the ropes, explain to them exactly what they were expected to do and when, and spell these expectations out clearly at the start. Students described eloquently how hard the transition from undergraduate to graduate student was, especially for those coming from small colleges.

Students described, sometimes in detail, their struggles with figuring out how to do research, how to think, how to organize knowledge, and whether and whom to ask for advice. In a culture that privileges being knowledgeable, they felt they had to pretend, remain silent, or fudge on questions raised. "Fake it till you make it," as it is commonly referred to. Many recounted their growing appreciation of posturing in scientific culture and learning how to posture themselves.

> I don't talk to many of my committee members about my future plans. If they ask me what my plans are, I tell them, "I want to be a professor just like

> you" because that's the answer I think they want! . . . I'm not sure what would happen if I were honest. Although I think that I would get less respect from some of my committee members, as if I were not a serious student—which of course I am.

But students identified posturing as a problem systemic in the culture of science and academia. They described the many times faculty would pretend to know something that they clearly did not, unconsciously modeling ways to posture. In summary, students described graduate school as a phase that they knew little about when they entered it. The early years of graduate school were spent figuring out the system and what was expected of them—slowly muddling their way through it. There seemed few departmental or university-wide resources that could help them through it. Student experiences were often idiosyncratic, dependent on the advisor/lab director's investment in the wellbeing of students. There seemed a huge variance in the quality of graduate mentoring. In laboratory situations, students' experiences with lab cultures depended on the laboratory directors and the degree to which they were willing to create a supportive climate for women students and intervene when necessary. Through the discussions of these anecdotes students were calling for more sustained departmental and university-wide support structures for graduate students and recourse for students who were working with unsupportive advisors or labs. Also, they felt faculty should be held accountable for their mentoring practices, and that institutions should reward good mentoring.

In this project, students were not asking for blanket approval or encouragement when it was not deserved. Students stressed that faculty who were not honest in their feedback and criticism were not good mentors either. Rather, they were asking that graduate orientation and socializing become a more uniform and explicit part of the curriculum.

FACULTY MENTORING

My colleagues and I talked to faculty about their part in the graduate school relationship and asked for their views on mentoring. We asked them to describe for us what they considered good mentoring, how they trained their students, what their expectations of students were, and how they had learned to mentor students. A point that all faculty underscored was that they did not receive any training in mentoring to equip them for the job at hand, and they tried to do the best they could.

> I don't know about you all, but going and getting a Ph.D. and going into industry and post-docing for a few years after that never prepared me to teach

> undergraduate classes of eighty people and serve as a mentor to graduate students. I mean, just because you have a Ph.D. in science doesn't provide any training at all for the job. There can't be any other profession in the world where you receive less training for what you actually do.

Faculty seemed hard pressed to name particular models that they used except to generally state that they gave students feedback, encouraged them, and did the best job they could.

The faculty narratives speak to the challenge of both encouraging and nurturing an individual while at the same time being the "disciplinarian." Mentoring models in faculty accounts seemed deeply impoverished. Faculty received little training, and these issues were seldom discussed in the departments in our project. In response, faculty tried to cope the best way they could—by following their instincts and experiences from their own lives, such as with children or their own graduate experience. One of the faculty remarked that many of the "jerks" in academia were the product of their advisors being "jerks." The only overt mentoring model faculty named was mentoring as "parenting"—a very problematic model for education.

Graduate students responded to faculty participants' admission of lack of training with little sympathy! They insisted that faculty were hiding behind these admissions to dodge responsibility, and that good mentoring could be learned and developed over the years, if a faculty member cared. But both faculty and students felt strongly that academic culture does not value mentoring. It carries little weight during promotion and tenure reviews, and there were few rewards to faculty who spent a great deal of time with students, often mentoring students that were not their own. Both groups felt that these were often women faculty. Students in each department could readily name these faculty members.

There were two main overall differences between faculty and students in their discussion of graduate education. First, faculty and graduate students used different languages to characterize mentoring relationships and graduate school experiences. Students stressed commonalties among graduate women's experience. They tended to have talked with others about graduate education, and to share with each other a language for it. Indeed, by the end of the second meeting, both student groups had developed a coherent and cogent narrative about gender and graduate education and about the process of graduate education in general. While there was some variation in student experiences, often shaped by lab groups and departments, students seemed to be able to agree on a general framework for graduate education remarkably quickly. Students were often astonished at how similar some experiences were across departments.

Discussion sessions provided a process to organize individual and seemingly idiosyncratic experiences of students into a collective analysis of graduate education. Student analyses characterized graduate education as a self-replicating and well-reinforced system.

Faculty, in contrast, generally viewed their relationships with students as individual, idiosyncratic, and particular. Anecdotes that students had offered in their meetings with each other as symptomatic of larger currents within graduate education, when relayed at faculty meetings, were usually analyzed as reflective of problems of individuals. Faculty did not seem to share a language with each other about the nature of graduate education, and when they analyzed difficulties they tended to use psychological terms like *self-esteem, self-motivation,* and *self-direction.* In contrast, graduate students preferred an idiom of culture, politics, and power. For faculty, it seemed less true that they saw themselves participating in a system; rather, they tended to see individual relationships and problems. Perhaps this is in large part because of an overriding importance of signaling expertise areas—where faculty habitually distinguished themselves from one another by different areas of expertise, and did not appreciate that they shared in a common enterprise. For graduate students, on the other hand, the process of being a protégé temporarily trumped field or expertise differences and put them in solidarity with others. Moreover, while graduate student narratives themselves seemed quite systematized, and were coherent and cogent, faculty understandings held considerable contradictions. For example, in a discussion of the student-produced list of "Rules for Graduate Students," faculty found it easy to understand this list simultaneously as a sign that students were immature and to express the idea that it was a good thing that students were cynical because academic life is hard. Indeed, as Barbara Lovitts's research on dissertation committee criteria across the disciplines repeatedly points out, there is a lack of consensus and considerable contradiction among faculty in how they develop standards of evaluation. This fits with the individualized, privatized model that the faculty were themselves trained in—they had no sense of collective or systemic agreement of the "rules" (Lovitts 2007).

Second, and in a related phenomenon, faculty and students had very different understandings of what happens over the course of a graduate school experience. Students tended to view becoming a scientist or mathematician as a constructed and somewhat arbitrary process. Faculty, in contrast, saw the process as natural, involving the growth and maturation of something already inside students. Faculty expressed an ideology of meritocracy, believing that there were those who are good at science and those who were not, those who had "it" and those who did not.

For example, the notion of a "spark" was an important part of faculty's description of their ideal students. Faculty largely believed that there was not a whole lot they could do to create this spark, but that once it was there, it was their role to mentor students toward successful scientific careers. Implicit in their narrative was the notion that the making of scientists was a natural process, involving the growth and maturation of something already inside students in incipient form. They felt this growth was idiosyncratic—something that sometimes happened for students and sometimes did not. Faculty felt they had only limited influence on it. Their understanding of what happens left little room for criticism of them as mentors, in the sense that it emphasized a "stay if you fit in, leave if you don't" perspective. That is, in faculty participants' accounts, a student should be able to tell that she or he is really "cut out to be a scientist" if the graduate education process comes to seem easy, reasonable, and rational. If it does not, the student was not meant to be a scientist. This framework (conveniently in the eye of students) short-circuited the possibility of critique. It also completely obscured the possibility of an inclusive pedagogy—an ethics that anyone can learn if taught!

Students in contrast were interested in challenging and reinterpreting the question of who could be a good scientist. They believed that good scientists could be made, and faculty should not rely on their perceptions on whether a student had a "spark." Students reacted strongly to the notion, suggesting that faculty were creating a mystical, nebulous quality to education ("like a mystic talking about attaining a plane of consciousness"), rather than addressing their role as educators in the making of scientists. Students interpreted faculty descriptions of mentoring as a disengaged process.

> It's very much a sink or swim attitude. As though they're just standing on the shore with their arms folded saying, "Yeah, that one is not going to make it. And that one is."
>
> If they told you that's what they were doing at least then while you're drowning you know. You could look up and see who was watching. Or you could say, "Oh! I am supposed to get out of this myself," but you know it's never made explicit and that's a problem.

THE "UNWRITTEN" RULES OF GRADUATE EDUCATION

What are the "written" rules of graduate education? There seemed to be surprisingly few. There were stipulations on required courses (with waivers if the committee agreed), deadlines on picking a dissertation advisor and topic, time for the comprehensive exam, the format of the exams, the dissertation committee and its composition, and limits on the length of time to the graduate degree (after which one had to retake comprehensive exams). But in each of these and in the rest of graduate student life, there was a surprising amount of flexibility. And indeed, there was tremendous variation in graduate student progress and performance. This sets up students to succeed in a very individualized model of achievement with a great deal of uncertainty around one's performance. I remember that when I was in graduate school, we were evaluated each year, and if nothing was amiss, we received a letter saying, "We have not found any evidence that you are not progressing satisfactorily toward your degree." Hardly a resounding endorsement!

One of the more powerful exercises we did was the rules exercise where we invited graduate student and faculty groups to name the unwritten rules governing graduate education. This exercise clearly encapsulated the fundamentally different standpoints that emerged among faculty and students. For this exercise we asked the four groups to name the unwritten rules of academia. Were there any rules that were not explicitly written down but widely believed to be rules or good practices by graduate students? If so, what were they? Each of the four groups was asked to generate the list without seeing the rules generated by the other groups. Anyone was allowed to suggest a rule. The rule was then discussed. If everyone agreed, it went on the list. Even if one person disagreed, the rule was left out. The wording was often carefully and painstakingly negotiated. The four lists of rules appear in the appendix to this chapter.

The rules exercise was much easier to do with the graduate students than the faculty groups. The students seemed to quickly grasp the point of the exercise, and there was easy agreement. There was more disagreement about the rules with the faculty group, and often this meant that rules were "worded down." Some of the faculty were uncomfortable with the strong wording and successfully argued to make them more general and open-ended. For example, "Do not have children" after a lengthy discussion was changed to "having children is an issue." Similarly, "pregnancy is a liability" was changed to "pregnancy and other time-consuming personal decisions are problematic." There was a striking difference in the level of consensus and the degree to which individuals felt there were clear rules. The differing tone and emotions in the room were palpable. If one faculty member suggested a rule, some other faculty member would counter it with an example or felt that was too rigid a stipulation. In the student group, by contrast, when a rule was suggested, they often felt like moments of revelation—that people had articulated something that had organized their lives for many years but no one had named until that point.

REACTIONS AND ANALYSIS

The faculty-student (via facilitators) conversations in our project highlighted the unique standpoints by which faculty and students experienced and interpreted scientific culture. Power implicitly shaped these differences in perspective just as power has demonstrably shaped the practice of science and the production of knowledge more generally. Ironically, and not surprisingly, initially neither group felt they had any power![3] One of the outcomes of the project was to get both groups to appreciate the power that they held, a power that could allow them to rethink and reshape the cultures they inhabited. As Foucault suggests, power is "exercised" in relation to others rather than a quality merely possessed (Foucault 1980). In our project, faculty and students experienced their power, or lack thereof, through relationships experienced through the graduate education process. The significance of the rules is not in whether these rules were empirically accurate or not. The significance lies in the implicit belief of the students in the rules, an implicitness absent among faculty narratives.

The rules exercise illustrates how gender norms and roles are operationalized in specific settings within graduate science and math education. The sets of discussions around the rules exercise provided a rich set of interpretations of the culture of science from participants. The culture of science, like all cultures, represents sets of rules and expectations of all individuals who participate in that culture. Social relations, and the rules and expectations that govern these relations in fields like science and engineering, have a long history of male domination and are particularly problematic for women. One of the strongest tenets in the culture of science is a studied silence about social relations, a silence designed to distinguish scientists from others. In our project, the resistance of faculty to talking about and recognizing larger patterns in their daily interactions and behaviors seemed a symptom of a more general resistance to their own cultural embeddedness. It suggested that those with power and privilege cultivate an exaggerated commitment to individuality as a form of resistance to cultural change (Subramaniam and Wyer 1998).

The expressions of privileged perspectives we found in our project were consistent with the concept of the "invisible knapsack" often used in introductory women's studies courses (McIntosh 2003). The nature of privilege, it is argued, is unmarked—an unweighted and invisible capacity to move without burden in a structure in which you are the norm, versus the burdens others carry of being marked. Scientific culture, students suggest, has unwritten rules and unacknowledged notions of who makes a good scientist. These characteristics of good scientists and science's "no culture" are deeply aligned through history with the west, masculinity, whiteness, and heterosexuality. It is precisely the privilege of *being* an individual that is critical, rather than a situation in which one's "individuality" is eclipsed by the markings of subaltern identity. One could argue that scientific claims to being culture free are not unlike a racist system's claims to color blindness—whereby the insistence that color doesn't matter is precisely what allows racial privilege to continue unchecked.

While many graduate women in our study were articulate about these institutional practices, it was apparent that initially they were reluctant to speak up, precisely because they felt surrounded by a culture of science that was silent on the importance of social categories of gender, race, class, and sexuality in the doing of science—in interpersonal dynamics, mentoring relations, and the role of personal issues in professional life. Students and faculty alike repeatedly remarked how much they learned purely by talking to one another about social relations, a practice they had seldom engaged within their home departments. The rules exercise provided a useful tool by which students and faculty examined the culture they worked in and in highlighting how similar and different those views can sometimes be. The exercise produced an interesting moment in knowledge generation as an actual *intervention* in the culture and practice of science.

In retrospect, the list of faculty rules more closely resembles the goals of graduate education than codes of conduct—they stressed the learning of the social and intellectual climate of one's field, learning to formulate and carry out research projects, and learning to communicate about one's work. As one student remarked, "Who could disagree with that?" Students interpreted their set of rules as the means and mechanisms by which they could attain the faculty's list of "goals." For example, faculty rules stress the following:

- Graduate education should be the major focus of your life.
- Personal lives and extracurricular activities are okay so long as they do not interfere with reasonable progress.
- Become the "expert" in your field.
- Show initiative and independence.

In order to attain the above goals, students pointed to rules in their lists that allowed them to meet the above expectations. These rules include:

- Work all the time with no break.
- Be visible around the department.
- Don't show fear, self- doubt, insecurity.
- Don't have a life outside the department (or don't talk about it).

- Be assertive, confident, good communicator.
- Be happy.
- Be busy.
- Don't complain even about real problems.
- Don't cry in public.

Thus students responded to the sets of unwritten expectations of faculty by a set of behaviors that best presented them as "serious" students. Students elaborated how gender was an important factor in their graduate school experiences and in their everyday lives. The rules exercise points to the perceptions women students have of their place in science. Women face an additional burden in having to follow the rules to be seen seriously as scientists. In their personal lives outside scientific culture, women face the pressure of conforming to our culture's notions of "femininity." But once they come to work, within scientific culture, they have to leave that identity behind, to assume another, namely, a form of "masculinity." Often, this takes the form of dissembling behaviors. Within the culture of science, women students participate in these rules in public, such as being happy or not crying in public. Students respond with their own set of rules in private, however—"cry in your office" or "leave your lab if you think you are going to explode." Student narratives unfold a world of performances, where they learn to perform a set of behaviors and practices that will allow them to be taken seriously. A "serious scientist" is single minded, dedicated, emotionless, happy, well connected, intellectually curious, the expert, and ready to follow in the footsteps of the mentor to reproduce the system. It is not that the students themselves change. The contradictions around emotions are always interesting—that emotions are always read as undesirable and feminine; hence it is possible to hold the contradiction of being emotionless and happy! Instead, they quickly learn to segment their lives—they cry in their offices or the bathroom, they learn to leave the lab to release their frustration or tension outside. And yet as the rules tell us, women scientists are not accepted completely if they become masculine. For no matter how much they give up femininity, they can never be men. Narratives that have emerged from the recent rise of the transgender movement, such as Ben Barres's, emphasize this point. This, I believe, is the double bind students describe and what Keller argues when she talks about women scientists being bounded by inauthenticity and subversion (E. Keller 1987).

The faculty were of course deeply disturbed by these rules. Their responses were threefold. One was complete refusal to believe that these rules exist. Some specifically talked about how they work against these rules especially by being models themselves. It remained a consistent struggle in discussions to have faculty elaborate a role of a collective culture even though they may have individually resisted specific rules. The second response was that students were missing the point. They thought that if students did not enjoy doing science, did not enjoy participating in these rules, that rather than dissembling or pretending or play acting, they ought not to pursue this line of work. A third response (especially around the persistence of the long work hours, the importance of connections, and the chilly climate for women) was that students had gotten it right and life would not get any easier when students joined the faculty ranks! Again, while there were differences between men and women faculty, especially when many women faculty felt marked by their own gender, there were far more similarities between men and women faculty than between women faculty and students.

With respect to gender, it seemed that students had a more unitary and cogent understanding of how gender impacted their lives as students and scientists than faculty did. Yet women faculty were quite explicit and active in enforcing gender norms, particularly with respect to dress (not too feminine, not too casual). Even so, it seemed that departments did evolve their own idiosyncratic cultures, and overall, gender norms and expectations were more coherent within a single department than across departments or the university as a whole. This, in turn, provided a clue as to why student and faculty interpretations might have differed. For students, power tended to be concentrated in the single form of the faculty mentor. For faculty, in contrast, power was much more diffused—in department heads, senior colleagues, administration, journal editors, professional colleagues at conferences, and so forth. Thus, faculty indeed experienced gendered pressures and inequalities in more diffuse contexts. Even pretenure faculty, who are perhaps the most vulnerable, recognize the need to impress all their faculty colleagues in contrast to students who are primarily accountable to their individual dissertation advisors.

It seemed to us, in listening to the two groups describe graduate education and the day-to-day life of graduate student-faculty mentor interaction, that both significantly misapprehended their relation to the other. Graduate students tended to overestimate faculty power, control, and investment in the day-to-day life of graduate students. They also reported that it was hard for them to understand their place in the wider academic system. They had difficulty understanding how graduate education fit within the larger set of constraints and obligations experienced by their faculty mentors. Faculty, in turn, tended to underestimate their power in relation to graduate students, their impact on the quality of life of their students, and their role as gatekeepers to professional careers in the science and mathematics.

As feminist scholarship reminds us, gender is something we learn to "do," not an attribute one "has." For the students in our project, graduate education was a contested exercise in enculturating students into scientific culture and teaching students how to "be" scientists. What women students in the project were often challenging and objecting to in their descriptions of the rules of scientific culture was not an unwillingness to follow the rules to do science. Rather, they challenged the notion that these rules did produce good science and, more important, that *not* following these rules would produce bad science. They weren't questioning the need for structures themselves, but critiquing the values attributed to those rules and the ways they dispersed privilege unevenly. Empirical work on the qualities of good scientists would suggest that there is no ideal constellation of cognitive abilities that shape the ideal scientist. Successful scientists demonstrate a range of characteristics such as deductive reasoning abilities, verbal skills, quantitative reasoning, intuition, and social skills (Handelsman et al. 2005). Narratives also highlight creativity, imagination, effective communication, and thinking outside the box as useful skills (Handelsman et al. 2005).

If the current rules did not necessarily produce "good" science, then we were free to imagine alternate frameworks. This re-envisioning allows us to draw on models that are diverse, flexible, and informed by difference. It was exciting to see that by the end of our project, participants were imagining alternate frameworks for graduate education in the sciences. By examining and critiquing their own culture, by the end of the project, faculty and students began imagining and exploring and ultimately even implementing alternate models for a culture of science, one that did not define science and scientists exclusively in terms that our culture has relegated to the "masculine."

Such an understanding of the culture of science can be immensely powerful for the transformation of institutions as well as those of individual students and faculty. Rather than train graduate women to perform successfully in the existing culture of science as most mentoring programs do, we can instead asked men and women in science, faculty and students, to examine the world in which they live. Through the articulation of what we do and how we do it, we can ask ourselves how we can do it differently—incorporating and celebrating the diversity of genders, ethnicities, classes, nationalities, and sexualities that make up our world. Perhaps, in so doing, one day the diversity of scientists will mirror the composition of that diverse world, and in so doing, produce very different—fuller, multifaceted, interdisciplinary science.

Appendix. NSF Model Project

FACULTY GROUP I

Some Unwritten Rules of Graduate School/Academia

1. Think of yourself as an independent researcher as early as possible.
2. Show initiative and independence.
3. Graduate education should become the major focus of your life.
4. You should decide early if you enjoy research and if not find something else.
5. Having children is an issue, although it's getting better.
6. Pedigree is still important.
7. Read more, think more than you are required to.
8. Go to seminars.

9a. Develop communication skills—give talks/seminars.

9b. Learn to write.

FACULTY GROUP II

Unwritten Rules of Graduate School

1. Attend seminars/colloquia.
2. Participate actively in seminars (at least eventually).
3. Be intellectually curious—take initiative, ask questions, visibly discuss science with faculty/students.
4. Take responsibility, be professional (you are not an undergraduate anymore)—work hard, make progress.
5. Take ownership of your research project—become an "expert" in your field.
   - Learn the background.
   - Be independent.
   - Take intellectual responsibility.
   - Learn how to locate information.
6. At least begin to learn the social and intellectual climates of your field (including outside the home institution).
7. Research/academic positions are construed as "best."
8. Exhibit integrated, clear thinking.
9. Cooperation and collegiality (exhibit it).
10. Don't procrastinate too long—realize that graduate school is a life phase, not a lifestyle.
11. Practice and develop communication skills—oral and written.
12. Learn to communicate about your own work.
13. Personal lives and extracurricular activities are okay as long as they do not interfere with reasonable progress.

14. Pregnancy and time-consuming personal decisions are problematic.
15. Personal and professional needs have to be balanced in a constructive way.
16. Don't expect your research advisor to be your personal and professional mentor.

STUDENT GROUP I

Unwritten Rules of Graduate School

1. Work all the time with no breaks.
2. The only work that counts is what is publishable.
3. Students on the fast track will get more resources and faculty time.
4. Emotions permitted; excitement, "controlled" emotions not allowed: being upset, overwhelmed, insecurity, apathy, and lack of interest.
5. Don't go public with career plans outside Research 1 university jobs.
6. Be like your advisor.
7. Connections count (good and bad).
8. Don't show fear, self-doubt, insecurity.
9. Don't have a life outside the department (or don't talk about it).
10. Be assertive, confident, good communicator.
11. Know your place in the hierarchy.
12. Students ought to be available whenever faculty want, faculty will be available to students whenever faculty want.
13. Faculty time is more important than student time.
14. Students should be more visible around the department (not true in Math).
15. Know the local landscape about rules.
16. If things aren't going well, you may not get direct feedback.

Rules for Women Graduate Students

1. If you get pregnant, you may be perceived as less serious but if you surmount that, kid(s) must never interfere with work.
2. Double binds: know your place and don't be timid.
3. You will be either a bitch or a doormat (more or less true depending on how many women are around).
4. Not okay emotions—being upset, feeling defeated, overwhelmed, insecurity, apathy, self-doubt (more true for women than men).

How do you get around the rules?

1. Ask permission.
2. Ignore the rules (can't always get busted because they are unwritten).

3. Pick and choose rules you want to follow (get help from other graduate students).

STUDENT GROUP II

Unwritten Rules for Graduate School

1. The official rules are flexible.
2. Early success helps.
3. Be visible at departmental functions—be a "good" citizen.
4. Be a colleague to faculty, but know your place.
5. Don't complain, even about real problems.
6. Don't make waves—don't be a troublemaker—you will be punished (this extends to junior–low-status faculty).
7. You don't have input, even on decisions that affect graduate school (even when asked).
8. Faculty will not interfere in other faculty's business.
9. Connections count.
10. Be happy.
11. Be busy.
12. Be visible in the department.
13. Don't cry in public.
14. Work all the time.
15. Be just like your advisor.
16. There is an "expected" track, but the funding crunch is modifying it.

Rules for women graduate students

1. Act happy.
2. Cry in your office.
3. Don't complain (especially true for women).
4. Dressing "feminine" elicits less respect, more help; your legs are public property.
5. Being a strong woman, a strong advocate for women, or notably a feminist is not good either.
6. Being a woman is a liability. You have less leeway with unwritten rules if you are a woman (varies with department).
7. Don't be feminine: don't exhibit—dress, intonation of voice (feminine speech), don't wear makeup, don't show feminine emotions (i.e., women more vocal about emotions, menstruation).
8. Pretend to be like your advisor or be quiet about it.
9. Pregnancy is a liability (pulls you off the expected career track).
10. Don't have a personal life.

CHAPTER NINE

The Emperor's New Clothes

Revisiting the Question of Women in the Sciences

> "The boy is right! The Emperor is naked! It's true!" The Emperor realized that the people were right but could not admit to that. He thought it better to continue the procession under the illusion that anyone who couldn't see his clothes was either stupid or incompetent. And he stood stiffly on his carriage, while behind him a page held his imaginary mantle.
>
> —Hans Christian Anderson, *The Emperor's New Clothes*

> The Lesson
>
> And after the Emperor had appeared naked and no one had disturbed the solemn occasion, one little girl went home in silence, and took off her clothes. Then she said to her mother, "Look at me, please, I am an Emperor." To which her mother replied, "Don't be silly, darling. Only little boys grow up to be Emperors. As for little girls, they marry Emperors; and they learn to hold their tongues, particularly on the subject of the Emperor's clothes."
>
> —Suniti Namjoshi, *Feminist Fables*

In the famous Hans Christian Anderson fable, *The Emperor's New Clothes,* the emperor commissions a cloak. Under pressure to produce ever finer cloaks, the cloth maker and tailor invents a deception—the finest fabric ever, a fabric so fine and fantastical that it is visible only to the truly intellectually deserving and is invisible to all else. While the emperor himself cannot see the cloak, he pretends to be able to and wears his imaginary cloak in a procession through town. The "emperor's new clothes" has become a metaphor for power, pretentiousness, social hypocrisy, and collective denial. I use the emperor's cloak as a useful metaphor to discuss "cloaks" that frame the literature on women in the sciences. As the famous fable unfolds, the cloak turns out to be nothing but an illusion—nonexistent and fantastical. It is only given visibility through the power of the emperor and the complicity of his subjects; the emperor is, in fact, quite naked.

Having examined variation's genealogy in evolutionary biology and traced its ecological geographies, in part III of this book I have been looking at how the question of variation has haunted our ideas of who can be a scientist and has shaped demographic patterns of scientific practitioners. Ideological debates about the importance of variation for evolutionary theory seeped into social relations within science in unexamined and disquieting ways. Women, as "outsiders" who might have provided variation in perspectives within science, have been in some historical periods expressly excluded, in others belittled and discounted, and in more recent times silenced and implicitly excluded through the micro-dynamics of "cumulative disadvantage" (Etzkowitz et al. 1994).

Nonetheless, women and other marginalized groups have fought long and hard to enter the hallways of science. In the past three decades activists have moved through the courts with antidiscrimination laws and affirmative action programs (such as Title IX in the United States) as well as through women in science and engineering (WISE) initiatives, projects, and programs. By WISE initiatives, I include programs literally named WISE at most universities but also women in science initiatives at professional societies and institutions, that is, funded programs and projects on "women in science and engineering." WISE initiatives have their political and intellectual roots in the women's movement and its academic arm, women's studies. Indeed, we can trace the origins of what is now called feminist science studies to early feminist work in the late 1970s and early 1980s when women scientists began to articulate how questions of sex, gender, race, sexuality, and nationality were inextricably interconnected and implicated in the institution of science, its history, cultures, people, practices, and knowledges. Despite these origins in the experiences of women scientists, feminist work within women's studies has gradually moved away from questions of the "problem" of women *in* the sciences to ask questions about gender *and* science. The literature on gender and science focused not on the practitioners of science but on the ways in which scientific knowledge was gendered; work shifted to the important project of exploring the production of knowledge in science. The move from women *in* the sciences to gender *and* science divorced questions of the practitioners and producers of scientific knowledge from questions about the scientific knowledge they produced. Scholars focused on the content of science, while activist faculty and staff focused on providing a supportive environment for students who were majoring in science, technology, engineering, and math (STEM) fields. The activist efforts resulted in the growth of WISE programs, largely housed within the administrative wings of science and engineering colleges. These programs have been broadly charged to increase the recruitment, retention, and overall status of women in the sciences. While their charge is broad, the focus is usually more instrumental than academic, in providing women with networking and mentoring opportunities, professional development activities, and leadership training as part of an equity feminist approach. They seldom articulate a critique of androcentric bias in academic practices, and instead offer a "you can do it" cheerleader approach to women students touched by the program. The programs are also usually evaluated by the popularity of events rather than the institutional change they have effected.

I am myself deeply indebted to WISE programs. They were an important part of my early beginnings and I have since worked in and continue to support WISE programs. Yet, as a woman in science who works in feminist science studies, I am struck by how little interaction there is between the two. This chapter is written in the spirit of the tremendous possibilities of coalitions and collaboration that I see between women's studies and WISE initiatives. At this time, with few exceptions, WISE programs tend not to be centrally located in feminist science studies or women's studies programs and departments since they are not engaged with questions of scholarship or curriculum. Given their broad mission, WISE programs could have developed a robust analysis. As I elaborate later in this chapter, it should not surprise us that they have not. Women and gender studies programs, in turn, have too few allies in science and engineering fields. While we could debate the details of institutional dynamics, one thing is clear. While all three fields—WISE programs, feminist science studies, and women/gender/sexuality studies—can arguably be said to be thriving, there is little interaction between them. The considerable funds invested in WISE programs and initiatives could, I believe, yield impressive results if WISE programs engaged more fully with feminist work in the social sciences and the humanities. If we are truly interested in diversity and addressing the question of variation on all levels—genealogical, geographic, and biographic—it must involve bringing the project of women *in* science back together with the question of gender *and* science.

Between the Matildas and Curies: A Brief History of Women *in* the Sciences

In challenging the claims of scientific objectivity, feminists have uncovered the systematic ways in which power and privilege have shaped science and scientific practice. They have detailed the systemic exclusion and discrimination against women.[1] These exclusions were based on a complex stew of ideologies about rationality, androcentrism, white privilege, eurocentrism, and biology. It is now apparent that white males dominate the scientific elite through no accident of history, as scientists were powerful cultural and institutional voices in sustaining assumptions about white male superiority in science—hardly disinterested or objective. In this section, I briefly review the history of women in the sciences in the United States—the dynamics of exclusion, as well as the social movements by women scientists to challenge these exclusions and demand their right to full participation in the scientific enterprise. There are two predominant concepts in this history of women in the sciences: the "Matilda effect" and the "Curie effect." The Matilda effect was named by Margaret Rossiter as a corollary to the Matthew effect (Rossiter 1993). The Matthew effect contends that privilege matters within science and that the work of highly ranked scientists is more likely to be noticed and recognized than that of lesser-ranked scientists (Merton 1968). Rossiter coined a term for the inverse effect, the *Matilda effect,* which argues that in contrast to the Matthew effect, the contributions of women scientists are usually denied or ignored. After their deaths, they are forgotten, their contributions lost and invisible in the historical records of science. The effect is named after Matilda Gage, an important American suffragist who argued that the inventor of the cotton gin was a woman (Hess 1997). In contrast to the Matilda effect, we have its converse, the Curie effect, which recognizes the popular conception of Marie Curie as *the* woman scientist (Des Jardins 2010). The Curie effect points out that in order to be famous, women scientists have to be unusually brilliant like Marie Curie, who won a Nobel Prize in the sciences not once but twice. These two effects represent two phenomena in the history of women in the sciences—invisibility and extraordinariness. Somewhere between the forgotten Matildas and the extraordinary Curies lies the worlds of most women in the sciences.

Nonetheless, historians have unearthed the long buried contributions of women to science, mostly focused on the United States and Western Europe (Gates and Shteir 1997, Shteir 1999). This is not a linear history where women have been accepted into the profession with increasing and growing respect.

Rather, political shifts have influenced science and women's contributions in complex ways (Rossiter 1982, 1995, Eisenhart and Finkel 1998). Historians have unearthed biographical details of "women of distinction" (Kohlstedt 2004: 3) such as Rachel Carson (Hynes 1989, Lear 1997), Anna Botsford Comstock (Henson 1987, Kohlstedt 2004), Marie Curie (Pycior, Slack, and Abir-Am 1996, Pycior 1997, Jardins 2010, Quinn 1995), Rosalind Franklin (Sayre 1975, Maddox 2002), Anna Mani (Sur 2001, 2011), Mileva Maric (Renn and Schulmann 2000), Barbara McClintock (E. Keller 1983), Lise Meitner (Sime 1997), Anna Kouzeletsova (Koblitz 1983), and Ellen Swallow Richards (Lippincott 2003), among others. Questions of nation, race, and class complicate these narratives (Koblitz 1983, Sur 2001, Johnson 2011, Bilimoria and Liang 2011).

Margaret Rossiter's now classic two volumes on women in science as well the work of other historians demonstrate that the systematic marginalization of women has kept the numbers of "great women" to relatively small numbers, that is, exceptions rather than the rule (Rossiter 1982, 1993, Abir-Am and Outram 1987, H. Rose 1994, Pycior et al. 1996, Henrion 1997). A wide variety of studies have revealed a consistent pattern of discrimination. Several influential phrases frame these perspectives (Wylie 2011)—women in academe face a "chilly climate" (Hall and Sandler 1982, 1984, Sandler 1986), women find themselves in the "outer circle" (Zuckerman and Cole 1991), or women are "outsiders in the sacred grove" of the prestigious academy (Aisenberg and Harrington 1988); if they do succeed, they quickly hit the "science glass ceiling" (Rosser 2004a). Programs aimed at women in the sciences are predicated on the idea that women will make excellent scientists if only given the opportunity.

The national statistics showing an increase in Ph.D.s awarded to women in most STEM fields provide concrete evidence that women have been determined to become scientists, despite obstacles. The success of women in the sciences in the United States owes a great deal to their persistent organization and activism. Kohlstedt (2004) notes key successful strategies developed in the late eighteenth and early nineteenth centuries, which continue to this date: (1) documenting discrimination against women scientists (Rossiter 1982, S. Levine 1995, Kohlstedt 2004); (2) identifying practices and standards used to exclude women (Rossiter 1982, 1993, Kohlstedt 2004); (3) determining practices used to contain women in socially appropriate areas of work such as home economics or some fields in the biological sciences (Nerad 1999, Kohlstedt 2004, Veit 2011); and (4) developing remedies to redress the marginalization through public recognition of women, collective action, and leadership development (Kohlstedt 2004). While these strategies facilitated women receiving advanced degrees, they often failed to get jobs, promotion, and recognition. By the 1920s, women's advancement seemed to plateau or even lose ground (Kohlstedt 2004). The growing professionalization of scientific fields developed cultures of growing masculinity that further marginalized women (Nye 1997). As old barriers fell, new barriers and forms of stratifications were put in place (Rossiter 1982). During and after World War II, women in the sciences continued to be paid less than men and had restricted and even reduced access to laboratories and resources compared to wartime levels. The push to return women back to the home in the postwar years further exacerbated these patterns as mainstream ethos highlighted the role of women inside the home as wives and mothers (Kohlstedt 2004, Rossiter 1982).

The consistent activism and strategies had a cumulative effect, however. The postwar baby boom led to an increase in the number of women scientists although most were employed as researchers and faculty in the poorest and least prestigious institutions. Those in more prestigious universities were clustered in home economics, though a select few had appointments in genetics and anatomy (Kohlstedt 2004, Rossiter 1995).

The post-*Sputnik* support for science along with the rise of the women's movement and the emergence of the academic field of women's studies facilitated the organization of women's groups in science.[2] Social transformations in the workplace through the Equal Pay Act, push for childcare, and legal challenges to discrimination were important for all women, including scientists (Freedman 2002). Scholars began to document how precipitously women's presence in science declined with increasing educational achievement and used these statistics to push for programs at NSF and other institutions (Rossiter 1995). Equally striking was the unevenness in the presence of women—clustered in particular fields like psychology and the life sciences but severely underrepresented in others like the physical sciences and engineering. The metaphor of the "pipeline" emerges in this era, a metaphor about the participation of women in the sciences that has endured in the United States ever since.

By the 1990s, feminists gathered data through various federal agencies to show that the gains of previous eras were not being sustained (Kohlstedt 2004). Scientists and activists pushed for legislative and institutional commitments to change. Congress passed the Women in Science and Technology Equal Opportunity Act in 1980 extending Title IX law (which forbids sex discrimination at educational institutions that receive federal funding) to the fields of science and engineering (Handelsman et al. 2005). Regular surveys of the scientific workforce and yearly compiling of data on undergraduate and graduate degrees (organized by gender, race, nationality, and ability) by NSF gave rise to constant quantitative and qualitative monitoring of women and minorities in science (Mervis 2000).

Alongside promoters of women and minorities in science have been critics of such programs. They have argued that women are not good at science; that women don't want to do science; that such initiatives shortchange boys; or that affirmative action initiatives are inherently unfair (Cole 1979, Lawrence 2006, Pinker 2002, Baron-Cohen 2003, Mansfield 2006). These debates have often pitted men against women, whites against students of color, or U.S. nationals against foreign nationals. Questions of parity and fairness have been presented as being at odds with each other, and it is claimed that diversity is being promoted at the cost of excellence (Mervis 2000). These debates have often created a backlash against initiatives to increase the representation of women and other marginalized groups (Holden 2000). The 1990s also revealed a plateau in the increasing number of women in science, while the number of men declined. Women remained concentrated in the psychological and behavioral sciences (Kohlstedt 2004), and continued to be clustered in less prestigious institutions, outside the ranks of tenure and in institutions that focused on teaching rather than research (Long 2001). In fact, in the 1990s women lost ground in mathematics and computer science, fields that emerged as the "hot fields" of that decade (Kohlstedt 2004, National Science Foundation 2002). The 1999 Massachusetts Institute of Technology (MIT) report "Women in the School of Science" reframed the public debate about the status of women in science by powerfully demonstrating the continued discrimination in the post–civil rights era (Wylie 2011, MIT 1999). Discrimination, the report argued, was created by a pattern of "exclusion and invisibility" affecting women's workplaces, their quality of work life, and their careers and career trajectories compared to similarly well-trained male scientists (MIT 1999: 8).

While the above brief history has focused on the United States, underrepresentation of women in the sciences persists throughout the world, although the historical patterns vary considerably across continents and national boundaries. Women are underrepresented in the sciences across the globe, especially in the more prestigious and well-paying fields. In national contexts where mathematics and physics appear to be less important, women are better represented and more visible. Cross-national analyses make a strong case against biologically determinist arguments and demonstrate powerfully the complexities of the demography of the scientific workforce (Mellström 2009, Kumar 2009, Subramanyam 1998).

Thus, women's marginalization in science is best understood as one of political and systemic exclusion. These practices of exclusion have been well documented by women scientists and their biographers, historians and social scientists, lawyers and university administrators, and feminist science studies scholars. The practices and processes that have discouraged, marginalized, and diverted women from their scientific interests and talents are alarmingly enduring, evident in the research that spans over two centuries of U.S. history. The ideological thematic that underwrites this durability is the recurring debate about women's intellectual and physical difference from men, a difference constructed against a western-white-male-as-norm backdrop. The touchstone points in this debate echo and reassert twentieth-century eugenic arguments about the importance of variability in evolution. Of most concern are the ways in which today's discussions about women's abilities are voiced by those who intend to be advocates for women in science, even while their arguments evade confronting the deep structural foundations of masculinist science. It is an unavoidable fact that elite white western male scientists embedded notions of "superior" and "inferior" human biology into early formulations of evolution. Contemporary versions of these ideas are evident in resorting to discussion of demands of pregnancy, motherhood, and family as particular burdens that women in science do (or do not) or should (or should not) embrace. The case for biological difference endures in these formulations. Do women have the intellectual ability to be scientists? Can women's brains grasp the advanced mathematical modeling necessary in science? Can women have careers in science and be mothers too? Discussing and answering these questions is much-tread territory within programs for women in the sciences. Unfortunately, such emphases, in the absence of critiques of cultural norms in science, sustain the unspoken assumptions that variability is of questionable value for humanity.

Women in Science and Engineering Programs and the Politics of Inclusion

Considering that women were once seen as incapable of rational thought and were barred from higher education and the hallowed halls of science, the success of women in STEM fields today is remarkable. The proportion of women in science and engineering occupations has increased considerably (Bilimoria and Liang 2011). For example, there has been a thirty-fold increase in the proportion of Ph.D.s granted to women in engineering. This period corresponds with women's persistent activism; this shift is best explained by the removal of cultural and structural impediments rather than innate differences (Handelsman et al. 2005). Indeed, we should not underestimate these changes. Yet, as we shall see, the increase has plateaued in recent years or at times, even declined. Why?

Over the past three decades, government agencies such as NSF, the National Institutes of Health (NIH), as well as private organizations such as the Association of Women in Science (AWIS), the American Association of University Women (AAUW), the Women in Engineering Professional Action Network (WEPAN), the Society of Women Engineers (SWE), as well as women's task-forces and caucuses within professional societies of scientific disciplines have all funded projects and programs to increase the recruitment and retention of girls and women in the sciences. There is an enormous literature that explores the successes and challenges of various intervention efforts at K–12, undergraduate, graduate, postgraduate, and faculty levels.[3] The pattern of underrepresentation of women persists for women in most science settings in contemporary times (National Academy of Sciences 2006, Settles et al. 2006), and studies repeatedly find that bias persists (Kenneth Chang 2012, Moss-Racusin et al. 2012). Targeted programs at NSF have worked to intervene in the institutional culture of departments (Rosser 2004b): working to reduce discriminatory practices in hiring and promotion (Fuller and Meiners 2005, Martinez et al. 2007, Rosser 2004a), as well as to empower and mentor women for a successful life in science and engineering.

To the often-asked question on whether feminism has changed science, most have answered in the affirmative but with qualifications (Bug 2003, Kass-Simon and Farnes 1990, Rosser 1990, Schiebinger 1999, 2008). Women's presence has not changed science because they have brought feminine or feminist values but because their very presence has helped erode traditional gender labels (E. Keller 2001). Indeed, despite well-developed theories and analyses on feminism and science, initiatives for women in the sciences have been grounded in and reduced to a narrow liberal idea of equity or equality of representation as the primary intellectual, political, and ideological strategy (Cinda-Sue Davis 1996). This comes from a larger cultural adoption of a principle of equality of *opportunity* rather than equality of *outcome*—which is the more radical commitment to social justice. It is well worth pondering why demographic numbers are the cornerstone of focus rather than cultural or structural transformation. Despite equity feminism that most of these efforts promote, parity for women in the sciences remains a distant goal—the numbers seem to have plateaued, and in some fields and subfields they have declined (National Science Foundation 2011). The professoriate remains male dominated, especially at prestigious institutions and senior ranks. Why? At the heart of the problem here is the very worldview that frames women in science discourse. The problem, as the pioneers of feminist science studies articulated it, is an exclusive focus on women instead of on the gendered and racialized nature of the institution of science itself. The exclusive focus on increasing numbers of women has led to a myopic and singular focus on equity and parity and a narrow vision of feminism and what feminist theories can offer. The women *in* science discourse has been haunted by a path of "relentless linear progressivism" as the result of particular discursive formations (Garforth and Kerr 2009). There have been some changes in the predominant language in the past three decades, but interrogating the underlying assumptions about women is rare. The shift from exclusion to inclusion is marked by a shift from exclusion based on claims of the innate biological inferiority of women's scientific abilities to a politics of inclusion dominated by policies that address women's biological bodies and gendered roles as wives, daughters, and mothers. Consistent with other segments of the labor market, in considering the various barriers facing women, women scientists most often cite the need to balance career and family (Rosser 2004b). Interviews, case studies, and statistical research consistently find that individuals report that family/work balance discriminates against women scientists at structural, institutional, and individual levels (Rosser 2004b, Rosser and Taylor 2009). The personal choices, relationships, and responsibilities of women outside the halls of science (especially as wives, mothers, and daughters) have nurtured and supported women's scientific work but also stymied and curtailed careers (Kohlstedt 2004, Zuckerman and Cole 1991, Laslett and Thorne 1997). Pregnancies, childcare, and housework have always largely fallen within the domain of women's work and women's roles as wives and mothers, and so have been consistently highlighted as a reason for women's lack of equal participation (Long 2001, Mason and Goulden 2004, Xie and Shauman 2003).

These concerns have led to a push in "female-friendly" policies. Despite their progressive ambitions, emphasizing issues of reproduction and family, advocates of "family-friendly" policies reassert women's reproductive potential as a central concern, marking "female difference" as hypervisible while leaving the worlds of masculine epistemic cultures untouched. To be sure, it is neither desirable nor persuasive to articulate an "anti-family-friendly" perspective at this moment in time, but as a strategy for inclusion, the consistent emphasis on family and women reinforces essentialist ideas about women. What has remained unchallenged is the normative model of the male as *the* ideal scientist, which insists on a productivity that can only be achieved by very long hours, a singular dedication to work, and an exclusive focus on one's profession. Solutions have included mentoring women to negotiate the normative model, promoting those who accept the normative model, retaining women through "special accommodations" that increase their workplace flexibility (part-time appointments with administrative or teaching responsibilities), automatic pregnancy leave, and family leave. The original standards for excellence are never challenged; the solution is about helping women conform to them. In extolling feminine virtues, the women in science discourse endlessly reinscribes women in relation to femininity and the domestic reproductive sphere (Garforth and Kerr 2009). The idea that the inclusion of women means first and foremost "family friendly" creates a universal woman and positions all women in relation to a heteronormative reproductive economic model (L. Morley 1999, 2006). In addition to essentializing women as different from men and without variability from one another in commitments to family and children, such strategies leave the mythology of an objective, value-free masculine world of science and technology untouched (Garforth and Kerr 2009). Indeed, as Louise Morley (2003) has shown, at higher levels of science, such a focus on women and family has created a hypervisibility around women, gender, and family, rendering women more vulnerable in their careers as their success is often seen as political rather than meritorious. In all of this, women get marked as separate and their bodies are marked as different and ones that need to be endlessly monitored, while men remain invisible and unmarked. After all, as feminists have long argued, gender is not what one "has" but what one "does" (Morland 2011). Challenging the elision of norms of masculinity and scientific masculinity has to be a central project for feminism and science.

Finally, initiatives related to women in science and engineering (WISE) have created an endless structure of vigilance and monitoring (Garforth and Kerr 2009). There are constant audits, grounded in identity categories of gender and race, and these evaluations constantly reinforce and reinscribe rather than challenge a rather narrow vision of race and gender. Again and again, gender is discussed in relation to work/life balance, childcare, harassment, bullying, human resources, recruitment, promotion, and so on. At the same time large swathes of epistemic, organizational, and personal lives of institutions and researchers remain unquestioned, including normative expectations, research policies, male networks, and privilege (Garforth and Kerr 2009). When inequities persist and the initiatives fail, they are seen as "implementation gaps." Affirmative action and equal opportunity policies become audits and an "inspecting body" rather than a challenge to the micropolitics of institutional power. These efforts have ended up making the identity of marginalized individuals more visible and their presence endlessly surveilled. Through all of this, masculinity and whiteness remain the unmarked and normative categories that escape scrutiny. As Garforth and Kerr (2009: 398) eloquently summarize: "Perhaps the problem with women in science is not women or science, but a women in science problem in its own right."

In examining the history of initiatives for women in the sciences, I consider *five* key problems that haunt efforts on behalf of women in the sciences.

METAPHORS THAT FRAME THE DISCOURSE ON WOMEN IN SCIENCE

One of the key insights of feminist science studies is that language matters (E. Keller 1985). Language embeds, expresses, and enacts the concepts that frame an issue, thereby shaping resulting policies and solutions. The WISE literature is rich with governing metaphors. Catch phrases and pithy summaries reveal a great deal about the ways in which interventions have been framed and characterized. Two related metaphors govern the field—the process of producing talented scientists as a pipeline that "leaks" and the process of career development as a "path" that leads necessarily to success albeit with some bumps along the way.

The "pipeline" metaphor creates a visual flow-chart about those entering and leaving the sciences and is widely used in discussions of the recruitment and retention of girls and women in science (Subramaniam 2009). The metaphor invokes a long pipe leading from kindergarten to the scientific laboratory with "leaks" representing the attrition of women and students of color. Solutions are conceptualized as "plugs" that keep more women and students of color within the institutional machinery of science. The pipeline model is predicated on the notion that if more women enter the pipeline and stay in it, more will become scientists and academics. Low participation in science and attrition from science are portrayed as resulting from individual choices rather than a process of discrimination. Such a portrayal fails to account for structures of institutions or practices of science (Schiebinger 1999, Bystydzienski and Bird 2006) or social influences and contexts (Xie and Shauman 2003).

While the metaphor of the pipeline describes flows in the machinery of science, it also evokes other visions. We could describe the pipes as long, dark, dingy, impenetrable tubes and masses of metal crisscrossing the terrain of industrial capital; pipes contain, constrain, limit, and cut off the oxygen from the travelers within. Imagining the regimented travels in pipes that give the travelers no agency in their journey, we might start cheering for the leaks and for those who escape the drudgery of pipe travel! Such a picture also describes the experiences of many women in the sciences, although, I hasten to add, not their thrill of discovery and exploration (Subramaniam 2009). And this, I believe, has been the crux of the difference between the literatures on women *in* versus women/gender *and* sciences. In the former the leaks are seen as a problem, and in the latter the problem is the pipe itself. If science is indeed in the business of laying down pipes that are inhospitable to marginal groups, why encourage young girls and women to enter them or stay in them? Why not rejoice at the leaks as a symbol of escape?

The second metaphor represents a scientific career as the final destination along a clear path of steps, aka "career ladder." Here, I draw on the work of Garforth and Kerr (2009), whose incisive and insightful critique of such programs in the United Kingdom translates well to the United States. In their analyses, career development in science is presented as a linear and hierarchical progression in the growth of expertise, experience, and vision, all in a timely fashion. There are clear "stages" that are most often imagined as a "ladder" or "path" along which scientists need to move toward excellence. Resting, stopping, or not progressing "on time" is evidence of failure (Garforth and Kerr 2009). The focus is thus on counting bodies at every stage of the career path—K–12, undergraduate, graduate, postdoctorate, assistant professor, associate professor, professor, senior membership in elite professional societies, and recipient of awards and accolades. There are also references to challenges confronting marginalized groups along the way, describing an obstacle course—bumps, hurdles, hoops, and pitfalls as well as institutional obstacles such as glass ceilings.

As Garfield and Kerr (2009) argue, the WISE literature is organized around two related narratives—one about progressive change over time and another about science as a meritocracy. In the progressive change narrative, women in the sciences march toward equality within institutions that are becoming more modern with equal opportunity and greater diversity. In this narrative, historical change is "progress," where "much has already been done," inspiring advocates for women in the science to continue the work because "more needs to be done for full equity." In this narrative, a "more equitable future is always in sight" (Garfield and Kerr 2009: 389). In the second narrative about science as a meritocracy, scientific institutions and their policies are seen as well structured and fair. It is individuals within them who are sometimes discriminatory. Through adequate training and mentoring, women can be taught to be more productive and assertive and the evaluators more objective and fair. Programs are largely focused on identifying women and people of color who "pass" or qualify for normative standards for insider status (Iverson 2007). Thus, it is assumed, that individuals with talent will inevitably (with hard work and dedication) be recognized and rewarded for their career achievements.

Both narratives eerily echo themes in the debates about variation in nineteenth-century evolutionary biology that asserted evolution as a historical progression toward excellence in a world filled with natural (and therefore desirable) hierarchies, with white males perching at the top. Evolutionary biology has yet to resolve the core debate between those who argue that variability promotes vitality and those who argue that variability dilutes it. The narrative of women *in* the sciences replicates science's own narrative of scientific progress, when in fact rather than linear progress, the history reveals circularity because the core problem of variation and difference is never solved. Thus the analogies of evolutionary biology's genealogy of variation mirror its biographic problems. Change-oriented activists on behalf of women in science would benefit from engaging with feminists who study the social world of science if they are to confront the narrative frameworks that contain their initiatives. For despite good intentions, strategies to include women have been governed by uninspired, regimented, and conformist notions about the conditions that foster a career in science. The programs work hard to enculturate women to survive the dingy recesses of the pipeline armed with strategies of self-motivation, networking, mentoring support, and superficial antidiscrimination policies to soldier on in a prescribed, timely fashion. Programs are inevitably focused on supporting women and people of color; the onus is always on women and people of color, not the structural forces of sexism, or racism, or other forms of discrimination (Iverson 2007). The western-white-male-heterosexual normative standards of science remain uninterrogated. If science developed as a world without women, women in science programs continue this by defining the goals of women's "success" in terms modeled on male scientific careers and female reproductive capacity.

THE CRISIS IN HIGHER EDUCATION AND THE QUESTION OF WOMEN IN SCIENCE

Recent headlines declare that there is a crisis in higher education in the United States (Hacker and Dreyfus 2011, Taylor 2010). We hear of deep cuts in public funding of higher education, challenges to collective bargaining of faculty, a steep reduction of tenure track faculty among the professoriate, escalating debt for students, poor learning outcomes (Arum and Roksa 2011), and high unemployment for graduates. Only 44.7 percent of students in the graduating class of 2011 held jobs that even needed an undergraduate degree—students with majors such as education, teaching, and engineering fared best in finding jobs that used their skills to continue to work in their area of specialization (Rampell 2011). Similarly, there has been a steady decline in graduate education, and in the numbers of those who go on to join the professoriate. "We are producing too many PhDs," says Mark C. Taylor (2011). The proportion of people with science Ph.D.s who get tenured academic positions in the sciences has declined steadily. The problem is most acute in the life sciences, where only 15 percent find a tenure track position after six years of a Ph.D. in contrast to 55 percent in 1973 (Cyronoski et al. 2011). Private industry does not have enough positions that require a doctorate to absorb the surplus. Some researchers argue that there is an overproduction of Ph.D.s and that it is "scandalous" that politicians and educators continue to speak of a Ph.D. shortage. Yet, as we saw in the previous two chapters, graduate education is geared to a future in academia and academic research was the top career choice for Ph.D. students in 2010; most Ph.D. programs train students specifically for such a future (Cyronoski et al. 2011). Given this overproduction, we need to reassess academe (Taylor 2011).

Predictions of an upcoming boom of jobs in STEM fields have persisted for decades, but as a share of doctoral-level employment opportunities across all sectors, tenure track academic jobs have shrunk dramatically—from 75 percent of the professoriate in the 1970s to an average of 25 percent in recent years (Bousquet 2008). Academic employment opportunities in short-term contract faculty positions with far less job and economic security have increased; women are overrepresented in such positions (Feldman and Turnley 2004). Similarly, in the private sector outsourcing and globalization have created significant wage reductions. Women are concentrated in lower-paying technical jobs rather than in positions as leaders or heads of research labs and departments. It is rather remarkable that despite these shrinking opportunities, there is a continuing effort to recruit women to the ranks of those awarded Ph.D.s. The overproduction of women Ph.D.s in a dwindling job market creates an overqualified but underpaid workforce. In this context, the relentless pursuit of WISE programs is astonishing. There is little discussion or preparation for this reality in WISE-related initiatives. As a result, WISE-related initiatives it would seem are complicit in the creation of a highly skilled but underpaid, underappreciated, underrecognized, and uncertain (largely female) workforce.

THE LIMITED MODELS OF DEFICIT VERSUS DIFFERENCE

Why are women less likely to stay in STEM fields and why are they less likely to be successful in scientific careers (Sonnert and Holton 1996)? Two competing models predominate in analyses of gender differences in educational and career outcomes in STEM disciplines (Sonnert and Holton 1995, Barbercheck 2001). The deficit model assumes that men and women are similar and are motivated by the same goals and aspirations. What accounts for the disparity is that women are disadvantaged by structural obstacles, lower status positions, lower pay, and limited access to resources and reward networks (Sax 2001). Thanks to the long and persistent activism of feminists, many of the formal barriers have been removed, but many subtle informal barriers remain. Those using a deficit model perspective focus on removing the barriers and training women to overcome them. The second model, the difference model, assumes that women and men have inherently different preferences, aspirations, and goals—either through genetics, socialization, or gendered cultural values (Sonnert and Holton 1995). From this perspective, attitudes about science define it as a male domain and, as a result, rigid gender roles discourage girls and women from expressing and developing their interests in science (Valian 1999, Barbercheck 2001). Women are socialized into attitudes and behaviors that deemphasize qualities such as competitiveness, ambition, and aggressiveness, which are central to science today; thus women are disadvantaged because they cannot conform to gender norms as scientists. Those using the difference model argue that rather than training women into a "male" model of science, we need to engage in a reform of science to accommodate a more diverse range of styles, behaviors, and epistemologies (Malcolm 1999, Barbercheck 2001). Using the same arguments, other scholars counter that women's difference explains their continuing marginalization. A well-publicized study by Ceci and Williams (2011) argued that the glass ceiling has been broken, that discrimination no longer existed, and that women "chose" the life in science they wanted (Zakaib 2011). In the difference model, the reason for gender disparities lies squarely in women's differential abilities or aspirations, desires, and goals.

It is unfortunate that for twenty years these two models that take an either/or approach have dominated frameworks for understanding the underrepresentation of women in the sciences. Grounded in equity feminism, the focus of both approaches is on increasing the numbers of "bodies" of women. The deficit model argues for a focus on overcoming structural barriers for girls and women. The actual structure of scientific education and training remains unscathed—rather, the focus has been on getting women to excel *within* that structure. The difference model focuses on developing a more diverse and plural institution of science, but draws on essentialist assumptions about what girls and women are taught, are good at, and enjoy. Again, the main institution of science remains uninterrogated and marked as "scientific," while essentially remaining male. Interventions within the difference model open up new spaces for the feminine and "female-friendly" science, as an accommodation to girls and women, often misunderstood as feminist efforts. Such ideas of difference mark female researchers and femininity as "visibly different" and male researchers and masculine epistemic subjectivity as invisible and unmarked (Harding 1991, Garforth and Kerr 2009). As Helen Longino (1989) eloquently argues, the conflation of feminist and feminine is not useful. Science needs the whole repertoire of human potential, not a pigeonholed vision of two mutually exclusive ways of being—either masculine or feminine.

What is unfortunate in both models is that mainstream science as a gendered institution remains entirely unexamined (Rolin 2004). The very rules of the game in the history of science are grounded in a history of a racialized heteronormative masculinity that precludes the participation of women; feminine science is a contradiction in terms. Furthermore, the seemingly benign vocabulary of "difference" obfuscates deeper claims of essential sex differences. Feminists have pointed out that the relegation of "feminine" and "masculine" traits is neither innocent nor accidental. When marginalized groups are marked as "different," this difference translates into "deficient" in the language of power (Fausto-Sterling 1985, Shiebinger 1989, Hubbard 1990). For example, women may be celebrated as having great "nurturing" talent; unsurprisingly, jobs that require nurturing talent are less prestigious and pay poorly. Furthermore, such "feminine" talents are rarely heralded as important to science and the scientific temperament. Rather than reinscribing traditional gender roles of masculine and feminine, feminism must be a project about troubling gender, about elaborating how gender might be done as well as undone and redone in new and radical ways or even not at all (Morland 2011).

WOMEN IN THE SCIENCES AND THE EMERGENCE OF THE NEOLIBERAL UNIVERSITY SYSTEM

Over the past two decades, scholars have noted deep erosion in support for public higher education. With progressive cuts to public education, many see this as an ideological turn in universities and the emergence of an increasingly neoliberal university (Slaughter and Leslie 2007, Slaughter and Rhoades, 2000, 2004, Aihwa Ong 2006, Wendy Brown 2011). As Aihwa Ong (2006: 140) argues, there is a double movement in U.S. higher education: "a shift from a national to a transnational space for producing knowledgeable subjects, and a shift from a focus on political liberalism and multicultural diversity at home to one on neoliberalism and borderless entrepreneurial subjects abroad." With growing neoliberalism across the globe, public higher education has increasingly put markets as central to social value. Universities have becomes sites for private entrepreneurship, and technoscience is increasingly the central focus of the university. The neoliberal university thus stresses training students close to the technoscience core of knowledge economies, embracing research that creates high-tech products and processes for private capital, and preparing the future workers for a global technoscience economy (Slaughter and Rhoades 2004).

This neoliberal turn has shifted the focus from the mission of higher education for the benefit of all citizens, to a focus on business and privatization of public goods and services.

WISE projects and programs have continued to thrive through the neoliberal turn and are increasingly well integrated into the neoliberal university. They have accomplished this by adopting the language of economic efficiencies, evaluation, and accountability as their universities move toward market/business models of organization and a rhetoric of transparency and public accountability (Deem and Morley 2005, Garforth and Kerr 2009). Programs are tasked with identifying "high-quality" women and people of color who show scholarly distinction and talent, and working to promote them (Iverson 2007). Program annual reports, evaluations, publications, and activities are now available via program websites that celebrate program innovations as opening new horizons for women in science. These are accomplished either through mentoring women to overcome their "deficit" in training or in accommodating women's "difference" in the evaluative logics of the university. In this trend, the language of equity is accompanied by new political and economic language. Programs tout both economistic (arguments about efficiency and the importance of "tapping" into hidden and unused talent) and democratic (liberal sentiments of equal opportunity and inclusion) goals.

Garforth and Kerr (2009) argue that these changes in language reposition discussions about inequalities through a rhetoric of social inclusion that rests on the logic of market economics: how we can produce more women innovators and entrepreneurs and women's startups, innovate in producing female-friendly toys and computer games, and enable women scientists to produce more papers, grants, awards, and patents? Thus left and right, labor and capital, can be brought together in this rhetoric. These two are presumed but not explained. Indeed, scientific practices are recouched in new terms. For example, Louise Morley (1999) argues that the new quality audit produces new forms of macho competitive individualist masculinities with an emphasis on outputs and targets, while women are made responsible for communal caretaking and the domestic elements of academic labor such as teaching and administration. We are thus moving from an old "collegial fraternity" to a new culture of "technocratic patriarchy" (Hearn 2001). To be sure, some women have benefited from these new managerial practices since celebrated "excellent" ones thrive, but these practices also have had profound effects on the ways in which gender inequality has been reproduced, reinforced, and reinvented through gendered employment insecurities. The successful woman scientist is one who exemplifies a new model of efficiency, producing the right number of publications and receiving the requisite number of grants and awards. But this is not true of the vast majority of women. Indeed, women are overrepresented in teaching faculty and teaching colleges. Women and faculty from marginalized groups are overrepresented in administrative tasks as they get recruited into various committees in the name of diversity. The rules of the new "audit" of who is productive, who receives "merit" money, and who is released from teaching reinscribe gendered differences that have been sutured into the underlying assumptions and logic of modern organizations. The discourse about discrete barriers or pipeline leaks severely misidentifies the problem. We need to recognize that the very logic of academic careers is gendered (Garforth and Kerr 2009).

Acker (2000), and Meyerson and Kolb (2000) argue that while the rhetoric of equal opportunity, equality, and fairness may continue, any work that is not in an organization's interest will be resisted. Demands of individual performativity always overshadow economic concerns, so financial implications are always subordinated to gender initiatives. As Garforth and Kerr (2009: 397) summarize:

> The dominant de-individualized economic discourse of women as resource, the reiteration of the linear career path, the dual rhetorics of institutional and policy progress, the re-inscription of domesticity and subjectivity onto visible women, and the proliferation of gender audit and experts that make up the package of "equality works" cloaks male researchers and the masculinity of science in its traditional robes of objectivity, neutrality, and normality.

Indeed, if science and academia itself is gendered, if the very calculus and logic is gendered, attempts to include women in this logic without transforming it merely reinscribes a gendered logic. Those men and women who fit into the neoliberal logic indeed do well, rising in the hierarchy and being richly rewarded. The rise of these successful women of "excellence" provides an insidious logic in privatizing the "failure" or "choice" of others who do not or choose not to embody such narrow definitions of "excellence." This fails to challenge the structure of advantage and the gendering of science and the division of labor (Wajcman 1991). Recent innovations in women in science initiatives as a result reinscribe gendered ideas that always put the burden on women and the housekeeping of administration, helping the soul of the university and national science initiatives. The deeply embedded rhetoric of nation and nationalism is striking (Garforth and Kerr 2009).

IMPOVERISHED MODEL OF FEMINISM

The idea of feminism in the contemporary United States provokes both backlash as a radical agenda and a "special interest" political perspective, as well as an idea that has been simultaneously and thoroughly mainstreamed. Equity is a case in point. Polls repeatedly show that the majority of Americans believe in gender equity, and indeed bringing gender equity to the world has been a reason to go to war! Yet equity has been thoroughly critiqued within feminist studies as reinforcing ideas of femininity and feminine difference rather than challenging the gender binary itself (Garforth and Kerr 2009).

The goals of equity, while laudable, have severe limitations. Drawing on the second wave of feminism, equity efforts have attempted to give women equal access to science. As feminists have demonstrated, however, equity efforts have not caught up with personal lives. The whole enterprise of programs for women and minorities is predicated on women and minorities as "always already" victims of oppression (Iverson 2007). Efforts of programs are focused on recruiting and retaining them. Yet these efforts are rarely in tune with the realities of gendered lives outside science. Women continue to be burdened by the "second shift" and with mainstream ideologies of femininity. Indeed, in order for science or mainstream culture to be "de-gendered," we need to concurrently "re-gender" relations between men and women (Lorber 2000). These efforts have been slow. An intransigent scientific culture coupled with a feminism that has reinforced essentialist ideas have been less than ideal in transforming lives for women or science. It should come as no surprise that faculty least likely to use mechanisms such as stopping tenure clocks for pregnancy are women faculty in STEM disciplines. Moving beyond the equity model of overcoming overt discrimination has been exceedingly difficult. Many women scientists have themselves been reluctant to embrace strategies that highlight their gender or identity categories. The exclusive focus on equity has thus been a limiting strategy in addressing the underrepresentation of women in science.

Learning to Count Past Two: Shifting Our Focus from Gender to Science

Alongside concerns about the overproduction of Ph.D.s, there is a crisis of confidence about science and engineering education in the United States. Demographics of student populations show increases in international and "foreign" students (Marklein 2012, Matthews 2010). There are similar increases in working international and "foreign" scientists and engineers (National Science Foundation 2012, Matthews 2010). It is argued that we are failing to train more "native" U.S. scientists and engineers (Bennett 2012). The United States is losing its competitive edge (Malone 2012) to European countries, Japan, and, in more recent years, China and India. Here, some of the literature moves into xenophobic territory, bemoaning the rise in foreign-born scientists and students rather than highlighting the significant disinvestment of public education in the United States and the changing economic contexts that have shifted the flows of labor. Two sets of factors propel these arguments. First, for more than two decades U.S. schoolchildren have consistently performed poorly in science and math tests when compared to students in the rest of the world. This educational deficit is not only in science and mathematics but is systemic and includes reading and writing. As the U.S. secretary of education said, "we're being out-educated" (Kobert 2011: 72). Second, there is much anxiety about the scientific workforce. Fewer U.S. citizens are opting for science and math careers, and this deficit is being filled by a greater reliance on foreign students and foreign-trained workers (Cyranoski et al. 2011, Nelson 2007). What was once dubbed the "brain drain" of a talented pool of foreign scientists and engineers who flocked to the United States from other countries has now been reversed. Increasingly, foreign students with U.S. degrees are leaving the United States after their training and returning to their homelands. In response, Congress and U.S. policy have sought to find ways to keep this talent in the United States. Again, a narrow model of equity in women in science and engineering seems detached from the complex political and demographic practices that are fueling contemporary flows of labor and international capital.

It is in within these larger shifts in national and global contexts that we need to evaluate WISE initiatives. WISE programs have been incredibly important in opening and promoting openings for women in science and engineering. I do not want to understate these efforts or the striking shift in numbers they have engendered. At the same time, the strategies have occasionally elided with larger political and economic shifts that are not always progressive. An exclusive focus on equity has led to a focus on the numbers of women and students of color instead of science. Feminist scholars have spent a great deal of time studying such patterns but there is little engagement with this literature in many WISE-related initiatives. Instead, sex, gender, and race are talked about as stable, ahistorical categories. What if we take the question of variation seriously and attend to our changing understandings of variation, diversity, and difference?

Studies of diversity, rather than questioning racial or other differences, often "mirror" and reinforce them (Baez 2004). For example, as Omi and Winant argue, identity categories like race do not have content by themselves. These categories emerged at particular historical moments and were mobilized toward clear political and economic ends. Race is a "formation" through the complex and shifting sets of political projects that organize human bodies and social structures in the aid of particular agendas (Omi and Winant 1994). The challenge then is to explore how and why our discourses of difference organize the category of difference within particular configurations of power and how they shape individual experiences and attitudes (Baez 2004). Gender and other identity categories are not qualities we "have" or possess but rather what we learn to "do"—it is an active process of construction in relation to historical, political, and economic contexts (Morland 2011).

These are also relational categories, always co-constructed with ideas of race, class, sexuality, and nation. In her classic essay nearly thirty years ago, "How Gender Matters, or, Why It's So Hard for Us to Count Past Two," Evelyn Fox Keller (1992) warned us about the slippage between sex and gender. She recounts that when she told people that she worked on "gender and science," she was usually asked, "So what have you found out about women in the sciences?" She reminded us that our inability to move beyond the binary frames of male/female elides sex and gender where "gender" often reinscribes ideas of femininity and women. And indeed the WISE literature is testament that we have been constantly and consistently caught up in a binary world. Thirty years since the emergence of the early feminist critiques of science, WISE initiatives continue to extol the importance, virtues, and joys of women and femininity while the mainstream culture of science has been left largely uncontested except in making accommodations to marginalized groups. The onus has been on marginalized groups making their case and fighting for inclusion rather than an onus on the scientific enterprise being held responsible for its exclusions.

While many researchers have explored the attrition of women from science, there has been limited work on understanding the *problem* of the underrepresentation of women in sciences (Gonsalves 2011). To redefine the problem of women in science, we need to decenter the category of "woman" and instead shift the center of the problem to that of science. It is science rather than women that needs to be interrogated. Rather than merely accommodating women's different biologies, we need to interrogate the narrow frameworks of "good scientists" and "good science" that continue to be valorized within scientific culture. As the previous two chapters demonstrate, individual scientists are limited by definitive yet narrow notions of what it means to be a good scientist. By excavating the histories that have shaped the narrow vision of scientists and science, we open up the possibilities of new and more inclusive futures. We may yet learn how to count past two.

Feminist efforts for women in the sciences must move beyond just increasing the numbers of women in science. Recent efforts of women in science initiatives that buy into promoting women into narrow definitions of efficiency and productivity elide with eugenic scripts of decreasing rather than increasing variation. Instead, a feminist project must include exploring and transforming scientific culture to include not only different bodies but also different visions and cultures, as well as different epistemologies, methodologies, and methods. The project of bringing more women and people of color into science must occur in concert with examining the knowledge the same academy produces about these groups. I am constantly astonished as my students in the sciences—who have been recruited in an effort to diversify science—tell me about the problematic work on biological differences (sex, race, sexuality, ability, nation) that are presented in their biology classes as "truths." What does it mean to recruit a group into an enterprise that simultaneously teaches them about their own biological inferiority? As parts I and II of this book have argued, scientific knowledge production is deeply implicated in its cultures and practitioners. Feminist transformation of the sciences has not embraced the radical potentials of its own critiques. We need to move beyond the frame of women *in* the sciences to embrace feminist analyses of science in its holistic frames. While women in science initiatives have clothed the emperor in new cloaks of femininity, science, like the emperor, is still quite naked. As the story about the emperor and his clothes go, we don't need to reclothe the emperor in feminine or feminist garb. We need to end monarchies!

CONCLUSION

New Cartographies of Variation

The Future of Feminist Science Studies

> I wish for all this to be marked on my body when I am dead. I believe in such cartography—to be marked by nature, not just to label ourselves on a map like the names of rich men and women on buildings. We are communal histories, communal books. We are not owned or monogamous in our taste or experience.
>
> —Michael Ondaatje, *The English Patient*

> If you want to build a ship don't gather people together to collect wood and don't assign task and work but rather teach them to long for the endless immensity of the sea.
>
> —Antoine de Saint-Exupéry, *The Wisdom of the Sands*

During the summer I was revising this manuscript, I decided to grow morning glories in my garden. An obeisance of sorts to this glorious creature that had captured my imagination two decades ago and has ever since made me long for the immensity of its naturecultural possibilities. It has been a glorious journey. Among the bewildering varieties now available, my favorite is the Japanese morning glory, *Ipomoea nil,* Imperial Star of India! As I ponder the flowers each morning, the naturecultural genealogies of morning glories I have recently discovered come tumbling into my mind. I contemplate the genus with its ancient origins in Pangaea (Mann 2011), migrating with the moving continents, domesticated, spreading globally through colonial expansion. It is at once a beloved cultivar and a noxious weed, technologically manipulated, commercially ebullient, and named with imperial pride. What a story! Morning glories, through their extraordinary naturecultural genealogies, their geographies mapped to global travels of colonialism and commerce, and, like their thigmotropic tendrils, their biographies woven into so many lives, including mine, embody this book on the question of variation.

When I started exploring the genealogies of the history of invasive species and the question of women in the sciences, similar complex and global naturecultural histories came tumbling out as well. The world, it would seem, and its many naturecultural objects are deeply interconnected and intertwined in their myriad histories. For these objects, and others in the world, no solitary, narrow history will do. And yet our contemporary cartography of knowledge maps this immense interconnected world into an impoverished, narrow, fragmented, hyperspecialized academia, where knowledge has been vivisected into narrow disciplinary and subdisciplinary formations. Interdisciplinarity today is most heralded in the interdisciplinary sciences that remain contained within the natural sciences, firmly ensconced in uncontested epistemologies and ontologies. Naturecultural histories and genealogies suggest that we need new cartographies of study. We need to replace these insular and narrowly focused areas of study with communal histories and communal storytelling. This to me is the profound gift the feminist studies of science has given me—a move from the myopic view of a morning glory field as a controlled experimental object to the expansive vision of a global traveler par excellence, its migrations and evolutions intertwined in a global history. How can I be bound by the myopia of a discipline when feminist studies of science has handed me the world and set me free?

Inspired by my morning glory philosophies, this book explores what is lost when parochial commitments to conventional disciplinary hyperspecialization prevail. I am making an argument for reinventing the ways in which we practice science and feminism, for reimagining nature and culture as naturecultures. Beginning with a central concept in evolutionary biology, variation, I have argued that this biological concept has been deeply intertwined with cultural ideas about diversity and difference since its very inception. Put starkly, evolutionary theories and models of variation owe their formulation to cultural debates around diversity and difference, culminating in their *eugenic scripts* that have haunted us ever since. This has not been a linear, strategic, or overt process. It is not as though scientists developed their theories, and then society enacted them. Rather, science and society have been co-produced and co-constituted through narratives about culture and nature that contain what can and cannot be said. Ignoring these naturecultural connections, I argue, produces ill-informed science and society, as well as impoverished, ill-informed understandings and models of nature and culture. Naturecultural visions show us that individual disciplines are each imbued with cultural norms and histories while being blind to those influences. Instead of these narrow genealogies of variation, I suggest new cartographies of variation, embedded in global circuits of knowledge.

Throughout the book I have used ghosts and the metaphor of haunting to highlight how a disciplinary academy and its knowledge traditions have consigned a brutal history to the invisible world of ghostly hauntings. In my experience, in a traditional academy, scientists learn little of the humanities and humanists learn little of science. As a young evolutionary biologist, I did not formally study the deep eugenic roots of my field and the cost of the horrendous eugenic policies they enabled. As a young feminist, I did not formally study the biological theory that grounds the histories of eugenics and educational policies about women. Biology was always removed from culture, and culture from biology. Women's studies has not been an innocent bystander in this dynamic, as feminist science studies exists only at the margins. From my naturecultural perch, as I pulled down the walls that separated the worlds of natures and cultures, the ghosts emerged, and through their stories I came to understand how thoroughly I had been educated in ignorance.

The metaphor of ghosts comes from Bollywood and Indian movies I grew up with. In this genre, ghosts wish to leave the earthly world and go home to the ghostly world but cannot do so because of injustices they have encountered or issues that remain unresolved. Their ghostly spirits linger in the earthly world, seeking justice and resolution. For me, the hauntings of ghosts represent the silenced eugenic history of ecology and evolutionary biology and our consequent refusal to suitably acknowledge the horrors of eugenics. In the refusal to acknowledge this past, the question of variation and its attendant implications surfaces again and again. These are the ghostly hauntings. Contrary to the obligations and responsibilities of higher education, disciplinary configurations, allegiances, and knowledges promote "agnotology," Robert Proctor's wonderful term that denotes the study of culturally induced ignorance or doubt. Disciplinary formations, I argue, obfuscate the inconvenient, avoid the uncomfortable, and promote ignorance about the profoundly powerful insights of interdisciplinary thinking. In studying the parts, nature and culture as ontologically separate zones, we lose sight of the whole, the naturecultures. Disciplines' narrow formulations of meaningful questions and definitions of objects of study preclude examining connections to other spheres. I have come to these reflections on disciplinary knowledge after years of doing discipline-constrained work and feeling the deep limitations of trying to do interdisciplinary work in a disciplinary academy. I have crafted an interdisciplinary intellectual life, and have worked with and through the institutional structures that disciplines accord in the contemporary university.

This book is modest evidence of the possibilities of taking that particular career path; no doubt it will be too adventurous for many and not adventurous enough for some. At least perhaps it will encourage disciplinary transgressions that are less timid than humble, less conventional than imaginative, and less about teams than about genuine collaboration. We are, I believe, paying too high a price in creativity and innovation for our loyalties and allegiances to our disciplinary groundings.

In particular, I elaborate the concept of naturecultural knowledge, that is, the thoroughly entwined, inseparable synergies of the natural world with humanity. I argue that in a truly collaborative academy that recognizes the connections between natures and cultures, scientists and feminist scholars could acknowledge and accept our culpability in the continuing injustices in the world and then join efforts to disrupt them. Ignoring the historical backdrop of eugenics debates dooms scientists (and I speak here specifically of evolutionary biologists as a particularly influential group) to a future as co-conspirators in the production of inequality. The ghosts cannot be chased off unless they are listened to and heard. In tracing the genealogy of variation, I maintain that the same questions about whether diversity is good or bad have fueled long-enduring debates on nature or nurture. The disciplinary boundaries that dictate professional training mask the continuing salience of these debates. Evolutionary biologists would benefit from realizing that these are not just political questions, they are at the core of biology. Humanists, including those in women's studies, would benefit from realizing that these are not just biological questions, they are at the core of our humanity. Nor are these questions unique to evolutionary theory. The same politics of diversity and difference haunted (and still haunt) the geographies of variation. The same questions about diversity emerged as nation-states built imaginary borders and imaginary communities filled with natives and nationalisms, now sustained in contemporary immigration politics and environmentalisms. Finally, perhaps most sadly, the ghosts of eugenics haunt the very lives and biographies of those who would, or have, become scientists. Until science educators recognize that science developed as a world that *by design* excluded women and people of color and third world scientists, until contemporary science educators grasp and disrupt the historical construction of science as a reification of white male intellectual supremacy, scientific training will reproduce and reward traditional conformist models of science and scientists. By examining the contemporary implications of the nineteenth-century eugenics debates, scientists can begin to confront challenges to cultivating genuine diversity. Understanding the meaning and value of variation, diversity, and difference has been and will continue to be the key challenge of our world unless or until the ghosts of eugenics are exposed, appeased, and put to rest. My argument is that silence about the co-construction of nature and culture, in particular, enables the proliferation of sexism, racism, classism, casteism, homophobia, ableism, and other forms of oppressions. What the ghosts want, in classic Bollywood fashion, is to be listened to, understood, acknowledged, recognized, and resolved. Making visible the connection of science and politics, of science and power, of science and culture, is the beginning of such a project.

The book demonstrates the deep and broad implications of taking a naturecultural perspective. Many of these implications challenge the structures, practices, and processes of creating new knowledge in higher education. A naturecultural vision is at the heart of the feminist studies of science and technology, an interdisciplinary field of study that does not neatly fit into any single disciplinary studies area because it embraces variation and thrives on intellectual adventure. In my case, feminist science studies was still relatively new, so I forged a way to my enduring commitments to science by building bridges to and through women's studies and the feminist critiques of science. While interdisciplinary, I found that women's studies, then and now, leans heavily on the humanities and social sciences for most major works; feminist science and technology studies remain relatively marginal. Within biology, except for the question of women in the sciences, feminist scholarship remains remote and irrelevant. In this book, I have made a plea for the critical importance of bringing these fields into conversation with each other. The stakes are high. The future of the Planet, its ecologies and evolutions, the futures of humanity, its inequalities and injustices, are all just too urgent for us to remain silent in our disciplinary silos. A feminist studies of science and technology—robustly interdisciplinary and supported by the humanities, social sciences, and the natural sciences—offers an antidote to the willful agnotology that disciplinary formations have created. It is this space beyond disciplines, freed from disciplinary shackles of rigor, which I imagine and long for. With such a vision and unbridled imagination, forever cognizant and wise to our sexist, racist, classist, and homophobic past, we can chart new academic practices in making possible new genealogies, geographies, and biographies.

Notes

Preface

Parts of the preface are drawn from Subramaniam (1998).

1. I am deeply indebted to my dear friends in graduate school, especially Rebecca Dunn, Peggy Schultz, Jim Bever, and Mary Malik, for their unflagging support and own forays into the politics of science and feminist and critical race scholarship; without them, my own engagement would have been difficult, if not impossible.

2. I am forever indebted to Jean O'Barr, director of the Women's Studies Program when I was a graduate student, and Mary Wyer, who was managing editor of the journal *Signs* while it was housed at my institution. They were instrumental in steering me through the literature in feminist theory and feminist science studies and worked patiently with me.

Introduction. Interdisciplinary Hauntings: The Ghostly World of Naturecultures

1. I am grateful to my colleague Angela Willey for many hours of discussions on the multiple genealogies of feminist science and technology studies. Her astute observations have considerably shaped my historical understanding of the emergence and development of the field.

Chapter One. Thigmatropic Tales: On the Politics and Social Lives of Morning Glories

1. I do realize that the term *eugenics* has historically specific origins. I'm using the term loosely here to indicate how this history translates into modern-day sensibilities.

2. Morning glory flower color variation is well studied. See, for example, Baucom et al. 2011, Barbara A. Brown and Clegg 1984, Epperson 1986, Epperson and Clegg 1983, Fine-

blum and Rausher 2002, Glover et al. 1996, Zufall and Rausher 2003, Shu-Mei Chang and Rausher 1999.

3. In 1911, Wilhelm Johannsen introduced the terms *genotype* and *phenotype* to refute the "transmission view" of genetics, where it was believed that personal qualities could influence the passing down of traits. He introduced these terms as way to signal the shift from an old science of "heredity" to a new science of "genetics." See Gudding (1996).

4. From a feminist standpoint, the language of genetics is fascinating. Alleles are said to be "dominant" and "recessive," even though these are not binary choices since alleles can be "co-dominant," as in the case of the w locus. While individuals may carry a particular allele, whether this is reflected in the emerging phenotype is not always self-evident—the proportion of individuals that express the phenotype is reflected in the "penetrance" of the allele or mutation. The field is rife with gendered metaphors and imagery.

Chapter Two. A Genealogy of Variation: The Enduring Debate on Human Differences

1. I entered the world of the history of eugenics expecting to find a clear link to the work I was doing. There was no singular history of "variation" that traced the idea through the history of eugenics, however. I also expected a well-researched field but was ill prepared for the sheer volume of history and historiography on eugenics! Eugenics, it would appear, is a discipline of its own. What I imagined would be a footnote grew into chapters 2 and 3 of this book along with a healthy respect for the painstaking work of historians!

2. My readings of history surprised me. What I found were distinct genres—internalist histories, critical histories, controversies in biology, biographies, period histories, and so forth. Each of these is a distinct genre, each rich and vibrant. But the internalist (told from within science) versus externalist (told from outside) are two broad frames that shape the history of science. See Allen (1986) for an internalist/externalist history of eugenics. See Ruse (2005), Grene (1983), Bashford and Levine (2010), and P. Levine and Bashford (2010) for their analyses on shifts in the fields of the history and philosophy of science from internalist to externalist explanations, and a shift from analyses of "pure" science to science as a set of practices located and shaped by particular histories, cultures, and politics.

3. For more on the debates on essentialism and evolutionary biology, see Winsor (2003) and Walsh (2006).

4. Several tenets of the modern synthesis led to debates once the molecular era began (Depew and Weber 1995). For example, questions about the units of selection and whether selective process can act at any and every level of scale continue to generate considerable debate (Depew and Weber 1995, Gould 2002).

5. The "new genetics" usually refers to the growth of genetic research and technology arising out of recombinant DNA in the 1970s.

Chapter Three. Singing the Morning Glory Blues: A Fictional Science

1. This character is inspired by Prashanta Mahalonobis's description of the philosophy of the Indian mathematician Srinivasa Ramanujam and Ashis Nandy's (1995) own description and analysis. I have, of course, embellished and taken many liberties with their descriptions.

Chapter Four. Alien Nation: A Recent *Bio*graphy

Some parts of Act I were previously published in Subramaniam (2001).

1. Connections between the body as fortress and the nation as fortress in late capitalism can be seen in Martin (1997). For thinking about the persistence of national states in an age of globalization, see Comaroff and Comaroff (2000, 2002).

2. For provocative thoughts about the production of ethnicity and civil society/nation, botanical taxonomies, and immigration policies in the United States, see Moallem and Boal (1999). For some recent work on ethnocentrism and nationalism produced as a certain politics, see Paola Bachetta's article on xenophobia and the Hindu right in India (1999). For national cultures, cultures, and questions of cultural nationalism, see Sahlins (2000). For cultural nationalism and new modes of citizenship, see Aihwa Ong (1999).

3. While the distribution of political and religious beliefs in science may not parallel mainstream culture, multiple and diverse political and ideological beliefs are indeed represented.

4. The campaigns are, of course, not by the same groups. Many ecologists have expressed reservations about genetically modified food. My point is about the rhetoric that circulates in the mainstream United States.

5. Anna Tsing (1994) makes a similar point in her analysis of native and exotic bees.

Chapter Five. My Experiments with Truth: Studying the Biology of Invasions

I offer this chapter title with due apologies to M. K. Gandhi's *The Story of My Experiments with Truth*. As an experimental biologist, I could not resist it!

This work was supported by an NSF grant to the author for "Impact of Soil Communities on Invasive Plant Species in Southern California" (NSF 0075072)

1. Jim Bever and Peggy Schultz were at the University of California, Irvine, when this project began and subsequently moved to Indiana University.

2. The native species were *Lotus purshianus, Salvia apiana, Nassella pulchra, Achillea millefolium, Galium angustifolium,* and *Isocoma menziesii,* and the naturalized species were *Rumex crispus, Bromus diandrus, Lactuca serriola, Medicago polymorpha, Lolium multiflorum, Marrubium vulgare, Salsola tragus,* and *Sonchus asper.*

3. A survey of how biologists categorize various species would be fascinating.

Chapter Six. Aliens of the World Unite! A Meditation on Belonging in a Multispecies World

1. This pattern is not universal and there are many exceptions. Not all invasives are named with their geographical origins. Why and how this happens is a fascinating future project.

2. On this issue, see also Pollan (1994) and Olwig (2003).

3. For a discussion of Denmark, see Olwig (2003), and of Britain, see Smout (2003).

4. Some do contest such a reading of Jensen; see Grese (2011).

5. The cookbook they have produced is Wong and Leroux (2012).

Chapter Seven. Through the Prism of Objectivity: Dispersions of Identity, Culture, Science

1. I was deeply influenced by the autobiographical narratives of E. Keller (1977), Weinstein (1977), and Hammonds (1993).

2. I do recognize that the experience of cultural isolation is field dependent. Fields such as engineering or business have a large number of South Asian students who came from elite and prestigious universities in South Asia, and their experiences may be very different.

3. This also pervades the workforce in this country with respect to salary. It is a common joke among Indians here that you can ask another Indian how much she or he earns but not about his or her sex life. Americans will tell you all about their sex life but salaries are too personal!

Chapter Eight. Resistance Is Futile! You Will Be Assimilated: Gender and the Making of Scientists

This work was supported by an NSF grant to the author for "Breaking the Silences: A Faculty-Student Action Project for Graduate Women in Science and Mathematics." (HRD9553439)

1. Perhaps one of the most interesting species on *Star Trek: The Next Generation,* the Borg have captured the imagination of many "trekkies." A number of websites have proliferated, the most interesting of which is the Borg Institute of Technology (BIT) with the maxim "Graduation is futile!" (previously http://grove.ufl.edu/~locutus/Bit/bit.html, now http://www.skippypodar.net/Bit/.

2. I am deeply indebted to Ann Gerber and Laura Briggs, who were co-facilitators with me during these sessions. Their observations and rich insights have profoundly shaped the interpretation and analysis of this work.

3. It is also important to note that there are strict hierarchies within faculty, especially within the same department, and faculty themselves are sometimes disempowered by the structures of academe, factors often not always visible to graduate students.

Chapter Nine. The Emperor's New Clothes: Revisiting the Question of Women *in* the Sciences

1. In connecting scientific knowledge to structures of power, the feminist literature suggests systematic discrimination and exclusion of the "nonnormative"—women scientists, scientists of color, working-class scientists, third world scientists, queer scientists, non-able-bodied scientists. "Women," after all, are not a universal group, and there are considerable differences among women within and between countries. These differences, however, are not that well studied, and what remains is largely a history of elite white women's efforts to gain entry into the hallways of science.

2. While white women have been the largest beneficiaries of affirmative action in the United States, other underrepresented groups have also pushed for equity measures in STEM disciplines, including groups promoting "minority," queer, disabled, and third world communities.

3. For a discussion of K–12 innovations, see Brotman and Moore (2008), Buck et al. (2012), Hill et al. (2010), Nosek et al. (2009), Scher and O'Reilly (2009), and Valla and Williams (2012). For innovations at the undergraduate level, see M. J. Chang, Cerna, et al. (2008), M. J. Chang, Eagan, et al. (2011), Cheryan and Plaut (2010), Else-Quest et al. (2010), Fox et al. (2011), Sonnert and Fox (2012), Hanson (2004), Hill et al. (2010), Nassar-McMillan et al. (2011), and Sax (2001). For research and programs at the graduate, postgraduate, faculty, and institutional levels, see Jill Adams (2008), Bhatacharjee (2004), Bilimoria and Liang (2011), Brickhouse et al. (2006), Bystydzienski and Bird (2006), Cech and Blair-Loy (2010), Hill et al. (2010), Gonsalves (2011), Intemann (2009), Leonard (2003), Maria Ong et al. (2011), Robinson (2011), Rosser (1990, 1995, 1997, 2004b), Scantlebury (2010), Schibeinger (1999, 2003), Stewart et al. (2007), and Wylie (2011).

References

Abir-Am, Pnina, and Dorinda Outram, eds. 1987. *Uneasy Careers and Intimate Lives: Women in Science, 1789–1979*. New Brunswick, NJ: Rutgers University Press.

Acker, Joan. 2000. "Gendered Contradictions in Organizational Equity Projects." *Organization* 7 (4): 625–32.

Action News 6, Philadelphia. 2002. "Killer Chinese Fish Surfaces in Maryland," July 10.

Adams, Jill. 2008, February 8. "Nurturing Women Scientists." *Science* 319: 831–36.

Adams, Mark. 2000. "Last Judgment: The Visionary Biology of J. B. S. Haldane." *Journal of the History of Biology* 33: 457–91.

Adams, Vincanne, and Stacy Leigh Pigg. 2005. *Sex in Development: Science, Sexuality and Morality in Global Perspective*. Durham, NC: Duke University Press.

Ahmed, S. 2008. "Open Forum Imaginary Prohibitions: Some Preliminary Remarks on the Founding Gestures of the 'New Materialism.'" *European Journal of Women's Studies* 15 (1): 23–39.

———. 2012. *On Being Included: Racism and Diversity in Institutional Life*. Durham, NC: Duke University Press.

Aisenberg, Nadya, and Mona Harrington. 1988. *Women of Academe: Outsiders in the Sacred Grove*. Amherst: University of Massachusetts Press.

Alaimo, S., and S. Hekman, eds. 2008. *Material Feminisms*. Bloomington: Indiana University Press.

Allen, Garland. 1986. "The Eugenics Record Office at Cold Spring Harbor, 1910–1940: An Essay in Institutional History." *Osiris* (Second Series) 2: 225–64.

———. 1996. "Science Misapplied: The Eugenics Age Revisited." *Technology Review* 99 (6): 22–31.

———. 2001. "Is a New Eugenics Afoot?" *Science* 294 (5540): 59–61. DOI: 10.1126/science.1066325.

Alonso, W. 1995. "Citizenship, Nationality and Other Identities." *Journal of International Affairs* 48: 585–99.

Alsop, Peter. 2009, November. "Invasion of the Longhorn Beetles." *Smithsonian Magazine,* http://www.smithsonianmag.com/science-nature/Invasion-of-the-Longhorns.html.

Alyokin, Andrei. 2011. "Non-Natives: Put Biodiversity at Risk." *Nature* 475: 36.

Anandhi, S. 1998. "Reproductive Bodies and Regulated Sexuality: Birth Control Debates in Early Twentieth-Century Tamilnadu." In *A Question of Silence: Sexuality Economics in Modern India,* edited by Mary E. John and Janaki Nair. Delhi: Kali for Women.

Anderson, Warwick. 2006. *The Cultivation of Whiteness: Science, Health, and Racial Destiny in Australia.* Durham, NC: Duke University Press.

———. 2008. *Colonial Pathologies: American Tropical Medicine, Race, and Hygiene in the Philippines.* Durham, NC: Duke University Press, 2006.

Arum, Richard, and Josipa Roksa. 2011. "Your So-Called Education," *The New York Times,* May 14.

Auld, B. A., and R. W. Medd. 1987. *Weeds: An Illustrated Botanical Guide to the Weeds of Australia.* Melbourne: Inkata.

Babbitt, Bruce. 1998. "Statement by Secretary of the Interior Bruce Babbitt on Invasive Alien Species." Science in Wildland Weed Management Symposium, Denver, CO, April 8, http://www.nps.gov/plants/alien/pubs/bbstat.htm.

Bachetta, Paola. 1999. "When the (Hindu) National Exiles Its Queers." *Social Text* 17 (4): 141–66.

Baez, Benjamin. 2004. "The Study of Diversity: The 'Knowing of Difference' and the Limits of Science." *Journal of Higher Education* 75 (3): 285–306.

Ball, George. 2006. "Border War," *The New York Times,* March 19.

Barad, Karen. 2007. *Meeting the Universe Halfway: Quantum Physics and the Entanglement of Matter and Meaning.* Durham, NC: Duke University Press.

Barbercheck, Mary. 2001. "Science, Sex, and Stereotypical Images in Scientific Advertising." In *Women, Science, and Technology: A Reader in Feminist Science Studies,* edited by Mary Wyer, Mary Barbercheck, Donna Giesman, Hatice Orun Ozturk, and Marta Wayne. New York: Routledge.

Barkley, T. M., ed. 1986. *Flora of the Great Plains.* Lawrence: University Press of Kansas.

Barlow, Connie, ed. 1995. *Evolution Extended: Biological Debates on the Meaning of Life.* Cambridge, MA: MIT Press.

Barnes, Jeff. 2002. "Juvenile Frankenfish Raise the Odds of Alien Invasion," *Washington Times,* July 13.

Baron-Cohen, Simon. 2003. *The Essential Difference: Men, Women, and the Extreme Male Brain.* London: Allen Lane.

Barres, Ben. A. 2006, July 13. "Does Gender Matter?" *Nature* 442: 133–36.

Barringer, Felicity. 2012. "New Rules Seek to Prevent Invasive Stowaways," *The New York Times,* April 7.

Barton, Angela Calabrese. 2001. "Capitalism, Critical Pedagogy and Urban Science Education: An Interview with Peter McLaren." *Journal of Research in Science Teaching* 38 (8): 847–59.

Bashford, Alison. 2010. "Internationalism, Cosmopolitanism, and Eugenics." In *The Oxford Handbook of the History of Eugenics,* edited by Alison Bashford and Philippa Levine, 154–72. Oxford: Oxford University Press.

Bashford, Alison, and Philippa Levine, eds. 2010. *The Oxford Handbook of the History of Eugenics.* Oxford: Oxford University Press.

Baskin, Y. 2002. *A Plague of Rats and Rubbervines: The Growing Threat of Species Invasions.* Washington, DC: Island.

Batten, Katherine M., Kate M. Scow, and Erin Espeland. 2008. "Soil Microbial Community Associated with an Invasive Grass Differentially Impacts Native Plant Performance." *Microbial Ecology* 55: 220–28.

Baty, Phil. 2010. "What's Sex Got to Do with It?" *Times Higher Education,* September 30.

Baucom, R. S., S-M. Chang, J. M. Kniskern, M. D. Rausher, and J. R. Stinchcombe. 2011. "Morning Glory as a Powerful Model in Ecological Genomics: Tracing Adaptation through Both Natural and Artificial Selection." *Heredity* 107: 377–85.

Bean, A. R. 2007. "A New System for Determining Which Plant Species are Indigenous in Australia." *Australian Systematic Botany* 20: 1–43.

Beatty, John. 1987. "Weighing the Risks: Stalemate in the Classical/Balance Controversy." *Journal of the History of Biology* 20 (3): 289–319.

Begley, Sharon. 2006. "He, Once a She, Offers Own View on Science Spat," *Wall Street Journal,* July 13, B1.

Benjamin, Rich. 2012. "The Gated Community Mentality," *New York Times,* March 29, 2012.

Bennett, William J. 2012. "U.S. Lag in Science, Math a Disaster in the Making," February 9, http://www.cnn.com/2012/02/09/opinion/bennett-stem-education.

Beoku-Betts, Josephine. 2004. "African Women Pursuing Graduate Studies in the Sciences: Racism, Gender Bias, and Third World Marginality." *NWSA Journal* 16 (1): 116–35.

Bever, James D. 2003. "Soil Community Dynamics and the Coexistence of Competitors: Conceptual Frameworks and Empirical Tests." *New Phytologist* 157: 465–73.

Bhattacharjee, Y. 2004. "Family Matters: Stopping Tenure Clock May Not Be Enough." *Science* 306: 2031–33.

Bilimoria, Diana, and Xiangfen Liang. 2011. *Gender Equity in Science and Engineering: Advancing Change in Higher Education.* London: Taylor & Francis.

Birke, Linda. 1999. *Feminism and the Biological Body.* Edinburgh: Edinburgh University Press.

Blair, Robert B. 2008. "Creating a Homogenous Avifauna." In *Urban Ecology: An International Perspective on the Interaction between Humans and Nature,* edited by John Marzluff, 405–24. New York: Springer.

Bleier, Ruth. 1984. *Science and Gender: A Critique of Biology and Its Theories on Women.* New York: Pergamon.

Bliss, Catherine. 2012. *Race Decoded: The Genomic Fight for Social Justice.* Stanford, CA: Stanford University Press.

Bombardieri, Marcella. 2005. "Harvard Women's Group Rips Summers," *Boston Globe,* January 19, http://www.boston.com/news/education/higher/articles/2005/01/19/harvard_womens_group_rips_summers/?page=full.

Bousquet, Marc. 2008. *How the University Works: Higher Education and the Low Wage Nation.* New York: NYU Press.

Brabazon, Tara. 1998, May. "What's the Story Morning Glory? Perth Glory and the Imagining of Englishness." *Sporting Traditions* 14 (2): 53–66.

Bradshaw, G. A., and M. Bekoff. 2001. "Ecology and Social Responsibility: The Re-embodiment of Science." *Trends in Ecology and Evolution* 16 (8): 460–65.

Brandon, Robert. N. 1990. *Adaptation and Environment.* Princeton, NJ: Princeton University Press.

Brickhouse, Nancy W., Margaret A. Eisenhart, and Karen Tonso. 2006. "Forum Identity Politics in Science and Science Education." *Cultural Studies of Science Education* 1: 309–24.

Briggs, Laura. 2002. *Reproducing Empire: Race, Sex, Science, and U.S. Imperialism in Puerto Rico.* Berkeley: University of California Press.

Bright, Christopher. 1998, November/December. "Alien Threat." *World Watch* 11 (6): 8.

———. 1999. "Invasive Species: Pathogens of Globalization." *Foreign Policy* 116 (51): 14.

Brinckman, Jonathan. 2001. "Creepy Strangler Climbs Oregon's Least-Wanted List," *Oregonian,* February 28, 1A.

Brockway, Lucile. 2002. *Science and Colonial Expansion: The Role of the British Royal Botanic Gardens.* New Haven, CT: Yale University Press.

Brotman, Jennie S., and Felicia M. Moore. 2008. "Girls and Science: A Review of Four Themes in the Science Education Literature." *Journal of Research in Science Teaching* 45 (9): 971–1002.

Brown, Barbara A., and Michael T. Clegg. 1984, July. "Influence of Flower Color Polymorphism on Genetic Transmission in a Natural Population of the Common Morning Glory, Ipomoea purpurea." *Evolution* 38 (4): 796–803.

Brown, J. H., and D. F. Sax. 2004. "An Essay on Some Topics Concerning Invasive Species." *Australian Ecology* 29: 530–36.

———. 2005. "Biological Invasions and Scientific Objectivity: Reply to Cassey et al." *Australian Ecology* 30: 481–83.

Brown, Wendy. 2011, Summer. "Neoliberalized Knowledge." *History of the Present* 1 (1): 113–29.

Brundrett, M. C. 2002. "Coevolution of Roots and Mycorrhizas of Land Plants." *New Phytologist* 154: 275–304.

Buchanan, Gale A., and Earl R. Burns. 1971, September. "Weed Competition in Cotton. I. Sicklepod and Tall Morningglory." *Weed Science* 19 (5): 576–79.

Buck, Gayle A., Nicole M. Beeman-Cadwallader, and Amy E. Trauth-Nare. 2012. "Keeping the Girls Visible in K–12 Science Education Reform Efforts: A Feminist Case Study on Problem Based Learning." *Journal of Women and Minorities in Science* 18 (2): 153–78.

Bug, Amy. 2003. "Has Feminism Changed Physics?" *Signs: Journal of women in culture and society* 28 (3): 881–900.

Burdick, Alan. 2005, May. "The Truth About Invasive Species: How to Stop Worrying and Learn to Love Ecological Intruders." *Discover Magazine* 26 (5).

Burelli, Joan. 2010, July. "Foreign Science and Engineering Students in the United States." InfoBrief, NSF 10–324, Arlington, VA.

Burian, R. M. 1983. "Adaptation." In *Dimensions of Darwinism*, edited by M. Grene, 286–314. Cambridge: Cambridge University Press.

Butler, Judith. 2006. *Gender Trouble: Feminism and the Subversion of Identity*. New York: Routledge.

Bystydzienski, Jill, and Sharon R. Bird, eds. 2006. *Removing Barriers: Women in Academic Science, Technology, Engineering, and Mathematics.* Bloomington: Indiana University Press.

Calautti, R. I., and J. J. MacIsaac. 2004. "A Natural Terminology to Define 'Invasive' Species." *Diversity and Distributions* 10: 135–41.

Campbell, Nancy. 2009. "Reconstructing Science and Technology Studies: Views from Feminist Standpoint Theory." *Frontiers* 30 (1): 1–29.

Canel, Annie, Ruth Oldenziel, and Karin Zachmann. 2000. *Crossing Boundaries, Building Bridges: Comparing the History of Women Engineers, 1870s–1990s.* Amsterdam: Harwood Academic.

Caplan, Arthur. L., Glenn McGee, and David Magnus. 1999. "What Is Immoral about Eugenics?" *British Medical Journal* 319 (7220): 1284.

Cardozo, Karen, and Banu Subramaniam. 2013, February. "Assembling Asian/American Naturecultures: Orientalism and Invited Invasion." *Journal of Asian American Studies* 16 (1).

Carey, Toni Vogel. 1998. "The Invisible Hand of Natural Selection and Vice Versa." *Biology and Philosophy* 13: 427–42.

Carthey, Alexandra J. R., and Peter B. Banks. 2012, February. "When Does an Alien Become a Native Species? A Vulnerable Native Mammal Recognizes and Responds to Its Long-Term Alien Predator." *PLoS One* 7 (2): 1–4.

CBS Evening News. 2002a. "Freak Fish Found in Two More States," August 3.

———. 2002b. "Wanted Dead: Voracious Walking Fish," July 3, www.cbsnews.com/stories/2002/07/03/eveningnews/main514182.shtmil.

Cech, Erin A., and Mary Blair-Loy. 2010. "Perceiving Glass Ceilings? Meritocratic versus Structural Explanations of Gender Inequality among Women in Science and Technology." *Social Problems* 57 (3): 371–97.

Ceci, Stephen J., and Wendy M. Williams. 2011, February 7. "Understanding Current Causes of Women's Underrepresentation in Science." *Proceedings of the National Academy of Sciences*. DOI 10.1073/pnas.1014871108.

Chaney, Lindsay, and Regina S. Baucom. 2012. "The Evolutionary Potential of Baker's Weediness Traits in the Common Morning Glory, *Ipomoea purpurea* (Convolvulaceae)." *American Journal of Botany* 99 (9): 1524–30.

Chang, Gordon. 2001. *Morning Glory, Evening Shadow: Yamato Ichihashi and His Internment Writings.* Stanford, CA: Stanford University Press.

Chang, Kenneth. 2012. "Bias Persists for Women of Science, a Study Finds." *The New York Times*, September 24.

Chang, M. J., O. Cerna, J. Han, and V. Saenz. 2008. "The Contradictory Roles of Institutional Status in Retaining Underrepresented Minorities in Biomedical and Behavioral Science Majors." *Review of Higher Education* 31: 433–64.

Chang, M. J., M. K. Eagan, M. H. Lin, and S. Hurtado. 2011. "Considering the Impact of Racial Stigmas and Science Identity: Persistence among Biomedical and Behavioral Science Aspirants." *Journal of Higher Education* 82: 564–96.

Chang, Shu-Mei, and Mark D. Rausher. 1999, October. "The Role of Inbreeding Depression in Maintaining the Mixed Mating System of the Common Morning Glory, *Ipomoea purpurea*." *Evolution* 53 (5): 1366–76.

Chaudhary, Bala, and Margot Griswold. 2001, Spring. *Ecesis: Newsletter of the Society for Ecological Restoration* (SERCAL), http://www.newfieldsrestoration.com/PDFs/Mycorrhizal_Fungi_Ecesis.pdf.

Cheater, Mark. 1992, September/October. "Alien Invasion: They're Green, They're Mean, and They May Be Taking Over a Park or Preserve Near You." *Nature Conservancy* 42 (5): 24–29.

Cheryan, Sapna, and Victoria C. Plaut. 2010. "Explaining Underrepresentation: A Theory of Precluded Interest." *Sex Roles* 63: 475–88.

Chew, Matthew, and Andrew L. Hamilton. 2011. "The Rise and Fall of Biotic Nativeness: A Historical Perspective." In *Fifty Years of Invasion Ecology: The Legacy of Charles Elton*, edited by David M. Richardson. Oxford: Blackwell.

Chew, Matthew, and M. D. Laubichler. 2003. "Natural Enemies—Metaphor or Misconception?" *Science* 301: 52–53.

"Chinese Snakehead Not Maryland's Only Foreign Problem." 2002. *Maryland Daily Record*. July 19, http://thedailyrecord.com/2002/07/19/chinese-snakehead-not-maryland8217s-only-foreign-problem/.

Clegg, Michael T., and Mary L. Durbin. 2000. "Flower Color Variation: A Model for the Experimental Study of Evolution." *Proceedings of the National Academy of Sciences USA* 97 (13): 7016–23.

CNN News Online. 2002. "Wanted: Snakehead," July 12, http://fyi.cnn.com/2002/fyi/news/07/12/news.for.you.

Coase, Ronald. 1998. "The New Institutional Economics." *American Economic Review* 88 (2).

Coates, Peter. 2003. "Editorial Postscript: Naming Strangers in the Landscape." *Landscape Research* 28: 131–37.

———. 2006. *American Perceptions of Immigrant and Invasive Species: Strangers on the Land*. Berkeley: University of California Press.

Cole, J. R. 1979. *Fair Science: Women in the Scientific Community*. New York: Collier Macmillan.

Collins, Patricia Hill. 1999. "Moving beyond Gender: Intersectionality and Scientific Knowledge." In *Revisioning Gender*, edited by Myra M. Ferree, Judith Lorber, and Beth B. Hess, 261–84. Thousand Oaks, CA: Sage.

Comaroff, Jean, and John Comaroff. 2000. "Millennial Capitalism: First Thoughts on Second Coming." *Public Culture* 12 (2): 291–343.

———. 2002. "Alien-Nation: Zombies, Immigrants, and Millennial Capitalism." *South Atlantic Quarterly* 104 (4): 779–805.

Conlon, Michael. 2010. "Flowering Cherry Trees—A Gift from Japan." USDA Foreign Agricultural Service Global Agricultural Information Network, GAIN Report Number: JA0507, http://www.usdajapan.org/en/reports/20100330_Japanese%20Flowering%20Cherry%20Trees.pdf.

Cooke, Kathy. J. 1997. "From Science to Practice, or Practice to Science? Chickens and Eggs in Raymond Pearl's Agricultural Breeding Research, 1907–1916." *Isis* 88: 62–86.

Coronado, Ty. 2006. "LA Immigrant Rights March," *ZNet*, March 31, http://www.zmag.org/content/showarticle.cfm?SectionID=72&ItemID=10022.

Cowan, Ruth Schwartz. 1997. *A Social History of American Technology.* New York: Oxford University Press.

———. 2008. *Heredity and Hope: The Case for Genetic Screening.* Cambridge, MA: Harvard University Press.

Cowherd, Kevin. 2002. "Soon Ghastly Fish Will Walk Off into the Sunset," *Baltimore Sun*, August 8.

CQ Researcher. 2000. "Bio-invasions Spark Concerns." 9 (37): 856.

Cravens, Hamilton. 1978. *The Triumph of Evolution: American Scientists and the Heredity-Environment Controversy, 1900–1940*. Philadelphia: University of Pennsylvania Press.

Creager, Angela, Elizabeth Lunbeck, and Londa Schiebinger, eds. 2001. *Feminism in Twentieth-Century Science, Technology, and Medicine.* Chicago: University of Chicago Press.

Crenshaw, Kimberle. 1991. "Mapping the Margins: Intersectionality, Identity Politics, and Violence against Women of Color." *Stanford Law Review* 43 (6): 1241–99.

Cronon, William. 1996a. "The Trouble with Wilderness; or, Getting Back to the Wrong Nature." In *Uncommon Ground: Rethinking the Human Place in Nature*, edited by William Cronon, 69–90. New York: Norton.

———. 1996b. "Introduction: In Search of Nature." In *Uncommon Ground: Rethinking the Human Place in Nature*, edited by William Cronon, 23–56. New York: Norton.

Crosby, Alfred W. 1986. *Ecological Imperialism: The Biological Expansion of Europe, 900–1900*. New York: Cambridge University Press.

Crow, James. F. 1987. "Dobzhansky and Overdominance." *Journal of the History of Biology* 20 (3): 351–80.

Crowley, R. H., and G. A. Buchanan. 1982. "Variations in Seed Production and Response to Pests of Morningglory (*Ipomoea*) Species and Smallflower Morningglory (*Jacquemontia tamnifolia*)." *Weed Science* 30: 187–90.

Crutzen, P. J., and E. F. Stoerme. 2000. "The Anthropocene." *Global Change Newsletter* 41: 17–18.

Cyranoski, David, Natasha Gilbert, Heidi Ledford, Anjali Nayar, and Mohammed Yahla. 2011. "Education: The PhD Factory." *Nature* 472: 276–79.

Dahl, Linda. 2001. *Morning Glory: A Biography of Mary Lou Williams.* Berkeley: University of California Press.

Darity, William, Jr. 1995. Introduction. In *Economics and Discrimination,* edited by William Darity Jr. Brookfield, VT: Edward Elgar.

Dart, Bob. 2002, July 24. "Invasive 'Walking' Fish Found Across U.S. South." *National Geographic.*

Darwin, Charles. 1922. *Descent of Man and Selection in Relation to Sex.* New York: D. Appleton and Co.

———. 1959. *The Origin of Species by Means of Natural Selection or The Preservation of Favoured Races in the Struggle for Life.* Middlesex: Penguin.

Daston, Lorraine, and Katherine Park. 2001. *Wonders and the Order of Nature, 1150–1750.* Zone.

Davis, Cinda-Sue. 1996. *The Equity Equation: Fostering the Advancement of Women in the Sciences, Mathematics, and Engineering.* San Francisco: Jossey-Bass.

Davis, Kathy. 2008. "Intersectionality as Buzzword: A Sociology of Science Perspective on What Makes a Feminist Theory Successful." *Feminist Theory* 9 (1): 67–85.

Davis, M. A. 2009. *Invasion Biology.* Oxford: Oxford University Press.

Davis, M. A., and K. Thompson. 2000. "Eight Ways to Be a Colonizer; Two Ways to Be an Invader." *Bulletin of the Ecological Society of America* 81: 226–30.

Davis, M. A., et al. 2011a, June. "Don't Judge Species on Their Origins." *Nature* 474: 153–54.

———. 2011b. "Researching Invasive Species 50 years after Elton: A Cautionary Tale." In *Fifty Years of Invasion Ecology: The Legacy of Charles Elton,* edited by David M. Richardson, 269–74. Oxford: Blackwell.

Davis, Noela. 2009. "New Materialism and Feminism's Anti-Biologism: A Response to Sara Ahmed." *European Journal of Women's Studies* 16 (1): 67–80.

Dawkins, Richard. 1986. *The Blind Watchmaker: Why the Evidence of Evolution Reveals a Universe Without Design.* New York: Norton.

De Laey, J. J., S. Ryckaert, and A. Leys. 1985. "The 'Morning Glory' Syndrome." *Ophthalmic Genetics* 5 (1/2): 117–24.

De Laplante, K. 2004. "Toward a More Expansive Conception of Ecological Science." *Biology and Philosophy* 19: 263–81.

Deem, Rosemary, and Louise Morley. 2005. "Negotiating Equity in Higher Education Institutions." Paper presented at the Economic and Social Research Council (ESRC) Seminar on Equalities in Education, University of Cardiff, http://dera.ioe.ac.uk/5886/1/rd10_05.pdf.

Defelice, Michael. 2001. "Tall Morningglory, *Ipomoea purpurea* (L.) Roth—Flower or Foe?" *Weed Technology* 15: 601–6.

Demeritt, D. 1998. "Science, Social Constructivism and Nature." In *Remaking Reality: Nature at the Millennium,* edited by N. Castree and B. Braun, 173–93. London: Routledge.

Department of Natural Resources. n.d. http://dnr.maryland.gov/fisheries/bass/docs/snakehead_contest_20110420.pdf.

Depew, David J. 2010. "Darwinian Controversies: An Historiographical Recounting." *Science and Education* 19: 323–66.

Depew, David, and Bruce Weber. 1995. *Darwinism Evolving—Systems Dynamics and the Genealogy of Natural Selection.* Cambridge, MA: MIT Press.

Derr, P. G., and E. M. McNamara. 2003. "Have You Seen This Fish?" In *Case Studies in Environmental Ethics.* Lanham, MD: Rowman & Littlefield.

Des Jardins, Julie. 2010. *The Madame Curie Complex: The Hidden History of Women in Science.* New York: Feminist Press.

Desrosières, Alain. 1998. *The Politics of Large Numbers: A History of Statistical Reasoning.* Cambridge, MA: Harvard University Press.

Devine, Robert. 1999. *Alien Invasion: America's Battle with Non-native Animals and Species.* Washington, DC: National Geographic Society.

Dietrich, Michael R. 1994. "The Origins of the Neutral Theory of Molecular Evolution." *Journal of the History of Biology* 25: 21–59.

———. 2006. "From Mendel to Molecules: A Brief History of Evolutionary Genetics." In *Evolutionary Genetics: Concepts and Case Studies,* edited by Charles W. Fox and Jason B. Wolf, 3–13. New York: Oxford University Press.

Dietz, T., and P. C. Stern. 1998. "Science, Values, and Biodiversity." *BioScience* 48: 441–44.

Dikötter, Frank. 1998. "Race Culture: Recent Perspectives on the History of Eugenics." *American Historical Review* 103 (2): 467–78.

Dobzhansky, Theodosius. 1955. "A Review of Some Fundamental Concepts and Problems in Population Genetics." *Cold Spring Harbor Symposium in Quantitative Biology* 20: 1–15.

———. 1962. *Mankind Evolving.* New Haven, CT: Yale University Press.

———. 1973. "Nothing in Biology Makes Sense Except in the Light of Evolution." *American Biology Teacher,* vol. 35: http://www.phil.vt.edu/Burian/NothingInBiolChFina.pdf.

Dobzhansky, Theodosius, and L. C. Dunn. 1952 [1946]. *Heredity, Race, and Society.* Rev. ed. New York: Pelican.

Doggett, Tom. 2002. "Crawling Snakehead Fish Scaring Washington," Reuters, July 28.

Dudley, Rachel. 2012. "Toward An Understanding of the 'Medical Plantation' as a Cultural Location of Disability." *Disability Studies Quarterly* 32 (4).

Duke Homestead Tour, 1990. Information from a personal tour of Duke Homestead, http://www.nchistoricsites.org/duke/.

Dumit, Joseph. 2012. "Prescription Maximization and the Accumulation of Surplus Health in the Pharmaceutical Industry: The BioMarx Experiment." In *Lively Capital: Biotechnologies, Ethics and the Governance in Global Markets,* edited by Kaushik Sunder Rajan. Durham, NC: Duke University Press.

Duster, Troy. 2003. *Backdoor to Eugenics.* 2nd ed. New York: Routledge.

Egan, Timothy. 2011. "Frankenfish Phobia," *The New York Times,* March 17, http://opinionator.blogs.nytimes.com/2011/03/17/frankenfish-phobia/.

Eisenhart, Margaret A., and Elizabeth Finkel. 1998. *Women's Science: Learning and Succeeding from the Margins.* Chicago: University of Chicago Press.

Ekberg, Merryn. 2007. "The Old Eugenics and the New Genetics Compared." *Social History of Medicine* 20 (3): 581–93.

Elmore, C. D., H. R. Hurst, and D. F. Austin. 1990. "Biology and Control of Morningglory (*Ipomoea* spp.)." *Journal Review of Weed Science* 5: 83–114.

Else-Quest, N. M., J. S. Hyde, and M. C. Linn. 2010. "Cross-national Patterns of Gender Differences in Mathematics: A Meta-Analysis." *Psychological Bulletin* 136: 103–27.

Elton, Charles C. 1958. *The Ecology of Invasion by Animals and Plants.* London: Methuen.

Emery, David. 2002. "Attack of the Frankenfish: Northern Snakehead Fish Invades North America," About.com Urban Legends, http://urbanlegends.about.com/od/fish/a/snakehead.htm.

Enserink, Martin. 1999. "Biological Invaders Sweep In." *Science* 285 (5435): 1834–36.

Epperson, Bryan K. 1986, December. "Spatial-Autocorrelation Analysis of Flower Color Polymorphisms within Sub-structured Populations of Morning Glory (*Ipomoea purpurea*)." *The American Naturalist* 128 (6): 840–58.

Epperson, Bryan K., and Michael T. Clegg. 1983. "Flower Color Variation in the Morning Glory, *Ipomoea purpurea." Journal of Heredity* 74 (4): 247–50.

Eriksen, Thomas Hylland. 2006. "Diversity versus Difference: Neo-liberalism in the Minority Debate." In *The Making and Unmaking of Difference,* edited by Richard Rottenburg, Burkhard Schnepel, and Shingo Shimada, 13–36. Bielefeld, Ger.: Transaction.

Escobar, Arturo. 1997. "Cultural Politics and Biological Diversity: State, Capital, and Social Movements in the Pacific Coast of Colombia." In *The Politics of Culture in the Shadow of Capital,* edited by Lisa Lowe and David Lloyd. Durham, NC: Duke University Press.

Etzkowitz, Henry, Carol Kemelgor, Michael Neuschatz, and Brian Uzzi. 1994. "Barriers to Women in Academic Science and Engineering." In *Who Will do Science? Educating the Next Generation,* edited by Willie Pearson Jr. and Irwin Fechter. Baltimore: Johns Hopkins University Press.

Faulkner, William. 1951. *Requiem for a Nun.* New York: Random House.

Fausto-Sterling, Anne. 1985. *Myths of Gender: Biological Theories About Men and Women.* New York: Basic.

———. 1987. "Society Writes Biology/Biology Constructs Gender." *Daedalus* 116 (4): 61–76.

———. 1995. "Gender, Race, and Nation: The Comparative Anatomy of 'Hottentot' Women in Europe, 1815–1817." In *Deviant Bodies: Critical Perspectives on Difference in Science and Popular Culture,* edited by Jennifer Terry and Jacqueline Uria, 19–48. Bloomington: Indiana University Press.

———. 2003. "Science Matters, Culture Matters." *Perspectives in Biology and Medicine* 46 (1): 109–24.

Feldman, Daniel C., and William H. Turnley. 2004. "Contingent Employment in Academic Careers: Relative Deprivation among Adjunct Faculty." *Journal of Vocational Behavior* 64 (2): 284–307.

Fields, Helen. 2005, February. "Invasion of the Snakeheads." *Smithsonian Magazine.*

Fineblum, Wendy L., and Mark D. Rausher. 2002, October 12. "Tradeoff between Resistance and Tolerance to Herbivore Damage in a Morning Glory." *Nature* 377: 517–20.

Fisher, Jill. 2011. *Gender and the Science of Difference: Cultural Politics of Contemporary Science and Medicine.* New Brunswick, NJ: Rutgers University Press.

Fisher, R. A. 1941. "Average Excess and Average Effect of Gene Substitution." *Annals of Eugenics* 11: 53–63.

Forrest, G. I., and A. F. Fletcher. 1995. "Implication of Genetic Research for Pinewood

Conservation." In *Our Pinewood Heritage*, edited by J. R. Aldhous, 97–106. Farnham: Royal Society for the Protection of Birds.

Foucault, Michel. 1977. *Discipline and Punish: The Birth of the Prison*. New York: Pantheon.

———. 1980. *Power/Knowledge: Selected Interviews and Other Writings, 1972–1977*. New York: Pantheon.

Fox, Mary Frank. 2000. "Organizational Environments and Doctoral Degrees Awarded to Women in Science and Engineering Departments." *Women's Studies Quarterly* 28: 47–61.

Fox, Mary Frank, Gerhard Sonnert, and Irina Nikiforova. 2011, October. "Programs for Undergraduate Women in Science and Engineering: Issues, Problems, and Solutions." *Gender & Society* 22 (5): 589–615.

Frazier, Ian. 2010, October 25. "Fish Out of Water: The Asian Carp Invasion." *New Yorker*, 66–73.

Freedman, Estelle. 2002. *No Turning Back: The History of Feminism and the Future of Women*. New York: Ballantine.

Fuller, Laurie, and Erica R. Meiners. 2005. "Reflections: Empowering Women, Technology, and (Feminist) Institutional Changes." *Frontiers: A Journal of Women Studies* 26 (1): 168–80.

Fullwiley, D. 2007. "The Molecularization of Race: Institutionalizing Human Difference in Pharmacogenetics Practice." *Science as Culture* 16 (1): 1–30.

Galton, Francis. 1865. "Hereditary Talent and Character." *Macmillan's Magazine* 12: 163.

———. 1909. *Memories of My Life*. 3rd ed. London: Methuen.

Gannett, Lisa. 2001. "Racism and Human Genome Diversity Research: The Ethical Limits of 'Population Thinking.'" *Philosophy of Science* 68 (3): S479–S492.

Garforth, Lisa, and Anne Kerr. 2009, Fall. "Women and Science: What's the Problem?" *Social Politics: International Studies in Gender, State and Society* 16 (3): 379–403.

Gates, Barbara T., and Ann B. Shteir. 1997. *Natural Eloquence: Women Reinscribe Science*. Madison: University of Wisconsin Press.

Gere, Cathy. 1999. "Bones That Matter: Sex Determination in Paleodemography, 1948–1995." *Studies in the History and Philosophy of Biology and the Biomedical Sciences* 30 (4): 455–71.

Ghiselin, Michael. T. 1971. "The Individual in the Darwinian Revolution." *New Literary History* 3 (1): 113–34.

———. 1995, December. "Darwin, Progress, and Economic Principles." *Evolution* 49 (6): 1029–37.

———. 2005. " The Darwinian Revolution as Viewed by a Philosophical Biologist." *Journal of the History of Biology* 38 (1): 123–36.

Gigerenzer, Gerd, Zeno Swijtink, Theodore Porter, Lorraine Daston, John Beatty, and Lorenz Kruger. 1989. *The Empire of Chance—How Probability Changed Science and Everyday Life*. Cambridge: Cambridge University Press.

Gilbert, Scott, Jan Sapp, and Alfred Tauber. 2012. "A Symbiotic View of Life: We Have Never Been Individuals." *Quarterly Review of Biology* 87 (4): 325–41.

Gilbert, S. F. 2002. "Genetic Determinism: The Battle between Scientific Data and Social

Image in Contemporary Developmental Biology." In *On Human Nature: Anthropological, Biological, and Philosophical Foundations,* edited by A. Grunwald, M. Gutmann, and E. M. Neumann-Held, 121–40. New York: Springer-Verlag.
Gilham, Nicholas. 2001. *A Life of Sir Francis Galton: From African Exploration to the Birth of Eugenics.* Oxford: Oxford University Press.
Gilman, Sander. 1985. *Difference and Pathology: Stereotypes of Sexuality, Race, and Madness.* Ithaca, NY: Cornell University Press.
Glover, Deborah, Mary L. Durbin, Gavin Huttley, and Michael T. Clegg. 1996. "Genetic Diversity in the Common Morning Glory." *Plant Species Biology* 11: 41–50.
Gobster, P. H. 2005. "Invasive Species as Ecological Threat: Is Restoration an Alternative to Fear-based Resource Management?" *Ecological Restoration* 23: 261–70.
Goonatilake, Susantha. 1984. *Aborted Discovery: Science and Creativity in the Third World.* London: Zed.
Gonsalves, Allison J. 2011. "Gender and Doctoral Physics Education: Are We Asking the Right Question?" In *Doctoral Education: Research Based Strategies for Doctoral Students, Supervisors, and Administrators,* edited by L. McAlpine and C. Amundsen. Springer Science+Business Media B.V.
Gordon, Avery. 1997. *Ghostly Matters: Haunting and the Sociological Imagination.* Minneapolis: University of Minnesota Press.
Gould, Stephen Jay. 1979, December. "Darwin's Middle Road." *Natural History* 88 (10): 27–31.
———. 1980. "Wallace's Fatal Flaw." *Natural History* 89 (1): 26–40.
———. 1981. *The Mismeasure of Man.* New York: Norton.
———. 1982. *The Structure of Evolutionary Theory.* Cambridge, MA: Harvard University Press.
———. 1991, December. "The Smoking Gun of Eugenics." *Natural History* 100 (12), http://homepages.dcc.ufmg.br/~assuncao/pgm/aulas/TheSmokingGunOfEugenics.pdf.
———. 1997. "An Evolutionary Perspective on Strengths, Fallacies, and Confusions in the Concept of Native Species." In *Nature and Ideology: Natural Garden Design in the Twentieth Century,* edited by Joachim Wolschke-Bulmahn, vol. 18: 11–19. Washington, DC: Dumbarton Oaks Research Library and Collection.
———. 2002. "An Evolutionary Perspective on the Concept of Native Species." In *I Have Landed: The End of a Beginning in Natural History,* 335–46. New York: Harmony.
Greco, M. 1993. "Psychosomatic Subjects and the 'Duty to Be Well': Personal Agency within Medical Rationality." *Economy and Society* 22 (3): 357–72.
Gregory, Richard. 1998. "Brainy Mind." *British Medical Journal* 317: 1693–95.
Grene, Marjorie. 1990, July. "Evolution, 'Typology' and 'Population Thinking.'" *American Philosophical Quarterly* 27 (3): 237–44.
———, ed. 1983. *Dimensions of Darwinism: Themes and Counterthemes in Twentieth-Century Evolutionary Theory.* Cambridge: Cambridge University Press.
Grese, Robert E. 2011. "Introduction." In *The Native Landscape Reader,* 3–22. Amherst: University of Massachusetts Press.

Gromley, Melinda. 2009. "Scientific Discrimination and the Activist Scientist: L. C. Dunn and the Professionalization of Genetics and Human Genetics in the United States." *Journal of the History of Biology* 42: 33–72.

Groning, G., and J. Wolschke-Bulmahn. 2003. "The Native Plant Enthusiasm: Ecological Panacea or Xenophobia?" *Landscape Research* 28: 75–88.

Grosz, Elizabeth. 1994. *Volatile Bodies: Toward a Corporeal Feminism.* Sydney: Allen and Unwin.

———. 2011. *Becoming Undone: Darwinian Reflections on Life Politics and Art.* Durham, NC: Duke University Press.

Gudding, Gabriel. 1996. "The Phenotype/Genotype Distinction and the Disappearance of the Body." *Journal of the History of Ideas* 57 (3): 525–45.

Haack, Robert A., Frank Herard, Jianghua Sun, and Jean J. Turgeon. 2010. "Managing Invasive Populations of Asian Longhorned Beetle and Citrus Longhorned Beetle: A Worldwide Perspective." *Annual Review of Entomology* 55: 521–46.

Hacker, Andrew, and Claudia Dreyfus. 2011. *Higher Education: How Colleges are Wasting Our Money and Failing Our Kids—and What We Can Do about It.* New York: St. Martin's Griffin.

Hall, Roberta M., and Bernice R. Sandler. 1982. *The Classroom Climate: A Chilly One for Women?* Washington, DC: Association of American Colleges.

———. 1984. *Out of the Classroom: A Chilly Climate for Women?* Washington DC: Association of American Colleges.

Haller, Mark. 1984. *Eugenics: Hereditarian Attitudes in American Thought.* New Brunswick, NJ: Rutgers University Press.

Hammonds, Evelynn. 1993. "Never Meant to Survive: A Black Woman's Journey." In *The "Racial" Economy of Science: Towards a Democratic Future,* edited by Sandra Harding, 239–48. Bloomington: Indiana University Press.

———. 1994, Summer/Fall. "Black (W)holes and the Geometry of Black Female Sexuality." *differences: A Journal of Feminist Cultural Studies* 6 (2/3): 126–46.

———. 1999. "The Logic of Difference: A History of Race in Science and Medicine in the United States." Presentation at the Women's Studies Program, UCLA.

Hammonds, Evelynn, and Rebecca M. Herzig. 2009. *The Nature of Difference: Sciences of Race in the United States from Jefferson to Genomics.* Cambridge, MA: MIT Press.

Hancock, Ange-Marie, and Nira Yuval-Davis, eds. 2011. *The Politics of Intersectionality.* Palgrave-Macmillan.

Handelsman, Jo, et al. 2005. "More Women in Science." *Science* 309 (5738): 1190–91.

Hansen, Nancy, Heidi L Janz, and Dick J. Sobsey. 2008. "21st Century Eugenics." *The Lancet* 372: 104–7.

Hanson, Sandra L. 2004, Spring. "African American Women in Science: Experiences from High School through the Post-Secondary Years and Beyond." *Feminist Formations* 16 (1): 96–115.

Haraway, Donna J. 1988. "Situated Knowledges: The Science Question in Feminism and the Privilege of Partial Perspective." *Feminist Studies* 14 (3): 575–99.

———. 1989. *Primate Visions: Gender, Race, and Nature in the World of Modern Science.* New York: Routledge.
———. 1991. *Simians, Cyborgs, and Women: The Reinvention of Nature.* New York: Routledge.
———. 1997. *Modest_Witness@Second_Millennium: FemaleMan__Meets_Onco Mouse.* New York: Routledge.
———. 1999. *How Like a Leaf: An Interview with Thyrza Nichols Goodve.* New York: Routledge.
———. 2007. *When Species Meet.* Minneapolis: University of Minnesota Press.
Harding, Sandra. 1986. *The Science Question in Feminism.* Milton Keynes, UK: Open University Press.
———. 1991. *Whose Science? Whose Knowledge?* Milton Keynes, UK: Open University Press.
———. 1993. *The "Racial" Economy of Science: Toward a Democratic Future.* Bloomington: Indiana University Press.
———. 1998. *Is Science Multicultural? Postcolonialisms, Feminisms, and Epistemologies.* Bloomington: Indiana University Press.
———. 2006. *Science and Social Inequality: Feminist and Postcolonial Issues.* Urbana: University of Illinois Press.
———. 2008. *Sciences From Below: Feminisms, Postcolonialities, and Modernities.* Durham, NC: Duke University Press.
———. 2011. *The Postcolonial Science and Technology Studies Reader.* Durham, NC: Duke University Press.
Harley, J. L., and S. E. Smith. 1983. *Mycorrhizal Symbiosis.* London: Academic.
Harmon, Amy. 2007. "Prenatal Test Puts Down Syndrome in Hard Focus," *The New York Times,* May 9.
Harrison, Carlos. 2002. "Something from the X-Files," *Miami New Times,* August 29.
Harrison, Faye V. 1996. "Anthropology, Fiction, and Unequal Relations of Intellectual Production." In *Women Writing Culture,* edited by Ruth Behar and Deborah Gordon, 233–45. Berkeley: University of California Press.
Hart, Roger. 1999. "On the Problem of Chinese Science." In *The Science Studies Reader,* edited by Mario Biagioli. New York: Routledge.
Hartmann, Betsy. 2010. "Rethinking Climate Refugees and Climate Conflict: Rhetoric, Reality and the Politics of Policy Discourse." *Journal of International Development* 22: 233–46.
———. 2011, Spring. "The Return of Population Control: Incentives, Targets and the Backlash against Cairo." *Different Takes,* no. 70, http://popdev.hampshire.edu/sites/default/files/uploads/u4763/DT%2070%20Hartmann.pdf.
Hartmann, Betsy, Banu Subramaniam, and Charles Zerner. 2005. *Making Threats: Biofears and Environmental Anxieties.* Lanham, MD: Rowman & Littlefield.
Hartmann, Elizabeth. 1995. *Reproductive Rights and Wrongs: The Global Politics of Population Control.* Brooklyn: South End.
Harwood, Jonathan. 1989. "Genetics, Eugenics and Evolution." *British Journal for the History of Science* 22 (3): 257–65.
Hattingh, Johan. 2011. "Conceptual Clarity, Scientific Rigour and 'the Stories We Are': Engaging with Two Challenges to the Objectivity of Invasion Biology." In *Fifty Years of*

Invasion Ecology: The Legacy of Charles Elton, edited by David M. Richardson, 359–75. Oxford: Blackwell.

Head, Lesley, and Jennifer Atchison. 2009. "Cultural Ecology: Emerging Human-Plant Geographies." *Progress in Human Geography* 33 (2): 236–45.

Hearn, Jeff. 2001. "Academia, Management, and Men: Making the Connections, Exploring the Implications." In *Gender and the Restructured University: Changing Management and Culture in Higher Education*, edited by Anne Brooks and Alison Mackinnon, 69–89. New York: Open University Press.

Hebert, Josef. 1998. "Feds to Fight Invaders." *ABC News*, February 3.

Helmreich, Stefan. 2009. *Alien Ocean: Anthropological Voyages in Microbial Seas*. Berkeley: University of California Press.

Hemmings, Clare. 2011. *Why Stories Matter: The Political Grammar of Feminist Theory*. Durham, NC: Duke University Press.

Henrion, Claudia. 1997. *Women in Mathematics: The Addition of Difference*. Bloomington: Indiana University Press.

Henson, Pamela M. 1987. "The Comstocks of Cornell: A Marriage of Interests." In *Creative Couples in the Sciences*, edited by Helena Pycior, Nancy Slack, and Pnina Abir-Am, 112–25. New Brunswick, NJ: Rutgers University Press.

Hess, David. 1997. *Science Studies: An Advanced Introduction*. New York: NYU Press.

Hettinger, Ned. 2011, May. "Exotic Species, Naturalization, and Biological Nativism." *Environmental Values* 10 (2): 193–224.

Hierro, Jose, Diego Villare, Ozkan Eren, Jon M. Graham, and Ragan M. Callaway. 2006, August. "Disturbance Facilitates Invasion: The Effects Are Stronger Abroad than at Home." *The American Naturalist* 168 (2): 145–56.

Hill, Constance, Christianne Corbett, and Andresse St. Rose. 2010. "Why So Few? Women in Science, Technology, Engineering and Mathematics," American Association of University Women, http://www.aauw.org/learn/research/upload/whysofew.pdf.

Hird, M. J. 2004. *Sex, Gender, and Science*. London: Palgrave.

Hobbs, Richard J., and David M. Richardson. 2011. "Invasion Ecology and Restoration Ecology: Parallel Evolution in Two Fields of Endeavour." In *Fifty Years of Invasion Ecology: The Legacy of Charles Elton*, edited by David M. Richardson, 61–70. Oxford: Blackwell.

Hodges. Sarah. 2008. *Contraception, Colonialism and Commerce*. Farnham, UK: Ashgate.

———. 2010. "South Asia's Eugenic Pasts." In *The Oxford Handbook of the History of Eugenics*, edited by Alison Bashford and Philippa Levine, 228–42. New York: Oxford University Press.

Hodgson, Dennis. 1991. "The Ideological Origins of the Population Association of America." *Population and Development Review* 17 (1): 1–34.

Holden, Constance. 2000. "Parity as a Goal Sparks Bitter Battle." *Science* 289 (5478): 380.

Hubbard, Ruth. 1983. "Have Only Men Evolved?" In *Discovering Reality: Feminist Perspectives on Epistemology, Metaphysics, Methodology, and Philosophy of Science*, edited by Sandra Harding and Merrill B. Hintikka, 45–69. Dordrecht, Neth.: Kluwer Academic.

———. 1990. *The Politics of Women's Biology*. New Brunswick, NJ: Rutgers University Press.

Hubbard, Ruth, and Elijah Wald. 1999. *Exploding the Gene Myth: How Genetic Information Is Produced and Manipulated by Scientists, Physicians, Employers, Insurance Companies, Educators, and Law Enforcers*. Boston: Beacon.

Hulme, Philip E. 2011. "Biosecurity: The Changing Face of Invasion Biology." In *Fifty Years of Invasion Ecology: The Legacy of Charles Elton*, edited by David M. Richardson, 301–14. Oxford: Blackwell.

Hurston, Zora Neale. n.d. "Letter to Countee Cullen." PBS American Masters, http://www.pbs.org/wnet/americanmasters/episodes/zora-neale-hurston/introduction/93/.

Huslin, Anita. 2002a. "Biologists on Mission to Kill," *Washington Post*, August 18, C03.

———. 2002b. "Freaky Fish Story Flourishes," *Washington Post*, July 3, B01.

———. 2002c. "U. S. Moves to Ban Import of Snakeheads," *Washington Post*, July 23, B03.

Huslin, Anita, and Michael Ruane. 2002. "Spawn of Snakehead?" *Washington Post*, July 10, B01.

Hyde, J. S., S. M. Lindbert, M. C. Linn, A. B. Ellis, and C. C. Williams. 2008. "Gender Similarities Characterize Math Performance." *Science* 321: 494–95.

Hynes, Patricia H. 1989. *The Recurring Silent Spring*. New York: Pergamon.

Intemann, Kristen. 2009. "Why Diversity Matters: Understanding and Applying the Diversity Component of the National Science Foundation's Broader Impacts Criterion." *Social Epistemology* 23 (3/4): 249–66.

Iverson, Susan VanDeventer. 2007. "Camouflaging Power and Privilege: A Critical Race Analysis of University Diversity Policies." *Educational Administration Quarterly* 43: 586–611.

Jacob, Margaret C. 1998. "Latour's Version of the Seventeenth Century." In *A House Built on Sand: Exposing Postmodernist Myths About Science*, edited by Noretta Koertge, 240–54. New York: Oxford University Press.

Jamieson, D. 1995. "Ecosystem Health: Some Preventive Medicine." *Environmental Values* 4: 333–44.

Jansson, Stefan, and Carl J. Douglas. 2007. "Populus: A Model System for Plant Biology." *Annual Review of Plant Biology* 58: 435–58.

Johnson, Dawn R. 2011, Winter. "Women of Color in Science, Technology, Engineering, and Mathematics (STEM)." *New Directions for Institutional Research* 2011 (152): 75–85.

Kahn, Jonathan. 2005, November/December. "BiDil: False Promises." *Gene Watch* 18 (6): 6–9.

Kaiser, Jocelyn. 1999. "Stemming the Tide of Invading Species." *Science* 285 (5435): 1836–40.

Kamin, Leon J. 1974. *The Science and Politics of IQ*. New York: Halsted.

Kaplan, Kim. 1991, July. "USDA Plant Hunters: Bring 'Em Back Alive and Growing." *Agricultural Research*, 4–12, http://pubpages.unh.edu/~gec/IA401/USDA%20Plant%20Hunters.PDF.

Kass-Simon, G., and Patricia Farnes. 1990. *Women of Science: Righting the Record*. Bloomington: Indiana University Press.

Kaufman, Sharon R., and Lynn M. Morgan. 2005. "The Anthropology of the Beginnings and Ends of Life." *Annual Review of Anthropology* 34: 317–41.

Keller, David R., and Frank B. Golley. 2000. *The Philosophy of Ecology: From Science to Synthesis.* Athens: University of Georgia Press.

Keller, Evelyn Fox. 1977. "An Anomaly of a Woman in Physics." In *Working It Out: 23 Women Writers, Artists, Scientists, and Scholars Talk About Their Lives and Work,* edited by Sara Ruddick and Pamela Daniels. New York: Pantheon.

———. 1983. *A Feeling for the Organism: The Life and Work of Barbara McClintock.* San Francisco: Freeman.

———. 1985. *Reflections on Gender and Science.* New Haven, CT: Yale University Press.

———. 1987. "Is Gender/Science System: or, Is Sex to Gender as Nature is to Science?" *Hypatia* 2 (3): 37–49.

———. 1992. "How Gender Matters, Or, Why It's So Hard for Us to Count Past Two." In *Inventing Women: Science, Technology and Gender,* edited by Gill Kirkup and Laurie Smith Keller, 42–56. Cambridge: Polity, in association with the Open University.

———. 1996. "The Biological Gaze." In *Future Natural: Nature/Science/Cultural,* edited by G. Roberson, M. Mash, L. Tickner, J. Bird, B. Curtis, and T. Putnam, 107–21. London: Routledge.

———. 2001. "Making a Difference: Feminist Movement and Feminist Critiques of Science." In *Feminism in Twentieth-Century Science, Technology, and Medicine,* edited by Andrea N. H. Creager, Elizabeth Lunbeck, and Londa Schiebinger. Chicago: University of Chicago Press.

———. 2010. *The Mirage of a Space between Nature and Nurture.* Durham, NC: Duke University Press.

Keller, Evelyn Fox, and Helen Longino, eds. 1996. "Introduction." In *Feminism and Science,* 1–14. Oxford: Oxford University Press.

Kerr, Anne. 1998. "Eugenics and the New Genetics in Britain: Examining Contemporary Professionals' Accounts." *Science, Technology, & Human Values* 23 (2): 175–98.

Kevles, Daniel, J. 1985. *In the Name of Eugenics: Genetics and the Uses of Human Heredity.* New York: Knopf.

———. 1998. *In the Name of Eugenics: Genetics and the Uses of Human Heredity.* Cambridge, MA: Harvard University Press.

Keynes, John Maynard. 1926. *The End of Laissez-faire,* http://www.panarchy.org/keynes/laissezfaire.1926.html.

Khalil, Elias. 2000. "Beyond Natural Selection and Divine Intervention: The Lamarckian Implication of Adam Smith's Invisible Hand." *Journal of Evolutionary Economics* 10: 373–93.

Kim, Kyung-Man. 1994. *Explaining Scientific Consensus: The Case of Mendelian Genetics.* New York: Guilford.

Kirby, V. 1997. *Telling Flesh: The Substance of the Corporeal.* New York: Routledge.

Kirksey, S. Eben, and Stefan Helmreich. 2010. "The Emergence of Multispecies Ethnography." *Cultural Anthropology* 25 (4).

Kitcher, Philip. 1997. *The Lives to Come: The Genetic Revolution and Human Possibilities.* New York: Simon and Schuster.

Klausen, Susanne, and Alison Bashford, 2010. "Eugenics, Feminism and Fertility Control." In *The Oxford Handbook of the History of Eugenics,* edited by Alison Bashford and Philippa Levine. Oxford: Oxford University Press.

Klein, H. 2002. *Weeds, Alien Plants and Invasive Plants.* PPRI Leaflet Series: Weeds Biocontrol, no. 1.1. Pretoria: ARC-Plant Protection Research Institute.

Kline, Wendy. 2005. *Building a Better Race: Gender, Sexuality, and Eugenics from the Turn of the Century to the Baby Boom.* Berkeley: University of California Press.

Klinkenborg, Verlyn. 2013. "Hey, You Calling Me an Invasive Species?" *The New York Times,* September 8.

Kluchin, Rebecca. 2009. *Fit to Be Tied: Sterilization and Reproductive Rights in America, 1950–1980.* New Brunswick, NJ: Rutgers University Press.

Kluger, Jeffrey. 2002, July 29. "Fish Tale." *Time Magazine,* http://www.time.com/time/magazine/article/0,9171,332042,00.html.

Knorr Cetina, Karin. 1999. *Epistemic Cultures: How the Sciences Make Knowledge.* Cambridge, MA: Harvard University Press.

Kobell, Rona, and Candus Thomson. 2002. "State Shares Victory over Snakehead Fish," *Baltimore Sun,* November 21.

Kobert, Elizabeth. 2011, January 31. "America's Top Parent: What's Behind the 'Tiger Mother' Craze?" *The New Yorker,* 70–73.

Koblitz, Anne Hibner. 1983. *A Convergence of Lives: Sofia Kovalevskaia, Scientist, Writer, Revolutionary.* New Brunswick, NJ: Rutgers University Press.

Kohlstedt, Sally Gregory. 2004. "Sustaining Gains: Reflections on Women in Science and Technology in 20th Century United States." *Feminist Formations* 16 (1): 1–26.

———, ed. 1999. *History of Women in the Sciences: Readings from Isis.* Chicago: University of Chicago Press.

Kolata, Gina. 2011. "Women Atop Their Fields Dissect the Scientific Life," *The New York Times,* June 6.

Kreitman, Martin. 2000. "Methods to Detect Selection in Populations with Applications to the Human." *Annual Review of Genomics and Human Genetics* 1: 539–59.

Kumar, Neelam, ed. 2009. *Women and Science in India: A Reader.* New Delhi: Oxford University Press.

Kyung-Man, Kim. 1994. *Explaining Scientific Consensus: The Case of Mendelian Genetics.* New York: Guilford.

Larson, Brendon. 2005. "The War of the Roses: Demilitarizing Invasion Biology." *Frontiers in Ecology and the Environment* 3: 495–500.

———. 2007a. "An Alien Approach to Invasive Species: Objectivity and Society in Invasion Biology." *Biological Invasions* 9: 947–56.

———. 2007b. "Entangled Biological, Cultural and Linguistic Origins of the War on Invasive Species." In *Body, Language and Mind,* vol. 2, *Sociocultural Situatedness,* edited by R. Frank, R. Dirven, J. Zlatev, and T. Ziemke. New York: Mouton de Gruyter.

———. 2007c. "Thirteen Ways of Looking at Invasive Species." In *Invasive Plants: Inven-*

tories, Strategies and Action. Topics in Canadian Weed Science, vol. 5, edited by D. R. Clements and S. J. Darbyshire, 131–46. Quebec: Canadian Weed Science Society.

Laslett, Barbara, and Barrie Thorne, eds. 1997. *Feminist Sociology: Life Histories of a Movement.* New Brunswick, NJ: Rutgers University Press.

Latour, Bruno. 1993a, Winter. "Pasteur on Lactic Acid Yeast: A Partial Semiotic Analysis." *Configurations* 1: 129–46.

———. 1993b. *We Have Never Been Modern.* Cambridge, MA: Harvard University Press.

Latour, Bruno, and Steven Woolgar. 1986. *Laboratory Life: The Construction of Scientific Facts.* Princeton, NJ: Princeton University Press.

Lawrence, Peter A. 2006. "Men, Women, and Ghosts in Science." *PLoS Biology* 4 (1): 13–15.

Lear, Linda. 1997. *Rachel Carson: Witness for Nature.* Boston: Mariner.

Lederman, Muriel, and Ingrid Bartsch, eds. 2001. *The Gender and Science Reader.* New York: Routledge.

Lemke, Thomas. 2007. "Susceptible Individuals and Risky Rights: Dimensions of Genetic Responsibility." In *Biomedicine as Culture: Instrumental Practices, Technoscientific Knowledge, and New Modes of Life.* London: Routledge.

Leonard, Eileen B. 2003. *Women, Technology, and the Myth of Progress.* Upper Saddle River, NJ: Prentice Hall.

Lerdau, Manuel, and Jacob D. Wickham. 2011. "Non-Natives: Four Risk Factors." *Nature* 475: 36–37.

Levine, Philippa, and Alison Bashford. 2010. "Introduction: Eugenics and the Modern World." In *The Oxford Handbook of the History of Eugenics,* edited by Alison Bashford and Philippa Levine, 3–26. Oxford: Oxford University Press.

Levine, Susan. 1995. *Degrees of Equality: The American Association of University Women and the Challenge of Twentieth-Century Feminism.* Philadelphia: Temple University Press.

Levy, David, and Sandra Peart. 2004. "Statistical Prejudice: From Eugenics to Immigrants." *European Journal of Political Economy* 20: 5–22.

Lewontin, Richard. 1970. "The Units of Selection." *Annual Review of Ecology and Systematics* 1:1–18.

———. 1974. *The Genetic Basis of Evolutionary Change.* New York: Columbia University Press.

———. 1978. "Adaptation." *Scientific American* 239: 212–30.

———. 1987. "Polymorphism and Heterosis: Old Wine in New Bottles and Vice Versa." *Journal of the History of Biology* 20 (3): 337–49.

Lippincott, Gail. 2003. "Rhetorical Chemistry: Negotiating Gendered Audiences in Nineteenth-Century Nutrition Studies." *Journal of Business and Technical Communication* 17: 10–49.

Lockwood, Julie L., Martha F. Hoopes, and Michael P. Marchetti. 2011. "'Non Natives' Plusses of Invasion Ecology." *Nature* 475: 36.

Lohmann, Larry. 2005. "Malthusianism and the Terror of Scarcity." In *Making Threats: Biofears and Environmental Anxieties,* edited by Betsy Hartmann et al., 81–98. New York: Rowman & Littlefield.

Lombardo, Paul. 2002. "The American Breed: Nazi Eugenics and the Origins of the Pioneer Fund." *Albany Law Review* 65 (3): 743–830.

Long, Scott. J., ed. 2001. *From Scarcity to Visibility: Gender Differences in the Careers of Doctoral Scientists and Engineers.* Washington, DC: National Academy of Sciences.

Longino, Helen E. 1989. "Can There Be a Feminist Science?" In *Feminism and Science,* edited by Nancy Tuana. Bloomington: Indiana University Press.

———. 1990. *Science as Social Knowledge: Values and Objectivity in Scientific Inquiry.* Princeton, NJ: Princeton University Press.

Lorber, Judith. 2000. "Using Gender to Undo Gender: A Feminist Degendering Movement." *Feminist Theory* 1 (1): 79–95.

Lorde, Audre. 1984. "Age, Race, Class and Sex: Women Redefining Difference." In *Sister Outsider,* edited by Audre Lorde. Freedom, CA: The Crossing Press.

Louçã, Francisco. 2008. "Emancipation through Interaction—How Eugenics and Statistics Converged and Diverged." *Journal of the History of Biology* 42 (4): 649–84.

Lovett, Laura. 2007. *Conceiving the Future: Pronatalism, Reproduction, and the Family in the United States, 1890–1930.* Gender and American Culture Series. Chapel Hill: University of North Carolina Press.

Lovitts, Barbara. 2007. *Making the Implicit Explicit: Creating Performance Expectations for the Dissertation.* Sterling, VA: Stylus.

Mabberley, D. J. 1997. *The Plant-Book: A Portable Dictionary of the Vascular Plants.* 2nd ed. Cambridge: Cambridge University Press.

Mabey, R. 2005. "From Corn Poppies to Eagle Owls." *ECOS* 26 (3/4): 41–46.

MacAusland, Carol, and Christopher Costello. 2004, September. "Avoiding Invasives: Trade Related Policies for Controlling Unintentional Exotic Species Introductions." *Journal of Environmental Economics and Management* 45 (2): 954–77.

Macdougall, A. S., and R. Turkington. 2005. "Are Invasive Species the Drivers or Passengers of Change in Degraded Ecosystems?" *Ecology* 86: 42–55.

Mack, Richard N., et al. 2000, Spring. "Biotic Invasions: Causes, Epidemiology, Global Consequence and Control." *Issues in Ecology* 15 (5): 12.

MacKenzie, Donald A. 1981. *Statistics in Britain, 1865–1930: The Social Construction of Scientific Knowledge.* Edinburgh: Edinburgh University Press.

Maddox, Brenda. 2002. *Rosalind Franklin: The Dark Lady of DNA.* New York: HarperCollins.

Magubane, Zine. 2012. "Which Bodies Matter? Feminism, Poststructuralism, Race and the Curious Theoretical Odyssey of the 'Hottentot Venus.'" *Gender and Society* 26 (6): 816–34.

Malcolm, Shirley M. 1999. "Fault Lines." *Science* 284: 1271.

Malone, Scott. 2012. "U.S. Economy Losing Competitive Edge: Survey," Reuters, January 18, http://www.reuters.com/article/2012/01/18/us-corporate-competitiveness-idUSTRE80H1HR20120118.

Mann, Charles C. 2011, May/June. "The Dawn of the Homogenocene: Tracking Globalization Back to Its Roots." *Orion Magazine.*

Mansfield, Harvey. 2006. *Manliness*. New Haven, CT: Yale University Press.
Marais, Robert de. 1974. "The Double-Edged Effect of Sir Francis Galton: A Search for the Motives in the Biometrician-Mendelian Debate." *Journal of the History of Biology* 7: 141–74.
Margulis, Lynn. 1998. *Symbiotic Planet: A Look at Evolution*. Amherst, MA: Perseus.
Margulis, Lynn, and Dorion Sagan. 2002. *Acquiring Genomes: A Theory of the Origins of Species*. New York: Basic.
Marinelli, Janet, and John M. Randall. 1996. *Invasive Plants: Weeds of the Global Garden*. Brooklyn: Brooklyn Botanic Garden.
Marklein, Mary Beth. 2012. "Record Number of Foreign Students in U.S.," *USA Today*, November 12.
Markowitz, Sally. 2001. "Pelvic Politics: Sexual Dimorphism and Racial Difference." *Signs* 26 (2): 389–414.
Marler, Marilyn, Catherine Zabinksi, and Ragan Callaway. 1999, June. "Mycorrhizae Indirectly Enhance Competitive Effects of an Invasive Forb on a Native Bunchgrass." *Ecology* 80 (4): 1180–86.
Martin, Emily. 1992. *The Woman in the Body*. Boston: Beacon.
———. 1997. "The End of the Body?" In *The Gender Sexuality Reader*, edited by Roger N. Lancaster and Micaela di Leonardo, 543–59. New York: Routledge.
———. 2006. "The Pharmaceutical Person." *BioSocieties* 1: 273–87.
Martinez, Elisabeth D., et al. 2007. "Falling Off the Academic Bandwagon." *European Molecular Biology Organization Reports* 8 (11): 977–81.
Marvier, Michelle, Peter Kareiva, and Michael Neubert. 2004. "Habitat Destruction, Fragmentation, and Disturbance Promote Invasion by Habitat Generalists in a Multispecies Metapopulation." *Risk Analysis* 24 (4): 869–78.
"Maryland Fears Carnivore Fish Invasion." *Edmonton Journal*. 2002. July 13.
Mason, M. A., and M. Goulden. 2004, November/December. "Do Babies Matter Part II? Closing the Baby Gap." *Academe*: 11–15.
Matt, Susan J. 2012. "The New Globalist is Homesick," *The New York Times*, March 21.
Matthews, Christine M. 2010. "Foreign Science and Engineering Presence in U.S Institutions and the Labor Force." Congressional Research Service 97–746, http://www.fas.org/sgp/crs/misc/97-746.pdf.
Mayberry, Maralee, Banu Subramaniam, and Lisa Weasel, eds. 2001. *Feminist Science Studies: A New Generation*. New York: Routledge.
Mayell, Hillary. 2002, July 2. "Maryland Wages War on Invasive Walking Fish." *National Geographic*.
Mayr, Ernst. 1973. "The Recent Historiography of Genetics." *Journal of the History of Biology* 6 (1): 125–54.
———. 1976. *Evolution and the Diversity of Life*. Cambridge, MA: Harvard University Press.
———. 1982. *The Growth of Biological Thought*. Cambridge, MA: The Belknap Press of Harvard University Press.
Mazumdar, Pauline M. 1991. *Eugenics, Human Genetics and Human Failings: The Eugenics Society, its Sources and its Critics in Britain*. London: Routledge.

———. 2002. "'Reform' Eugenics and the Decline of Mendelism." *Trends in Genetics* 18 (1): 48–52.
McCall, Leslie. 2005, Spring. "The Complexity of Intersectionality." *Signs* 30 (3).
McCann, Carole R. 2009. "Malthusian Men and Demographic Transitions." *Frontiers* 30 (1): 142–72.
McDonald, Andrew. 1991. "Origin and Diversity of Mexican Convolvulaceae." *Anales del Instituto de Biologia. Serie Botanica* 62 (1): 65–82.
McDonald, Kim A. 1999. "Biological Invaders Threaten U.S. Ecology." *Chronicle of Higher Education* 45 (23): A15.
McIntosh, Peggy. 2003. "White Privilege: Unpacking the Invisible Knapsack." In *Understanding Prejudice and Discrimination,* edited by Scott Plous, 191–96. New York: McGraw Hill.
McNeely, J. A. 2001. *The Great Reshuffling: Human Dimensions of Invasive Alien Species.* Gland, Switzerland: IUCN.
Mellström, Ulf. 2009. "The Intersection of Gender, Race and Cultural Boundaries, or Why is Computer Science in Malaysia Dominated by Women?" *Social Studies of Science* 39 (6): 885–907.
Merchant, Carolyn. 1980. *The Death of Nature: Women, Ecology, and the Scientific Revolution.* San Francisco: Harper & Row.
Merton, R. 1968. "The Matthew Effect in Science." *Science* 159 (3810): 56–63.
Mervis, Jeffrey. 2000. "Diversity: Easier Said Than Done." *Science* 289 (5478): 378–79.
Meyerson, Debra E., and Deborah M. Kolb. 2000. "Moving out of the Armchair: Developing a Framework to Bridge the Gap Between Feminist Theory and Practice." *Organization* 7(4): 553–72.
Mills, E. L. , J. H. Leach, J. T. Carlton, and C. L. Secor. 1994. "Exotic Species and the Integrity of the Great Lakes: Lessons from the Past." *Bioscience* 33 (10): 666–76.
Mills, E. L., M. D. Scheurell, D. L. Strayer, and J. T. Carlton. 1996. "Exotic Species in the Hudson River Basin: A History of Invasions and Introductions." *Estuaries* 19 (4): 814–23.
MIT (Massachusetts Institute of Technology). 1999. *A Study of the Status of Women Faculty in Science at MIT,* Special Issue of the MIT Faculty Newsletter 9 (March), http://web.mit.edu/fnl/women/women/html.
Moallem, Minoo, and Iain A. Boal. 1999. "Multicultural Nationalism and the Poetics of Inauguration." In *Between Women and Nation,* edited by Caren Kaplan, Norma Alarcon, and Minoo Moallem, 243–63. Durham, NC: Duke University Press.
Mohanty, Chandra. 2003. *Feminism Without Borders: Decolonizing Theory, Practicing Solidarity.* Durham, NC: Duke University Press.
Moi, Barbara. 2009. "A Dawn Tea Ceremony for the First Blooming of the Morning Glory." *Moebius* 7 (1): 31–38.
Moore, James. R. A. 2007. "Fisher: A Faith Fit for Eugenics." *Studies in the History and Philosophy of Science* 38 (1): 110–35.
Morland, Iain. 2011. "Intersex Treatment and the Promise of Trauma." In *Gender and the*

Science of Difference: Cultural Politics of Contemporary Science and Medicine, edited by Jill Fisher, 147–63. New Brunswick, NJ: Rutgers University Press.

Morley, B. D., and H. R. Toelken. 1983. *Flowering Plants in Australia*. Adelaide, Australia: Rigby.

Morley, Louise. 1999. *Organizing Feminisms: The Micropolitics of the Academy*. London: Macmillan.

———. 2003. *Quality and Power in Higher Education*. Berkshire, UK: Society for Research into Higher Education/Open University Press.

———. 2006. "Hidden Transcripts: The Micropolitics of Gender in Commonwealth Universities." *Women's Studies International Forum* 29: 543–51.

Morrison, Toni. 1992. *Playing in the Dark: Whiteness and the Literary Imagination*. New York: Vintage.

Moss-Racusin, Corinee A., John F. Dovidio, Victoria L. Brescoll, Mark J. Graham, and Jo Handelsman. 2012, September 17. "Science Faculty's Subtle Gender Biases Favor Male Students." *Proceedings of the National Academy of Science*. DOI 10:1073/pnas.1211286109

Mottier, Véronique. 2008. "Eugenics, Politics and the State: Social Democracy and the Swiss 'Gardening State.'" *Studies in History and Philosophy of Biology and Biomedical Sciences* 39: 263–69.

Muller, H. J. 1984. *Out of the Night: A Biologist's View of the Future*. New York: Garland.

Murray, T. R., D. A. Frank, and C. A. Gehring. 2010, March. "Ungulate and Topographic Control of Arbuscular Mychorrhizal Fungal Spore Community Composition in a Temperate Grassland." *Ecology* 91 (3): 815–27.

Nandy, Ashis. 1995. *Alternative Sciences: Creativity and Authenticity in Two Indian Scientists*. 2nd ed. Oxford: Oxford University Press.

Nash, Jennifer. 2008. "Re-thinking Intersectionality." *Feminist Review* 89: 1–15.

———. 2009. "Un-disciplining Intersectionality (Review Essay)." *International Feminist Journal of Politics* 11 (4): 587–93.

Nassar-McMillan, Mary Wyer, Maria Oliver-Hoyo, and Jennifer Schneider. 2011. "New Tools for Examining Undergraduate Students' STEM Stereotypes: Implications for Women and Other Underrepresented Groups." *New Directions for Institutional Research* 2011 (152): 87–98.

National Academy of Sciences. 2006. *Beyond Bias and Barriers: Fulfilling the Potential of Women in Academic Science and Engineering*. Washington, DC: National Academy Press.

National Science Foundation. 2002. "Demographics of the S&E Workforce." In *Science and Engineering Indicators 2002*. Arlington, VA.

———. 2011. "Demographics of the S&E Workforce." In *Science and Engineering Indicators 2011*. Arlington, VA.

———. 2012. "Demographics of the S&E Workforce." In *Science and Engineering Indicators 2012*. Arlington, VA.

Nelkin, Dorothy, and Susan Lindee. 1995. *The DNA Mystique: The Gene as Cultural Icon*. New York: W. H. Freeman.

Nelson, Donna. 2007. "An Analysis of Minorities in Science and Engineering Faculties in Research Universities," http://faculty-staff.ou.edu/N/Donna.J.Nelson-1/diversity/Faculty_Tables_FY07/07Report.pdf.

Nerad, Maresi. 1999. *The Academic Kitchen: A Social History of Gender Stratification at the University of California's Berkeley.* Albany: State University of New York Press.

New York Times, Editorial. 2012. "Dream Act for New York," March 12.

Neyhfakh, Leon. 2011. "The Invasive Species War," *Boston Globe,* July 31.

Noble, David F. 1992. *A World Without Women: The Christian Clerical Culture of Western Science.* Oxford: Oxford University Press.

Norton, Bernard J. 1978. "Karl Pearson and Statistics: The Social Origins of Scientific Innovation." *Social Studies of Science* 8 (1): 3–34.

———. 1983. "Fisher's Entrance in Evolutionary Science: The Role of Eugenics." In *Dimensions of Darwinism,* edited by Marjorie Grene, 19–29. Cambridge: Cambridge University Press.

Nosek, B. A., F. L. Smyth, N. Sriram, N. M. Lindner, T. Devos, A. Ayala, et al. 2009. "National Differences in Gender-Science Stereotypes Predict National Sex Differences in Science and Math Achievement." *Proceedings of the National Academy of Sciences* 106: 10593–97.

Novas, Carlos, and Nikolas Rose. 2000. "Genetic Risk and the Birth of the Somatic Individual." *Economy and Society* 29 (4): 484–513.

Nye, Robert A. 1997. "Medicine and Science as Masculine 'Fields of Honor.'" *Osiris* 12: 60–79.

O'Brien, Dennis. 2004. "A Crusade to Stop the Voracious Fish," *Baltimore Sun,* April 30.

O'Brien, W. 2006. "Exotic Invasions, Nativism and Ecological Restoration: On the Persistence of a Contentious Debate." *Ethics, Place and Environment* 9: 63–77.

Odum, E. P. 1997. *Ecology: A Bridge between Science and Society.* Sunderland, MA: Sinauer Associates.

Ogilvie, Marilyn B., and Joy D. Harvey, eds. 2000. *The Biographical Dictionary of Women in Science: Pioneering Lives from Ancient Times to the Mid-20th Century.* New York: Routledge.

Oliver, L. R., R. E. Frans, and R. E. Talbert. 1976. "Field Competition between Tall Morningglory and Soybean. I. Growth Analysis." *Weed Science* 24 (5): 482–88.

Olwig, R. 2003. "Natives and Aliens in National Landscape." *Landscape Research* 28 (1): 61–74.

Omi, Michael, and Howard Winant. 1994. "Racial Formation." In *Racial Formation in the United States: From the 1960's to 1990's.* New York: Routledge.

Ong, Aihwa. 1999. *Flexible Citizenship: The Cultural Logics of Transnationality.* Durham, NC: Duke University Press.

———. 2006. "Higher Learning in Global Space." In *Neoliberalism as Exception: Mutations in Citizenship and Sovereignty,* 139–56. Durham, NC: Duke University Press.

Ong, Maria, Carol Wright, Lorelle L. Espinosa, and Gary Orfield. 2011, June. "Inside the Double Bind: A Synthesis of Empirical Research on Undergraduate and Graduate Women of Color in Science, Technology, Engineering, and Mathematics." *Harvard Educational Review* 6: 172–209.

Onishi, Norimitsu. 2012. "Crayfish to Eat, and to Clean the Water," *The New York Times*, July 12.

Orben, Robert. 2006. Quoted in Robert Byrne, *The 2,548 Best Things Anybody Ever Said*. New York: Simon & Schuster.

Paretti, Jonah. 1998. "Nativism and Nature: Rethinking Biological Invasions." *Environmental Values* 7 (2): 183–92.

Paul, Diane B. 1987. "'Our Load of Mutations' Revisited." *Journal of the History of Biology* 20 (3): 321–35.

———. 1995. *Controlling Human Heredity: 1865 to the Present*. Atlantic Highlands, NJ: Humanities Press.

———. 1998. "Did Eugenics Rest on an Elementary Mistake?" In *The Politics of Heredity: Essays on Eugenics, Biomedicine and the Nature-Nurture Debate*, 117–32. Albany: State University of New York Press.

———. 2001. "History of Eugenics." In *International Encyclopedia of Social and Behavioral Sciences*, edited by Neil Smelser and Paul Baltes, 4896–4901. Amsterdam: Elsevier.

———. 2007. "On Drawing Lessons from the History of Eugenics." In *Reprogenetics: Law, Policy, and Ethical Issues*, edited by Lori P. Knowles and Gregory E. Kaebnick. Baltimore: Johns Hopkins University Press.

Paul, Diane, and H. G. Spencer. 1995. "The Hidden Science of Eugenics." *Nature* 374 (6529): 302–4.

Pauly, Philip. 1993. "The Eugenics Industry: Growth or Restructuring?" *Journal of the History of Biology* 26 (1): 131–45.

———. 1996. "The Beauty and Menace of the Japanese Cherry Trees: Conflicting Visions of American Ecological Independence." *Isis* 87 (1): 51–73.

———. 2002, July. "Fighting the Hessian Fly: American and British Responses to Insect Invasion: 1776–1789." *Environmental History* 7 (3): 485–507.

PBS Newshour. 2012. "An Old Fashioned Strategy to Keep Asian Carp at Bay in the Great Lakes: Eat Them," July 26, http://www.pbs.org/newshour/bb/science/july-dec12/carp_07-26.html.

Philip, Kavita. 2004. *Civilizing Natures: Race, Resources, and Modernity in Colonial South India*. New Brunswick, NJ: Rutgers University Press.

Pigliucci, Massimo. 2012. "The One Paradigm To Rule Them All: Scientism and the Big Bang Theory." In *The Big Bang Theory and Philosophy: Rock, Paper, Scissors, Aristotle, Locke*. Oxford: Blackwell.

Pinker, Steven. 2002. *The Blank Slate: The Modern Denial of Human Nature*. New York: Penguin.

Pollan, Michael. 1994. "Against Nativism," *The New York Times Magazine*, May 15.

Porter, Theodore. 2005. *Karl Pearson: The Scientific Life in a Statistical Age*. Princeton, NJ: Princeton University Press.

Prakash, Gyan. 1999. *Another Reason: Science and the Imagination of Modern India*. Princeton, NJ: Princeton University Press.

Preston, Christopher D. 2009. "The Terms 'Native' and 'Alien'—A Bio-geographical Perspective." *Progress in Human Geography* 33: 702–13.

Pringle, Anne, James Bever, Monique Gardes, Jeri Parrent, Matthias Rillig, and John Klironomos. 2009. "Mycorrhizal Symbioses and Plant Invasions." *Annual Review of Ecology, Evolution, and Systematics* 40: 699–715.

Provine, W. 1971. *The Origins of Theoretical Population Genetics.* Chicago: University of Chicago Press.

———. 1988. *Sewall Wright and Evolutionary Biology.* Chicago: University of Chicago Press.

Puar, Jasbir. 2007. *Terrorist Assemblages: Homonationalism in Queer Times.* Durham, NC: Duke University Press.

———. 2012. "'I Would Rather Be a Cyborg Than a Goddess': Becoming-Intersectional of Assemblage Theory." *PhiloSOPHIA: A Journal of Feminist Philosophy* 2: 49–66.

Pycior, Helena. 1997. "Marie Curie: Time Only for Science and Family." In *A Devotion to Their Science: Pioneer Women of Radioactivity,* edited by Marelene F. Rayner-Canham and Geoffrey W. Rayner-Canham, 31–50. Philadelphia: Chemical Heritage Foundation.

Pycior, Helena, Nancy Slack, and Pnina Abir-Am, eds. 1996. *Creative Couples in the Sciences.* New Brunswick, NJ: Rutgers University Press.

Pysek, P., D. M. Richardson, J. Pergl, V. Jaros, Z. Sixtova, and E. Weber. 2008. "Geographical and Taxonomic Biases in Invasion Ecology." *Trends in Ecology and Evolution* 23: 237–44.

Pysek, P., D. M. Richardson, M. Rejmanek, G. L. Webster, M. Williamson, and J. Kirschener. 2004. "Alien Plants in Checklists and Floras: Towards Better Communication between Taxonomists and Ecologists." *Taxon* 53: 131–43.

Quinn, Susan. 1995. *Marie Curie: A Life.* New York: Simon & Schuster.

Rabinow, Paul, and Nikoas Rose. 2006. "Biopower Today." *BioSocieties* 1: 195–217.

Rader, Karen. 2004. *Making Mice: Standardizing Animals for American Biomedical Research, 1900–1955.* Princeton, NJ: Princeton University Press.

Raffles, Hugh. 2011. "Mother Nature's Melting Pot," *The New York Times,* April 2.

Raina, Dhruv, and S. Irfan Habib. 2004. *Domesticating Modern Science: A Social History of Science and Culture in Colonial India.* New Delhi: Tulika.

Rampell, Catherine. 2011. "Many with New College Degrees Find the Job Market Humbling," *New York Times,* May 18.

Ramsden, Edmund. 2002. "Carving Up Population Science: Eugenics, Demography and the Controversy over the 'Biological Law' of Population Growth." *Social Studies of Science* 32 (5/6): 857–99.

———. 2006. "Confronting the Stigma of Perfection: Genetic Demography, Diversity and the Quest for a Democratic Eugenics in the Post-war United States." Working Paper No. 12/06. On the Nature of Evidence: How Well Do "Facts" Travel? http://eprints.lse.ac.uk/22536/.

———. 2008. "Eugenics from the New Deal to the Great Society: Genetics, Demography and Population Quality." *Studies in the History and Philosophy of Biology and Biomedical Sciences* 39: 391–406.

———. 2009. "Confronting the Stigma of Eugenics: Genetics, Demography and the Problems of Population." *Social Studies of Science* 39 (6): 853–84.

Ramusack, Barbara. 1989. "Embattled Advocates: The Debate over Birth Control in India, 1920–1940." *Journal of Women's History* 1 (2): 34–64.

Raver, Anne. 2012. "Finding Flavor in the Weeds," *The New York Times*, July 6.

Reardon, Jenny. 2001. "The Human Genome Diversity Project: A Case Study in Coproduction." *Social Studies of Science* 31 (3): 357–88.

———. 2005. *Race to the Finish: Identity and Governance in an Age of Genomics.* Princeton, NJ: Princeton University Press.

———. 2011. "The Democratic, Anti-Racist Genome? Technoscience and the Limits of Liberalism." *Science as Culture* 21 (1).

Renn, Jürgen, and Robert Schulmann. 2000. *Albert Einstein, Mileva Maric: The Love Letters.* Princeton, NJ: Princeton University Press.

Resnick, Michael. 1997. *Turtles, Termites and Traffic Jams: Explorations in Massively Parallel Microworlds.* Cambridge, MA: MIT Press.

Reuters. 2002. "Remains of Abused South African Woman Given Final Resting Place," *The New York Times*, August 9, http://www.nytimes.com/2002/08/09/international/09RTRS-VENU.html.

Rich, Benjamin. 2012. "The Gated Community Mentality," *The New York Times*, March 29.

Richardson, D. M. 2011. "Invasion Science: The Roads Travelled and the Roads Ahead." In *Fifty Years of Invasion Ecology: The Legacy of Charles Elton*, edited by David M. Richardson, 397–407. Oxford: Blackwell.

Richardson, D. M., P. M. Holmes, K. J. Esler, et al. 2007. "Riparian Zones—Degradation, Alien Plant Invasions and Restoration Prospects." *Diversity and Distributions* 13: 126–39.

Richardson, D. M., and P. Pysek. 2008. "Invasion Ecology and Restoration Ecology: Parallel Evolution in Two Fields of Endeavor." In *Fifty Years of Invasion Ecology: The Legacy of Charles Elton*, edited by David M. Richardson, 61–70. Oxford: Blackwell.

Richardson, Sarah. 2012. "Sexing the X: How the X Became the 'Female Chromosome.'" *Signs* 37 (4): 909–33.

Ridley, Mathew. 2000. *Genome: The Autobiography of a Species in 23 Chapters.* New York: Harper Perennial.

Ringle, Ken. 2002. "Stop That Fish!" *Washington Post*, July 3, C01.

Roberts, Dorothy. 1997. *Killing the Black Body: Race, Reproduction, and the Meaning of Liberty.* New York: Pantheon.

———. 2008. "Is Race Based Medicine Good for Us? African American Approaches to Race, Biomedicine, and Equality." *Journal of Law, Medicine and Ethics* 36 (3): 537–45.

———. 2009. "Race, Gender, and Genetic Technologies: A New Reproductive Dystopia?" *Signs* 34 (4): 783–803.

———. 2012. *Fetal Invention: How Science, Politics, and Big Business Re-create Race in the Twenty-first Century.* New York: The New Press.

Roberts, Philip D., Hilda Diaz-Soltero, David J. Hemming, Martin J. Parr, Nicola H. Wakefield, and Holly J. Wright. 2013. "What Is the Evidence That Invasive Species Are a

Significant Contributor to the Decline or Loss of Threatened Species? A Systematic Review Map." *Environmental Evidence,* 2 (5), http://www.environmentalevidencejournal.org/content/2/1/5.

Robichaux, Mark. 2000. "Alien Invasion: Plague of Asian Eels Highlights Damage from Foreign Species," *Wall Street Journal,* September 27, 12A.

Robinson, Carol. 2011. "Women in Science: In Pursuit of Female Chemists." *Nature* 476: 273–75.

Rodger, D., J. Stokes, and J. Ogilvie. 2003. *Heritage Trees of Scotland.* London: The Tree Council.

Rodman, J. 1993: "Restoring Nature: Natives and Exotics." In *In the Nature of Things: Language, Politics and the Environment,* edited by J. Bennett and W. Chaloupka, 139–53. Minneapolis: University of Minnesota Press.

Rolin, Kristina. 2004, Winter. "Three Decades of Feminism in Science: From 'Liberal Feminism' and 'Difference Feminism' to Gender Analysis of Science." *Hypatia* 19 (1): 292–96.

Roll-Hansen, Nils. 2010. "Eugenics and the Science of Genetics." In *The Oxford Handbook of the History of Eugenics,* edited by Alison Bashford and Philippa Levine, 80–97. Oxford: Oxford University Press:.

Rone, Adam. 2008, July. "Nature Wars, Culture Wars: Immigration and Environmental Reform in the Progressive Era." *Environmental History* 13 (3): 432–53.

Roosth, Sophia, and Astrid Schrader, eds. 2012. "Feminist Theory Out of Science." Special issue, *differences* 23 (3).

Rose, Hilary. 1994. *Love, Power and Knowledge: Towards a Feminist Transformation of the Sciences.* Cambridge: Polity.

———. 1997. "Good-bye Truth, Hello Trust: Prospects for Feminist Science and Technology Studies at the Millennium?" In *Science and the Construction of Women,* edited by Mary Maynard. New York: Routledge.

———. 2007. "Eugenics and Genetics: The Conjoint Twins? New Formations," March 22, http://business.highbeam.com/150795/article-1G1-162578234/eugenics-and-genetics-conjoint-twins.

Rose, Nikolas. 2006. *The Politics of Life Itself: Biomedicine, Power, and Subjectivity in the Twenty-First Century.* Princeton, NJ: Princeton University Press.

Rosser, Sue V. 1990. *Female-Friendly Science: Applying Women's Studies Methods and Theories to Attract Students.* New York: Pergamon.

———, ed. 1995. *Teaching the Majority: Breaking the Gender Barrier in Science, Mathematics, and Engineering.* New York: Teachers College Press.

———. 1997. *Re-engineering Female Friendly Science.* New York: Teachers College Press.

———. 2004a. *The Science Glass Ceiling: Academic Women Scientists and the Struggle to Succeed.* New York: Routledge.

———, ed. 2004b. "Using POWRE to ADVANCE: Institutional Barriers Identified by Women Scientists and Engineers." *NWSA Journal* 13 (1): 139–49.

———, ed. 2008. *Women, Science, and Myth: Gender Beliefs from Antiquity to the Present.* Santa Barbara, CA: ABC-CLIO.

Rosser, Sue V., and Mark Zachary Taylor. 2009. "Why Are We Still Worried about Women in Science?" *Academe* 95 (3): 7–10.
Rossiter, Margaret W. 1982. *Women Scientists in America: Struggles and Strategies to 1940*. Baltimore: Johns Hopkins University Press.
———. 1993. "The Matilda Effect in Science." *Social Studies of Science* 23: 325–41.
———. 1995. *Women Scientists in America: Before Affirmative Action, 1940–1972*. Baltimore: Johns Hopkins University Press.
Roy, Deboleena. 2004. "Feminist Theory in Science: Working Toward a Practical Transformation." *Hypatia* 19 (1): 255–79.
———. 2008. "Asking Different Questions: Feminist Practices for the Natural Sciences." *Hypatia* 23 (4): 134–57.
Rubin, David. 2012. "'An Unnamed Blank That Craved a Name': A Genealogy of Intersex as Gender." *Signs* 37 (4): 883–908.
Ruse, Michael. 2005. "The Darwinian Revolution, as Seen in 1979 and as Seen Twenty-Five Years Later in 2004." *Journal of the History of Biology* 38 (1): 3–17.
Rushdie, Salman. 1988. *The Satanic Verses*. London: Viking.
Sagoff, Mark. 2000, June. "Why Exotic Species Are Not as Bad as We Fear." *Chronicle of Higher Education* 46 (42): B7.
———. 2005. "Do Non-native Species Threaten the Natural Environment?" *Journal of Agricultural Environmental Ethics* 18: 215–36.
———. 2006. "Environmental Ethics and Environmental Science." In *Environmental Ethics and International Policy*, edited by Henk ten Have, 145–61. Paris: UNESCO.
Sahlins, Marshall. 2000. "'Sentimental Pessimism' and Ethnographic Experience: Or, Why Culture Is Not a Disappearing Object." In *Biographies of Scientific Objects*, edited by Lorraine Daston, 158–202. Chicago: University of Press.
Said, Edward. 1979. *Orientalism*. New York: Vintage.
Samerski, Silja. 2009. "Genetic Counseling and the Fiction of Choice: Taught Self-Determination as a New Technique of Social Engineering." *Signs* 32 (4): 735–61.
Sandler, Bernice R. 1986. "The Campus Climate Revisited: Chilly for Women Faculty, Administrators, and Graduate Students." In *Project on the Status and Education of Women*, edited by Bernice R. Sandler. Washington, DC: Association of American Colleges.
Sax, Linda J. 2001, Winter. "Undergraduate Science Majors: Gender Differences in Who Goes to Graduate School." *The Review of Higher Education* 24 (2): 153–72.
Sayre, Anne. 1975. *Rosalind Franklin and DNA*. New York: Norton.
Sayres, Janet. 1982. *Biological Politics: Feminist and Anti-feminist Perspectives*. London: Tavistock.
Scantlebury, Kate, ed. 2010. *Revisioning Science Education from Feminist Perspectives: Challenges, Choices, and Careers*. New York: Sense.
Scher, Lauren, and Fran O'Reilly. 2009. "Professional Development for K–12 Math and Science Teachers: What Do We Really Know?" *Journal of Research on Educational Effectiveness* 2 (3): 209–49.
Schiebinger, Londa. 1989. *The Mind Has No Sex? Women in the Origins of Modern Science*. Cambridge, MA: Harvard University Press.

———. 1993. *Nature's Body: Gender in the Making of Modern Science.* Boston: Beacon.

———. 1999. *Has Feminism Changed Science?* Cambridge, MA: Harvard University Press.

———. 2003. "Introduction: Feminism Inside the Sciences." *Signs* 28 (3): 859–66.

———. 2004. "Feminist History of Colonial Science." *Hypatia* 19 (1): 233–53.

———, ed. 2008. *Gendered Innovations in Science and Engineering.* Stanford, CA: Stanford University Press.

Schweber, S. S. 1978. "The Genesis of Natural Selection—1838: Some Further Insights." *BioScience* 28 (5): 321–26.

Scott, Joan. 1991. "The Evidence of Experience." *Critical Inquiry* 17 (4): 773–97.

Scott, Timothy Lee, and Stephen Buhmer. 2010. "Invasive Plant Medicine: The Ecological Benefits and Healing Abilities of Invasives." Rochester, VT: Inner Traditions.

Searle, G. R. 1976. *Eugenics and Politics in Britain, 1900–1914.* Leiden, Neth.: Noordhof International.

Selden, Steven. 2005. "Transforming Better Babies into Fitter Families: Archival Resources and the History of the American Eugenics Movement, 1908–1930." *Proceedings of the American Philosophical Society* 149 (2): 199–225.

Settles, Isis H., Lilia M. Cortina, Janet Malley, and Abigail J. Stewart. 2006. "The Climate for Women in Academic Science: The Good, the Bad, and the Changeable." *Psychology of Women Quarterly* 30: 47–58.

Sharpley-Whiting, Denean T. 1996. "The Dawning of Racial-Sexual Science: A One Woman Showing, a One Man Telling." *Ethnography in French Literature* 33: 115–28.

Shteir, Ann. B. 1999. *Cultivating Women, Cultivating Science: Flora's Daughters and Botany in England, 1760–1860.* Baltimore: Johns Hopkins University Press.

Simberloff, Daniel. 2003. "Confronting Invasive Species: A Form of Xenophobia?" *Biological Invasions* 5: 179–92.

Simberloff, Daniel, et al. 2011. "Non-natives: 141 Scientists Object." *Nature* 475 (36). DOI: 10.1038/457036a

Sime, Ruth Lewin. 1997. *Lise Meitner: A Life in Physics.* Berkeley: University of California Press.

Sivaramakrishnan, K. 2011. "Thin Nationalism: Nature and Public Intellectualism in India." *Contributions to Indian Sociology* 45: 85–111.

Slaughter, Sheila, and Larry L. Leslie. 2007. "Expanding and Elaborating the Concept of Academic Capitalism." *Organization* 8 (2): 154–61.

Slaughter, Sheila, and Gary Rhoades. 2000, Spring/Summer. "The New Liberal University." *New Labor Forum* 6: 73–79.

———. 2004. *Academic Capitalism and the New Economy: Markets, State and Higher Education.* Baltimore: Johns Hopkins University Press.

Slobodkin, L. B. 2001. "The Good, the Bad and the Reified." *Evolutionary Ecology Research* 3:1–13.

Smocovitis, V. B. 1992. "Unifying Biology: The Evolutionary Synthesis and Evolutionary Biology." *Journal of the History of Biology* 25: 58–59.

Smout, Chris. T. 2003. "The Alien Species in 20th Century Britain: Constructing a New Vermin." *Landscape Research* 29 (1): 11–20.
Sober, Elliott. 1980. "Evolution, Population Thinking, and Essentialism." *Philosophy of Science* 47 (3): 350–83.
Somerville, Siobhan. 2005. "Notes Toward a Queer History of Naturalization." *American Quarterly* 57 (3): 659–75.
Sonnert, Gerhard, and Mary Frank Fox. 2012, January/February. "Women, Men, and Academic Performance in Science and Engineering: The Gender Difference in Undergraduate Grade Point Averages." *Journal of Higher Education* 83 (1): 73–101.
Sonnert, Gerhard, and Gerald J. Holton. 1995. *Who Succeeds in Science? The Gender Dimension*. New Brunswick, NJ: Rutgers University Press.
———. 1996. "Career Patterns of Men and Women in the Sciences." *American Scientist* 84: 63–71.
Soulé, M. E. 1990. "The Onslaught of Alien Species, and Other Challenges in the Coming Decades." *Conservation Biology* 4: 233–39.
Soulé, M. E., and G. Lease. 1995. *Reinventing Nature: Responses to Postmodern Deconstruction*. Washington, DC: Island.
Spelke, Elizabeth S. 2005. "Sex Differences in Intrinsic Aptitude for Mathematics and Science?: A Critical Review." *American Psychology* 60: 950–58.
Srivastava, R. C. 1983. "A Taxonomic Study of the Genus *Ipomoea L.* (*Convolvulaceae*) in Madhya Pradesh." *Journal of Economic and Taxonomic Botany* 4: 765–75.
Stampe, Elizabeth, and Curtis Daehler. 2003. "Mycorrhizal Species Identity Affects Plant Community Structure and Invasion: A Microcosm Study." *Oikos* 100: 362–72.
Steinem, Gloria. 2005. Acceptance speech cited in Taylor, http://articles.chicagotribune.com/2005-05-18/features/0505180315_1_minority-owned-firms-women-s-movement-anger.
Stengers, Isabelle. 2000. *The Invention of Modern Science*. Minneapolis: University of Minnesota Press.
———. 2011. *Thinking with Whitehead: A Free and Wild Creation of Concepts*. Cambridge, MA: Harvard University Press.
Stepan, Nancy. 1982. *The Idea of Race in Science: Great Britain, 1800–1960*. Hamden, CT: Anchor.
———. 1985. "Biological Degeneration: Races and Proper Places." In *Degeneration: The Dark Side of Progress*, edited by Edward Chamberlin and Sander L. Gilman. New York: Columbia University Press.
———. 1986. "Race and Gender: The Role of Analogy in Science." *Isis* 77 (2): 261–77.
Stern, Alexandra Minna. 2005. *Eugenic Nation: Faults and Frontiers of Better Breeding in Modern America*. Berkeley: University of California Press.
———. 2010. "Gender and Sexuality: A Global Tour and Compass." In *The Oxford Handbook of the History of Eugenics*, edited by Alison Bashford and Philippa Levine, 173–91. Oxford: Oxford University Press.

Stewart, Abigail J., Janet E. Malley, and Danielle Lavaque-Manty, eds. 2007. *Transforming Science and Engineering: Advancing Academic Women.* Ann Arbor: University of Michigan Press.

Stewart, Barbara. 2001. "The Invasion of the Woodland Soil Snatchers," *The New York Times*, April 24, 1B.

Stoler, Ann, C. McGranahan, and P. C. Perdue. 2008. *Imperial Formations.* Santa Fe, NM: School for Advanced Research Press.

Subrahmanyan, Lalita. 1998. *Women Scientists in the Third World: The Indian Experience.* New Delhi: Sage.

Subramaniam, Banu. 1994. "Maintenance of the Flower Color Polymorphism at the W Locus in the Common Morning Glory, *Ipomoea purpurea.*" Ph.D. diss., Duke University.

———. 1998. "A Contradiction in Terms: Life as a Feminist Scientist." *Women's Review of Books* 15 (5): 25–26.

———. 2001. "The Aliens Have Landed! Reflections on the Rhetoric of Biological Invasions." *Meridians: feminism, race, transnationalism* 2 (1): 26–40.

———. 2009. "Moored Metamorphoses: A Retrospective Essay on Feminist Science Studies." *Signs* 34 (4): 951–80.

Subramaniam, Banu, Rebecca Dunn, and Lynn Broaddus. 1992. "'Sir'vey or 'Her'vey." In *Engaging Feminism: Students Speak Up and Speak Out,* edited by Jean O'Barr and Mary Wyer. Chicago: University of Chicago Press.

Subramaniam, Banu, and Mark Rausher. 2000. "Balancing Selection on a Floral Polymorphism." *Evolution* 54 (2): 691–95.

Subramaniam, Banu, and Mary Wyer. 1998. "Assimilating the 'Culture of No Culture' in Science: Feminist Interventions in (De)Mentoring Graduate Women." *Feminist Teacher* 12 (1): 12–28.

Sur, Abha. 2001. "Dispersed Radiance: Women Scientists in C. V. Raman's Laboratory." *Meridians: Feminism, Race, Transnationalism* 1 (2): 95–127.

———. 2011. *Dispersed Radiance: Caste, Gender, and Modern Science in India.* New Delhi: Navanyana.

Swift, Katherine. 2008, May. "Sinister Science: Eugenics, Nazism, and the Technocratic Rhetoric of the Human Betterment Foundation." *Lore* 6 (2): 1–11.

Takeshita, Chikako. 2011 *The Global Biopolitics of the IUD: How Science Constructs Contraceptive Users and Women's Bodies.* Cambridge, MA: MIT Press.

Taylor, Mark. C. 2010. *Crisis on Campus: A Bold Plan for Reforming Our Colleges and Universities.* New York: Knopf.

———. 2011, April. "Reform the PhD System or Close It Down." *Nature* 472: 261.

Teresi, Dick. 2001. *Lost Discoveries: Ancient Roots of Modern Science—From the Babylonians to the Mayans.* New York: Simon & Schuster.

Terry, Jennifer. 1999. *An American Obsession: Science, Medicine, and Homosexuality in Modern Society.* Chicago: University of Chicago Press.

Theodoropoulos, D. I. 2003. *Invasion Biology: Critique of a Pseudoscience.* Blythe, CA: Avvar.

Todd, Kim. 2001. *Tinkering with Eden: A Natural History of Exotics in America*. New York: Norton.

Tomes, Nancy. 2000, February. "The Making of a Germ Panic, Then and Now." *American Journal of Public Health* 90 (2): 191–99.

Townsend, M. 2005. "Is the Social Construction of Native Species a Threat to Biodiversity?" *ECOS* 26: 1–9.

Traweek, Sharon. 1992. *Beamtimes and Lifetimes: The World of High Energy Physicists.* Cambridge, MA: Harvard University Press.

———. 1993. "Cultural Differences in High-Energy Physics: Contrasts between Japan and the United States." In *The "Racial" Economy of Science: Toward a Democratic Future,* edited by Sandra Harding, 398–407. Bloomington: Indiana University Press.

Tredoux, Gavan. 2001. *Two Geneticists: J.B.S. Haldane and C.D. Darlington,* http://www.mail-archive.com/ctrl@listserv.aol.com/msg73477.html.

Tsing, Anna Lowenhaupt. 1994. "Empowering Nature, or Some Gleanings in Bee Culture." In *Naturalizing Power: Essays in Feminist Cultural Analysis,* edited by Sylvia Yanagisako and Carol Delaney, 113–43. New York: Routledge.

Tuana, Nancy, ed. 1989. *Feminism and Science.* Bloomington: Indiana University Press.

Turda, Marius. 2010. *Modernism and Eugenics.* Basingstoke, UK: Palgrave Macmillan.

United Press International. 1998. "Native Species Invaded." *ABC News,* March 16.

U.S. Department of Agriculture. 2012. "Natives, Invasive, and Other Plant-Related Definitions," http://www.ct.nrcs.usda.gov/plant_definitions. html.

———. n.d. "Plants." http://www.invasivespeciesinfo.gov/plants/main.shtml.

U.S. Geological Survey Midcontinent Ecological Science Center. 1999, May 13. "USGS Research Upsets Conventional Wisdom on Invasive Species Invasions." News release.

Valent, Barbara. 1990. "Rice Blast as a Model System for Plant Pathology." *Phytopathology* 80 (1): 33–36.

Valentine, G. 2004. "Geography and Ethics: Questions of Considerability and Activism in Environmental Ethics." *Progress in Human Geography* 28: 258–64.

Valian, Virginia. 1999. *Why So Slow? The Advancement of Women.* Cambridge, MA: MIT Press.

Valla, Jeffrey M., and Wendy M. Williams. 2012. "Increasing Achievement and Higher-Education Representation of Under-Represented Groups in Science, Technology, Engineering, and Mathematics Fields: A Review of Current K–12 Intervention Programs." *Journal of Women and Minorities in Science* 18 (1): 21–53.

Van Driesche, Jason, and Roy Van Driesche. 2000. *Nature Out of Place: Biological Invasions in the Global Age.* Washington, DC: Island.

Veit, Helen Zoe. 2011. "Time to Revive Home Ec," *The New York Times,* September 5.

Venter, O., N. N. Brodeur, L. Nemiroff, B. Belland, I. J. Dolinsek, and J. W. A. Grant. 2006. "Threats to Endangered Species in Canada." *Bioscience* 56: 903–10.

Vermeij, G. J. 2005. "Invasion as Expectation: A Historical Fact of Life." In *Species Invasions: Insights into Ecology, Evolution, and Biogeography,* edited by D. F. Sax, J. J. Stachowicz, and S. D. Gaines, 315–36. Sunderland, MA: Sinauer Associates.

Verran, Helen. 2001. *Science and an African Logic.* Chicago: University of Chicago Press.

Verrengia, Joseph B. 1999. "When Ecologists Become Killers." *MSNBC News,* October 4, www.msnbc.com/news.

———. 2000. "Some Species Aren't Welcome." *ABC News,* September 27, www.abcnews.go.com.

Veuille, Michel. 2010. "Darwin and Sexual Selection: 100 Years of Misunderstanding." *C. R. Biologies* 333 (2): 145–56.

Vining, D. R. 1983. "Fertility Differentials and the Status of Nations: A Speculative Essay on Japan and the West." In *Intelligence and National Achievement,* edited by R. B. Cattell. Washington, DC: Cliveden.

Viswanathan, Shiv. 2002. "The Laboratory and the World: Conversations with C. V. Seshadri," *Economic and Political Weekly,* June 1, 2163–70.

Visweswaran, Kamala. 1994. *Fictions of Feminist Ethnography.* Minneapolis: University of Minnesota Press.

Waggoner, Miranda. R. 2012. "Motherhood Preconceived: The Emergence of the Preconception Health and Health Care Initiative." *Journal of Health Politics, Policy and Law* 38 (2): 345–71.

Wajcman, Judy. 1991. *Women Confront Technology.* University Park: Penn State University Press.

Walsh, Denis. 2006. "Evolutionary Essentialism." *British Journal of Philosophy of Science* 57 (2): 425–48.

Walters, Suzanna. 1996. "From Here to Queer." *Signs* 21 (4): 857.

Wanzo, Rebecca. 2009. "In the Shadows of Anarcha." In *This Suffering Will Not Be Televised.* New York: State University of New York Press.

Warren, Charles R. 2007. "Perspectives on the 'Alien' versus 'Native' Species Debate: A Critique of Concepts, Language and Practice." *Progress in Human Geography* 31 (4): 427–46.

Wayne, Marta L. 2000. "Walking a Tightrope: The Feminist Life of a *Drosophila* Biologist." *NWSA Journal* 12 (3): 139–50.

Weiner, H. 1996. "Congress Threatens Wild Immigrants." *Earth Island Journal* 11 (4): 19.

Weinstein, Naomi. 1977. "'How Can a Little Girl Like You Teach a Great Big Class of Men?' the Chairman Said, and Other Adventures of a Woman in Science." In *Working it Out: 23 Women Writers, Artists, Scientists, and Scholars Talk About Their Lives and Work,* edited by Sara Ruddick and Pamela Daniels. New York: Pantheon.

Wilson, Elizabeth A. 1998. *Neural Geographies: Feminism and the Microstructure of Cognition.* New York: Routledge.

———. 2004. *Psychosomatic: Feminism and the Neurological Body.* Durham, NC: Duke University Press.

Winsor, M. P. 2003. "Non-essentialist Methods in Pre-Darwinian Taxonomy." *Biology and Philosophy* 18: 387–400.

Woodhouse. Edward I. 2005. "Reconstructing Technological Society by Taking Social Construction Even More Seriously." *Social Epistemology* 19: 2–3.

Woods, Mark, and Paul Veatch Moriarty. 2001. "Strangers in a Strange Land: The Problem of Exotic Species." *Environmental Values* 10 (2): 163–91.

Wong, Tama Matsuoka, and Eddy Leroux. 2012. *Foraged Flavor: Finding Fabulous Ingredients in Your Backyard or Farmer's Market.* New York: Clarkson Potter.

Wright, S. F., and A. Upadhyaya. 1998. "A Survey of Soils for Aggregate Stability and Glomalin, a Glycoprotein Produced by Hyphae of Arbuscular Mycorrhizal Fungi." *Plant and Soil* 198: 97–107.

Wyer, Mary. 1993. *NSF Model Project, Women in Science and Engineering.* Durham, NC: Duke University Press.

Wyer, Mary, Mary Barbercheck, Donna Giesman, Hatice Orun Ozturk, and Marta Wayne, eds. 2008. *Women, Science, and Technology: A Reader in Feminist Science Studies.* 2nd ed. New York: Routledge.

Wylie, Alison. 2011. "What Knowers Know Well: Women, Work and the Academy." In *Feminist Epistemology and Philosophy of Science: Power in Knowledge.* New York: Springer.

Xie, Yu, and Kimberlee Shauman. 2003. *Women in Science: Career Progress and Outcomes.* Cambridge, MA: Harvard University Press.

Yeh, Ling-Ling. 1995. "U.S. Can't Handle Today's Tide of Immigrants." *Christian Science Monitor* 87 (81): 19.

Young, Robert M. 1990. "Darwinism and the Division of Labour." *Science as Culture* 9: 110–24.

Yuval-Davis, Nira. 2006. "Intersectionality and Feminist Politics." *European Journal of Women's Studies* 13 (3): 193–209.

Zachariah, Benjamin. 2001. "The Uses of Scientific Argument: The Case of 'Development' in India c. 1930–1950." *Economic and Political Weekly* 36 (39): 3695.

Zakaib, Gwyneth Dickey. 2011. "Science Gender Gap Probed." *Nature* 470: 153.

Zerner, Charles, ed. 2000. *People, Plants, and Justice: The Politics of Nature Conservation.* New York: Columbia University Press.

Zhang, Qian, Ruyi, Yang, Jianjun Tang, Haishui Yang, Shuijin Hu, and Xin Chen. 2010. "Positive Feedback between Mycorrhizal Fungi and Plants Influences Invasion Success and Resistance to Invasion." *PLos One* 5 (8): 1–10.

Zimmer, Oliver. 1998. "In Search of Natural Identity: Alpine Landscape and the Reconstruction of the Swiss Nation." *Comparative Studies in Society and History* 40 (4): 637–65.

Zuckerman, Harriet, and Jonathan R. Cole. 1991. *The Outer Circle: Women in the Scientific Community.* New York: Norton.

Zufall, R. A., and M. D. Rausher. 2003. "The Genetic Basis of Flower Color Polymorphism in the Common Morning Glory, *Ipomoea purpurea*." *Journal of Heredity* 94 (6): 442–48.

Index

Banu Subramaniam is an associate professor of women, gender, sexuality studies at the University of Massachussetts, Amherst, and a coeditor of *Feminist Studies: A New Generation and Making Threats: Biofears and Environmental Anxieties.*

The University of Illinois Press
is a founding member of the
Association of American University Presses.

Composed in 10.5/13 Arno Pro
with Avenir display
by Jim Proefrock
at the University of Illinois Press
Manufactured by Sheridan Books, Inc.

University of Illinois Press
1325 South Oak Street
Champaign, IL 61820-6903
www.press.uillinois.edu